Electric Transmission Line Fundamentals

Electric Transmission Line Fundamentals

Edwin M. Anderson, P.E.
Professor of Electrical and Electronics Engineering
North Dakota State University
Fargo, North Dakota

Reston Publishing Company, Inc.
A Prentice-Hall Company
Reston, Virginia

Library of Congress Cataloging in Publication Data
Anderson, Edwin M.
Electric transmission line fundamentals.

Bibliography: p.
Includes index.
1. Electric lines. 2. Telecommunication lines.
I. Title.
TK3226.A54 1985 621.319′2 84-11624
ISBN 0-8359-1597-2

10 9 8 7 6 5 4 3 2 1

Printed in the United States of America

Dedicated to my students—
past, present, and future

Contents

Preface

Transmission line theory has been offered in a number of different formats over the years as curriculums in electrical engineering departments have been rethought, restructured, and retreaded due to the pressures of new discoveries and an increasing amount of knowledge, along with higher quality students and constant changes in the job market. In the early 1950's students were offered a course primarily concerned with power transmission lines and perhaps some material on the principles of telephone and telegraph lines. These courses covered not only the basic concepts from the electrical point of view but often dealt with the mechanical construction of the line. Although the theoretical mathematical approach was used, the courses were often taught with a practical emphasis since the instructors were usually persons with some experience in industry and most of the graduates found employment in the power area. As time went on and younger teachers with more interest in communications and in electronics arrived on the scene, the emphasis swung towards communications-type courses; often the power material was reduced or deleted or the department ended up with two different transmission line courses.

The next trend was to reduce all the material concerned with transmission lines; often it ended up as an afterthought in a course in electromagnetic theory. In recent years there have been very few new texts on the theory of transmission lines and many students have graduated without an appreciation of the application of transmission line fundamentals to engineering problems that may not, at first glance, seem to lend themselves to such applications. A more appropriate title for this text might be *Circuits With Distributed Parameters*, since the engineer may need the theory to deal with many situations that have nothing to do with overhead lines stretching across the "back forty." Designers of high-speed, digital computers have found that they need a more complete knowledge of signals moving along a line that may be short in physical length but electrically long in an engineering sense. Communication engineers are constantly facing new challenges as their systems become more complex and encompass communication not only between human operators but between machines and computers, with transmission lines serving as the links between points that may be separated by distances varying from a few inches to hundreds of miles. When antennas are included in the system, the distances may stretch to millions of miles. And even today the traditional area of power transmission continues to need engineers trained in the basic concepts as the problems in the energy areas force new solutions to old problems. As technology advances at an increasing rate, graduates are needed who are strongly grounded in engineering fundamentals so that they will be able to continue their professional lives with a minimum of difficulty as the state-of-the-art technology changes in the years ahead.

In this text the emphasis is on developing a fundamental understanding of basic concepts that will help electrical engineers throughout their profesional lives, no matter what specialized area of engineering they may find themselves engaged in after leaving the university. A prerequisite for the course is a previous study of electromagnetic field theory so that the student will have a prior knowledge of the calculation of distributed parameters associated with the inductance and capacitance of two-wire lines, as well as an understanding of such concepts as intrinsic impedance, reflection coefficients, standing waves, traveling waves, and other ideas as they pertain to waves traveling through space.

Over the years, as this material was presented to students, it became apparent to the author that the use of rotating phasors was an aid to understanding, allowing students to solve a large number of problems without memorizing many equations. By understanding the rotating phasor concept the student can visualize what is taking place on the line and is often able to arrive at a solution that requires little or no mathematical manipulation. Since rotating phasors will not take care of all the engineering problems associated with transmission lines and distributed parameter circuits, the text presents the mathematical approach whenever it is needed. Since electrical engineering students branch out into a wide variety of areas when they graduate there has been no attempt to treat each topic in great detail. Rather, the approach has been to treat the fundamentals so that the graduate will be in a position to enter any area with enough understanding to allow specific problems to be solved with a minimum of further study and effort.

Since many students will have access to high frequency laboratories, the text, whenever it seems appropriate, includes material which will be helpful in making measurements on transmission lines operating at rather high frequencies since the wavelength is small enough to make measurements, observe standing waves, and, in general, to observe the items mentioned throughout this text.

Any writing project of this magnitude requires the help of a number of individuals. I would like to thank my wife, Kathleen, who has spent many hours alone during the several years this text was being written and rewritten. Dr. William Bares, Head of the Department of Electrical and Electronics Engineering at North Dakota State University, has helped not only by supplying material, secretarial help, and student help but who has offered every encouragement to the author. Dr. David Rogers and Dr. Donald Smith have used the preliminary text for several years and have been most helpful. A special thanks must go to Professor Donald Peterson, Director of our University Computer Center, who volunteered the use of the Center's equipment and provided instruction in its use. Mrs. Margaret Skolness, one of our secretaries, spent a great deal of time and effort on the early versions and deserves a big thanks for her help on a very difficult task. One person who has probably contributed the most through her help, comments, proofreading of the early manuscripts, and her suggestions is Mrs. Wilma Jean McDonald, listed as our departmental secretary, but in reality a valued professional colleague. Finally a big thanks must go to the thousands of students who have been a source of inspiration to the author over the past many years. Their contributions to our society in the "real world" away from the university have been a constant joy.

A number of publishers have allowed the author to reproduce material from other authors and for this the author is very grateful. A special thanks goes to Analog Instruments of New Providence, New Jersey for allowing the use of the Smith chart as an aid to the understanding of transmission line theory.

Electric Transmission Line Fundamentals

1

Fundamentals

1.1 Introduction to the Text

Thousands of miles of transmission lines span the countryside of this land connecting synchronous alternators to substations located near the centers of electrical loads. Supported by wooden, aluminum, steel or concrete towers, these lines are either three-phase alternating current or direct current and are operating at voltages that run from around 115 kilovolts to over 750 kilovolts. At the substations, step-down transformers reduce the voltage to lower values and many more miles of distribution lines transmit the energy to the consumer, either as three-phase or single-phase. If the line is d.c. there are converters at each end of the transmission line changing the a.c. to d.c. at the sending end and back to a.c. at the receiving end before the voltage is reduced for distribution. The transmission of electrical power from the point of generation to the point of utilization has been the concern of a large number of electrical engineers working for utility companies for many decades. The problems associated with the transmission of large amounts of energy over long distances at high efficiencies demanded a great deal of study and work and resulted in most electrical engineering students taking courses in 60 hertz power lines up to the time of World War II.

After the war the rapid advances in electronics and communications produced separate courses in power and high frequency transmission lines in many electrical engineering departments in this country. Today many electronic engineers are concerned with transmission lines that are operating at low power levels with physical lengths that may run down to a few inches, with efficiency usually of minor or no consideration but with maximum power transfer and accurate signal reproduction of major importance. Since transmission lines may consist of conductors stretched between towers spanning a countryside, an underground coaxial cable in a city, a pair of wires in an electronic computer, or a wave guide feeding a high-frequency antenna, this text will concentrate on the fundamentals of transmission line theory, which is common to all situations. While the transmission of electrical energy can be considered as an electromagnetic field phenomenon with the conductors serving only to guide the field, we will study transmission lines primarily from a circuits viewpoint except in the case of wave guides where field theory is not only extremely useful but necessary. An antenna system consisting of a sending antenna and a receiving antenna may be thought of as a transmission line without any physical connection between the sending and receiving ends except the atmosphere and as a special case of a transmission line. As you might suspect, field theory is once again a necessary tool.

Transmission line theory applies to all situations where electrical power is transmitted along guided paths by conductors, although in many practical cases the need for detailed analysis proves to be unnecessary. Only a knowledge of the theory, however, will allow the engineer to know when approximations may be made and a detailed study is not needed. In basic circuit courses, a voltage was assumed to be applied to a circuit with instantaneous action throughout the circuit since in many cases the time for the current and voltage to travel in the circuit was unimportant. However, the engineer designing a modern, high-speed, digital computer must be concerned with the time it takes a signal to move from one location to another location. The timing of several signals arriving at one location may be critical. When the signal the engineer is working with is nonsinusoidal, the time of transit of the different frequency components becomes important if one wishes to reproduce the input waveshape at the output of the circuit with faithful accuracy.

Maximum power transfer is important if the engineer starts with only a small fraction of a watt. The medical researcher studying the electrical signals generated by the human body or the communication engineer starting with extremely low levels of signal strength at an antenna represent cases of maximum power transfer importance. Efficiency becomes important if one begins with thousands of kilowatts and especially so as the cost of energy rises. From d.c. to the gigahertz range of frequencies, transmission line theory is very important in a large number of practical engineering problems. Electrical and electronic engineers have a wide spectrum of areas to work in and there are few specialties where a knowledge of transmission line theory does not enter from time to time. Today, even after decades of work and development, the utility engineer is still faced with new problems that require a fundamental knowledge of transmission line concepts. While the state-of-the-art technology changes from day to day, the fundamental concepts do not and so this text will concentrate on basic ideas so that the student will be prepared to solve engineering problems associated with the transmission of electrical energy throughout his or her working life. Major emphasis will be on having the student develop a feeling for the physical phenomenons associated with the voltage, current and power on the lines as well as being able to handle the mathematics associated with the study. The mathematical level is usually not difficult and many of the basic concepts may be handled by using old friends from previous course work.

1.2 Introduction to Chapter 1

As one starts the study of a new area of engineering it always becomes necessary to learn some new terminology and in our case to treat some old friends in a new way. Transmission line theory becomes necessary when the distributed parameters along the length of the line affect the voltage, current and power relationships on the line. Distributed parameters arise because each foot of the line has some resistance, some inductance, some capacitance and some leakage current. We find that because of distributed parameters the current entering the line is different from the current entering the load at the other end, and that the voltage across the line varies from point to point on the line. It is even possible that the voltage at the load end is higher than the input voltage at the sending end of the line.

Voltage and current waves travel down a transmission line much as the electric and magnetic fields traveled through space when we studied field theory. If the waves are reflected at the load end, the possibility exists of having two waves of both current and voltage traveling in opposite directions on the line. By terminating the line with the proper value, one can eliminate the reflected wave.

In this chapter we will take a small section of the line and replace it with an approximate circuit which will become more exact as the length of the small section is reduced towards zero. By writing the circuit equations for this small section we will end up with the equations necessary for an understanding of the physical effects on the transmission line. These basic equations will be used from time to time throughout the entire text with modifications to take into account different viewpoints needed when dealing with various practical cases.

One very important engineering use of transmission line theory occurs when one is dealing with the steady-state sinusoidal case. The introduction to this important engineering application gives rise in this chapter to two new concepts, characteristic impedance and propagation constant. Their counterparts in field theory were the intrinsic impedance and the propagation constant.

Throughout the text we will be discussing the two very important engineering cases of no losses on the transmission line, which is an approximation to the low-loss line, and the high-frequency line, which gives similar results. We will refer to both cases as the "lossless" case. Since Chapter 1 is the beginning and the foundation for the course, the student should treat it with special care. The student will find that the material is not difficult but only new and unfamiliar.

1.3 Distributed Parameters

In circuits, basic elements such as resistors, inductors and capacitors are treated as individual, two-terminal devices that can be picked up and handled and purchased as such from any electrical supply firm. They are referred to as lumped-circuit elements. However, a transmission line conductor has resistance along each inch of its length and there is no lumped resistor that we can point to. In the study of electromagnetic field theory it was found that a parallel, two-wire line has inductance and capacitance distributed along its length and calculated as henrys per meter and farads per meter. Field texts give these equations as:

$$\frac{L}{\ell} = \frac{\mu}{\pi} \ln \frac{D}{a} \text{ henrys/meter} \tag{1-1a}$$

where D is the separation between the centers of the two conductors, a is the radius of the conductor and μ is the permeability in henrys per meter.

$$\mu = \mu_r \mu_o. \tag{1-1b}$$

μ_r is the relative permeability and μ_o is the permeability of free space and has a value of $4\pi \times 10^{-7}$ henrys per meter.

$$\frac{C}{\ell} = \frac{\pi\varepsilon}{\ln D/a} \text{ farads/meter} \tag{1-2a}$$

where ε is the permittivity in farads per meter.

$$\varepsilon = \varepsilon_r\varepsilon_o. \tag{1-2b}$$

ε_r is the relative permittivity and is often called the dielectric constant and ε_o is the permittivity of free space and has a value of $(8.85415 \pm 0.00002) \times 10^{-12}$ farads per meter. A more convenient number to remember is $\frac{10^{-9}}{36\pi}$ and this is the number we will use throughout the text since it is approximately correct with a value of 8.84194×10^{-12} farads per meter. $\frac{L}{\ell}$ and $\frac{C}{\ell}$ are referred to as distributed parameters and are the first major differences between elementary circuit theory and transmission line theory.

In this text the rationalized M.K.S. system of units will normally be used so we usually will be dealing with the meter, kilogram, and second. However, when working with transmission lines in the United States, other units of length such as the foot and the mile are often encountered. The student should take some care to keep from making errors in his work when different units of length are being used. In (1-1a) as long as D and a are in the same units it doesn't make any difference, but if we desire the inductance in henrys per mile the equation must be properly modified.

Example 1-A

Let us anticipate some results that will come later in the text by making some manipulations which, on the surface, may make no sense at this stage of our learning process.

(a) What is the value of the product of $\frac{L}{\ell}$ and $\frac{C}{\ell}$ for a parallel-wire line in free space? (b) What is the value of the reciprocal? (c) What is the value of the square root of the reciprocal? (d) What are the units of the square root of the reciprocal?

Solution

$$\frac{L}{\ell} = \frac{\mu_o}{\pi} \ln \frac{D}{a}. \qquad \frac{C}{\ell} = \frac{\pi \varepsilon_o}{\ln \frac{D}{a}}.$$

$$\frac{L}{\ell} \times \frac{C}{\ell} = \mu_o \varepsilon_o = (4\pi \times 10^{-7}) \frac{(10^{-9})}{36\pi}$$

$$= \frac{10^{-16}}{9} \frac{\text{henry-farads}}{\text{meter}^2}. \qquad \text{(a)}$$

$$\frac{1}{\frac{10^{-16}}{9}} = 9 \times 10^{+16} \frac{\text{meter}^2}{\text{henry-farads}}. \qquad \text{(b)}$$

$$\sqrt{9 \times 10^{16}} = 3 \times 10^{8} \frac{\text{meters}}{\sqrt{\text{henry-farad}}}. \qquad \text{(c)}$$

Let us find the units for henrys and farads.

$$v = L \frac{di}{dt}. \qquad L = \frac{vdt}{di}. \qquad \frac{(\text{volt-sec})}{\text{ampere}}.$$

$$i = C \frac{dv}{dt}. \qquad C = \frac{idt}{dv}. \qquad \frac{(\text{amp-sec})}{\text{volt}}.$$

So our answer for (c) comes out to be a familiar quantity.

$$\frac{\text{meters}}{\sqrt{\text{henry-farad}}} = \frac{\text{meters}}{\sqrt{\frac{\text{volt-sec}}{\text{amp}} \frac{\text{amp-sec}}{\text{volt}}}} = \frac{\text{meters}}{\text{second}} = \text{velocity}. \qquad \text{(d)}$$

The student, hopefully, has noticed that, by manipulating the equations for the inductance and capacitance per unit length, one obtains the velocity of light if one uses the permeability and permittivity of free space. By retracing the steps we have taken we obtain:

$$\text{velocity} = \sqrt{\frac{1}{\frac{(L)}{\ell} \frac{(C)}{\ell}}}.$$

Besides distributed resistance, inductance, and capacitance, our work requires one additional distributed parameter. Wires must be insulated from each other to prevent large currents flowing directly from wire to wire. However, since there is no perfect insulation material there will always be some small leakage current between conductors when a voltage exists between them. This current may flow from a bare conductor up a string of insulators to a wooden beam, across the beam to a second string of insulators and down to a second wire. In this case the leakage current may be highly variable depending on dirt, humidity, materials involved and other factors which make the leakage current difficult or impossible to compute by using equations. In addition, this type of leakage current would occur only at a tower location which means it would have random spacing on the line. On the other hand, a twisted pair of conductors would have leakage current all along the line depending on the insulating properties of the wire covering. So one additional parameter must be added to take this leakage current into account—G, the shunt conductance of the line, which is in mhos per meter. It is the parameter that is the hardest to compute and often difficult to measure.

Thus in this text, R, L, C, and G will have different meanings than they did in elementary circuits. By definition they will be: R in ohms per meter, L in henrys per meter, C in farads per meter and G in mhos per meter. In the beginning part of the book we will assume that they are known. If it is a two-wire line the parameters will be for both wires. For a three-phase, three-wire line the parameters will be given on a per-conductor or a per-phase basis.

1.4 Transmission Lines from an Electromagnetic Field Viewpoint

All transmission of energy from a source to a point of utilization is basically a field phenomenon. The transmission line acts as a wave guide. In this text the emphasis will be on the circuit properties of transmission lines, but it is interesting to look at the field aspects briefly. Let us assume a two-wire parallel line with a generator at one end and a load at the other. At some instant of time the electric and magnetic field pattern will appear as shown in Figure 1-1.

If we apply Poynting's vector, $\vec{P} = \vec{E} \times \vec{H}$, at every point in space we find that the energy flow is always in the same direction; into the paper, as indicated in Figure 1-2.

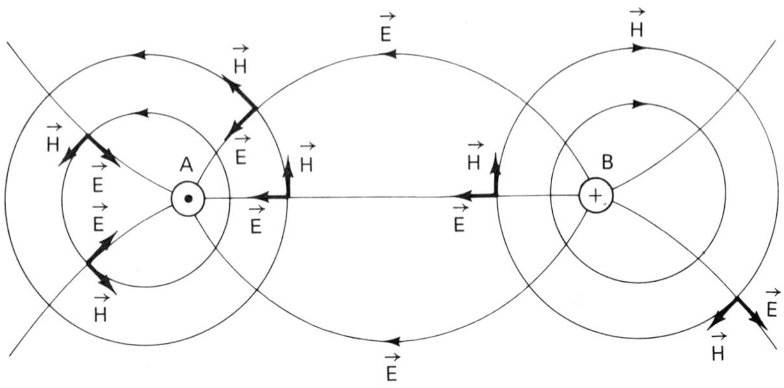

Figure 1-1. Electric and magnetic fields for parallel-wire line.

This indicates that the energy is actually going from the generator to the load in the space surrounding the wires. The wires are acting as a wave guide for the electric and magnetic fields traveling towards the load. Furthermore, if we consider what is happening at the surface of a wire an interesting result is obtained.

Let us assume that the conductor is located on the z axis with current going in the positive z direction. In the following equations, σ is the conductivity in mhos per meter, ρ is the distance in meters from the z axis in cylindrical coordinates, $\vec{J}$ is the current density in amperes per meter squared, $\vec{a}_z$, $\vec{a}_\phi$, and $\vec{a}_\rho$ are unit vectors in the cylindrical coordinate system, and S is the cross-sectional area of the conductor in meters squared.

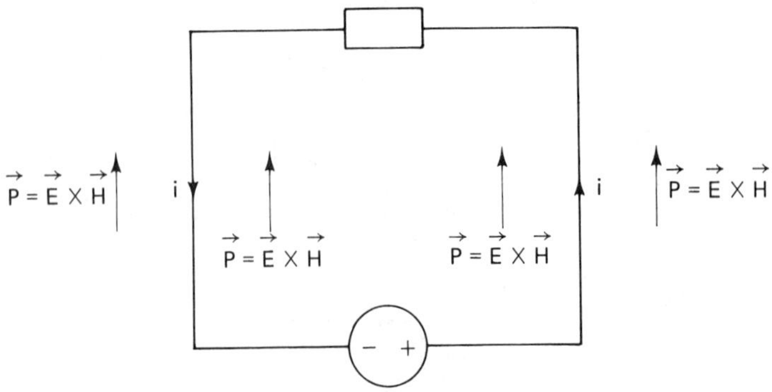

Figure 1-2. Poynting's vector around a two-wire line.

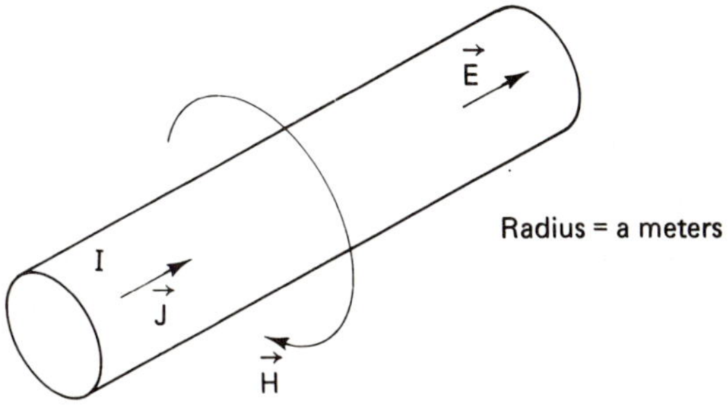

Figure 1-3. Fields at surface of conductor.

cylindrical coordinate system, and S is the cross-sectional area of the conductor in meters squared.

The student should realize that some Greek symbols are used to represent more than one physical item. For instance, ρ is used as one of the coordinates when working with the cylindrical coordinate system, but it is also used to represent the volume charge density in coulombs per meter cubed.

At the surface of the conductor,

$$\vec{J} = \sigma\vec{E} \text{ or } \vec{E} = \frac{J}{\sigma}\vec{a}_z.$$

$$\vec{H} = \frac{I}{2\pi\rho}\vec{a}_\phi. \quad \vec{P} = \vec{E} \times \vec{H} = \frac{J}{\rho}\frac{I}{2\pi a}(-\vec{a}_\rho) \text{ watts/meter}^2.$$

The power flow turns out to be perpendicular to the surface of the conductor in a direction towards the conductor.

The energy flow is from the external field into the conductor. Let us evaluate the power flow into the conductor by multiplying Poynting's vector by the surface area.

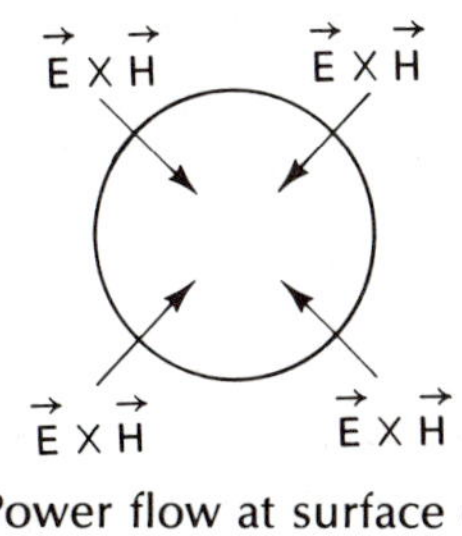

Figure 1-4. Power flow at surface of conductor.

$$|\vec{P}| = |(\vec{E} \times \vec{H})|(\text{area}) = \frac{J}{\sigma}\frac{I}{2\pi a}(2\pi a\ell) \text{ watts.}$$

$$J = \frac{I}{\pi a^2}. \quad P = \frac{I}{\sigma\pi a^2}\frac{I}{2\pi a}(2\pi a\ell) = I^2\frac{\ell}{\sigma\pi a^2} = I^2\frac{\ell}{\sigma S} = I^2R.$$

The I^2R loss of the conductor is supplied by the external field feeding the energy into the surface of the conductor. When dealing with transmission lines we do not have to worry about how the energy is transmitted since we are usually dealing with voltages, currents, and power and our normal circuit expressions will handle our calculations.

1.5 The Basic Transmission Line Circuits

As always when starting the study of a new area of engineering it is important that the student learn the terminology. Figure 1-5 gives the basic circuit and notation which will be used throughout this book.

Small letters will indicate functions of time and capital letters will indicate steady-state, r.m.s., a. c. values or, in some cases, d. c. values. The subscript $_g$ will be used to indicate that we are dealing with the generator (or source) voltage, current or impedance, (V_g, v_g,

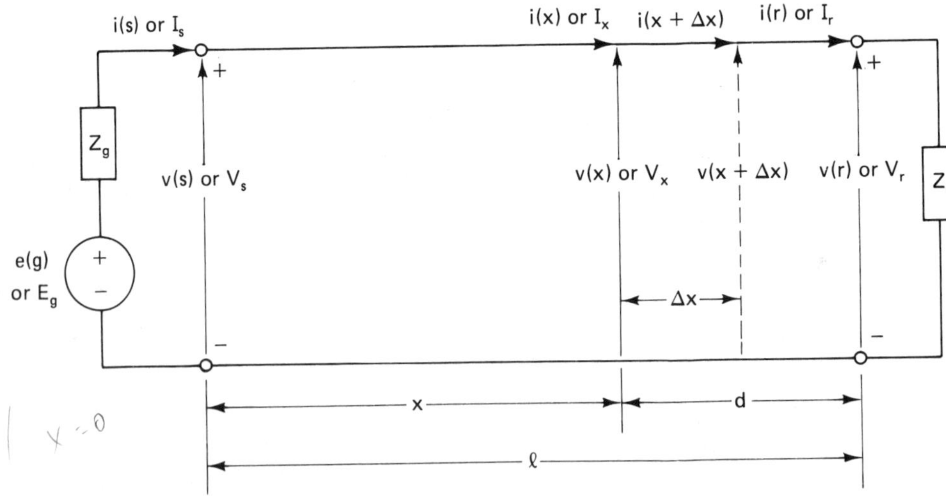

Figure 1-5. Basic circuit and notation.

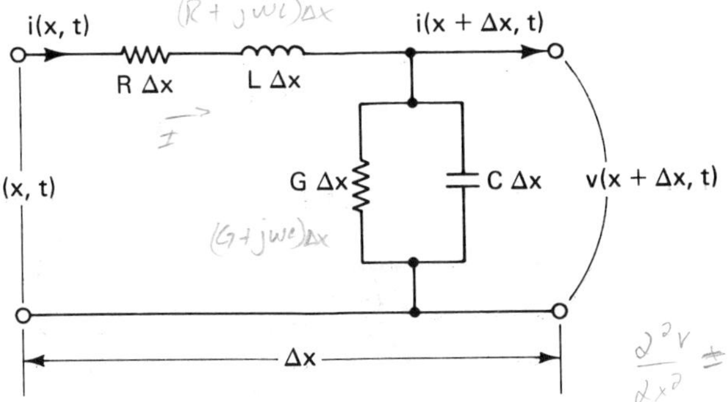

Figure 1-6. L-type equivalent circuit.

I_g, i_g, Z_g). The subscript s will be used at the input terminals of the transmission line, (V_s, v_s, I_s, i_s, Z_s). The subscript r will be used at the receiving end of the transmission line (load terminals), (V_r, v_r, I_r, i_r, Z_r). x will be used to designate the distance from the sending end to some point on the line whereas d will be the distance from the receiving end to the same point. ℓ represents the total length of the line from the sending end terminals to the receiving end terminals. We can write for all cases to be considered:

$$\ell = x + d. \tag{1-3}$$

Since the parameters are distributed along the line we can take an infinitesimal length of line, Δx, and represent it by several equivalent circuits which will approach the actual conditions as Δx approaches zero. Possible configurations are shown in Figures 1-6, 1-

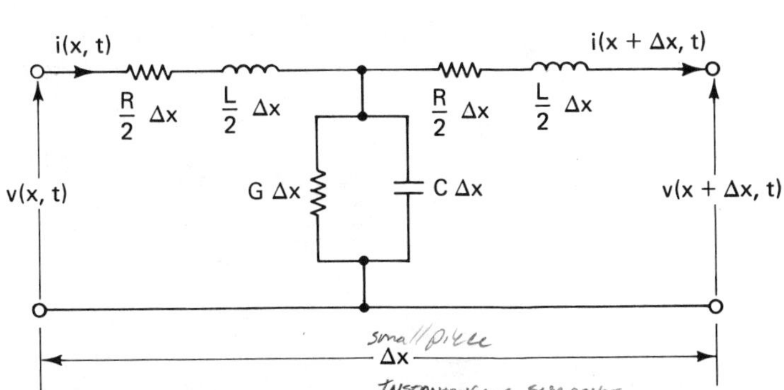

Figure 1-7. T-type equivalent circuit.

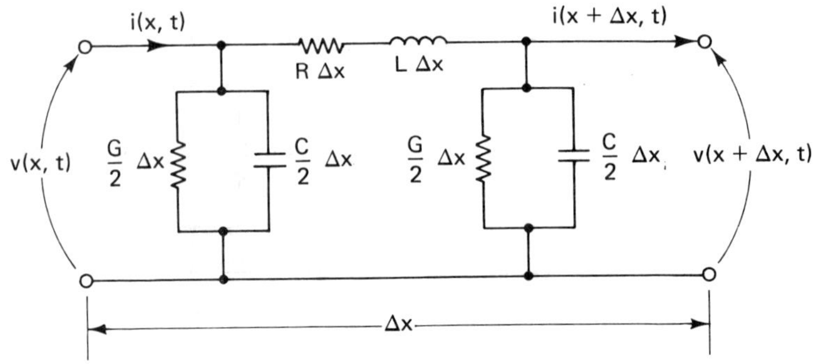

Figure 1-8. Pye-type equivalent circuit.

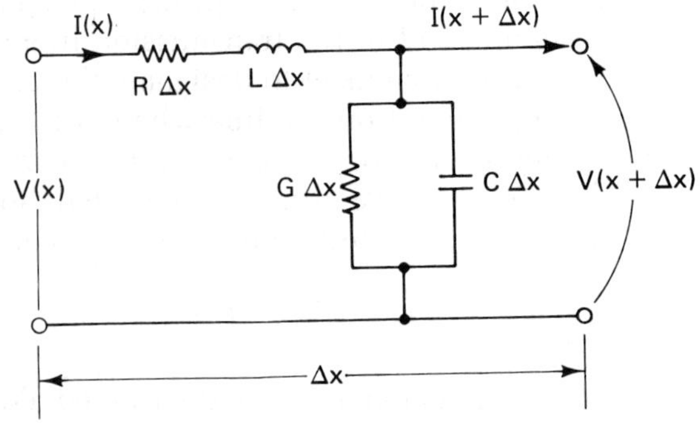

Figure 1-9. L-type equivalent circuit.

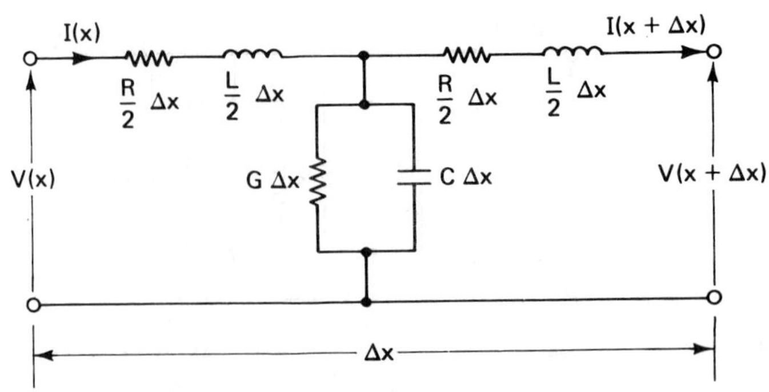

Figure 1-10. T-type equivalent circuit.

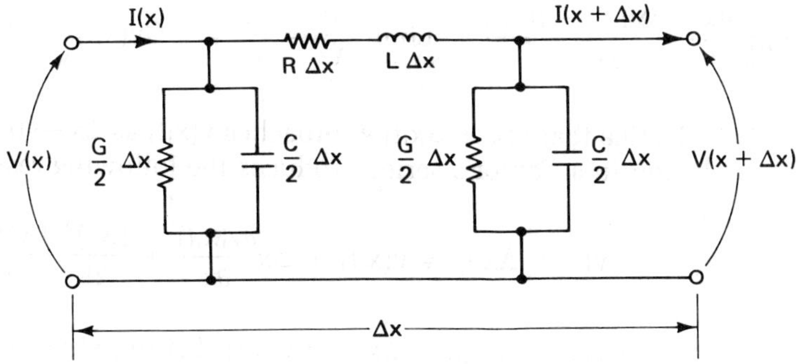

Figure 1-11. Pye-type equivalent circuit.

7, and 1-8 in the time domain and in Figures 1-9, 1-10 and 1-11 in the frequency domain.

The student should note that $R\Delta x$ gives ohms, $L\Delta x$ gives henrys, $C\Delta x$ gives farads and $G\Delta x$ gives mhos so that the elements in the equivalent circuits are back into familiar circuit terms. Let us apply Kirchhoff's laws to Figure 1-6.

$$v(x + \Delta x,t) = v(x,t) - R\Delta x\, i(x,t) - L\Delta x\frac{\partial i}{\partial t}(x,t). \tag{1-4}$$

$$i(x + \Delta x,t) = i(x,t) - G\Delta x\, v(x + \Delta x,t) - C\Delta x\frac{\partial v}{\partial t}(x + \Delta x,t). \tag{1-5}$$

or

$$v(x + \Delta x,t) - v(x,t) = \Delta v = - R\Delta x\, i(x,t) - L\Delta x\frac{\partial i}{\partial t}(x,t). \tag{1-6}$$

$$i(x + \Delta x,t) - i(x,t) = \Delta i = - G\Delta x\, v(x + \Delta x,t) - C\Delta x\frac{\partial v}{\partial t}(x + \Delta x,t) \tag{1-7}$$

This notation is used since at each point on the line the instantaneous value of the voltage or the current is a function of both position and time. If we divide by Δx and take the limit as $\Delta x \rightarrow 0$,

$$\underset{\Delta x \rightarrow 0}{\text{Limit}} \frac{v(x + \Delta x,t) - v(x,t)}{\Delta x} = \underset{\Delta x \rightarrow 0}{\text{Limit}} \frac{\Delta v}{\Delta x} = \frac{\partial v}{\partial x} = -Ri(x,t) - L\frac{\partial i}{\partial t}(x,t). \tag{1-8}$$

$$\underset{\Delta x \to 0}{\text{Limit}} \frac{i(x + \Delta x,t) - i(x,t)}{\Delta x} = \underset{\Delta x \to 0}{\text{Limit}} \frac{\Delta i}{\Delta x} = \frac{\partial i}{\partial x} = -Gv(x,t) - C\frac{\partial v}{\partial t}(x,t). \quad (1\text{-}9)$$

Notice that $v(x + \Delta x,t)$ approaches $v(x,t)$ as $\Delta x \to 0$ since an expansion in a Taylor's series produces the following result:

$$v(x + \Delta x,t) = v(x,t) + \Delta x\frac{\partial v(x,t)}{\partial x} + \frac{(\Delta x)^2}{2!}\frac{\partial^2 v(x,t)}{\partial x^2} + \ldots$$

so as $\Delta x \to 0$, $v(x + \Delta x,t) \to v(x,t)$. Letting $v(x,t)$ and $i(x,t)$ be represented by v and i to simplify the notation gives the following:

$$\frac{\partial v}{\partial x} = -Ri - L\frac{\partial i}{\partial t}. \quad (1\text{-}10)$$

$$\frac{\partial i}{\partial x} = -Gv - C\frac{\partial v}{\partial t}. \quad (1\text{-}11)$$

If we take the derivative of (1-10) with respect to x and (1-11) with respect to time we obtain (1-12) and (1-13).

$$\frac{\partial^2 v}{\partial^2 x} = -R\frac{\partial i}{\partial x} - L\frac{\partial^2 i}{\partial x \partial t}. \quad (1\text{-}12)$$

$$\frac{\partial^2 i}{\partial t \partial x} = -G\frac{\partial v}{\partial t} - C\frac{\partial^2 v}{\partial t^2}. \quad (1\text{-}13)$$

Since

$$\frac{\partial^2 i}{\partial x \partial t} = \frac{\partial^2 i}{\partial t \partial x}$$

we obtain:

$$\frac{\partial^2 v}{\partial x^2} = -R\left(-Gv - C\frac{\partial v}{\partial t}\right) - L\left(-G\frac{\partial v}{\partial t} - C\frac{\partial^2 v}{\partial t^2}\right)$$

or (1-14)

$$\frac{\partial^2 v}{\partial x^2} = RGv + RC\frac{\partial v}{\partial t} + LG\frac{\partial v}{\partial t} + LC\frac{\partial^2 v}{\partial t^2}.$$

By reversing the order of differentiation we obtain (1-15).

$$\frac{\partial^2 i}{\partial x^2} = RGi + RC\frac{\partial i}{\partial t} + LG\frac{\partial i}{\partial t} + LC\frac{\partial^2 i}{\partial t^2}. \tag{1-15}$$

Rearranging gives us the so-called transmission line equations.

$$\frac{\partial^2 v}{\partial x^2} = RGv + (RC + LG)\frac{\partial v}{\partial t} + LC\frac{\partial^2 v}{\partial t^2}, \tag{1-16}$$

$$\frac{\partial^2 i}{\partial x^2} = RGi + (RC + LG)\frac{\partial i}{\partial t} + LC\frac{\partial^2 i}{\partial t^2}. \tag{1-17}$$

These equations result no matter which of the approximate equivalent circuits is chosen.

Example 1-B

Develop equations (1-10) and (1-11) for the equivalent circuit given in Figure 1-8 which is repeated here as Figure 1-12.

Solution

Again we use Kirchhoff's laws.

$$v(x + \Delta x,t) = v(x,t) - R\Delta x\, i_1 - L\Delta x\frac{\partial i_1}{\partial t}.$$

$$i_1 = i(x,t) - \frac{C}{2}\Delta x\frac{\partial v(x,t)}{\partial t} - \frac{G}{2}\Delta x\, v(x,t).$$

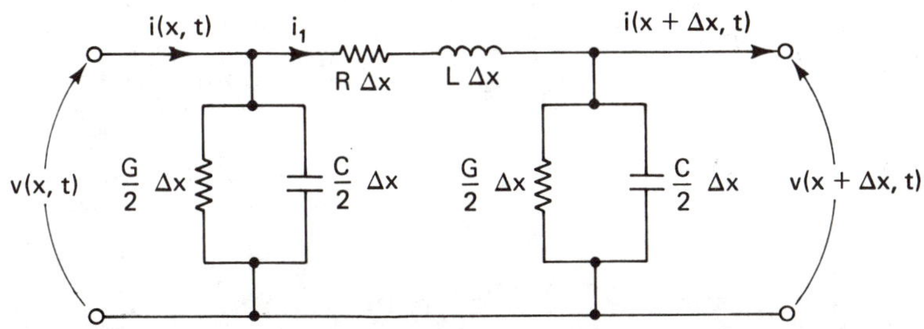

Figure 1-12. Pye-type equivalent circuit.

$$v(x + \Delta x,t) = v(x,t) - R\Delta x\left[i(x,t) - \frac{C}{2}\Delta x\frac{\partial v(x,t)}{\partial t} - \frac{G}{2}\Delta x\, v(x,t)\right]$$
$$-L\Delta x\left[\frac{\partial i}{\partial t}(x,t) - \frac{C}{2}\Delta x\frac{\partial^2 v}{\partial t^2}(x,t) - \frac{G}{2}\Delta x\frac{\partial v(x,t)}{\partial t}\right]$$

so

$$v(x + \Delta x,t) - v(x,t) = \Delta v = -R\Delta x\, i(x,t) + \frac{RC}{2}\Delta x^2\frac{\partial(x,t)}{\partial t} + \frac{RG}{2}\Delta x^2 v(x,t)$$
$$-L\Delta x\frac{\partial i(x,t)}{\partial t} + \frac{LC}{2}\Delta x^2\frac{\partial^2 v(x,t)}{\partial t^2} + \frac{LG}{2}\Delta x^2\frac{\partial v(x,t)}{\partial t}.$$

Dividing by Δx and taking the limit as $\Delta x \to 0$ gives the following:

$$\operatorname*{Limit}_{\Delta x \to 0}\frac{v(x + \Delta x,t) - v(x,t)}{\Delta x} = \operatorname*{Limit}_{\Delta x \to 0}\frac{\Delta v}{\Delta x} = \frac{\partial v}{\partial x} =$$
$$\operatorname*{Limit}_{\Delta x \to 0} -Ri(x,t) + \frac{RC}{2}\Delta x\frac{\partial v(x,t)}{\partial t} + \frac{RG}{2}\Delta x\, v(x,t) - L\frac{\partial i(x,t)}{\partial t}$$
$$+\frac{LC}{2}\Delta x\frac{\partial^2 v(x,t)}{\partial t^2} + \frac{LG}{2}\Delta x\frac{\partial v(x,t)}{\partial t} =$$
$$\frac{\partial v}{\partial x} = -Ri(x,t) - L\frac{\partial i(x,t)}{\partial t} = -Ri - L\frac{\partial i}{\partial t}$$

which is equation (1-10).

Also

$$i(x + \Delta x,t) = i(x,t) - \frac{G}{2}\Delta x\, v(x,t) - \frac{C}{2}\Delta x\frac{\partial v(x,t)}{\partial t}$$
$$-\frac{G}{2}\Delta x\, v(x + \Delta x,t) - \frac{C}{2}\Delta x\frac{\partial v(x + \Delta x,t)}{\partial t}.$$
$$\operatorname*{Limit}_{\Delta x \to 0}\frac{i(x + \Delta x,t) - i(x,t)}{\Delta x} = \operatorname*{Limit}_{\Delta x \to 0}\frac{\Delta i}{\Delta x} = \frac{\partial i}{\partial x} =$$
$$\operatorname*{Limit}_{\Delta x \to 0} -\frac{G}{2}v(x,t) - \frac{C}{2}\frac{\partial v(x,t)}{\partial t} - \frac{G}{2}v(x + \Delta x,t) - \frac{C}{2}\frac{\partial v(x + \Delta x,t)}{\partial t} =$$
$$-Gv(x,t) - C\frac{\partial v(x,t)}{\partial t} = \frac{\partial i}{\partial x} = -Gv - C\frac{\partial v}{\partial t}$$

which is equation (1-11) since in the limit $v(x + \Delta x,t)$ approaches $v(x,t)$ as $\Delta x \to 0$.

1.6 Solutions of the Transmission Line Equations for the Lossless Case

The transmission line equations have very complex solutions for the general case but if we consider a special case where $R = G = O$, we can obtain some insight into the fundamental physical phenomena on the line. Furthermore this so-called "lossless" case is very important since it turns out that in many practical engineering situations this case is approached and the "lossless" solutions give us usable answers. Equation (1-14) with R and G equal to zero gives us (1-18).

$$\frac{\partial^2 v}{\partial x^2} = LC\,\frac{\partial^2 v}{\partial t^2}. \tag{1-18}$$

and (1-15) becomes (1-19)

$$\frac{\partial^2 i}{\partial x^2} = LC\,\frac{\partial^2 i}{\partial t^2}. \tag{1-19}$$

Since these equations have the same form as some of the equations we saw in field theory it would seem logical that one could assume a similar solution for either of them. In field theory we had an equation of the type given in equation (1-20),

$$\frac{\partial^2 E_z}{\partial x^2} = \frac{\partial^2 E_z}{\partial t^2}. \tag{1-20}$$

which has a solution of:

$$E_z = E_{z\ max} \cos\,(\omega t - \beta x).$$

So for equation (1-18) we will assume the solution given in equation (1-21),

$$v = V_m \cos\,(\omega t - \beta x). \tag{1-21}$$

This, indeed, will be a solution if it satisfies the differential equation in equation (1-18). Let us find out if it does.

$$\frac{\partial v}{\partial x} = -V_m \sin(\omega t - \beta x)\,(-\beta) = \beta V_m \sin(\omega t - \beta x).$$

$$\frac{\partial^2 v}{\partial x^2} = \beta V_m \cos(\omega t - \beta x)(-\beta) = -\beta^2 V_m \cos(\omega t - \beta x).$$

$$\frac{\partial v}{\partial t} = -V_m \omega \sin(\omega t - \beta x) \text{ and } \frac{\partial^2 v}{\partial t^2} = -\omega^2 V_m \cos(\omega t - \beta x).$$

Substitution gives the following results:

$$\frac{\partial^2 v}{\partial x^2} = LC\frac{\partial^2 v}{\partial t^2}.$$

$$-\beta^2 V_m \cos(\omega t - \beta x) = LC[-\omega^2 V_m \cos(\omega t - \beta x)].$$

Thus we see that it is truly a solution if

$$\beta^2 = LC\omega^2 \text{ or } \beta = \pm\omega\sqrt{LC}. \tag{1-22}$$

Using the positive sign we have the following:

$$\frac{\omega}{\beta} = \frac{1}{\sqrt{LC}}. \tag{1-23}$$

If we investigate this equation dimensionally we find:

$$v = L\frac{di}{dt}. \qquad L = v\frac{dt}{di}. \qquad \text{henrys} = \frac{\text{volts-seconds}}{\text{ampere}}.$$

$$i = C\frac{dv}{dt}. \qquad C = i\frac{dt}{dv}. \qquad \text{farads} = \frac{\text{ampere-seconds}}{\text{volt}}.$$

The reader should note that in the above equations L and C are the traditional inductance and capacitance in henrys and farads whereas in equation (1-23) L and C are in henrys/meter and farads/meter.

$$\frac{\omega}{\beta} = \frac{1}{\sqrt{LC}} = \frac{1}{\sqrt{\frac{\text{henrys}}{\text{meter}}\frac{\text{farads}}{\text{meter}}}} = \frac{1}{\sqrt{\frac{\text{volt-seconds}}{\text{ampere-meter}}\frac{\text{ampere-second}}{\text{volt-meter}}}} =$$

$$\frac{1}{\sqrt{\frac{\text{second}^2}{\text{meter}^2}}} = \sqrt{\frac{\text{meter}^2}{\text{seconds}^2}} = \frac{\text{meters}}{\text{second}} = \text{velocity}.$$

We will define this velocity as the phase velocity, v_p.

$$v_p = \frac{\omega}{\beta} = \frac{1}{\sqrt{LC}} \frac{\text{meters}}{\text{second}}. \quad (1\text{-}24)$$

The solution, $v = V_m\cos(\omega t - \beta x)$, equation (1-21) can be rearranged as follows:

$$v = V_m\cos\omega(t - \frac{\beta x}{\omega}) = V_m\cos\omega(t - \frac{x}{\frac{\omega}{\beta}}) = V_m\cos\omega(t - \frac{x}{v_p}). \quad (1\text{-}25)$$

This is recognized as the equation for a traveling wave. If it is plotted as a function of x for different values of time we obtain Figure 1-13.

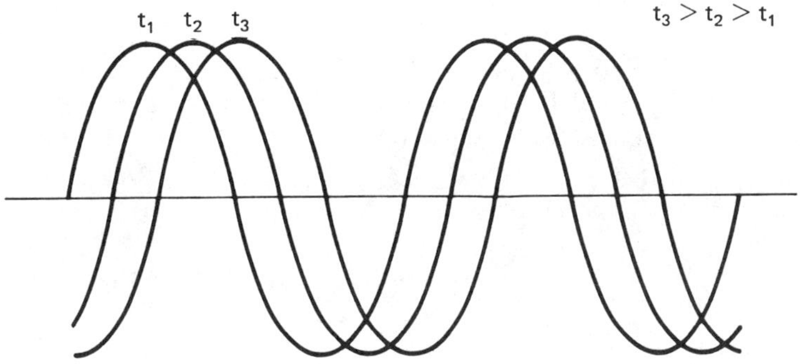

Figure 1-13. Voltage traveling waves on line.

The result is a voltage wave traveling in the positive x direction at a velocity of v_p. Notice that at a fixed value of x the voltage at that point varies as a cosine wave as a function of time.

Since the $\cos\theta = \cos(-\theta)$ we can rewrite our equation as follows:

$$v = V_m\cos(\omega t - \beta x) = V_m\cos(\beta x - \omega t) = V_m\cos\beta(x - \frac{\omega t}{\beta}) =$$
$$= V_m\cos\beta(x - v_p t). \quad (1\text{-}26)$$

It turns out that any function of $(x - v_p t)$ is a solution for our differential equation. Let us assume that $v = (x - v_p t)^3$.

$$\frac{\partial v}{\partial x} = 3(x - v_p t)^2. \qquad \frac{\partial^2 v}{\partial x^2} = 6(x - v_p t).$$

$$\frac{\partial v}{\partial t} = 3(x - v_p t)^2(-v_p).$$

$$\frac{\partial^2 v}{\partial t^2} = 6(x - v_p t)\,(-v_p)\,(-v_p) = 6v_p{}^2(x - v_p t).$$

$$\frac{\partial^2 v}{\partial x^2} = LC\frac{\partial^2 v}{\partial t^2}. \qquad (1\text{-}18)$$

$$6(x - v_p t) = LC\,[6v_p{}^2(x - v_p t)].$$

$$1 = LC\,v_p{}^2 \text{ or } v_p{}^2 = \frac{1}{LC}.$$

Thus we do have a solution of our differential equation if

$$v_p = \frac{1}{\sqrt{LC}}$$

which is the same condition that we had before.

Suppose a pulse of voltage is a function of $(x - v_p t)$. At $t = t_1$ it is located at x_1. At t_2 which is greater than t_1 the function will be constant if the quantity $(x - v_p t)$ is constant at its previous value. This can only happen if x has increased to x_2 (See Figures 1-14 and 1-15).

Therefore

$$x_1 - v_p t_1 = x_2 - v_p t_2 \text{ or } x_2 - x_1 = v_p(t_2 - t_1)$$

or

$$v_p = \frac{x_2 - x_1}{t_2 - t_1} = \frac{\Delta x}{\Delta t}$$

which will happen if the wave travels at a velocity of v_p; the wave travels Δx units in Δt seconds.

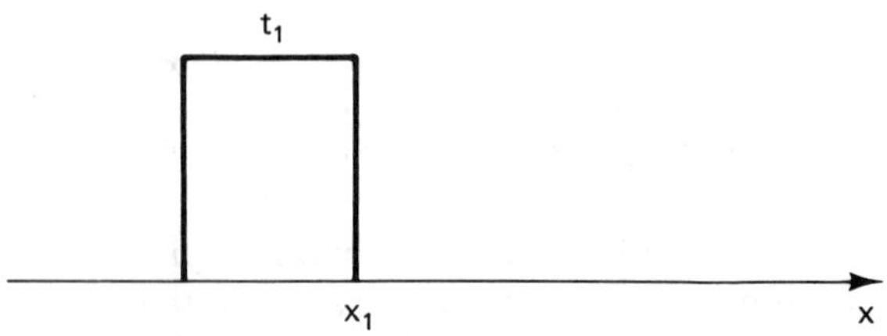

Figure 1-14. Pulse of voltage at $t = t_1$.

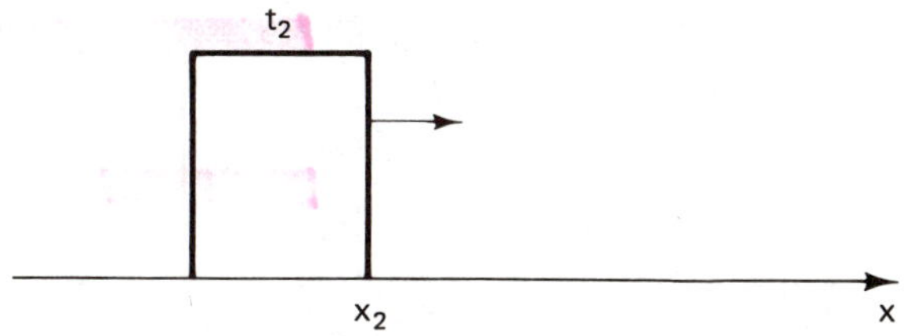

Figure 1-15. Pulse of voltage at $t = t_2$. $t_2 > t_1$.

1.7 Solutions of the Transmission Line Equations for the Steady-State, Sinusoidal Case. Propagation Constant and the Characteristic Impedance

Starting with (1-16) and (1-17) we can consider the case when the voltages and currents are sinusoidal.

$$\frac{\partial^2 v}{\partial x^2} = RGv + (RC + LG)\frac{\partial v}{\partial t} + LC\frac{\partial^2 v}{\partial t^2}. \tag{1-16}$$

$$\frac{\partial^2 i}{\partial x^2} = RGi + (RC + LG)\frac{\partial i}{\partial t} + LC\frac{\partial^2 i}{\partial t^2}. \tag{1-17}$$

Now we convert to the frequency domain and remember that when converting, the phasor is multiplied by $j\omega$ for the first derivative, and by $(j\omega)^2$ for the second derivative.

$$\frac{d^2V}{dx^2} = RGV + j\omega(RC + LG)V + LC(j\omega)^2V. \tag{1-27}$$

$$\frac{d^2I}{dx^2} = RGI + j\omega(RC + LG)I + LC(j\omega)^2I. \tag{1-28}$$

where V and I are now the phasor values. Notice, as before, that the equation for the voltage is identical to the equation for the current with V substituted for I. Let us look at the equation for the voltage. It can be rearranged to:

$$\frac{d^2V}{d^2x} = [RG + j\omega(RC + LG) + (j\omega L)(j\omega C)]\ V.$$

$$\frac{d^2V}{dx^2} = [(R + j\omega L)(G + j\omega C)]\ V. \tag{1-29}$$

Let us define a quantity, the propagation constant, γ, much as we do in field theory.

$$\gamma = \alpha + j\beta.$$

Anticipating future equations or remembering what happened in field theory, α will be defined as the attenuation constant in nepers per meter and β will be defined as the phase constant in radians per meter.

$$\gamma = \sqrt{(R + j\omega L)(G + j\omega C)} = \alpha + j\beta. \tag{1-30}$$

Notice that in general γ is a complex number. Equation (1-29) can now be written as (1-31).

$$\frac{d^2V}{dx^2} = \gamma^2 V.$$

or

$$\frac{d^2V}{dx^2} - \gamma^2 V = 0. \tag{1-31}$$

Using operational mathematics we are able to obtain a solution.

$$(D^2 - \gamma^2)V = 0. \qquad D^2 - \gamma^2 = 0. \qquad D^2 = \gamma^2. \qquad D = \pm\gamma.$$

The solution for V becomes:

$$V = Ae^{-\gamma x} + Be^{+\gamma x}. \tag{1-32}$$

Let us convert (1-10) to the phasor form.

$$\frac{\partial v}{\partial x} = -Ri - L\frac{\partial i}{\partial t} \text{ gives } \frac{dV}{dx} = -Ri - j\omega LI.$$

$$\frac{dV}{dx} = -(R + j\omega L)I.$$

or

$$I = -\frac{1}{R + j\omega L}\frac{dV}{dx}. \tag{1-33}$$

Using (1-32) we obtain:

$$I = -\frac{1}{R + j\omega L}[-\gamma A e^{-\gamma x} + \gamma B e^{\gamma x}] = \frac{\gamma}{R + j\omega L}[Ae^{-\gamma x} - Be^{\gamma x}]$$

$$= \frac{\sqrt{(R + j\omega L)(G + j\omega C)}}{R + j\omega L}[Ae^{-\gamma x} - Be^{\gamma x}] =$$

$$\sqrt{\frac{(R + j\omega L)(G + j\omega C)}{(R + j\omega L)^2}}[Ae^{-\gamma x} - Be^{\gamma x}] = \frac{1}{\sqrt{\frac{(R + j\omega L)}{(G + j\omega C}}}[Ae^{-\gamma x} - Be^{\gamma x}] \tag{1-34}$$

The quantity $\sqrt{\frac{(R + j\omega L)}{(G + j\omega C)}}$ has the dimensions of $\sqrt{\frac{\text{ohms/meter}}{\text{mhos/meter}}} =$

$\sqrt{\text{ohms}^2}$ = ohms, and is defined as the Characteristic Impedance. Another name, often used by the power engineer, is the Surge Impedance. Defining the characteristic impedance as Z_o, we have:

$$Z_o = \sqrt{\frac{(R + j\omega L)}{(G + j\omega C)}} \text{ ohms.} \tag{1-35}$$

The reciprocal is defined as the Characteristic Admittance.

$$Y_o = \frac{1}{Z_o} = \sqrt{\frac{(G + j\omega C)}{(R + j\omega L)}} \text{ mhos.} \tag{1-36}$$

We now have the two fundamental equations:

$$V = Ae^{-\gamma x} + Be^{\gamma x}. \tag{1-32}$$

$$I = \frac{Ae^{-\gamma x}}{Z_o} - \frac{Be^{\gamma x}}{Z_o}. \tag{1-37}$$

The student should notice the minus sign in front of the second term for the current.

In many practical engineering situations it is useful to assume the lossless case, $(R = G = 0)$ or the high frequency case, $\omega L \gg R$ and $\omega C \gg G$.

$$\gamma = \sqrt{(R + j\omega L)(G + j\omega C)} \simeq \sqrt{(j\omega L)(j\omega C)}$$
$$\simeq j\omega\sqrt{LC} = \alpha + j\beta. \tag{1-38}$$

In this case,

$$\alpha = 0, \text{ and } \beta = \omega\sqrt{LC}.$$

This gives:

$$\frac{\omega}{\beta} = \frac{1}{\sqrt{LC}}.$$

This is the same result as equation (1-23) which we obtained in the time domain, and which was defined as the phase velocity, v_p.

$$v_p = \frac{\omega}{\beta} = \frac{1}{\sqrt{LC}}. \quad (1\text{-}24)$$

Under our assumed conditions the characteristic impedance also changes to a purely resistive quantity.

$$Z_o = \sqrt{\frac{(R + j\omega L)}{(G + j\omega C)}} \simeq \sqrt{\frac{j\omega L}{j\omega C}} \simeq \sqrt{\frac{L}{C}}. \quad (1\text{-}39)$$

The d. c. case may be thought of as a special case with the frequency set to zero. If we set the frequency to zero we have:

$$\gamma = \sqrt{(R + j\omega L)(G + j\omega C)} \simeq \sqrt{R\,G} = \alpha + j\beta. \quad (1\text{-}40)$$

$$\alpha \simeq \sqrt{R\,G}. \qquad \beta \simeq 0.$$

$$Z_o = \sqrt{\frac{(R + j\omega L)}{(G + j\omega C)}} \simeq \sqrt{\frac{R}{G}} \quad (1\text{-}41)$$

When taking the square root for the propagation constant or the characteristic impedance one could take it as plus or minus. We will take the positive sign for the characteristic impedance which gives a positive resistive term. We could take either sign for the propagation constant which would simply change the constants in (1-32) and (1-37), but it is common practice to take the plus sign and we shall continue this practice.

1.8 Ranges for the Propagation Constant and the Characteristic Impedance

Looking at the propagation constant and the characteristic impedance indicates that in general they are complex numbers with a magnitude and an angle.

$$\gamma = \sqrt{(R + j\omega L)(G + j\omega C)} = \alpha + j\beta = \gamma \angle \phi.$$

We have already seen the extreme cases.

If $\omega = 0$, or $R \gg \omega L$ and $G \gg \omega C$:

$$\gamma = \sqrt{RG} = \alpha + j0 = \sqrt{RG} \angle 0° \text{ so } \phi = 0°.$$

If $R = G = 0$ or $\omega L \gg R$ and $\omega C \gg G$

$$\gamma = \sqrt{(j\omega L)(j\omega C)} = 0 + j\omega\sqrt{LC} = \alpha + j\beta$$
$$= \omega \sqrt{LC} \angle 90° \text{ so } \phi = 90°.$$

So ϕ varies between 0° and 90°.

For Z_o we have:

$$Z_o = \sqrt{\frac{(R + j\omega L)}{(G + j\omega C)}} = Z_o \angle \theta°.$$

Looking at the extreme cases we have:

$$R = C = 0. \qquad Z_o = \sqrt{\frac{j\omega L}{G}} = \sqrt{\frac{\omega L}{G}} \angle 45°.$$

$$G = L = 0. \qquad Z_o = \sqrt{\frac{R}{j\omega C}} = \sqrt{\frac{R}{\omega C}} \angle -45°.$$

So we see that the angle of the characteristic impedance, θ, has a theoretical range from $-45°$ to $+45°$.

Example 1-C

A coaxial cable connects a radio transmitter in the university library with the antenna on the roof. It can be assumed lossless. If L equals 800×10^{-9} henrys per meter and C equals 320.0 picofarads per meter and the frequency is 970 kilohertz determine:

(a) The characteristic impedance.
(b) The propagation constant.
(c) The phase velocity.

Solution

$$\omega = 2\pi f = 2\pi(970 \times 10^3) = 6.09 \times 10^6 \text{ radian/sec.}$$

$$\omega L = (6.09 \times 10^6)(800 \times 10^{-9}) = 4.875 \text{ ohms.}$$

$$\omega C = (6.09 \times 10^6)(320 \times 10^{-12}) = 1.95 \times 10^{-3} \text{ mhos.}$$

$$Z_o = \sqrt{\frac{L}{C}} = \sqrt{\frac{800 \times 10^{-9}}{320 \times 10^{-12}}} = 50 \text{ ohms.} \quad \text{(a)}$$

$$\gamma = \sqrt{(0 + j4.875)(0 + j1.95 \times 10^{-3})} = 97.5 \times 10^{-3} \underline{/90^\circ} = 0 + j97.5 \times 10^{-3}$$

$$\alpha = 0, \beta = 97.5 \times 10^{-3} \text{ radians/meter.} \quad \text{(b)}$$

$$v_p = \frac{\omega}{\beta} = \frac{6.09 \times 10^6}{97.5 \times 10^{-3}} = 62.5 \times 10^6 \text{ meters/second.} \quad \text{(c)}$$

1.9 A Detailed Look at the Characteristic Impedance

Let us assume that the transmission line is terminated in its characteristic impedance, Z_o, and that we go to the receiving end where $x = \ell$, $V = V_r$ and $I = I_r$ (See Figure 1-16).

$$V = Ae^{-\gamma x} + Be^{\gamma x} \quad \text{(1-32)}$$

becomes

$$V_r = Ae^{-\gamma \ell} + Be^{\gamma \ell}$$

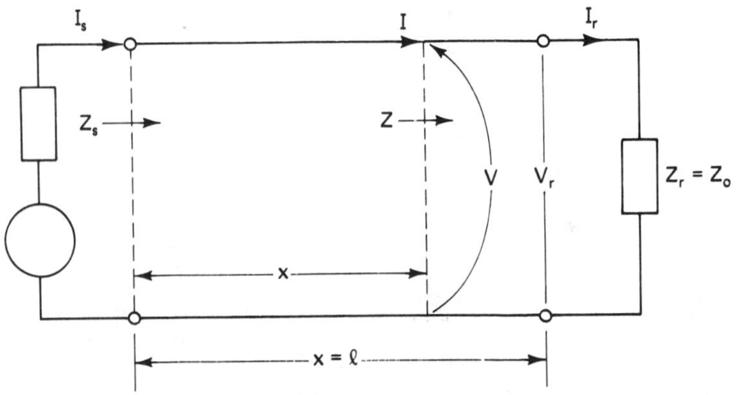

Figure 1-16. Line terminated in Z_o.

and

$$I = \frac{Ae^{-\gamma x}}{Z_o} - \frac{Be^{\gamma x}}{Z_o} \tag{1-37}$$

becomes

$$I_r = \frac{Ae^{-\gamma \ell}}{Z_o} - \frac{Be^{\gamma \ell}}{Z_o}.$$

But:

$$I_r = \frac{V_r}{Z_o} = \frac{V_r}{Z_r} = \frac{A\,e^{-\gamma \ell}}{Z_o} + \frac{B\,e^{\gamma \ell}}{Z_o} = \frac{A\,e^{-\gamma \ell}}{Z_o} - \frac{B\,e^{\gamma \ell}}{Z_o}.$$

So

$$B = 0 \text{ and } V_r = Ae^{-\gamma \ell} \text{ and } I_r = \frac{Ae^{-\gamma \ell}}{Z_o}.$$

$$Z_r = \frac{V_r}{I_r} = \frac{Ae^{-\gamma \ell}}{\frac{Ae^{-\gamma \ell}}{Z_o}} = Z_o$$

and

$$V = Ae^{-\gamma x}$$

and

$$I = \frac{A\,e^{-\gamma x}}{Z_o}.$$

One important consideration on transmission lines is the impedance at any point on the line looking towards the load. This is simply the voltage at that point (x) divided by the current at that same point. Here we find that:

$$Z = \frac{V}{I} = \frac{Ae^{-\gamma x}}{\frac{A\,e^{-\gamma x}}{Z_o}} = Z_o.$$

At the sending end the input impedance of the line becomes:

$$Z_s = \frac{V_s}{I_s}$$

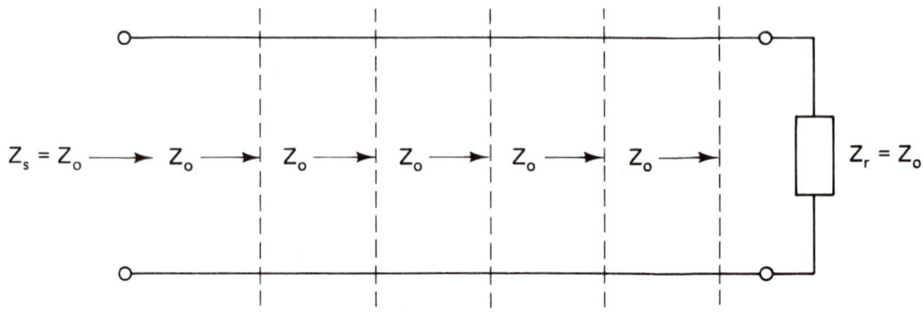

Figure 1-17. Line terminated in Z_o.

but at the sending end x = 0, so the input impedance becomes

$$Z_s = \frac{A}{\frac{A}{Z_o}} = Z_o.$$

We now have the interesting result that if the transmission line is terminated in its characteristic impedance, the impedance all along the line looking towards the load is the characteristic impedance, Z_o, and the input impedance is also Z_o (See Figure 1-17). Under these conditions the source sees an impedance of Z_o (See Figure 1-18). Under these conditions the source sees an impedance of Z_o.

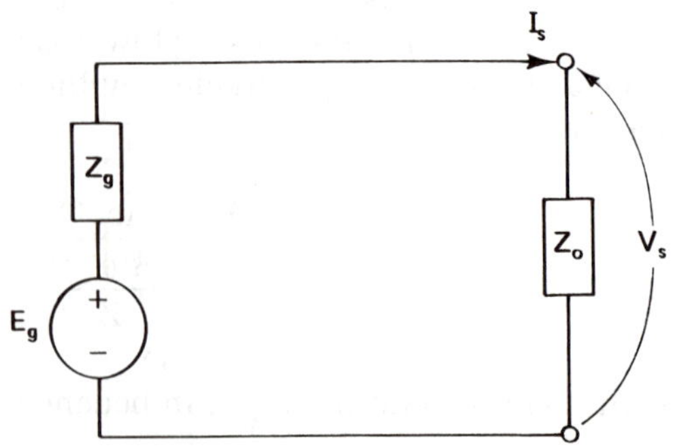

Figure 1-18. Line terminated in Z_o.

So:

$$I_s = \frac{E_g}{Z_g + Z_o} \tag{1-42}$$

and

$$V_s = I_s Z_o = \frac{E_g Z_o}{Z_g + Z_o}. \tag{1-43}$$

At $x = 0$, $I = I_s$ and $V = V_s$.

$$I = \frac{A}{Z_o} e^{-\gamma x}$$

so

$$I_s = \frac{A}{Z_o}.$$

$$A = I_s Z_o$$

and

$$I = \frac{I_s Z_o e^{-\gamma x}}{Z_o} =$$

$$I = I_s e^{-\gamma x}. \tag{1-44}$$

$$V = Ae^{-\gamma x}$$

so

$$V_s = A$$

or

$$V = V_s e^{-\gamma x}. \tag{1-45}$$

and

$$V = \frac{E_g Z_o\, e^{-\gamma x}}{Z_g + Z_o}. \tag{1-46}$$

$$I = \frac{E_g e^{-\gamma x}}{Z_g + Z_o}. \tag{1-47}$$

Let us now assume that the transmission line is infinite (See Figure 1-19).

(1-32) $V = Ae^{-\gamma x} + Be^{\gamma x}$.

(1-30) $\gamma = \alpha + j\beta$,

Figure 1-19. Infinite line.

so

$$V = Ae^{-(\alpha + j\beta)x} + Be^{(\alpha + j\beta)x} = Ae^{-\alpha x}e^{-j\beta x} + Be^{\alpha x}e^{j\beta x}.$$

$$I = \frac{A}{Z_o} e^{-\gamma x} - \frac{Be^{\gamma x}}{Z_o} = \frac{A}{Z_o} e^{-(\alpha + j\beta)x} - \frac{B}{Z_o} e^{(\alpha + j\beta)x} =$$

(1-37)

$$I = \frac{A}{Z_o} e^{-\alpha x}e^{-j\beta x} - \frac{B}{Z_o} e^{\alpha x}e^{j\beta x}.$$

With α positive the term $e^{-\alpha x}$ becomes smaller as x increases and the term $e^{\alpha x}$ becomes larger. Since $e^{-j\beta x} = \cos\beta x - j\sin\beta x = 1 \angle \beta x$ and $e^{j\beta x} = \cos\beta x + j\sin\beta x = 1 \angle \beta x$ these terms always have a magnitude of unity. Since V and I must remain finite as x increases it is obvious that B must be equal to zero. So we end up with the same conditions that we had when the line was terminated in Z_o.

$$V = Ae^{-\gamma x}$$

and

$$I = \frac{A}{Z_o} e^{-\gamma x}$$

and the impedance anywhere along the line looking towards the load is the characteristic impedance, Z_o. At the input of the line the results are the same whether the line is terminated in Z_o or the line is infinite. A line terminated in Z_o is said to be matched and, as we shall see shortly, has no standing wave on it. Energy received by the load is completely absorbed and no energy is reflected back towards the sending end of the line.

Example 1-D

Two lines, each with a characteristic impedance of Z_o, are each terminated in a load impedance of Z_o. What is the input impedance at the point where the two lines are connected in parallel?

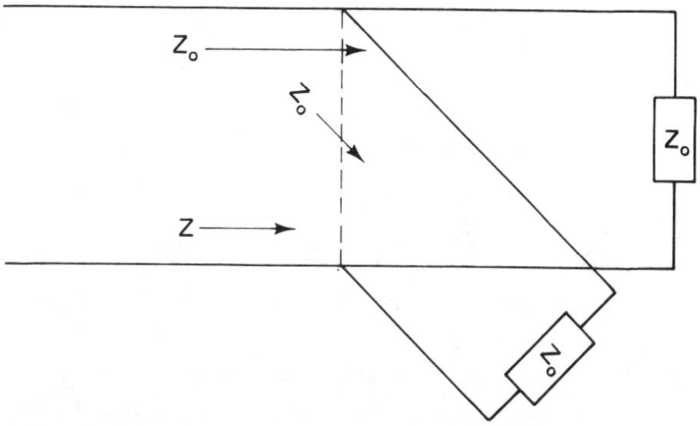

Solution

Since each line is terminated in Z_o, each line has an input impedance of Z_o at the point of paralleling. Therefore:

$$Z = \frac{(Z_o)(Z_o)}{Z_o + Z_o} = \frac{Z_o}{2} \text{ ohms.}$$

Notice that if the line to the left of the point of paralleling had a characteristic impedance of $Z_o/2$ that line would be matched with the two lines in parallel and the input impedance of that line would be $Z_o/2$.

1.10 Traveling Waves on a Transmission Line

If we substitute the propagation constant into (1-32) we have:

$$V = Ae^{-\alpha x}e^{-j\beta x} + Be^{\alpha x}e^{j\beta x}. \tag{1-48}$$

We can obtain the function of time by multiplying the $e^{j\omega t}$ and taking the real part.

$$
\begin{aligned}
v(t) &= \mathcal{R}[Ae^{-\alpha x}e^{-j\beta x}e^{j\omega t} + Be^{\alpha x}e^{j\beta x}e^{j\omega t}] = \\
&\mathcal{R}[Ae^{-\alpha x}e^{j(\omega t - \beta x)} + Be^{\alpha x}e^{j(\omega t + \beta x)}] = \\
&Ae^{-\alpha x}\cos(\omega t - \beta x) + Be^{\alpha x}\cos(\omega t + \beta x). \qquad (1\text{-}49)
\end{aligned}
$$

The term $Ae^{-\alpha x}\cos(\omega t - \beta x)$ represents a wave traveling in the positive x direction that is being attenuated at a rate determined by $e^{-\alpha x}$. The term $Be^{\alpha x}\cos(\omega t + \beta x)$ represents a wave traveling in the minus x direction that has its magnitude changed by the factor, $e^{\alpha x}$. We have already seen that the term $\cos(\omega t - \beta x)$ is a solution of the transmission line differential equations. The term $\cos(\omega t + \beta x)$ is also a solution for the differential equations. In Section 1-8 we saw that a line terminated in Z_o (or an infinite line) has the constant, B, equal to zero. Therefore a matched line has only one traveling wave moving towards the load. We shall call this the forward wave. If B is not equal to zero because the line is not matched then we will have the wave traveling in the minus x direction. We shall call this the backward, reflected, or reverse wave. We would then have two traveling waves moving in opposite directions.

As we shall see later, this results in a standing wave on the line and many other interesting results. If we consider the attenuation caused by $e^{\alpha x}$ we are aware that with α positive we really have negative attenuation since the term grows larger as x increases. Physically, we have a wave traveling from the sending end to the receiving end; its magnitude is decreasing as it moves towards the load. We also have a wave traveling from the load towards the sending end; its magnitude is also decreasing as it moves towards the sending end but is increasing as we see how its magnitude changes from the sending end to the receiving end.

If we insert the propagation constant into (1-37) for the current, we have:

$$
I = \frac{A}{Z_o}e^{-\alpha x}e^{-j\beta x} - \frac{B}{Z_o}e^{\alpha x}e^{j\beta x}
$$

Now we convert to the time function.

$$
\begin{aligned}
i(t) &= \mathcal{R}\left[\frac{A}{Z_o}e^{-\alpha x}e^{-j\beta x}e^{j\omega t} - \frac{B}{Z_o}e^{\alpha x}e^{j\beta x}e^{j\omega t}\right] = \\
&\frac{A}{Z_o}e^{-\alpha x}\cos(\omega t - \beta x) - \frac{Be^{\alpha x}}{Z_o}\cos(\omega t + \beta x). \qquad (1\text{-}50)
\end{aligned}
$$

We see that the current also, in general, consists of two traveling waves moving in opposite directions unless B equals zero.

If the characteristic impedance is not real; that is, if

$$Z_o = |Z_o|\underline{/\theta}.$$

$$i(t) = \mathcal{R}\left[\frac{A}{|Z_o|e^{j\theta}}e^{-\alpha x}e^{-j\beta x}e^{j\omega t} - \frac{B}{|Z_o|e^{j\theta}}e^{\alpha x}e^{j\beta x}e^{j\omega t}\right] =$$

$$\mathcal{R}\left[\frac{A}{Z_o}e^{-\alpha x}e^{j\omega t}e^{-j\beta x}e^{-j\theta} - \frac{B}{Z_o}e^{\alpha x}e^{j\omega t}e^{\beta x}e^{-j\theta}\right] =$$

$$\frac{A}{Z_o}e^{-\alpha x}\cos(\omega t - \beta x - \theta) - \frac{B}{Z_o}e^{\alpha x}\cos(\omega t + \beta x - \theta). \quad (1\text{-}51)$$

Under these conditions the forward voltage and current waves are out of phase with each other, with the current lagging the voltage by the angle, θ. The backward current is also lagging the backward voltage by the same angle. It is important to notice that the forward waves are associated with $Ae^{-\gamma x}$ and the backward waves are associated with $Ae^{\gamma x}$.

Let us assume a lossless line with $R = G = 0$, and consider only the forward wave. Under these conditions $\alpha = 0$ and $\beta = \omega\sqrt{LC}$ so

$$v(t) = A\cos(\omega t - \beta x) = A\cos(\omega t - \omega\sqrt{LC}\,x) = A\cos\omega(t - \sqrt{LC}\,x) = A\cos\omega\left(t - \frac{x}{\frac{1}{\sqrt{LC}}}\right) = A\cos\omega\left(t - \frac{x}{v_p}\right).$$

$v(t)$ will have a constant magnitude if $t - \frac{x}{v_p}$ is equal to a constant $= C$.

With

$$t - \frac{x}{v_p} = C$$

and taking the derivative we have:

$$dt - \frac{dx}{v_p} = 0 \text{ or } v_p = \frac{dx}{dt}$$

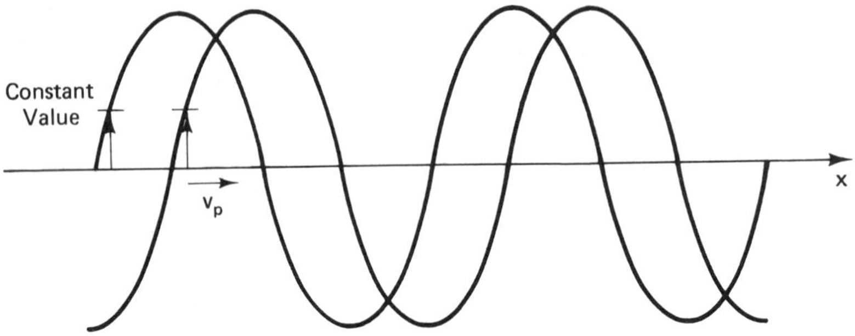

Figure 1-20. Constant value of traveling wave.

and the wave is moving in the positive x direction with a velocity of v_p. In other words, a constant value of $v(t)$ moves in the x direction at the phase velocity (See Figure 1-20).

If we plot our voltage wave as a function of x for a constant value of time we obtain results as shown in Figure 1-21.

A given value of the wave will repeat itself when $\beta x = 2\pi$ radians. This value of x between the two points on the wave is called the wavelength, λ, or $\beta\lambda = 2\pi$. So we have as the result:

$$\lambda = \frac{2\pi}{\beta}. \tag{1-52}$$

Since

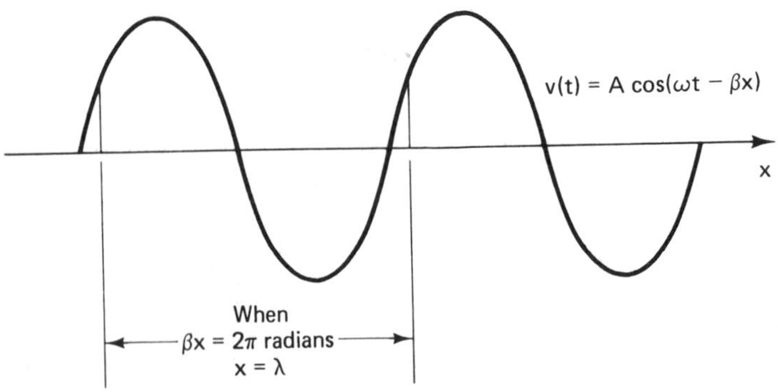

Figure 1-21. Traveling wave as a function of x.

$$v_p = \frac{\omega}{\beta} \qquad \lambda = \frac{2\pi}{\frac{\omega}{v_p}} = \frac{2\pi\, v_p}{\omega} = \frac{2\pi v_p}{2\pi f} = \frac{v_p}{f}.$$

$$v_p = \lambda f. \tag{1-53}$$

In many cases v_p is the velocity of light so we have:

$$\lambda = \frac{3 \times 10^8}{f}. \tag{1-54}$$

The velocity of light is $(2.997930 \pm 0.0000003) \times 10^8$ meters per second but we will use the value 3×10^8 which is close enough for most transmission line applications.

A convenient equation for the student to remember can be written in the following form:

$$\lambda \text{ (in meters)} = \frac{300}{f\text{(in megahertz)}}. \tag{1-55}$$

From this equation we can set up Table 1-1.

Table 1-1

Frequency (hertz)	*Frequency (megahertz)*	*Wavelength*
30×10^6	30	10.0 meter
300×10^6	300	1 meter
$3{,}000 \times 10^6$	3,000	0.1 meter

It is apparent that a transmission line physically short (several feet) may contain many wavelengths at 10 gigahertz and a line physically long (several hundred miles) would have only a fraction of a wavelength at 60 hertz with a wavelength of approximately 3,000 miles.

Example 1-E

The voltage on a transmission line is given as:

$$v(t) = 150 \cos(7.5 \times 10^{10} t - 500x)$$

Determine: (a) the frequency, (b) the phase velocity, (c) the wavelength, and (d) the phasor expression for v.

Solution

Compare

$$v(t) = 150 \cos(7.5 \times 10^{10}t - 500x)$$

with

$$v(t) = V_m \cos(\omega t - \beta x)$$

$$\beta = 500 \qquad \omega = 7.5 \times 10^{10} = 2\pi f$$

$$f = \frac{7.5 \times 10^{10}}{2\pi} = 1.19 \times 10^{10} \text{ hertz} \qquad \text{(a)}$$

$$v_p = \frac{\omega}{\beta} = \frac{7.5 \times 10^{10}}{500} = 150 \times 10^{6} \text{ meters/second} \qquad \text{(b)}$$

$$\lambda = \frac{v_p}{f} = \frac{150 \times 10^{6}}{1.19 \times 10^{10}} = 12.56 \times 10^{-3} \text{ meters} \qquad \text{(c)}$$

$$V = 150 \underline{/-500x} \text{ volts.} \qquad \text{(d)}$$

1.11 Chapter Summary

1. The assumption of "lumped" circuit elements may or may not provide solutions that are useful to the engineer. Distributed parameters along any transmission line system may have to be taken into account depending on a number of factors such as: (a) Length of line; (b) Frequencies involved; (c) Types of signals; (d) Waveform of signal and need for accurate reproduction at the receiving end; (e) Timing between signals; (f) Efficiency considerations; (g) Type of load; (h) Maximum power transfer to receiving end.
2. The distributed parameters, R(ohms/length), G(mhos/length), L(henrys/length), C(farads/length), must be multiplied by a unit of length in order to obtain the quantities we are most familiar with in elementary circuit analysis such as ohms, henrys, etc. The distributed parameters are used in different types of circuits to represent a differential length of line. By applying Kirchhoff's voltage and current laws and allowing the length of the line to approach zero, the circuit equations become the basic differ-

ential transmission line equations. These equations are difficult to solve in the general case but easy to handle for a number of important practical engineering cases.

3. The lossless line case with $R = G = 0$ gives a traveling wave type of solution with voltage and current waves traveling at a phase velocity of v_p. It accounts for the possibility of two waves, each traveling in opposite directions on the line.
4. One of the more important engineering solutions is the steady-state sinusoidal phasor case. This solution introduces two useful concepts:

(a) Characteristic Impedance

$$Z_o = \sqrt{\frac{R + j\omega L}{G + j\omega C}}.$$

(b) Propagation Constant

$$\gamma = \alpha + j\beta = \sqrt{(R + j\omega L)(G + j\omega C)}$$

where α is the attenuation constant in meters/length and β is the phase constant in radians/length.

Important engineering cases arise for the lossless and the high frequency case.

$$Z_o = \sqrt{\frac{L}{C}}, \ \alpha = 0 \text{ and } \beta = \omega \sqrt{LC}.$$

5. The basic phasor solutions are

$$V = Ae^{-\gamma x} + Be^{\gamma x}$$

and

$$I = \frac{Ae^{-\gamma x}}{Z_o} - \frac{Be^{\gamma x}}{Z_o}.$$

These phasor solutions may be converted to the time domain by multiplying by $e^{j\omega t}$ and taking the real part of the resulting equation. The first term represents a wave traveling in the forward $(+x)$ direction and the second term represents a wave traveling in the opposite (reverse) $(-x)$ direction. If B is zero there is only the one, forward traveling wave. B will be zero if the load impedance equals the characteristic impedance or if the line is infinite.

6. If the characteristic impedance is real the sinusoidal voltage and the sinusoidal current of both the forward and reverse waves are in phase with each other but if the characteristic impedance is complex there is an angle between the respective voltages and currents.
7. If we look at the distribution of either the voltage or the current on the line for the a. c. case at any instant of time there will be a sinusoidal distribution along the line; the distance between two adjacent like-values is called the wavelength, λ.
8. The waves travel at a velocity called the phase velocity.

$$v_p = \frac{\omega}{\beta}.$$

9. The wavelength and the phase velocity are related to each other through the frequency.

$$\lambda = \frac{v_p}{f}.$$

10. In the United States, the parameters of transmission lines are often expressed in units other than meters. Resistance may be given in terms of ohms per thousand feet and the inductance in terms of henrys per mile. This does not affect the characteristic impedance as long as all the parameters are given on the basis of the same unit of length, but the student must be careful with the propagation constant since one can obtain such things as nepers/foot, nepers/mile, etc.

1.12 Review and Study Questions

1. State Kirchhoff's voltage and current laws.
2. State Ampere's circuital law and Gauss's law.
3. Define capacitance.
4. Define flux linkages and inductance.
5. How is the inductance and capacitance of a parallel wire line affected by changing the distance between the two wires?
6. State the units for the distributed parameters, R, L, G and C.
7. Sketch the pye, T, and the two L-type representations of a differential length of transmission line giving the values of each element. Write the basic differential equations for each of these circuits. Then develop the transmission line equations by letting the length of the line approach zero.
8. Give the equations for the characteristic impedance and the propagation constant and give the units of each.

9. How is a cosinusoidal (or sinusoidal) function of time converted to the frequency domain (phasor form)? How is a phasor converted to a sinusoidal (cosinusoidal) function of time?
10. What is the range of values for the angle of the characteristic impedance and for the angle of the propagation constant?
11. Under what conditions will there be only a forward traveling wave on a transmission line?
12. How do we convert the following to hertz?
    (a) kilohertz (b) megahertz (c) gigahertz
13. Under what conditions will the attenuation constant be zero?
14. Under what conditions will the phase constant be zero?
15. What are the units of $\frac{1}{\sqrt{LC}}$?
16. How many solutions are there to the differential equation (1-18)?
17. What does the term dielectric strength mean?

Problems

1-1. Starting with the circuit in Figure 1-7 develop equations (1-10) and (1-11).

1-2. Determine if the following voltages are solutions for the differential equation given as equation (1-18).
(a) $Ae^{-\alpha x}\cos(\omega t - \beta x)$
(b) $B\sin(\omega t - \beta x)$
(c) $C(x - v_p t)^4$

1-3. The following shows a d. c. circuit "representation" of a transmission line.

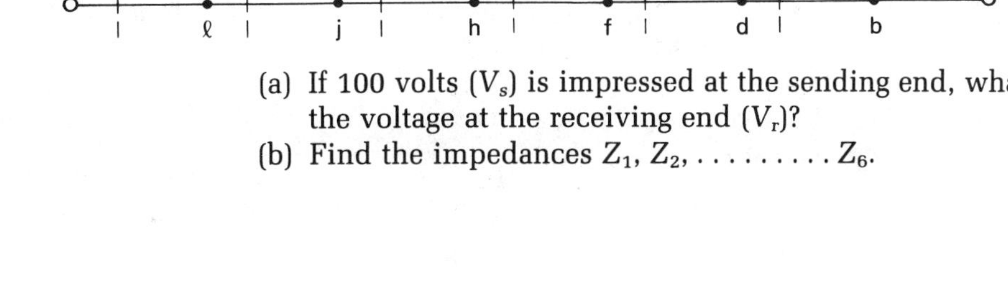

(a) If 100 volts (V_s) is impressed at the sending end, what is the voltage at the receiving end (V_r)?
(b) Find the impedances $Z_1, Z_2, \ldots\ldots\ldots Z_6$.

(c) If the sections continued on and on and on to the left do you suppose that $Z_7, \ldots\ldots\ldots Z_{100}, \ldots\ldots\ldots Z_{1,000}$ would approach some sort of a limit? If so, what is the limit?

1-4. Convert the following:
(a) 45 ohms per thousand feet to ohms per mile.
(b) 2 millihenrys per meter to henrys per mile.
(c) 2 picofarads per meter to farads per foot.
(d) 4.5×10^{-6} mhos per mile to mhos per meter.

1-5. A lossless transmission line has a characteristic impedance of 50 ohms. Its load consists of three identical lines connected in parallel. These lines have a characteristic impedance of Z_o and are terminated in a load impedance so that the load is matched with the characteristic impedance.
(a) If the 50 ohm line is to be matched with its load what should be the value of Z_o?
(b) If under the conditions of part (a), 10 volts r.m.s. is impressed on the sending end of the 50 ohm line, what will be the power delivered to each of the three loads?

1-6. If a transmission line has a characteristic impedance of $200 - j100$ ohms it is obviously a lossy line.
(a) What indicates it is lossy?
(b) What is its characteristic admittance?
(c) If we place a resistor and capacitance in parallel as a load on the line what should be the values of R and C to match the line if the frequency is 15 kilohertz?

1-7. If $R/L = G/C$ for a transmission line determine:
(a) α
(b) β
(c) Z_o

1-8. A certain transmission line has the following constants per mile. The frequency is 10 kilohertz.

$$R = 3.0 \text{ ohms/mile}, \; L = 2 \times 10^{-3} \text{ henrys/mile},$$
$$G = 0.02 \times 10^{-6} \text{mhos/mile}, \; C = 0.05 \times 10^{-6} \text{ farads/mile}.$$

Determine:
(a) α (c) Z_o (e) v_p
(b) β (d) λ

1-9. At high frequencies the inductance, capacitance and resistance per unit length of a coaxial cable transmission line are given in M.K.S. units as:

$$\frac{L}{\ell} = \frac{\mu}{2\pi} \ln \frac{b}{a} \qquad \frac{C}{\ell} = \frac{2\pi\varepsilon}{\ln \frac{b}{a}} \qquad \frac{R}{\ell} = \frac{1}{2}\sqrt{\frac{f\mu}{\pi\sigma}}\left[\frac{1}{a} + \frac{1}{b}\right]$$

where a is the radius of the inner conductor and b is the inner radius of the outer conductor. Assume that $\mu = \mu_o$, and $\varepsilon = \varepsilon_o$. For copper $\sigma = 5.8 \times 10^7$ mhos/meter. If $a = 1.0$ mm and $b = 2.7$ mm and the frequency is 2×10^{10} hertz determine: (a) α, (b) β, (c) Z_o, (d) v_p, (e) λ.

1-10. Design a coaxial cable for high frequency, lossless operation with a characteristic impedance of 50 ohms if the inner conductor has a radius of 0.1 inch and the dielectric constant is 1.0. See problem 1-9.

1-11. In the following circuit $L = 1 \times 10^{-3}$ henrys, $C = 0.1 \times 10^{-6}$ farads and $R = 100$ ohms.

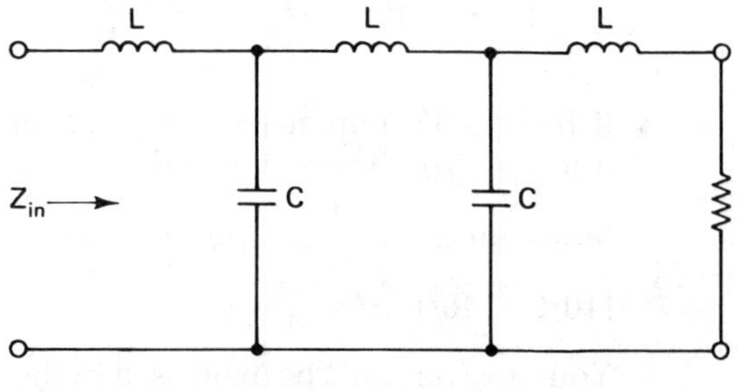

Find Z_{in} when
(a) $\omega = 100$ radians per second.
(b) $\omega = 100{,}000$ radians per second.
(c) $\omega = 100 \times 10^6$ radians per second.

1-12. Repeat problem 1-11 if $L = 40 \times 10^{-3}$ henrys, $C = 1 \times 10^{-6}$ farads and $R = 200$ ohms.

1-13. Give the lengths of the following lines in wavelengths if we assume the velocity of propagation is equal to the speed of light.
(a) Fargo, North Dakota to Minneapolis, Minnesota (240 miles) at 1,500 hertz.
(b) Minneapolis, Minnesota to Chicago, Illinois (405 miles) at 2,000 hertz.
(c) A power line from Bismarck, North Dakota to West Fargo, North Dakota (190 miles) at 60 hertz.
(d) A line 3 feet long at 8 gigahertz.

1-14. A parallel wire line consists of two No. 10 copper wires (solid) separated by a distance of 1 foot. Assuming that the resistance is equal to the d. c. resistance and that $G = 0$, determine the characteristic impedance at 60 hertz and 25° C. The d. c. resistance of a No. 10 copper wire is 0.102 ohm per hundred feet at 25° C. The diameter is 0.102 inch.

1-15. The voltage on a transmission line is given as:

$$v(t) = 150 \cos (4 \times 10^8 t - 2x)$$

Find: (a) Phase velocity, v_p.
(b) Frequency.
(c) Wavelength.
(d) The phasor expression for the voltage.

1-16. Two basic equations for a distributed parameter transmission line developed in the chapter are:

$$\frac{\partial v}{\partial x} = -Ri - L\frac{\partial i}{\partial t}. \qquad \frac{\partial i}{\partial x} = -Gv - C\frac{\partial v}{\partial t}.$$

If $R = 0.053$ ohm/meter, $L = 0.62$ microhenry/meter, $G = 950$ micromicromhos/meter and $C = 39.5$ micromicrofarads/meter; determine: $\frac{\partial i}{\partial x}$ at a point x on the line where $v(t) = 60 \cos (10^3 t + 45°)$.

Your answer must be expressed in the form $\frac{\partial i}{\partial x} = A \cos(\omega t \pm B)$ with B in degrees.

1-17. An infinite transmission line is lossless and has a characteristic impedance of 100 ohms. A signal generator with an internal voltage of $100 \angle 0°$ r.m.s. volts and an internal impedance of $50 + j50$ ohms is connected to the input of the line. Determine:
(a) The sending end voltage, V_s
(b) The sending end current, I_s
(c) The input impedance, Z_s.

2

Lossless Lines with Short-Circuit and Open-Circuit Loads

2.1 Introduction

We will define a lossless line as one where α equals zero. In Chapter 1 we found that this occurs if we assume that $R = G = O$, or if we assume that $\omega L \gg R$ and $\omega C \gg G$. A study of the lossless line is worthwhile since many practical engineering problems approximate this case. Furthermore the lossless case is rather simple to consider mathematically and physically and allows us to look at some basic concepts which can then be expanded to take the lossy case into account later after we have obtained a feeling for what is happening on our line. Even though the engineering situation is never completely lossless it may be that the losses are so minor that there is very little effect on the end results. For example, a "Ham" radio operator can assume a lossless case if the line from his transmitter is from his second-floor bedroom to the roof of his two-story home.

To make it even easier for us as we start out we will consider two special loads as terminations for our lossless line. The short-circuit load has zero impedance and the open-circuit load has infinite impedance so these two loads represent the two extremes of all possible loads. But more importantly the open and short-circuit loads have a practical use in many transmission line situations and are easy to handle on a mathematical basis. At high frequencies the short-circuited tuning stub can be made to look like a reactance, a short-circuit or an open-circuit depending on the length of the stub. Also when making measurements on a high frequency line the short-circuit load is often a useful tool.

We already know a great deal about what might happen with a

short-circuit or an open-circuit load. Neither of these loads can absorb energy so any energy that reaches the short or the open must be completely reflected back towards the source. This produces a reflected wave which, along with the forward or incident wave, establishes a standing wave on the line. This "standing" wave is easily observed in the laboratory and allows us to determine an unknown frequency. The efficiency of our line must be zero since the output is zero and, of course, the load current is zero for an open-circuit load and the load voltage is zero for the short-circuit load.

2.2 Basic Equations for the Lossless Transmission Line

In our work it is often convenient to change from the variable x to the variable d. We will then be concerning ourselves with what is happening on the line as we move from the load to the sending end. We will start with our basic equations.

$$\text{(1-32)}\ V = Ae^{-\gamma x} + Be^{\gamma x}.$$

$$\text{(1-37)}\ I = \frac{A}{Z_o} e^{-\gamma x} - \frac{B}{Z_o} e^{\gamma x}.$$

Under our assumed conditions, $\gamma = \alpha + j\beta = 0 + j\beta$ and Z_o is real. So we have:

$$V = Ae^{-j\beta x} + Be^{j\beta x}$$

and

$$I = \frac{A}{Z_o} e^{-j\beta x} - \frac{B}{Z_o} e^{j\beta x}.$$

Since $x = \ell - d$, we can substitute and obtain:

$$V = Ae^{-j\beta(\ell - d)} + Be^{j\beta(\ell - d)} = Ae^{-j\beta\ell}e^{j\beta d} + Be^{j\beta\ell}e^{-j\beta d}.$$

For a given line, $Ae^{-j\beta\ell}$ is a constant as is $Be^{j\beta\ell}$. Let us define these constants as follows:

$$V^+ = Ae^{-j\beta\ell}. \qquad (2\text{-}1)$$

$$V^- = Be^{j\beta\ell}. \qquad (2\text{-}2)$$

Now we can write our equations in terms of V^+ and V^-.

$$V = V^+e^{j\beta d} + V^-e^{-j\beta d}. \quad (2\text{-}3)$$

$$I = \frac{V^+}{Z_o}e^{j\beta d} - \frac{V^-}{Z_o}e^{-j\beta d}. \quad (2\text{-}4)$$

Notice that the forward wave which was associated with $Ae^{-\gamma x}$ is now associated with $V^+e^{j\beta d}$ and the backward wave is now associated with the term $V^-e^{-j\beta d}$.

When $d = 0$, we are at the receiving end and we have:

$$V = V^+ + V^- = V_r; \text{ and} \quad (2\text{-}5)$$

$$I = \frac{V^+}{Z_o} - \frac{V^-}{Z_o} = I_r = I^+ + I^-. \quad (2\text{-}6)$$

Where we are defining the quantities:

$$I^+ = \frac{V^+}{Z_o}; \text{ and} \quad (2\text{-}7)$$

$$I^- = -\frac{V^-}{Z_o}. \quad (2\text{-}8)$$

Note the minus sign associated with I^-.

2.3 The Short-Circuit Load

A short circuit load may not seem very important at first glance but as we shall see it has important uses on transmission lines at very high frequencies (See Figure 2-1).

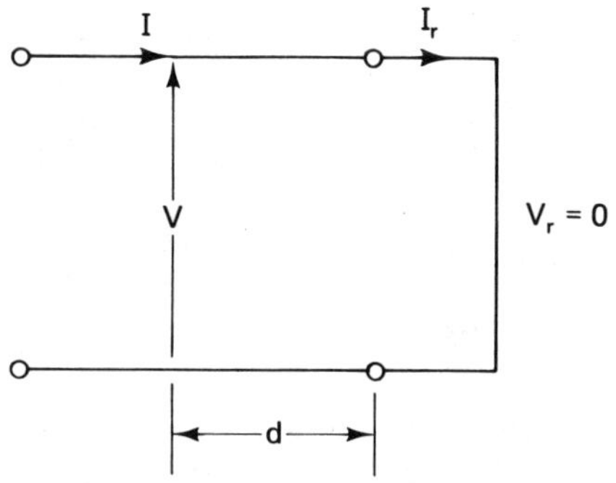

Figure 2-1. Short-circuit load.

With the line terminated in a short circuit we can write:

$$V_r = V^+ + V^- = 0$$

so

$$V^- = -V^+$$

$$V = V^+e^{j\beta d} + V^-e^{-j\beta d} = V^+e^{j\beta d} + (-V^+)e^{-j\beta d} = V^+ (e^{j\beta d} - e^{-j\beta d}) =$$

$$V^+ \frac{(e^{j\beta d} - e^{-j\beta d})}{2j} 2j = 2jV^+\sin\beta d = V.$$

$$I = \frac{V^+}{Z_o} e^{j\beta d} - \frac{V^-e^{-j\beta d}}{Z_o} = \frac{V^+e^{j\beta d}}{Z_o} - \frac{(-V^+)}{Z_o} e^{-j\beta d} = \frac{V^+}{Z_o} (e^{j\beta d} + e^{-j\beta d}) =$$

$$I = \frac{V^+}{Z_o} \frac{(e^{j\beta d} + e^{-j\beta d})2}{2} = \frac{2V^+}{Z_o} \cos\beta d. \tag{2-10}$$

At any point on the line, d units from the load, the impedance looking towards the load is:

$$Z = \frac{V}{I} = \frac{2jV^+\sin\beta d}{\frac{2V^+}{Z_o}\cos\beta d} = jZ_o\tan\beta d. \tag{2-11}$$

when $d = \ell$, $V = V_s$, $I = I_s$, and $Z = Z_s$.

$$V_s = 2jV^+\sin\beta\ell. \tag{2-12}$$

$$I_s = \frac{2V^+}{Z_o} \cos\beta\ell. \tag{2-13}$$

$$Z_s = jZ_o\tan\beta\ell. \tag{2-14}$$

We could solve either (2-12) or (2-13) for V^+, but let us use (2-12).

$$V^+ = \frac{V_s}{2j\sin\beta\ell} \tag{2-15}$$

Now we can substitute back into (2-9) and (2-10).

$$V = 2j \frac{V_s}{2j\sin\beta\ell} \sin\beta d = \frac{V_s\sin\beta d}{\sin\beta\ell}. \tag{2-16}$$

$$I = \frac{2}{Z_o} \frac{V_s}{2j\sin\beta\ell} \cos\beta d = \frac{-j\,V_s}{Z_o \sin\beta\ell} \cos \beta d. \tag{2-17}$$

$$\text{When } d = 0,\ I = I_r. \tag{2-18}$$

$$I_r = -j\frac{V_s}{Z_o \sin\beta\ell}.$$

Let us go back and solve equation (2-13) for V^+.

$$V^+ = \frac{I_s Z_o}{2\cos\beta\ell}. \tag{2-19}$$

Using this value in equations (2-9) and (2-10) results in the following equations.

$$V = 2j\frac{I_s Z_o}{2\cos\beta\ell}\sin\beta d = jI_s Z_o\frac{\sin\beta d}{\cos\beta\ell}. \tag{2-20}$$

$$I = \frac{2}{Z_o}\frac{I_s Z_o}{2\cos\beta\ell}\cos\beta d = \frac{I_s\cos\beta d}{\cos\beta\ell}. \tag{2-21}$$

Let us summarize our results.

$$\text{(2-16) } V_s\frac{\sin\beta d}{\sin\beta\ell} = V = jI_s Z_o\frac{\sin\beta d}{\cos\beta\ell}. \tag{2-20}$$

$$\text{(2-17) } -j\frac{V_s}{Z_o}\frac{\cos\beta d}{\sin\beta\ell} = I = I_s\frac{\cos\beta d}{\cos\beta\ell}. \tag{2-21}$$

An interesting result is obtained when the phasor is converted to a function of time. Let us start with equation (2-9).

$$V = 2jV^+ \sin\beta d$$

$$v(t) = \mathcal{R}\left[2jV^+ (\sin\beta d)e^{j\omega t}\right] =$$

$$\mathcal{R}\left[2jV^+ \sin\beta d(\cos\omega t + j\sin\beta t)\right] = -2V^+\sin\beta d\,\sin\omega t. \tag{2-22}$$

Instead of a traveling-wave type of equation we now have an equation which indicates a standing-wave pattern on the line. We have a function varying with time as a sine function but with a magnitude that is varying as a sine function of distance. It is customary to plot the absolute value of the magnitude as a function of the distance d, if we know β or βd if we do not know β (See Figure 2-2).

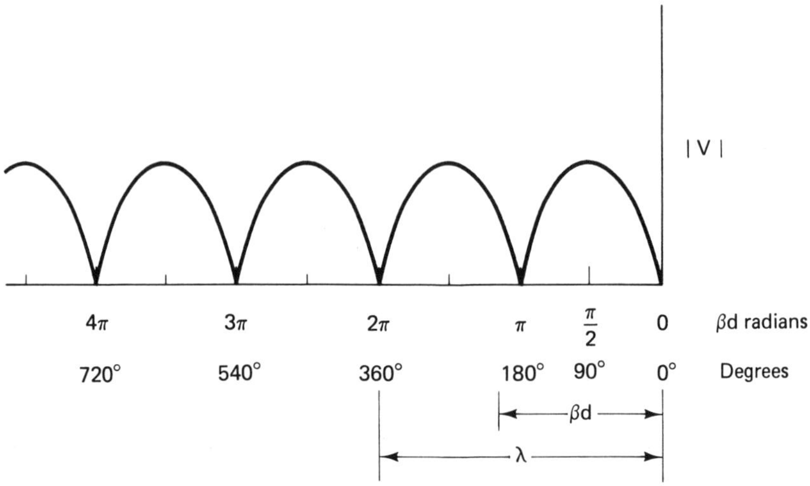

Figure 2-2. Standing wave for short-circuited line.

The voltage is zero at the load and every half wavelength down the line. The magnitude is a maximum at a quarter of a wavelength from the load and this repeats every half wavelength. At each point on the line we have a sinusoidal voltage varying with time but as we move down the line the magnitude of this voltage varies from zero to a maximum value. Since the original equations involved both a forward and backward wave with equal magnitudes, ($V^- = -V^+$, $|V^-| = |V^+|$) we see that a standing wave results whenever we have a forward and backward wave or two waves moving in opposite directions. Repeating the process for the current we will start with equation (2-10).

$$I = \frac{2V^+}{Z_o}\cos\beta d. \quad i(t) = \mathcal{R}\left[\frac{2V^+}{Z_o}(\cos\beta d)e^{j\omega t}\right] =$$

$$\mathcal{R}\left[\frac{2V^+}{Z_o}\cos\beta d(\cos\omega t + j\sin\omega t)\right] = \frac{2V^+}{Z_o}\cos\beta d\cos\omega t. \quad (2\text{-}23)$$

Again we have a standing wave but now the magnitude varies as a cosine wave. Since $Z = jZ_o \tan \beta d$, we can now give a complete graphical picture of the situation on the line in Figure 2-3. At the load the voltage is zero and the current is a maximum and the imped-

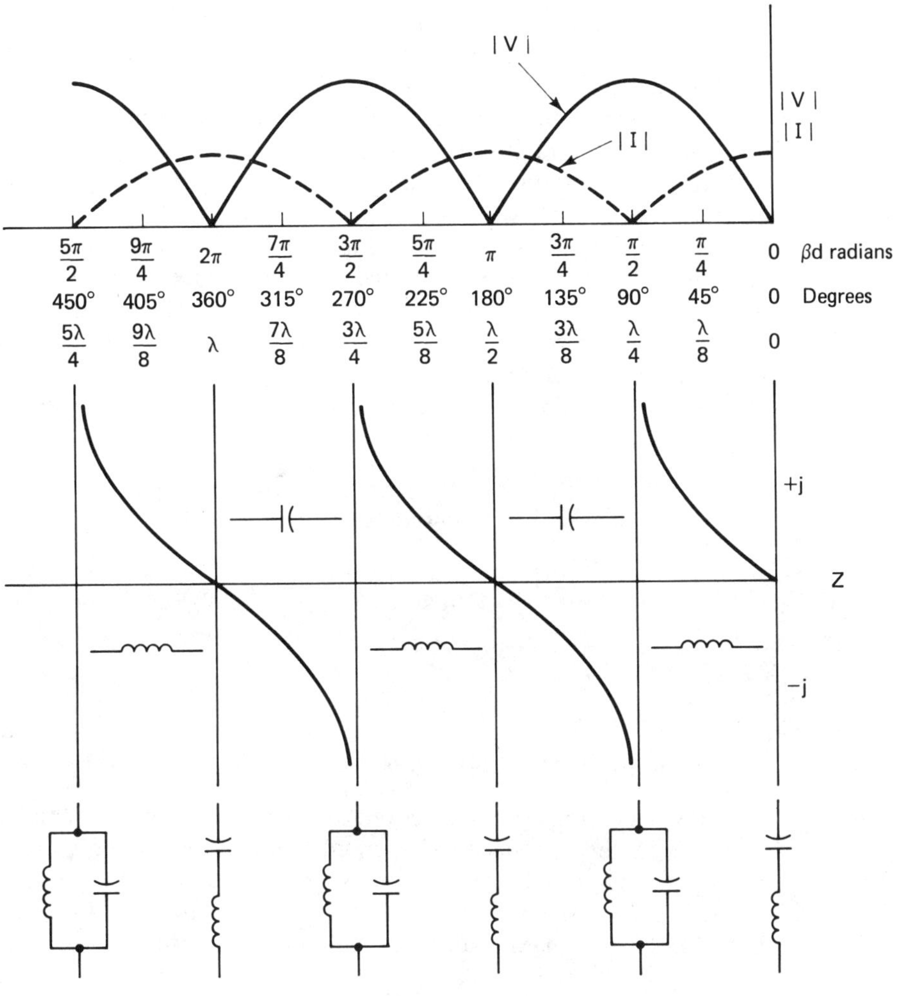

Figure 2-3. Short-circuit load patterns on line.

ance is zero which can be represented by a perfect series resonant circuit. As we move away from the load the impedance looking towards the load looks like a pure inductance for the first quarter of a wavelength. At 1/4 wavelength the impedance is infinite so it looks like a perfect parallel resonant circuit. For the second quarter wavelength, the short-circuited line looks like a pure capacitor and then the process repeats itself every half wavelength. Since a short cannot absorb any real power, all the real power is reflected back to the source and since the line was assumed lossless the source is supplying only reactive power or vars. The input impedance depends on the length of the line. If the line was exactly 1/4 wavelength long the short-circuit load would look like an open circuit. If it is 1/2 wavelength long the input impedance would again be a short circuit. If it was 7/8 wavelength long, the short circuit would look like a pure capacitor.

The student must be cautious in dealing with a.c. steady-state situations since there is sometimes confusion between the maximum and r.m.s. values of the sinusoidal a.c. voltage or current. Let us look at the first term of equation (2-3) which represents the forward or incident wave component.

$V = V^+e^{j\beta d}$. To obtain the time function we multiply by $e^{j\omega t}$ and take the real part.

$$\begin{aligned} v(t) &= \mathcal{R}\left[V^+e^{j\beta d}e^{j\omega t}\right] = R\left[V^+e^{j(\omega t + \beta d)}\right] \\ &= \mathcal{R}\left[V^+\cos(\omega t + \beta d) + jV^+\sin(\omega t + \beta d)\right] \\ &= V^+\cos(\omega t + \beta d). \end{aligned}$$

We can see from this that V^+ is really the maximum value of the voltage, $v(t)$ and the r.m.s. value would be $\frac{V^+}{\sqrt{2}}$ or 0.707 V^+.

However, since the square root of two is a constant, we can forget about it in most cases in our equations since it cancels out. For instance, if we take any of the previous equations such as (2-15) and divide by the square root of two we find the equations are good for both the phasor maximum value or the phasor r.m.s. value.

$$V^+ = \frac{V_s}{2j\sin\beta\ell}. \qquad \frac{V^+}{\sqrt{2}} = \frac{V_s}{\sqrt{2}\ 2j\sin\beta\ell}$$

or

$$V^+_{\text{r.m.s.}} = \frac{V_{s\ \text{r.m.s.}}}{2j\ \sin\beta\ell}.$$

The only time we have to be concerned is if we are given the r.m.s. value and have to find the actual value of the expression for the time function or if we go the other way and wish to find the r.m.s. value from the time function.

Example 2-A

(a) Convert $v(t) = 100 \cos(377t + 45°)$ to the phasor value of V_{max}.
(b) Convert $v(t)$ in (a) to the r.m.s. phasor value.
(c) The r.m.s. value of a current is $10 \angle -63°$. Convert this current to a function of time if the frequency is 60 hertz.

Solution

$$V_{phasor(max)} = 100 \angle 45° \text{ volts.} \quad \text{(a)}$$

$$V_{phasor(r.m.s.)} = 0.707(100) \angle 45° = 70.7 \angle 45° \text{ volts.} \quad \text{(b)}$$

$$i = 10\sqrt{2} \cos(377t - 63°) = 14.14 \cos(377t - 63°). \quad \text{(c)}$$

When we plot the absolute value of our quantities as a function of the distance from the load end, the maximum value of the voltage or the current on the plot could be either the actual maximum value or it could be the maximum value of the r.m.s. value. Since we are usually interested only in the shape of the curve rather than the actual numerical quantities it really doesn't matter which we use, maximum or r.m.s. Most of the time we will be talking about the r.m.s. values since these are the meter values and therefore of more use to us.

Example 2-B

A lossless line is terminated in a short circuit, has a characteristic impedance of 50 ohms, an input voltage of $100 \angle 0°$ volts, r.m.s., is 0.8 meters long and β is 2.5 radians per meter. Determine: I_s, I_r, V_r, Z_s, P_s, P_r, Q_s and Q_r.

$$V_s = 2jV^+ \sin \beta\ell. \qquad 100 \angle 0° = 2jV^+ \sin(2.5)(.8)$$

$$V^+ = \frac{100 \angle 0°}{2 \angle 90° \sin 2} = 54.9 \angle -90°.$$

Note: $\beta\ell$ is in radians.

$$V = (2 \angle 90°)(54.9 \angle -90°) \sin 2.5d = 109.97 \sin 2.5d.$$

$$I = \frac{2V^+}{Z_o} \cos \beta d = \frac{2}{50} (54.9 \angle -90° \cos 2.5d = 2.19 \cos 2.5d \angle -90°.$$

Solution

$$I_s = 2.19\cos(2.5)(.8)\angle{-90^\circ} = .915\angle{90^\circ} = j0.915 \text{ amperes.} \quad \text{(a)}$$

$$I_r = 2.19\cos(2.5)(0)\angle{-90^\circ} = 2.19\angle{-90^\circ} \text{ amperes.} \quad \text{(b)}$$

$$V_r = 0 \text{ (short)} \qquad Z_s = jZ_o\tan\beta\ell = j50\tan2 = -j109.25 \text{ ohms.} \quad \text{(c)}$$

$$P_s = V_sI_s\cos\theta_s = (100)(.915)\cos90^\circ = 0 \text{ watts.} \quad \text{(d)}$$

$$P_r = V_rI_r\cos\theta_r = (0)(2.19)\cos\theta = 0 \text{ watts.} \quad \text{(e)}$$

$$Q_s = V_sI_s\sin\theta_s = (100)(.915)\sin90^\circ = 91.5 \text{ vars.} \quad \text{(f)}$$

$$Q_r = V_rI_s\sin\theta_r = (0)(2.19)\sin 90^\circ = 0 \text{ vars.} \quad \text{(g)}$$

2.4 Further Discussion of a Short-Circuit Load

At very high frequencies, lumped, two-terminal circuit elements are very unpredictable. A high Q coil, for instance, always has some resistance and distributed capacitance between the turns. At very high frequencies the capacitive effects will appear and the resistance will change due to an effect which we call "skin effect". The coil may even appear capacitive in the circuit. Since a low-loss short-circuited line can appear as an L or a C, depending on the length, they are often used at high frequencies as circuit elements even though we would not have a perfect L or C when losses are taken into account. A short-circuited stub can be made from a coaxial cable with a movable short, as shown in Figure 2-4.

The device shown in Figure 2-4 is called a short-circuit stub tuner and the input impedance can approximate zero, infinity, L, or

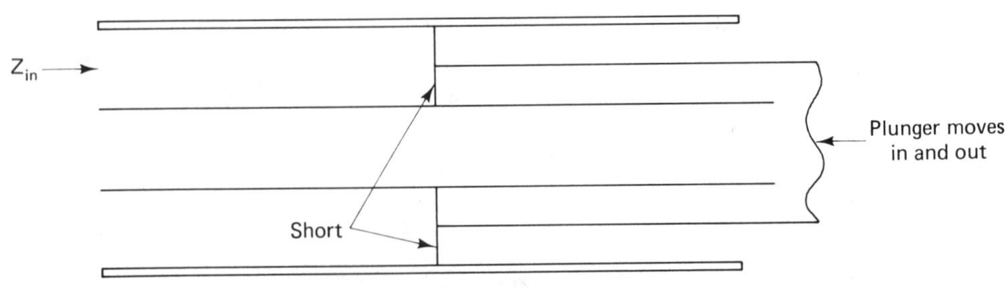

Figure 2-4. Coaxial cable with movable short.

C. In the practical case, with losses present the input impedance would be small, but not zero, for lengths of 1/2, 1, 3/2, 1/4, 3/4, and 5/4 wavelengths. In the C and L regions the input impedance looks like a C or L with a very small R in series or a very large R in parallel. As we will see later, a stub tuner can be used to "tune" a line so that the input impedance becomes the characteristic impedance even if the load impedance is different from the characteristic impedance.

2.5 The Open-Circuit Load

An open-circuit load, like the short-circuit load, has a useful role in our study of transmission lines. It cannot absorb any real power but can, like the short-circuit load, act as a circuit element which varies with the length of the line. Starting with the same equations as before will allow us to determine the relationship between the phasors V^+ and V^- at the receiving end. Throughout our preliminary work, this relationship will be of extreme importance for it determines what will happen on the line as we move towards the sending end (See Figure 2-5). We will proceed just as we did for the short-circuit load.

(2-4) $$I = \frac{V^+}{Z_o} e^{j\beta d} - \frac{V^-}{Z_o} e^{-j\beta d}.$$

at $d = 0$, $I_r = 0$.

$$I_r = 0 = \frac{V^+}{Z_o} - \frac{V^-}{Z_o} \text{ so } V^- = V^+.$$

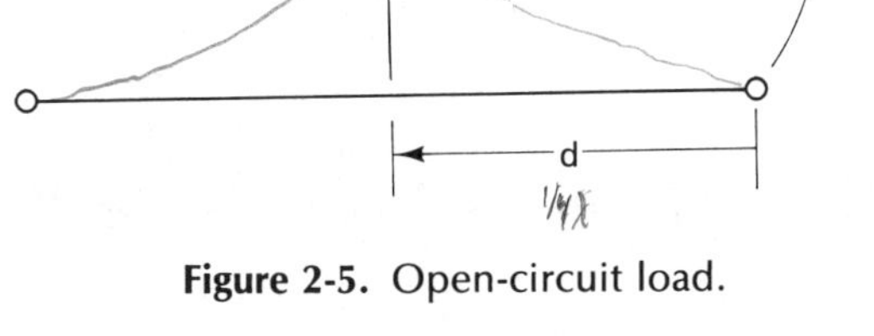

Figure 2-5. Open-circuit load.

$$I = \frac{V^+}{Z_o} e^{j\beta d} - \frac{V^+}{Z_o} e^{j\beta d} = \frac{V^+}{Z_o}(e^{j\beta d} - e^{-j\beta d}) = \frac{2V^+}{Z_o} j \frac{(e^{j\beta d} - e^{-j\beta d})}{2j} =$$

$$I = \frac{2V^+}{Z_o} j \sin\beta d. \tag{2-24}$$

$$\text{(2-3)}\ V = V^+e^{j\beta d} + V^-e^{-j\beta d} = V^+e^{j\beta d} + (V^+)e^{-j\beta d} =$$

$$V^+ \frac{(e^{j\beta d} + e^{-j\beta d})}{2} 2 = 2V^+\cos\beta d = V. \tag{2-25}$$

When $d = \ell$, $V = V_s$ and $I = I_s$.

$$V_s = 2V^+\cos\beta\ell. \tag{2-26}$$

$$I_s = 2j\frac{V^+}{Z_o}\sin\beta\ell. \tag{2-27}$$

Solving (2-26) for V^+ gives:

$$V^+ = \frac{V_s}{2\cos\beta\ell}$$

so

$$V = 2\frac{V_s}{2\cos\beta\ell}\cos\beta d = V_s\frac{\cos\beta d}{\cos\beta\ell}. \tag{2-28}$$

$$I = \frac{2}{Z_o}j\frac{V_s}{2\cos\beta\ell}\sin\beta d = j\frac{V_s}{Z_o}\frac{\sin\beta d}{\cos\beta\ell}. \tag{2-29}$$

If, instead, we solve (2-27) for V^+ we obtain:

$$V^+ = \frac{I_sZ_o}{2j\sin\beta\ell}$$

so

$$V = 2\frac{I_sZ_o}{2j\sin\beta\ell}\cos\beta d = -j\frac{I_sZ_o\cos\beta d}{\sin\beta\ell}. \tag{2-30}$$

$$I = \frac{2j}{Z_o}\frac{I_sZ_o}{2j\sin\beta\ell}\sin\beta d = I_s\frac{\sin\beta d}{\sin\beta\ell} \tag{2-31}$$

$$Z = \frac{V}{I} = \frac{-jI_sZ_o\dfrac{\cos\beta d}{\sin\beta\ell}}{I_s\dfrac{\sin\beta d}{\sin\beta\ell}} = -jZ_o\frac{\cos\beta d}{\sin\beta d} = -jZ_o\text{cotan}\beta d. \tag{2-32}$$

As in the previous case we are interested in the voltage and current as functions of time. The open circuit with an infinite impedance and the short circuit with zero impedance represent the extreme limits on impedance values. All other possible impedances exist between these two values. If we see what happens at each end we will be in better position to understand what happens when the impedance at the load end is somewhere between zero and infinity.

$$(2\text{-}25)\ V = 2V^{+}\cos\beta d \text{ so } v(t) = \Re\,[2V^{+}(\cos\beta d)e^{j\omega t}] =$$

$$\Re\,[2V^{+}\cos\beta d(\cos\omega t + j\sin\omega t)] = 2V^{+}\cos\beta d\cos\omega t = v(t). \quad (2\text{-}33)$$

$$(2\text{-}24)\ I = j\frac{2V^{+}}{Z_o}\sin\beta d \text{ so } i(t) = \Re\left[j\frac{2V^{+}}{Z_o}(\sin\beta d)e^{j\omega t}\right] =$$

$$\Re\left[j\frac{2V^{+}}{Z_o}\sin\beta d(\cos\omega t + j\sin\omega t)\right] = -2\frac{V^{+}}{Z_o}\sin\beta d\sin\omega t = i(t). \quad (2\text{-}34)$$

In Figure 2-6 we see the pattern for the open-circuit case.

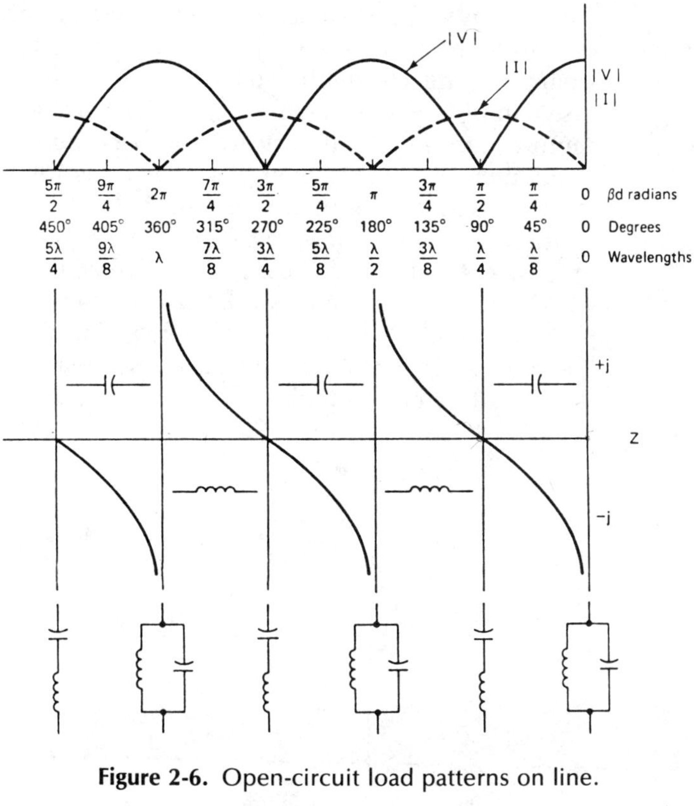

Figure 2-6. Open-circuit load patterns on line.

The pattern is similar to the one we obtained for the short-circuit case with everything reversed. When the length of the line is such that the input impedance is inductive for a short-circuit load it would be capacitive for an open-circuit load. Where the short circuit has zero impedance the open circuit has infinite impedance. Again everything repeats itself every half wavelength. At the receiving end $V^+ = V^-$ for the open circuit case and so these two phasors are in phase at the load and we have a maximum value of voltage at this point. The two current phasors I^+ and I^- are 180 degrees out of phase and equal in magnitude so the current is zero as it should be for an open-circuit load. At each point on the line we have an a.c. voltage or current, whose magnitude varies with the distance from the load. The input impedance of the line depends on the length of the line and in the lossless case would be either zero, infinite, purely inductive or purely capacitive. An open circuit could then be used as a circuit element but the short circuit is of more practical use since a movable short is physically easier to construct than a movable open and the open-circuit load may act as an antenna and radiate some real power out of the open and it then ceases to be a perfect open circuit.

While the actions of the short- and open-circuit loads may seem strange one must be reminded that we are dealing with a hypothetical case with $R = G = 0$, so that the line has only series inductance and shunt capacitance. The forward and backward traveling waves are equal in magnitude but their phase relations are such that a standing wave appears on the line. In the next chapter we shall study these phase relations in detail. In circuit theory we studied the action of inductances and capacitances under so-called ideal conditions ($R = 0$) which led to series and parallel resonant conditions that gave apparently strange results such as voltages in the circuit larger than the input voltage. It should not seem strange then to find our present results with a circuit looking like it does in Figure 2-7.

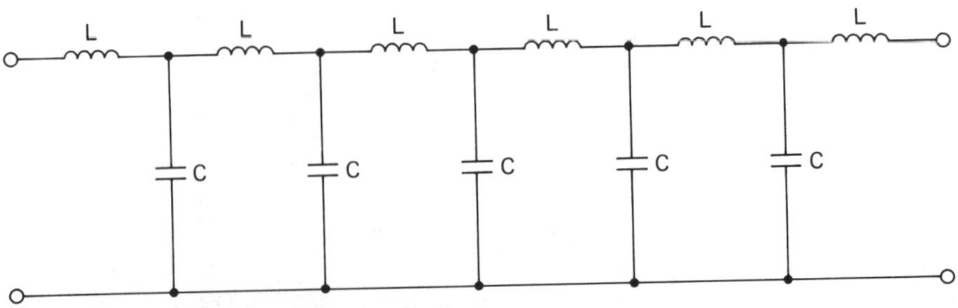

Figure 2-7. Lossless line representation.

Example 2-C

A lossless line is terminated in an open circuit, has a characteristic impedance of 50 ohms and input voltage of 100 $\angle 0°$ volts, r.m.s., is 0.8 meters long and β is 2.5 radians per meter. Determine: I_s, I_r, V_r, Z_s, P_s, P_r, Q_s and Q_r.

$$V = V_s \frac{\cos\beta d}{\cos\beta\ell}$$

so

$$V = \frac{100\angle 0° \cos 2.5d}{\cos(2.5)(0.8)} = -240.3 \cos 2.5d.$$

$$I = j\frac{V_s}{Z_o}\frac{\sin\beta d}{\cos\beta\ell}$$

so

$$I = j\frac{100\angle 0°}{50}\frac{\sin 2.5d}{\cos(2.5)(0.8)} = -j4.81 \sin 2.5d.$$

at $d = 0$, $V_r = -240.3 \cos 0° = 240.3 \angle 180°$ volts $I_r = 0$ (open circuit). With $d = 0.8$ meters,

Solution

$$I_s = -j4.81 \sin(2.5)(0.8) = -j4.37 = 4.37 \angle -90° \text{ amperes.} \quad \text{(a)}$$

$$Z_s = -jZ_o\text{cotan}\beta\ell = -j\frac{Z_o}{\tan\beta\ell} = -j\frac{50}{\tan(2.5)(0.8)} = j22.88 \text{ ohms.}$$

(b)

$$P_s = V_sI_s\cos\theta_s = (100)(4.37)\cos 90° = 0 \text{ watts.} \quad \text{(c)}$$

$$P_r = V_rI_r \cos\theta_r = (240.7)(0) \cos\theta = 0 \text{ watts.} \quad \text{(d)}$$

$$Q_s = V_sI_s\cos\theta_s = (100)(4.37)\sin 90° = 437 \text{ vars.} \quad \text{(e)}$$

$$Q_r = V_rI_s\cos\theta_r = (240.3)(0)\sin 90° = 0 \text{ vars.} \quad \text{(f)}$$

2.6 Chapter Summary

1. The "lossless" or "high frequency" cases may be considered together and represent the simplest "steady-state" a.c. transmission line situation.
2. With a short-circuit load or an open-circuit load the lossless transmission line has "standing" waves of voltage and current

with complete reflection of the energy which reaches the load end of the line. For such a line the efficiency is zero.

3. The input impedance of a lossless line terminated in an open-circuit or short-circuit load may be zero, infinite or a pure inductance or capacitance with the actual value determined by the length of the line and the wavelength or frequency of the input source.
4. The magnitude of the phasor values of the voltage and current on the lossless line terminated in either a short or an open circuit varies from zero to a maximum value depending on the location on the line. Everything repeats itself every half-wavelength.
5. Since no real power can be transmitted down a line terminated in a short or open circuit the voltage and the current must be ninety degrees out of phase at every point on the line.
6. It is customary to plot the situation on the line in terms of the distance from the receiving end and using the absolute magnitudes of the voltage and current. The sending end will be on the left and the receiving end on the right so the plot goes from right to left, starting at the receiving end.
7. In equations involving ωt it is mathematically correct only when using radians for ωt but it is customary to think in terms of electrical degrees with 2π radians equivalent to 360 degrees. In equations involving βd or βx or $\beta \ell$ it is mathematically correct only when using radians for the three items but it is normal to think in terms of degrees or wavelengths with 2π radians equivalent to one wavelength or 360 degrees.
8. Because of their properties, short- and open-circuit loads are useful at high frequencies as circuit elements and for making measurements.

2.7 Review and Study Questions

1. How is a "lossless" line defined?
2. How is a "high-frequency" line defined?
3. Under the conditions of this chapter what is the angle associated (a) with the characteristic impedance? (b) with the propagation constant?
4. Under the conditions of this chapter what is
(a) α? (b) β? (c) v_p?
5. The impedance looking towards the load of a lossless transmission line terminated in a short-circuit load is zero, infinite, inductive or capacitive if the line is (a) one-eighth wavelength

long? (b) one-quarter wavelength long? (c) two wavelengths long? (d) five-sixteenths wavelength long?

6. Give the answers for number 5 if the lossless line is terminated in an open circuit.
7. Under what conditions will a short-circuit load look like an open-circuit on a lossless line?
8. What conditions are required on a lossless line in order to have the voltage and current standing waves go to zero at some points on the line?
9. Why is the sending end power always equal to zero on a lossless line terminated in a short or open circuit?
10. Can the receiving end voltage be higher than the sending end on a lossless line terminated in an open circuit? If so, under what conditions?
11. What is the input impedance of a pure inductance and pure capacitance under resonant conditions (a) connected in parallel? (b) connected in series?
12. Where else in your college work have you had a situation where standing waves existed?
13. Sketch the standing waves on a lossless line terminated in (a) a short-circuit load, (b) an open-circuit load.
14. Define and give the units of (a) real power, (b) reactive power, (c) apparent power.
15. If the input impedance of a lossless line is capacitive is the reactive power input to the line considered positive or negative?
16. How do you expect that having a lossy line will affect the results obtained in this chapter?
17. If the input impedance of a lossless line is infinite when it is terminated in a short circuit what will be the input impedance if the short is replaced with an open circuit?
18. If we define a quantity called the voltage standing-wave ratio as the ratio of the maximum value of the voltage on the line divided by the minimum value of the voltage on the line, what is the voltage standing-wave ratio for the lossless line terminated in (a) a short circuit? (b) an open circuit?

Problems

2-1. A lossless transmission line has a characteristic impedance of 50 ohms. It is terminated in a short circuit. It is 1 meter long and β is 3 radians per meter. If the short-circuit current is 10 amperes, r. m. s., determine the following: V_r, V_s, I_s, Z_s, P_s, P_r, Q_r, Q_s.

2-2. In Problem 2-1 determine the impedance looking towards the load when the distance from the load is
(a) 0.1 meters.
(b) 0.25 meters.
(c) 0.5 meters.
(d) 0.8 meters.

2-3. Solve problem 2-1 if the short-circuit load is replaced with an open circuit.

2-4. Solve problem 2-2 if the short-circuit load is replaced with an open circuit.

2-5. Determine the output voltage of a 2.35 wavelength, open-circuited line, having a Z_o of 600 ohms, for an input voltage of 100 volts, r.m.s. Determine the current at the input to the line.

2-6. A lossless line with a Z_0 of 300 ohms is terminated in an open circuit. The line is 6.375 wavelengths long. If the input voltage is 100 volts, r.m.s., determine:
(a) the input current.
(b) the output voltage.
(c) the output current.
(d) the input impedance.

2-7. For the line of problem 2-6 sketch the absolute magnitude of the voltage and current for one wavelength down the line from the load. The ratio of the maximum value of the voltage to the minimum value of the voltage is defined as the voltage standing wave ratio. What is its value for this line?

2-8. Repeat problem 2-6 when the open circuit is replaced by a short circuit.

2-9. Repeat problem 2-7 for the conditions of problem 2-8.

2-10. A lossless transmission line is open-circuited at the load end. One and five-eighths wavelengths from the load end there is a short across the line. Exactly one-quarter wavelength beyond the short is the signal generator. What impedance does the signal generator see if the characteristic impedance of the line is 50 ohms?

2-11. A coaxial cable has a characteristic impedance of 50 ohms and may be considered lossless. The load is a short circuit. Determine: (a) the input impedance if the cable is 2 wavelengths long, (b) the input impedance if the cable is 3/4 wavelength long, (c) the input impedance if the cable is 0.86 wavelength long.

2-12. A two-wire line has a characteristic impedance of 100 ohms and is terminated in a short circuit. The line is 7.125 wavelengths long. It may be considered lossless. If the input voltage is 100 volts, r.m.s., determine: (a) the input current, (b) the output (load) voltage, (c) the load current, (d) the input impedance.

2-13. A line has a characteristic impedance of 300 ohms and may be considered lossless. It is terminated with an open circuit and the input voltage is 100 volts, r.m.s. It is 6.357 wavelengths long. Determine: (a) the input current, (b) the output voltage, (c) the load current, (d) the input impedance.

2-14. A short-circuited transmission line, three-eighths wave length long is to be used as a circuit element. What is the input impedance if the characteristic impedance is 50 ohms and the line may be assumed lossless?

2-15. An open-circuited lossless line is one-quarter wavelength long. $Z_o = 50$ ohms. It is fed from a signal generator with an internal impedance of $10 + j0$ ohms and an emf of 100 volts, r.m.s. Determine: (a) the sending end voltage, (b) the sending end current, (c) the receiving end voltage, (d) the receiving end current.

2-16. The voltage on a transmission line is given as

$$v(t) = 150 \cos(9 \times 10^{10}t - 600x).$$

Determine: (a) frequency, (b) phase velocity, (c) wavelength, (d) the phasor expression for v.

3

Lossless Lines and Rotating Phasors

3.1 Introduction

In the last chapter we were introduced to the lossless line terminated in some rather special loads such as an open circuit or a short circuit. These loads resulted in standing waves of voltage and current which are the result of traveling waves with equal magnitudes moving in opposite directions on the line. In Chapter 1 we found that we have both waves if the load impedance does not match the characteristic impedance of the line, so the possibility of standing waves exists even if we do not have a short- or open-circuit load. To investigate the possibility it becomes convenient to use an old friend from a. c. circuit analysis, the phasor. Its use in the study of transmission lines will allow us to solve many practical problems with a minimum of mathematical complexity, but more importantly, it will allow us to visualize what is happening on the line to the voltage, current, and impedances as we move down the line. Simple sketches of the phasor relationships will give us a great deal of information which does not appear without some thought when dealing with only the equations.

The engineer must understand what is happening before he makes decisions about what to do in a given situation. Simply going to a handbook and digging out equations may lead to serious engineering errors. The phasor is a simple tool to use, gives useful results and, if necessary, allows us to return to the time domain whenever it is necessary. In many cases the engineer can make a few simple sketches to determine exactly what is happening on the line and can come up with practical solutions without resorting to equations and extensive calculations. And if the line is lossy a simple modification of the lossless case gives a phasor solution that is easy to use without remembering any complicated equations. In fact a great number of real-world problems may be solved by using phasors and all one has to remember is a couple of simple equations.

At the start of our new approach it will be convenient to consider only the lossless and/or the high frequency case. By defining a new quantity known as the reflection coefficient it is possible to quickly relate the phasor values of the forward and reflected waves at the load end of the line to each other. By remembering this one simple equation for the voltage reflection coefficient, the engineer can approach many of the practical engineering problems encountered at higher frequencies with confidence and he is able to obtain useful solutions with a minimum of effort. Once the phase relationship of the two phasors is known at the load end of the line they are determined at every other point on the line. By tracing these phasor relationships from the load end back towards the sending end the engineer can obtain a picture of the voltage and current standing waves on the line and determine the impedances on the line as one looks towards the load end. The same information that we obtained in the last chapter for the open- and short-circuit loads by using equations may be obtained with less effort when we use the phasor solutions. Since the phasor solution involves two phasors rotating in opposite directions, this technique is often called the counter-rotating phasor method.

Once again we will start out with the simpler cases and work up to the more complicated examples. As we do so we will find that a pure resistive load can look like a complex impedance and a complex impedance load can look like a pure resistance depending on the length of the line. If we determine what happens for any type of load for the lossless case then it becomes a rather simple matter to extend our procedure to the lossy case.

The phasor technique is also valuable to us in understanding how to interpret measurements made either on an actual line or in the laboratory. By simply measuring the distance from the load to a voltage minimum or a voltage maximum of the standing wave on the line and by a simple measurement of the magnitudes of the voltage maximum and the voltage minimum, it is possible to determine an unknown load impedance. A simple sketch of the phasors involved will indicate how this is accomplished without the use of involved equations.

3.2 Basic Concepts of Rotating Phasors

Once again we have to return to our basic equations.

(1-32) $$V = Ae^{-\gamma x} + Be^{\gamma x}.$$

$$I = \frac{A}{Z_o} e^{-\gamma x} - \frac{B}{Z_o} e^{\gamma x}. \tag{1-37}$$

Since $x = \ell - d$

$$V = Ae^{-\gamma(\ell - d)} + Be^{\gamma(\ell - d)} = Ae^{-\gamma\ell}e^{\gamma d} + Be^{\gamma\ell}e^{-\gamma d} = V. \tag{3-1}$$

$$I = \frac{A}{Z_o} e^{-\gamma(\ell - d)} - \frac{B}{Z_o} e^{\gamma(\ell - d)} = \frac{A}{Z_o} e^{-\gamma\ell}e^{\gamma d} - \frac{B}{Z_o} e^{\gamma\ell}e^{-\gamma d} = I. \tag{3-2}$$

Let us define V^+ and V^- differently than we did in Chapter 2.

$$V^+ = Ae^{-\gamma\ell}. \tag{3-3}$$

$$V^- = Be^{\gamma\ell}. \tag{3-4}$$

It should be pointed out that these definitions are consistent with those given in Chapter 2 if the attenuation constant is zero.

We now obtain equations which are good for either the lossless or the lossy cases.

$$V = V^+e^{\gamma d} + V^-e^{-\gamma d} = V^+e^{(\alpha + j\beta)d} + V^-e^{-(\alpha + jb)d}$$
$$= V = V^+e^{\alpha d}e^{j\beta d} + V^-e^{-\alpha d}e^{-j\beta d}. \tag{3-5}$$

If α is zero we would obtain equation (2-3). Starting with equation (1-37) we can repeat our procedure for the current, I, and obtain equation (3-6).

$$I = \frac{V^+}{Z_o}e^{\alpha d}e^{j\beta d} - \frac{V^-}{Z_o}e^{-\alpha d}e^{-j\beta d}. \tag{3-6}$$

Equations (3-5) and (3-6) become our general equations good for all cases and for the lossless case if we simply set α equal to zero. Let us look at the various terms.

The $e^{\alpha d}$ and $e^{-\alpha d}$ both equal 1 at the receiving end when $d = 0$. As d is increased from zero, they simply change the magnitude of V^+ and V^-, increasing V^+ and decreasing V^-. When we assume that α is equal to zero the magnitudes of the two phasors, V^+ and V^- do not change on the line. $e^{j\beta d} = \cos\beta d + j\sin\beta d = 1\,\underline{/\beta d}$ is a complex operator that always has a magnitude of unity and has an angle of βd, which is added to the angle of V^+. At the receiving end V^+ has some phase angle and the angle of βd is zero so as we move towards the sending end, the adding of the angle βd simply rotates the phasor V^+ in the counterclockwise direction. In circuits when we multiplied a phasor by j the phasor was rotated by 90° in the counterclockwise direction. j is a complex operator with unity magnitude and adds 90° to the original angle of the phasor. The only difference in our case is that the angle added to the phasor is not constant but increases with distance from the load end. $e^{-j\beta d} = \cos \beta d - j \sin \beta d = 1 \angle{-\beta d}$ does the same thing except that it adds a negative angle to the phasor V^-. As we move from the receiving end towards the sending end this term rotates the phasor V^- in a clockwise direction. So V, in general, consists of a phasor, V^+ rotating counterclockwise and a phasor, V^-, rotating clockwise. Equation (3-6) shows that the current acts in the same way. One phasor, $I^+ = V^+/Z_o$ rotates in the counterclockwise direction and the other phasor $I^- = -V^-/Z_o$ rotates in the clockwise direction. It is important to remember that V^+ and I^+ are associated with forward traveling waves and V^- and I^- are associated with backward traveling waves. If the load is matched with the line so that the backward waves do not exist then we only have V^+ and I^+ phasors traveling in the counterclockwise direction since both V^- and I^- would be equal to zero.

Let us assume that α is zero and Z_o is real. Our equations become:

$$V = V^+ e^{j\beta d} + V^- e^{-j\beta d}. \tag{2-3}$$

$$I = I^+ + I^- = \frac{V^+}{Z_o} e^{j\beta d} - \frac{V^-}{Z_o} e^{-j\beta d}. \tag{2-4}$$

When $d = 0$, $V = V_r$ and $I = I_r$ so $V_r = V^+ + V^-$ and $I_r = \dfrac{V^+}{Z_o} - \dfrac{V^-}{Z_o}$.

We will now define the ratio of the phasor value of the backward (reflected) component to the phasor value of the forward (incident) component at the receiving end as the voltage reflection coefficient.

$$\Gamma_r = \frac{V^-}{V^+}. \tag{3-7}$$

$$Z_r = \frac{V_r}{I_r} = \frac{V^+ + V^-}{\dfrac{V^+}{Z_o} - \dfrac{V^-}{Z_o}} = Z_o \frac{(V^+ + V^-)}{(V^+ - V^-)} = Z_o \frac{1 + \dfrac{V^-}{V^+})}{(1 - \dfrac{V^-}{V^+})} =$$

$$Z_o \frac{(1 + \Gamma_r)}{(1 - \Gamma_r)} = Z_r. \qquad \frac{Z_r}{Z_o} = \frac{1 + \Gamma_r}{1 - \Gamma_r}. \tag{3-8}$$

Solving for Γ_r gives:

$$\Gamma_r = \frac{Z_r - Z_o}{Z_r + Z_o} = \frac{V^-}{V^+}. \tag{3-9}$$

Thus the reflected phasor value of voltage, in terms of the incident phasor value of voltage at the receiving end, is determined by the relationship between the load impedance and the characteristic impedance of the line.

$$V^- = \Gamma_r V^+. \tag{3-10}$$

In general the voltage reflection coefficient, Γ_r, is complex with a magnitude and angle, which means that V^- is in phase with V^+ at the receiving end only if the angle of Γ_r is zero degrees. This equation verifies our previous discussions since if $Z_r = Z_o$ (line is matched), $\Gamma_r = 0$ and V^- is zero and there is no backward or reflected wave. It would seem that Γ_r should vary between 0 and 1 since it is hard to conceive of a situation where the reflected component is larger than the incident component but one must be careful since we are dealing with hypothetical situations involving pure L and C in the lossless case. So we will leave the discussion of the possible values of Γ_r for later.

At this point the reader may well wonder what happens when the reverse wave returns to the sending end. An educated guess would be that it is again reflected and a new traveling wave starts towards the load end. As we shall see later we do have a reflection coefficient at the sending end which looks exactly like equation (3-9) with Z_g (source impedance) substituted for Z_r. From a practical standpoint it is much easier to match the characteristic impedance with the source impedance than to match the characteristic impedance with the load impedance since the load impedance is often fixed by factors that are not under the control of the engineer. If the wave returning to the sending end has been attenuated because the line is lossy then the reflection at the sending end may not be of much importance. If the source impedance matches the characteristic impedance there will be no reflection at the sending end and this will be our assumption for most of the discussion in this text.

3.3 The Short-Circuit Load Again

Equation (3-10) is important to remember since this simple equation is the starting point for our phasor solution and one can use it and equation (3-9) to solve many practical problems.

Let us go back and solve the problem where the line is terminated in a short-circuit load. At the receiving end $Z_r = 0$ so we obtain the following:

$$\Gamma_r = \frac{Z_r - Z_o}{Z_r + Z_o} = \frac{0 - Z_o}{0 + Z_o} = -1 = 1\,\angle 180^\circ$$

$$= \frac{V^-}{V^+} \text{ and } V^- = -V^+ = V^+\,\angle 180^\circ$$

at the receiving end. The starting point for our phasor solution is the complex plane. The solution will be easiest if we always assume that V^+ is located on the positive real axis at the receiving end. The phasor diagram given in Figure 3-1 shows the relationship between V^+ and V^- at the receiving end.

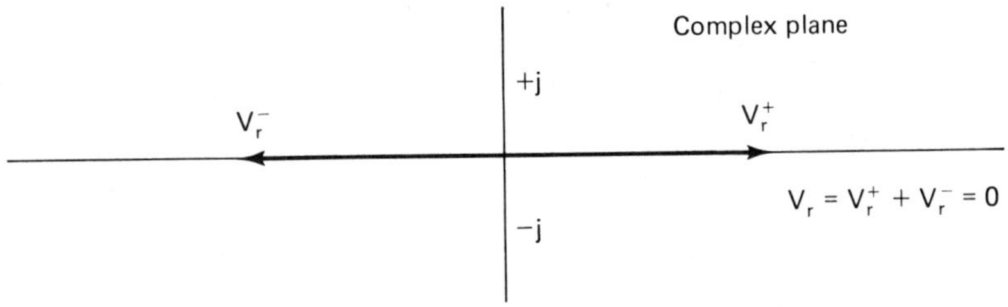

Figure 3-1. Phasor diagram at receiving end for short-circuit load.

What length should we use for V^+? Any convenient scale is satisfactory but once we choose a length for V^+ the length for V^- is determined by Γ_r as is the angle between V^+ and V^-. If Z_o is real (as we have assumed), then $I^+ = V^+/Z_o$ is in phase with V^+.

But $I^- = -V^-/Z_o$ so I^- is 180° out of phase with V^- when Z_o is real. In this case:

$$I^- = -\frac{V^-}{Z_o} = -\frac{(-V^+)}{Z_o} = \frac{V^+}{Z_o} = I^+$$

at the receiving end.

The phasor diagram for the currents is given in Figure 3-2.

The subscript r will always indicate the phasor at the receiving end. Again any convenient scale of I^+ should be picked and this will determine the length of I^-. Combining the two phasor diagrams gives us Figure 3-3.

Figure 3-3 shows the phasor diagram and the phasor values of

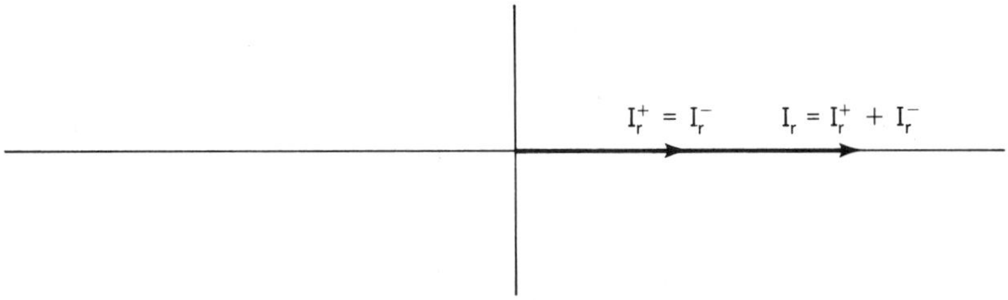

Figure 3-2. Phasor diagram for currents at the receiving end for short-circuit load.

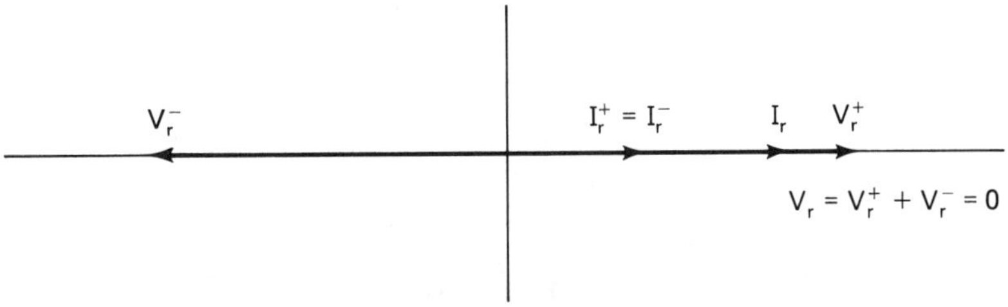

Figure 3-3. Phasor diagram for short-circuit load.

the voltage and current at the receiving end. Now we remember that as we move towards the sending end, the phasors V^+ and I^+ rotate in the counterclockwise direction and the phasors V^- and I^- rotate in the clockwise direction. Let us assume that $\beta d = \pi/4$ radians which is equivalent to 45 degrees or that we have moved 1/8 wavelength down the line from the short circuit. The new phasor diagram is given in Figure 3-4.

At this point on the line V^+ and V^-, as well as I^+ and I^- have the same magnitudes as they did at the load end but the angles have changed by 45 degrees in the counterclockwise direction for V^+ and I^+ and by 45 degrees in the clockwise direction for V^- and I^-. The total voltage, V, now has an angle of +90 degrees and the total current, I, is lagging the voltage by 90 degrees and is located on the positive real axis. In other words, the impedance at this point on the line, looking towards the load, is a pure inductance with an inductive reactance equal to the magnitude of the voltage divided by the magnitude of the current. For each point on the line we can sketch a new phasor diagram by rotating the phasors in the proper direction. In order to find the actual magnitudes of the voltage or current at any point on the line we would have to know the voltage at some

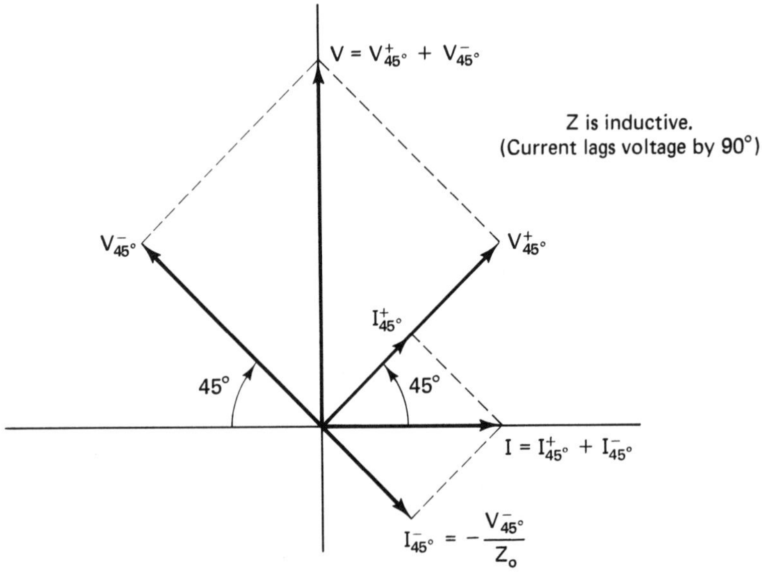

Figure 3-4. Phasor diagram for $\beta d = \pi/4$ radians (45°) with short-circuit load.

point in order to determine V^+, since knowing V^+ will allow us to find the magnitudes of the other quantities involved. Usually we would know the voltage, either at the sending or the receiving end. In this special case, knowing that the receiving end voltage is zero will not help us to find V^+ but we could find V^+ if we knew the actual value of the receiving end current and Z_o since $I_r = 2\,|V^+/Z_o|$. Z, the impedance, can be found, however, without knowing the actual magnitude of V^+. For instance, for the conditions of Figure 3-5 we can write:

$$|V| = 2\,|V^+|\cos 45^\circ = 1.414\,|V^+| \text{ and } |I| = 2\,|I^+|\cos 45^\circ = 1.414\,|I^+| \;\therefore\; |Z| = \frac{1.414\,|V^+|}{1.414\,|I^+|} = \frac{|V^+|}{\frac{|V^+|}{Z_o}} = Z_o$$

or we can express V and I as phasors.

$$V = |V^+|\,\underline{/45^\circ} + |V^-|\,\underline{/135^\circ}. \qquad |V^+| = |V^-|.$$

$$V = |V^+|\,\underline{/45^\circ} + |V^+|\,\underline{/135^\circ} = j\,1.414\,|V^+|.$$

$$I = \frac{|V^+|}{Z_o}\,\underline{/45^\circ} + \frac{|V^-|}{Z_o}\,\underline{/-45^\circ} = \frac{1.414\,|V^+|}{Z_o}\,\underline{/0^\circ},$$

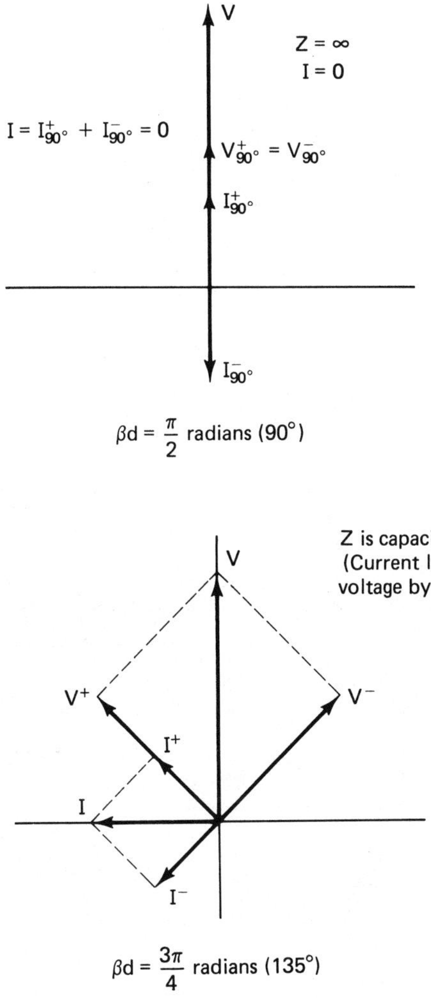

Figure 3-5. Phasor diagrams—short-circuit load.

so
$$Z = \frac{V}{I} = \frac{j\,|V^+|\,1.414}{\frac{|V^+|}{Z_o}\,1.414} = jZ_o.$$

Using phasors will give us the magnitude and angle of the impedance, Z (See Figure 3-6).

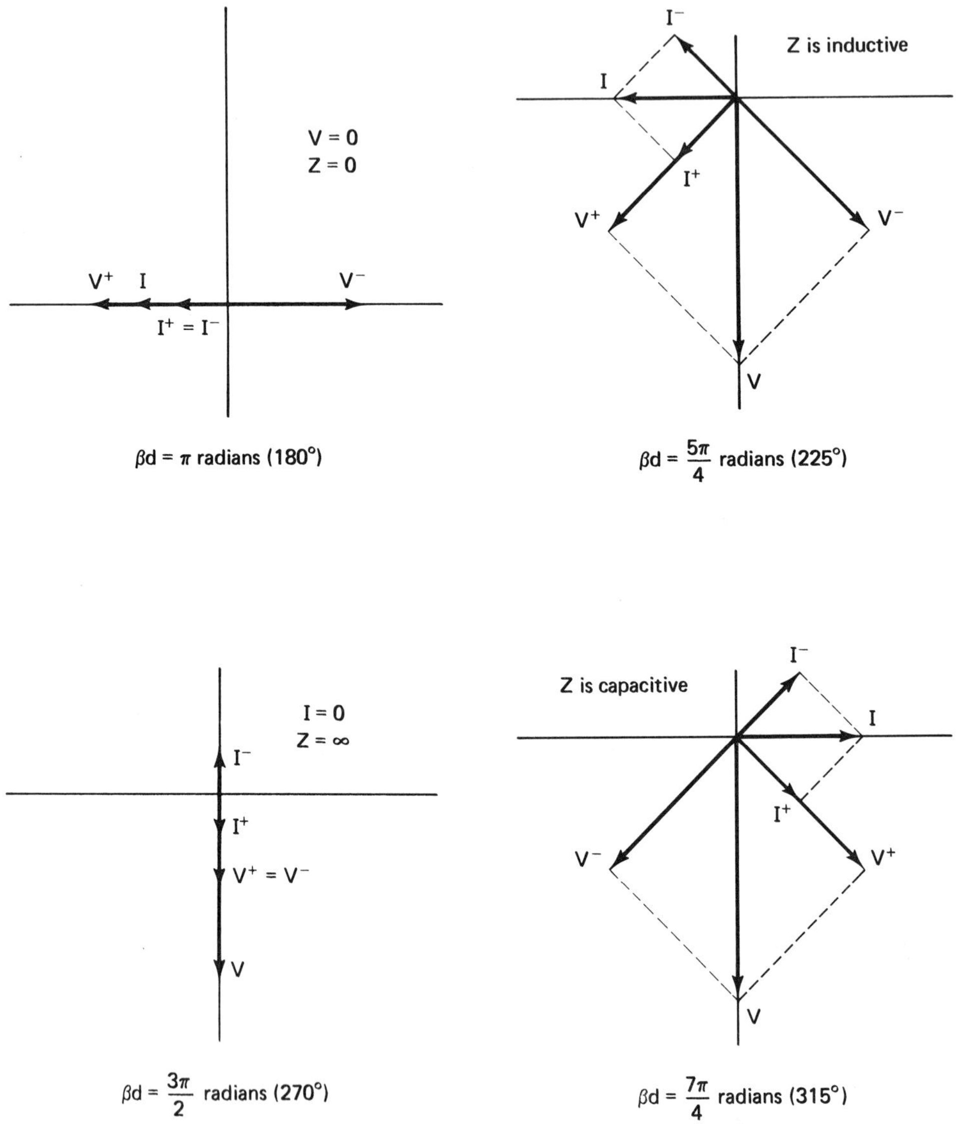

Figure 3-6. Phasor diagrams for short-circuit load.

Let us repeat the process for other points on the line. When $\beta d = 2\pi$ radian, (360°) or one wavelength, each phasor has rotated one complete revolution and will be back in the same position that it occupied at the load, and Figure 3-3 will again be the phasor diagram. As we continue down the line, away from the load, we continue to rotate the phasors and the phasor diagrams will repeat themselves. The phasor diagrams will be the same for 0°, 360°, 720°, etc. Notice, however, that the magnitudes of the voltage and current and the angles between them repeat every 180° or one-half wavelength. At 45°, the current lags the voltage by 90° and at 45° + 180° = 225° the current again lags the voltage by 90°. The magnitude of the voltage at 45° (1/8 wavelength) will be the same as its magnitude at 225° (5/8 wavelength) and the magnitudes of the currents will be the same at the two different angles so that the impedance magnitude will also be equal at both locations. Using our phasor diagrams we could then proceed to sketch the magnitudes of the voltage and current on the line, and knowing the phase relationships between the voltages and currents we could come up with the type of impedance we have at each point on the line. All we really need to start this process is the voltage reflection coefficient, Γ_r.

Figure 3-7 gives us the same information as Figure 2-3, which

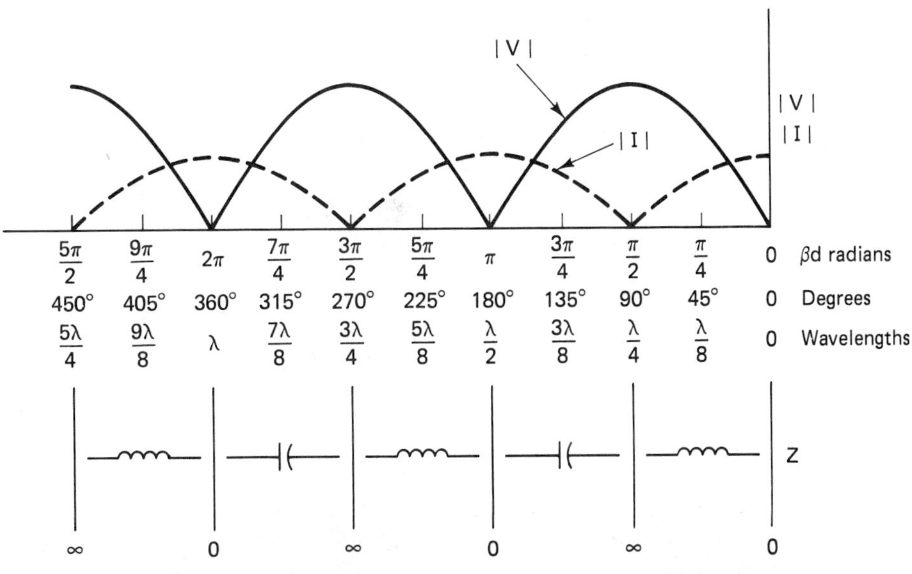

Figure 3-7. Pattern for short-circuit load from phasor diagrams.

you will remember was obtained by using equations. After becoming familiar with the technique one can rapidly sketch out what is happening on the line, determine the type of impedance for a given length of line and do it without ever scaling the diagrams. Let us repeat one of the examples in Chapter 2 using the phasor technique.

3.4 Example of the Phasor Solution for a Short-Circuit Load

A lossless line is terminated in a short-circuit, has a characteristic impedance of 50 ohms, an input voltage of 100 volts, r.m.s., is 0.8 meters long and β is 2.5 radians per meter. Using phasors determine I_s, I_r V_r, Z_s, P_s, and P_r.

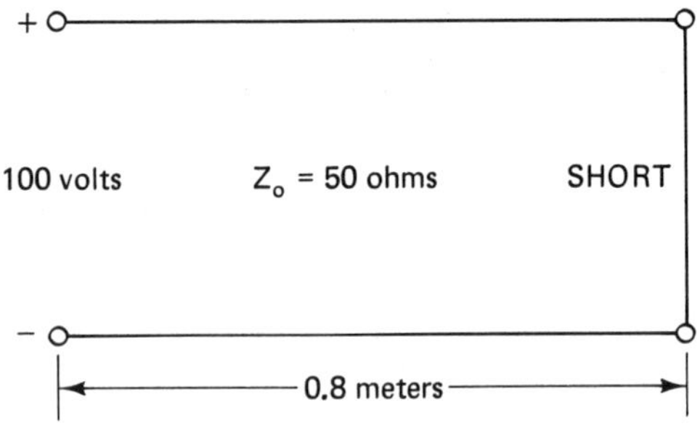

The starting point for the phasor solution is the voltage reflection coefficient.

$$\Gamma_r = \frac{z_r - z_o}{z_r + z_o} = \frac{0 - z_o}{0 + z_o} = -1$$
$$= \frac{V^-}{V^+}$$

So, as we already know, $V^- = -V^+$ at the receiving (load) end. Since the sending end voltage is given we must go to the sending end to determine the magnitude of V^+.

$\beta d = (2.5)(0.8) = 2$ radians (114.59°). The line is less than a wavelength long and the phasors will rotate 114.59° from their positions at the load end to arrive at the sending end. The receiving and sending end phasor diagrams will be as shown in Figure 3-4b.

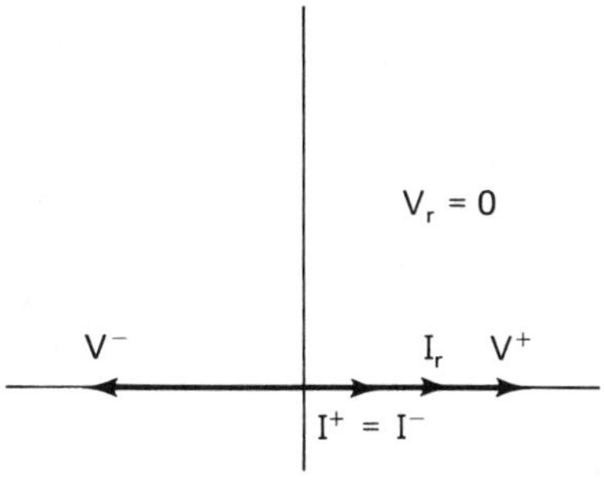

Receiving end

Notice that we really don't have to worry about the rotation of I^+ and I^-. After V^+ is rotated through the proper angle in the counterclockwise direction (114.59°), we place I^+ in phase with it and after V^- is rotated through the proper angle in the clockwise direction (114.59°), we place I^- 180° out of phase with it. Remember this only happens if Z_o is real. At the sending end we can find V^+ by using techniques from trigonometry or by using complex number methods.

$$|V_s| = 2\,|V^+|\cos 24.59° = 100.\ |V^+| = \left|\frac{100}{2\cos 24.59°}\right| = 55 \text{ volts.}$$

It is easy to work with just the magnitudes and to supply the angles from the phasor diagram. For instance, at the sending end, $V_s = 100\,\angle 90°$ just by glancing at the diagram. Remember that all angles are determined by the fact that we took the reference as V^+ with an angle of zero degrees at the load end. That is, our reference phasor is $V^+\,\angle 0°$.

$$|I^+| = \left|\frac{V^+}{Z_o}\right| = \frac{55}{50} = 1.09. \qquad |I_s| = 2\,I^+\cos 65.41° =$$

$2(1.09)\cos 65.41° = 0.915$ amperes.

Again glancing at the diagram we would write it as:

$$I_s = 0.915\,\angle 180°. \quad \therefore Z_s = \frac{V_s}{I_s} = \frac{100\,\angle 90°}{.915\,\angle 180°}$$

$$= 109.2\,\angle -90° = -j109.2 \text{ ohms}$$

Since the current is leading the voltage at the sending end the input impedance is capacitive.

$I_r = 2\ I^+ = 2(1.09) = 2.18$ amperes or $I_r = 2.18\ \underline{/0°}$ when we take the angle from the phasor diagram. $V_r = 0$ (short), $P_s = 0$ since V_s and I_s are 90 degrees out of phase and $P_r = 0$. ($V_r = 0$.)

Our solution gives the same magnitudes as in the example in Section 2.4. The angles are different because in the example in Chapter 2 we took the sending end voltage as the reference, ($V_s = 100\ \underline{/0°}$). In this example we used V^+ at the receiving end as the reference. We could have asked for the voltage, current, power or impedance anywhere along the line in this example. For instance, suppose we wish to find out what is happening 1/8 wavelength from the load. The first step is to construct the phasor diagram at the location; 1/8 wavelength is equivalent to 45 degrees.

$V = j(2)(V^+) \cos 45° =$
$j(2)(55) \cos 45° =$
$j77.76$ volts

$I = 2(I^+) \cos 45° =$
$2(1.09) \cos 45° =$
$1/55\ \underline{/0°}$ amperes

$Z = \dfrac{V}{I} = \dfrac{77.76\ \underline{/90°}}{1.55\ \underline{/0°}} =$
$50\ \underline{/90°} =$
$j50$ ohms

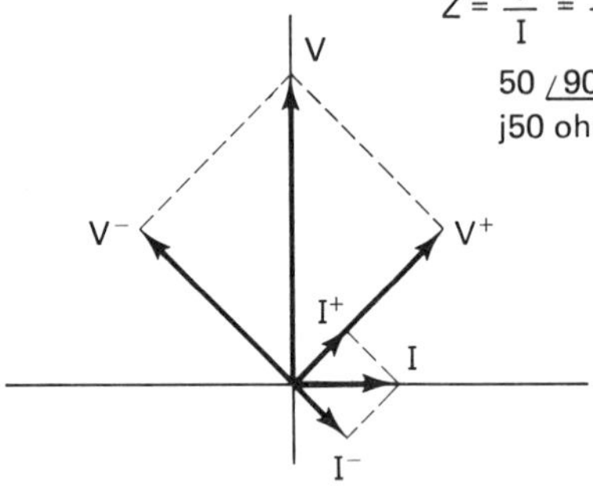

Equation (2-11) was $Z = j\ Z_o \tan \beta\ d = j\ 50 \tan 45° = j50$ ohms. Strictly speaking βd is in radians but since most of us are more familiar and comfortable with degrees it is common to think in terms of wavelength or degrees rather than in radians.

3.5 The Open-Circuit Load Again

In the last chapter we looked at the open-circuit case by using the basic equations. Let us rework our solution using the phasor method.

$$\Gamma_r = \frac{Z_r - Z_o}{Z_r + Z_o} = \frac{1 - \dfrac{Z_o}{Z_r}}{1 + \dfrac{Z_o}{Z_r}}. \tag{3-9}$$

$$Z_r = \infty, \text{ so } \Gamma_r = \frac{1}{1} = 1 = \frac{V^-}{V^+}.$$

Now the phasor diagram at the load end must indicate that the reflected component of the voltage must be in phase with the incident component. Once we know the phasor relationship at the load end we can rotate the phasors as we move down the line towards the sending end and see what is happening at each point on the line. The process is just a repeat of what we did for the short-circuit case with a different phasor diagram at the load end to start with.

When βd is equivalent to 360°, the phasor diagram is the same as it was at the load end.

The results of Figures 3-8, 3-9 and 3-10 can be quickly sketched as in Figure 3-11.

The results agree with those shown in Figure 2-6, which were obtained through the use of equations. Now we will resolve our previous problem of the open-circuit load using phasors.

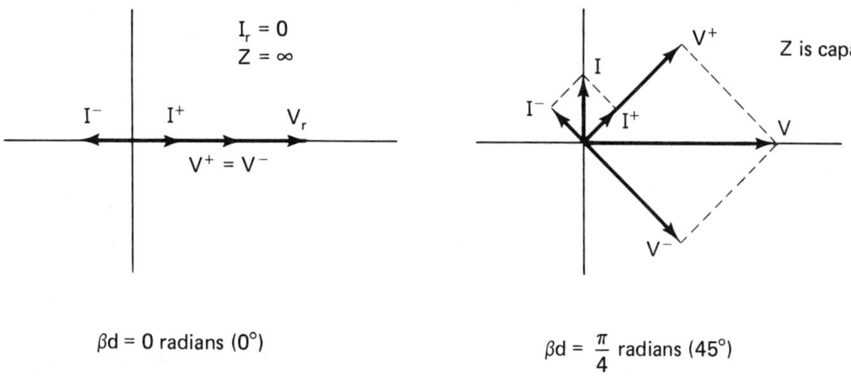

Figure 3-8.

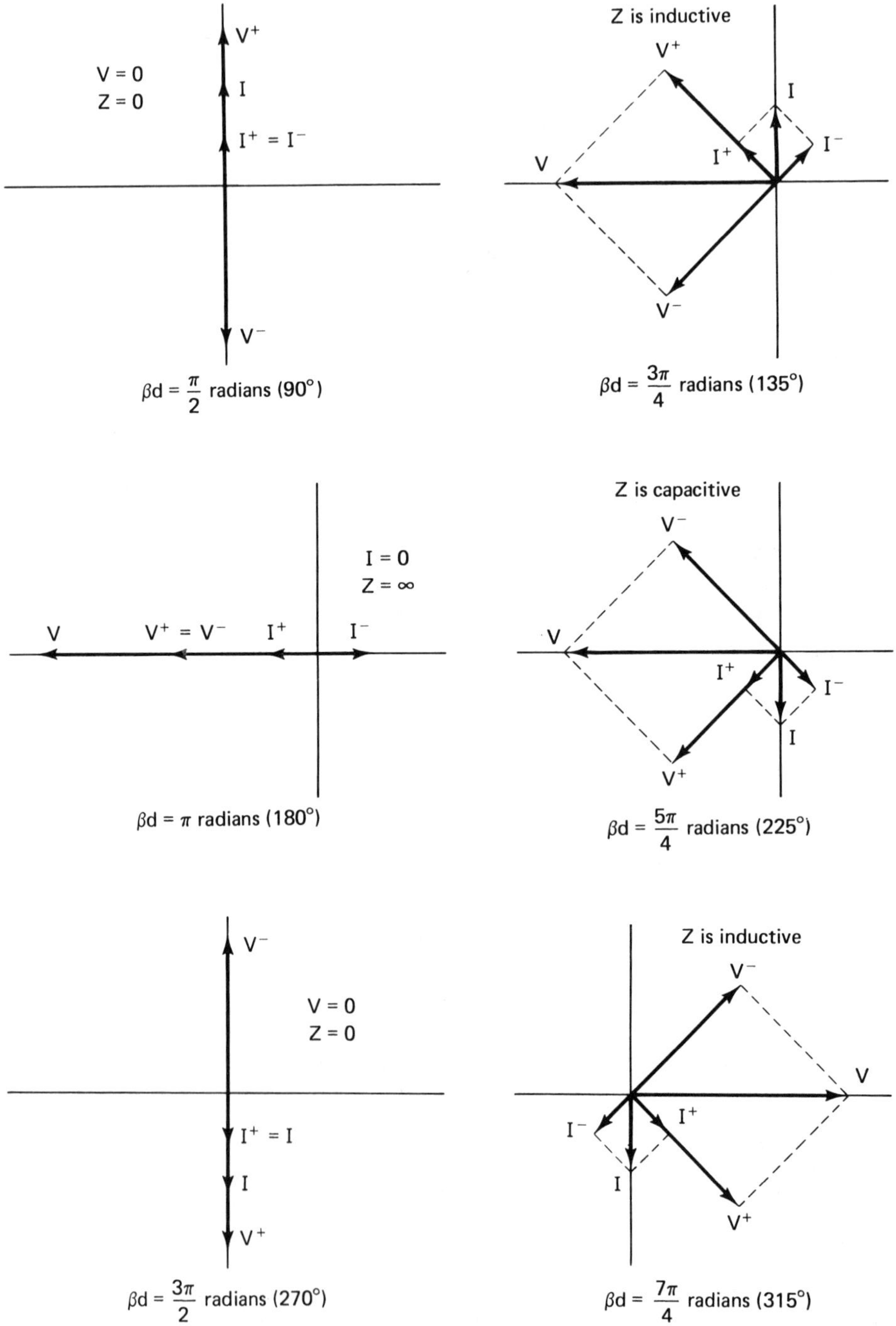

Figure 3-9. Open-circuit phasor diagrams.

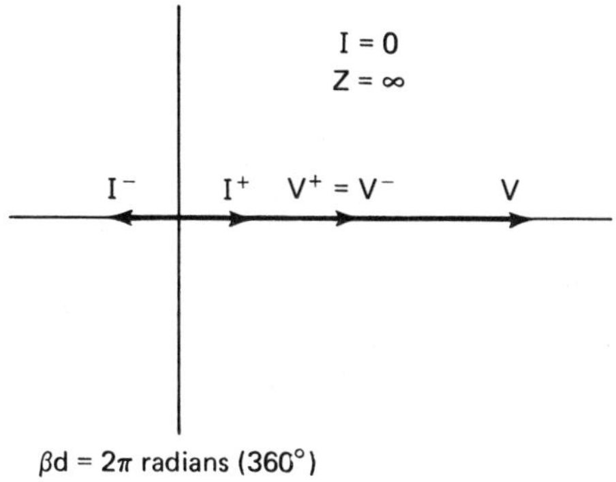

Figure 3-10. Open-circuit phasor diagram.

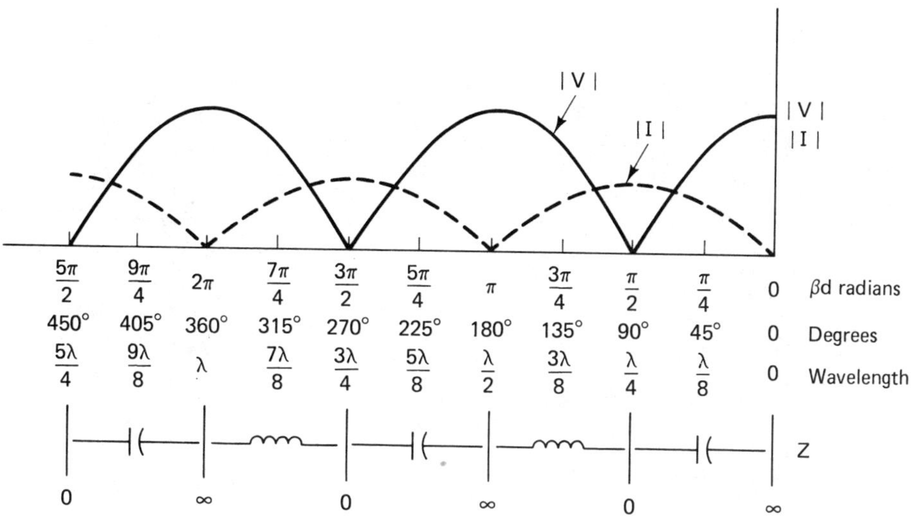

Figure 3-11. Open-circuit pattern on line.

3.6 Example of the Phasor Solution for an Open-Circuit Load

A lossless line is terminated in an open circuit, has a characteristic impedance of 50 ohms, an input voltage of 100 volts, r.m.s., is 0.8 meter long and β is 2.5 radians per meter. Using phasors, determine I_s, I_r, V_r, Z_s, P_s, and P_r.

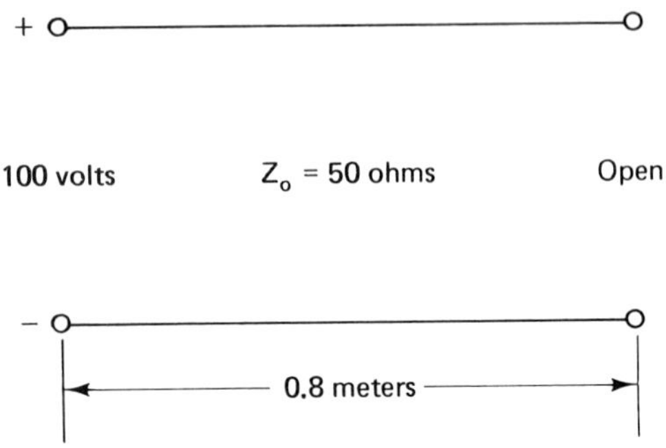

Our starting point is always the voltage reflection coefficient.

$$\Gamma_r = \frac{z_r - z_o}{z_r + z_o} = \frac{\infty - z_o}{\infty + z_o} = +1$$
$$= \frac{V^-}{V^+}$$

$$\beta\ell = (2.5)(0.8) = 2 \text{ radians } (114.59°).$$

Knowing Γ_r and $\beta\ell$ we can construct the phasor diagram at the load end and at the sending end.

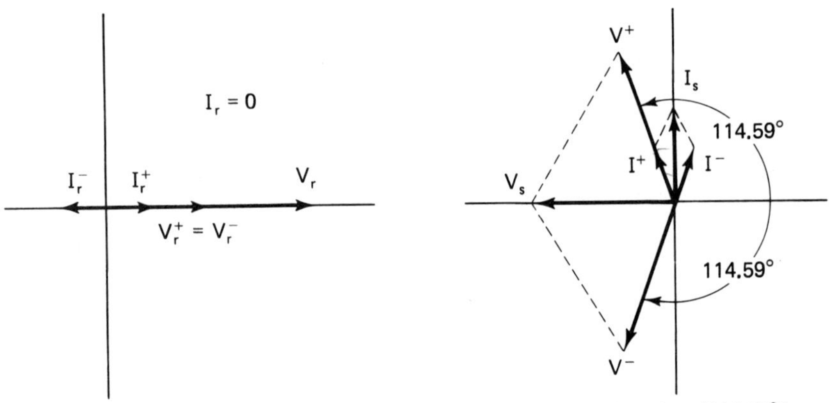

$\beta d = 0°$ (receiving end)

$\beta d = 2$ radians (114.59°) (sending end)

$$|V_s| = 2\,|V^+| \cos 65.410 = 100.$$

$$|V^+| = \frac{100}{2 \cos 65.41°} = 120.15 \text{ volts. } |I^+| = \left|\frac{V^+}{Z_o}\right| = \frac{120.15}{50} = 2.40 \text{ amperes.}$$

$$\therefore V_r = 2\,|V^+| \angle 0° = 2(120.15) \angle 0° = 240.3 \angle 0° \text{ volts.}$$

$$I_s = j2\,|I^+| \cos 24.59° = j2(2.4) \cos 24.59° = j4.37 = 4.37 \angle 90° \text{ amperes.}$$

$$V_s = 2\,|V^+| \cos 65.41° \angle 180° = 2(120.15) \cos 65.41° \angle 180° = 100 \angle 180° \text{ volts.}$$

$$Z_s = \frac{V_s}{I_s} = \frac{100 \angle 180°}{437 \angle 90°} = 22.88 \angle 90° = 0 + j22.88 \text{ ohms.}$$

$I_r = 0$ (open). $P_s = 0$ (current and voltage are 90° out of phase).

The input impedance is inductive (current lags the voltage by 90°). $P_r = 0$ since $I_r = 0$. We have the same magnitudes as in Section 2-7 but the angles are different since we had different reference phasors (V_s in Section 2.7 and V^+ at the load end here).

The voltage and current distribution on the line as shown in Figure 3-7 for the short-circuit load and in Figure 3-11 for the open-circuit load should be observed by the student. If we were moving a voltmeter down the line measuring these magnitudes and trying to pinpoint the exact spot for a maximum or minimum value we would find that near the voltage maximum the curve is rather flat, whereas at the voltage minimum there is a rapid change in the voltage as we move either way. If we were trying to determine the frequency by measuring the wavelength we could do it by measuring the distance between two voltage minimums, realizing that this distance is exactly one half wavelength. Because of the shape of the curves we would have more accurate measurements by taking the distance between two voltage minimums than by taking it between two voltage maximums.

We should also notice that for these two special cases the curves are exact duplicates for the magnitudes of pure sine and cosine waves. We shall see later that this is not true for loads other than a short or an open circuit. Let us look at the voltage for the short circuit. Our phasor diagram at the load end is as it was before. At the receiving end V^+ has an angle of 0 degrees and V^- an angle of 180 degrees. As we move down the line from the load end by a distance of d meters, V^+ rotates βd radians in the counterclockwise direction and V^- rotates βd radians in the clockwise direction. So V at any point on the line d meters from the load can be written as:

$$V = V^+ \underline{/\theta} + V^- \underline{/180} - \theta \text{ where } \theta = \frac{180}{\pi} \beta d \text{ degrees.}$$

(See Figure 3-12).

Since

$$|V^+| = |V^-|.\ V = V^+[1 \underline{/\theta} + 1 \underline{/180} - \theta] =$$

$$V^+ [\cos\theta + j\sin\theta + \cos(180 - \theta) + j\sin(180 - \theta)] =$$

$$V = V^+ [\cos\theta + j\sin\theta + \cos 180\cos\theta + \sin 180\sin\theta +$$

$$j(\sin 180\cos\theta - \cos 180\sin\theta)] =$$

$$V^+ [\cos\theta + j\sin\theta - \cos\theta + 0 + j[0 - (-1)\sin\theta]$$

$$= V^+ [2j\sin\theta].$$

$$|V| \text{ magnitude} = |V^+ (2j\sin\theta)| = |2V^+ \sin\theta.| \quad (3\text{-}11)$$

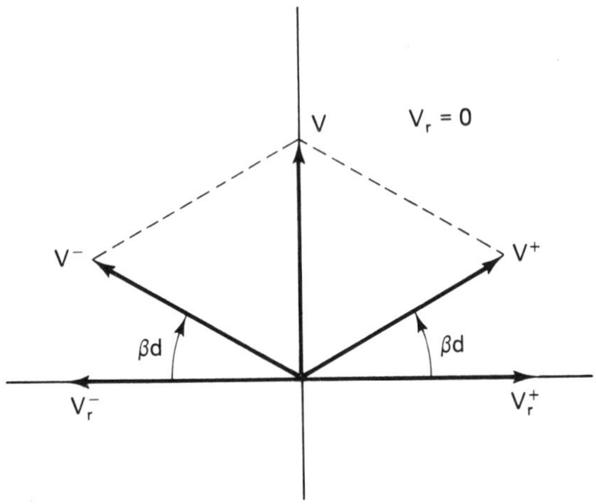

Figure 3-12. Phasor diagram for the short-circuit case.

If we did the same thing for the open-circuit load we would have:

$$|V| \text{ magnitude } = |2V^+ \cos \theta|. \tag{3-12}$$

3.7 Example of the Pure Resistance Load, $Z_r > Z_o$

Short- and open-circuit loads cannot absorb any real power in watts so in order to do any work the load impedance must have a resistive component. Let us first consider the case where the lossless line is terminated in a pure resistance that is larger than the characteristic impedance. Let us assume some simple values.

$$Z_r = 100 + j0 \text{ ohms and } Z_o = 50 + j0 \text{ ohms.}$$

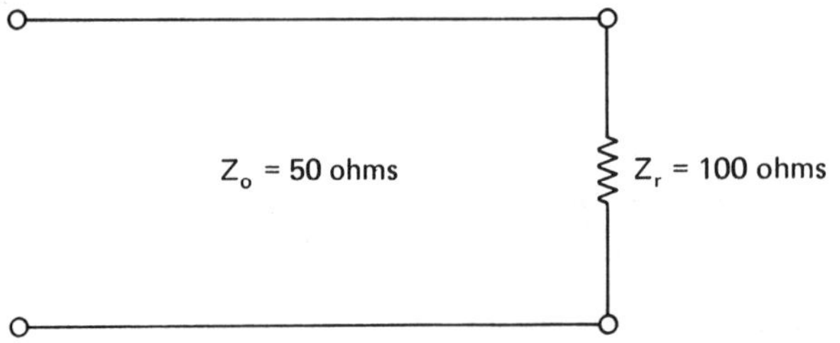

$$\Gamma_r = \frac{z_r - z_o}{z_r + z_o} = \frac{100 - 50}{100 + 50} = \frac{50}{150} = \frac{1}{3} = 0.333 \angle 0^\circ = \frac{V^-}{V^+}.$$

$$V^- = \frac{1}{3} V^+$$

To make it easy, let us assume that $V_r = 100$ r.m.s. volts. Our phasor diagram at the receiving end is shown in Figure 3-13.

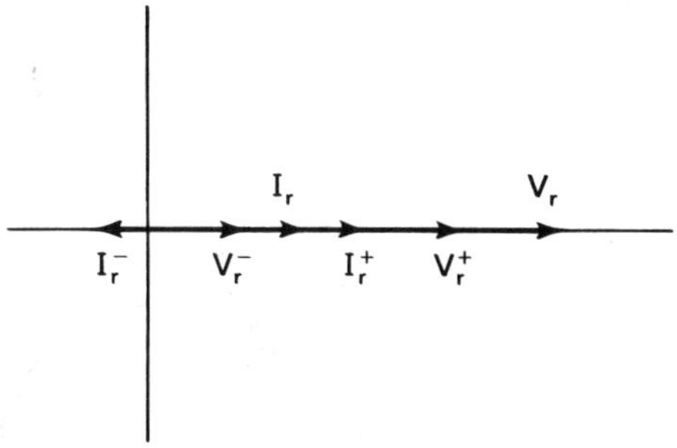

Figure 3-13. Phasor diagram for a pure resistance load at the receiving end when the load resistance is greater than the characteristic impedance.

At the receiving end V^+ and V^- are in phase with each other and we can write:

$$|V^+| + |V^-| = 100 = V_r. \qquad |V^+| + 1/3\,|V^+| = 100.$$

$$4/3\,|V^+| = 100. \qquad |V^+| = 3/4(100) = 75 \text{ volts.}$$

$$|V^-| = 1/3\,|V^+| = 1/3(75) = 25 \text{ volts.}$$

$$|I^+| = \left|\frac{V^+}{Z_o}\right| = \frac{75}{50} = 1.5 \text{ volts.}$$

$$|I^-| = \left|\frac{V^-}{Z_o}\right| = \frac{25}{50} = 0.5 \text{ amperes} = 1/3\,|I^-|.$$

It is important to point out that under these conditions with V^+ and V^- in phase at the load and I^+ and I^- 180° out of phase at the load that we have a voltage maximum and a current minimum at the load end. These are the maximum and minimum values of the r.m.s. values. If we go 1/8 wavelength (45°) towards the sending end the phasor diagram will be as depicted in Figure 3-14.

Now the current, I, is leading the voltage by some angle less than 90°.

Let us use complex numbers for our solution.

$$V = |V^+|\,\underline{/45^\circ} + |V^-|\,\underline{/-45^\circ} = 75\,\underline{/45^\circ} + 25\,\underline{/-45^\circ} = 79.05\,\underline{/26.56^\circ} \text{ volts.}$$

$$I = |I^+|\,\underline{/45} + |I^-|\,\underline{/135^\circ} = |I^+|\,\underline{/45^\circ} - |I^-|\,\underline{/-45^\circ} =$$

$$1.5\,\underline{/45^\circ} - 0.5\,\underline{/-45^\circ} = 1.58\,\underline{/63.44^\circ} \text{ amperes.}$$

$$Z = \frac{V}{I} = \frac{79.05\,\underline{/26.56^\circ}}{1.58\,\underline{/63.44^\circ}} = 50.0\,\underline{/-36.8^\circ} = 40 - j30 \text{ ohms.}$$

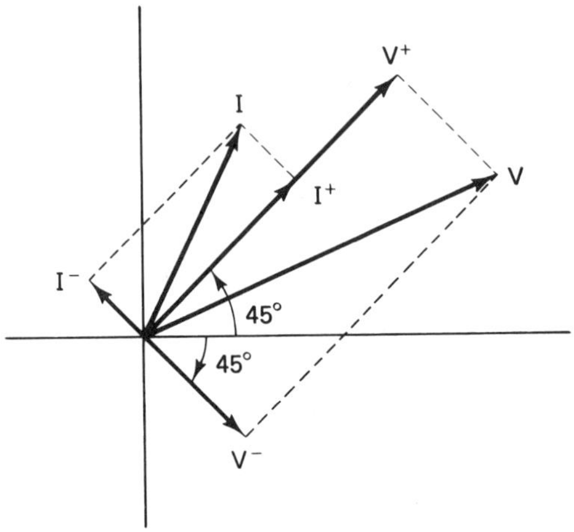

Figure 3-14. Phasor diagram for 1/8 wavelength from the load.

So 1/8 wavelength from the load we have an impedance looking towards the load, which looks like a 40-ohm resistor in series with 30 ohms of capacitive reactance.

This seems like a strange result. A pure resistance load looks like an R and C in series if we cut the line off exactly 1/8 wavelength from the load. But our intuition didn't prepare us for some of the things that can happen in lumped parameter circuits either. Let us move down the line so that we are 1/6 wavelength from the load, which would be equivalent to 60° or $\pi/3$ radians (See Figure 3-15).

The impedance looking towards the load is still capacitive looking, with a resistance component of 30.78 ohms.

Now we will look at the case where we are 1/4 wavelength from the load (90° or $\pi/2$ radians) (See Figure 3-16).

When we have moved 1/4 wavelength from the load, the 100-ohm resistance looks like a 25-ohm resistance. While this is a specific example, it will always be true that when the line is lossless with a real characteristic impedance and the load is a pure resistance greater than Z_o, that 1/4 wavelength down the line the impedance will be real and smaller than the characteristic impedance. It is also true that, in the region between the load and 1/4 wavelength from the load, the impedance looking towards the load will be equivalent to an R and C in series. After we pass the point where the voltage is a minimum and the current a maximum, the impedance will change its character for the next 1/4 wavelength. Let us take our next point

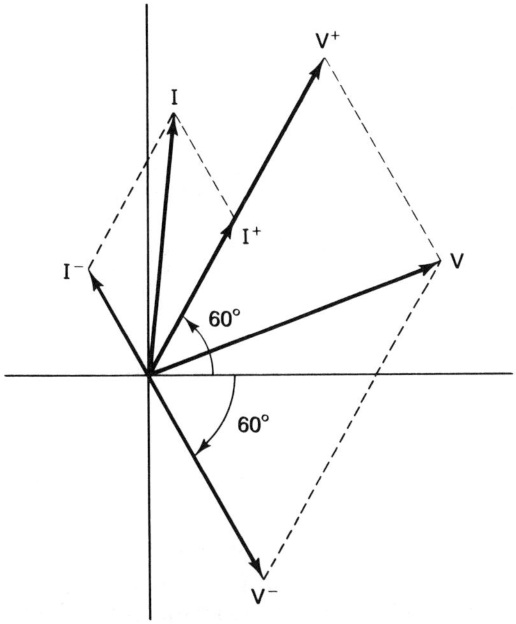

$V = 75 \angle 60^\circ + 25 \angle -60^\circ =$
$66.14 \angle 40.89^\circ$ volts.

$I = 1.5 \angle 60^\circ - 0.5 \angle -60^\circ =$
$1.80 \angle 73.89^\circ$ amperes.

$$Z = \frac{V}{I} = \frac{66.14 \angle 40.89^\circ}{1.80 \angle 73.89^\circ} =$$
$36.70 \angle -32^\circ = 30.78 - j20$ ohms.

Figure 3-15. Phasor diagram for 1/6 wavelength from the load.

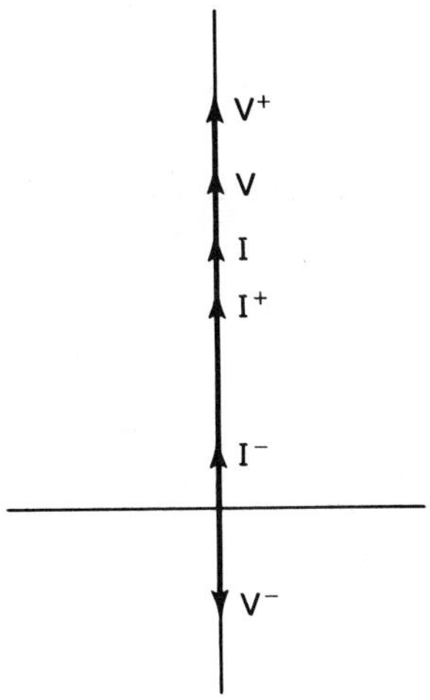

Now V^+ and V^- are exactly 180° out of phase and I^+ and I^- are in phase. Notice we now have a voltage minimum and a current maximum. Since V and I are in-phase, the impedance looking towards the load will be a pure resistance.

$V = 75 \angle 90^\circ + 25 \angle -90^\circ = 50 \angle 90^\circ = j50$ volts.

$I = 1.5 \angle 90^\circ - 0.5 \angle -90^\circ = 2 \angle 90^\circ = j2$ amperes.

$$Z = \frac{V}{I} = \frac{j50}{j2} = 25 \angle 0^\circ = 25 + j0 \text{ ohms.}$$

Figure 3-16. Phasor diagram for 1/4 wavelength from the load.

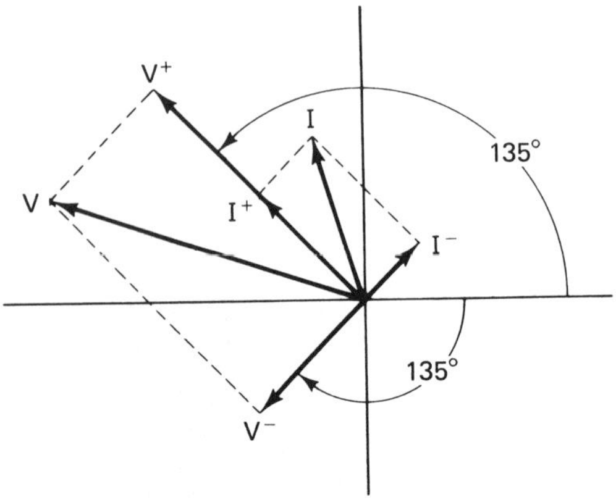

$$V = 75\,\underline{/135^\circ} + 25\,\underline{/-135^\circ} = 79.06\,\underline{/153.44^\circ} \text{ volts.}$$

$$I = 1.5\,\underline{/135^\circ} - 0.5\,\underline{/-135^\circ} = 1.58\,\underline{/116.56^\circ} \text{ amperes.}$$

$$Z = \frac{V}{I} = \frac{79.06\,\underline{/153.44^\circ}}{1.58\,\underline{/116.56^\circ}} = 50\,\underline{/36.87^\circ} = 40 + j30 \text{ ohms.}$$

Figure 3-17. Phasor diagram for 3/8 wavelength from the load.

as being 3/8 wavelength from the load ($\beta d = 3\pi/4$ radians, 135°) (See Figure 3-17).

So the impedance now looks like a resistor and inductance in series since I is lagging V by some angle less than 90°.

When we proceed down the line so that we are exactly 1/2 wavelength from the load we might guess by this time that the conditions at the load will repeat themselves; that is, the magnitudes of the voltage and current will be the same as at the load (but the angles will be different) and the impedance will be the same (See Figure 3-18).

So the magnitudes of the voltage, current, and impedance are the same as at the load. The impedance is equal to the load impedance value of 100 ohms. The only difference between the two locations

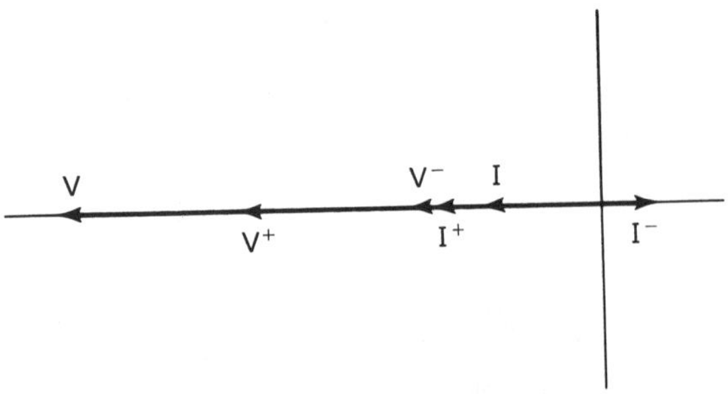

$$V = 75\,\underline{/180^\circ} + 25\,\underline{/180^\circ} = 100\,\underline{/180^\circ} = -100 + j0 \text{ volts.}$$

$$I = 1.5\,\underline{/180^\circ} - 0.5\,\underline{/-180^\circ} = 1\,\underline{/180^\circ} = -1 + j0 \text{ amperes.}$$

$$Z = \frac{V}{I} = \frac{-100}{-1} = 100 \text{ ohms.}$$

Figure 3-18. Phasor diagram for 1/2 wavelength from the load.

is that the voltage and current, one half wavelength down the line, are 180 degrees out of phase with the voltage and current at the load. Once again using the phasor diagrams we can rapidly sketch out the conditions on the line as shown in Figure 3-19.

Under these conditions the voltage and current never go to zero as they did for the short- and open-circuit cases. A voltage maximum occurs with a current minimum giving an impedance maximum (100 ohms) which is a pure resistance. A voltage minimum occurs with a current maximum giving an impedance minimum (25 ohms) which is also a pure resistance.

Conditions still repeat themselves every half wavelength. For the first quarter wavelength the impedance looks like a resistor and capacitor in series and for the next quarter wavelength the impedance looks like a resistor and inductance in series. Then for the next

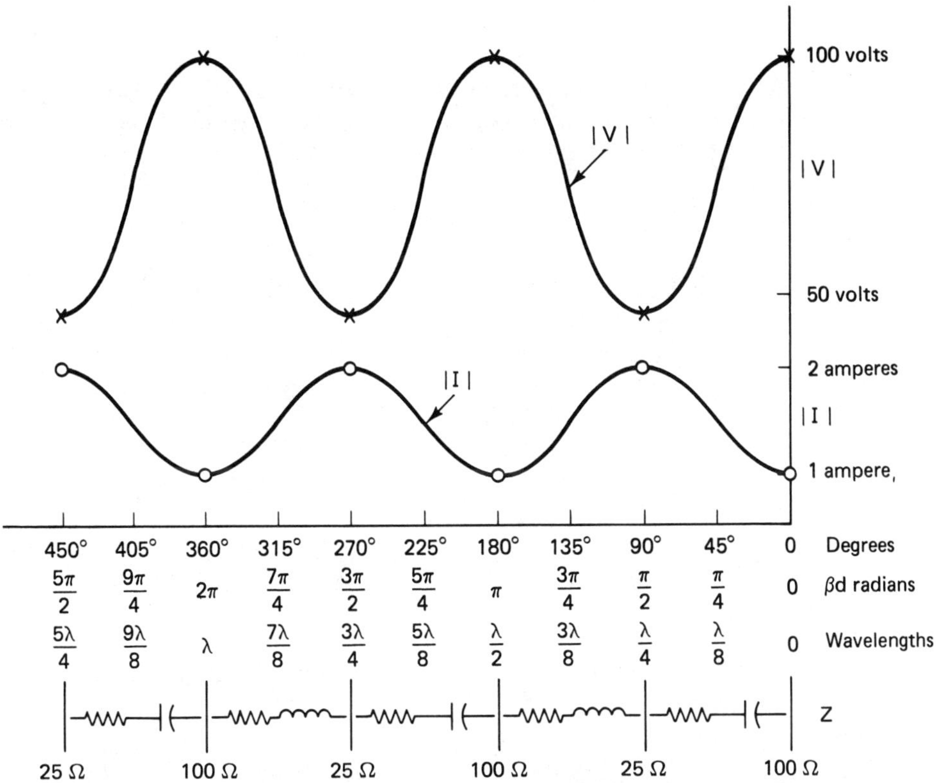

Figure 3-19. Patterns on line for resistive load.

quarter wavelength it goes back to the resistor and capacitor and so on down the line. If we took a pair of tin snips and cut the line at 225° from the load our sketch indicates that the input impedance to the line would look like a resistance in series with a capacitor. Its actual value could easily be determined with a phasor diagram and calculations such as we have just done for certain locations on the line. It seems obvious from our case that the magnitude of the impedance will vary from 25 to 100 ohms but that there are an infinite number of possible impedances between these two values.

3.8 Voltage Standing Wave Ratio

As the phasors, V^+ and V^-, rotate in opposite directions they become in phase with each other twice in each revolution or every half wavelength. This becomes a point of a voltage maximum.

$$V_{max} = |V^+| + |V^-|. \tag{3-13}$$

As they continue to rotate there will be two points in every wavelength where the two phasors are 180 degrees out of phase and we will obtain a voltage minimum.

$$V_{min} = |V^+| - |V^-|. \tag{3-14}$$

Let us define the Voltage Standing Wave Ratio (VSWR) as:

$$\text{VSWR} = \frac{V_{max}}{V_{min}} = \frac{|V^+| + |V^-|}{|V^+| - |V^-|} = \frac{|V^+| + |\Gamma_r V^+|}{|V^+| - |\Gamma_r V^+|} = \frac{1 + |\Gamma_r|}{1 - |\Gamma_r|}. \tag{3-15}$$

In Section 3.7 we had a case where the maximum voltage was 100 volts and the minimum value was 50 volts with $\Gamma_r = 1/3$. So the VSWR equals:

$$\text{VSWR} = \frac{V_{max}}{V_{min}} = \frac{100}{50} = 2 = \frac{1 + |1/3|}{1 - |1/3|} = \frac{4/3}{2/3} = 2.$$

The short- and open-circuit loads produce the maxium value of the VSWR.

Short-circuit case:

$$\Gamma_r = -1 = 1 \angle 180°. \ \text{VSWR} = \frac{1 + |\Gamma_r|}{1 - |\Gamma_r|} = \frac{1 + 1}{1 - 1} = \frac{2}{0} = \infty.$$

Open-circuit case:

$$\Gamma_r = +1 = 1\,\underline{/0^\circ}. \text{ VSWR} = \frac{1 + |\Gamma_r|}{1 - |\Gamma_r|} = \frac{1 + 1}{1 - 1} = \frac{2}{0} = \infty.$$

When the load is matched to the line, $\Gamma_r = 0$ so the VSWR becomes:

$$\text{VSWR} = \frac{1 + |\Gamma_r|}{1 - |\Gamma_r|} = \frac{1 + 0}{1 - 0} = 1.$$

Since the minimum value of the voltage cannot be larger than the maximum value, the VSWR must vary between one and infinity. As we shall see later the ideal situation is where we have no standing wave on the line and this occurs when $\Gamma_r = 0$ and the VSWR = 1.

3.9 Example of a Phasor Solution

A lossless line with a characteristic impedance of 50 ohms (real) is terminated in a pure resistance load of 100 ohms. It has a β of 2.5 radians per meter, is 0.8 meters long and the input voltage is 100 volts, r.m.s. Determine I_s, V_r, I_r, P_s, P_r, Q_r, and Q_s.

Solution

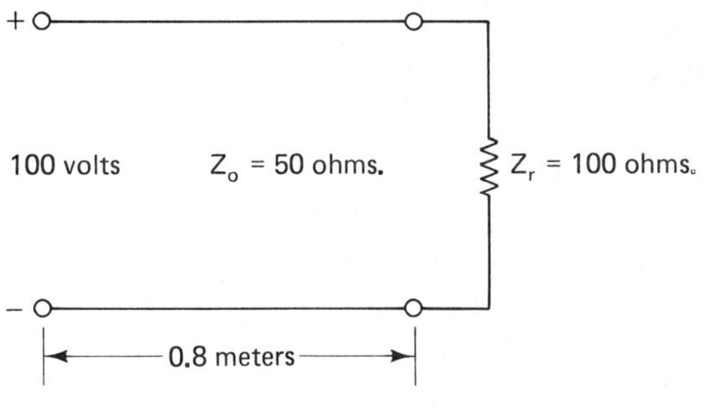

$$\Gamma_r = \frac{Z_r - Z_o}{Z_r + Z_o} = \frac{100 - 50}{100 + 50} = \frac{1}{3}\,\underline{/0^\circ}$$

$$\beta\ell = (2.5)\,(0.8) = 2 \text{ radians. } (114.59^\circ)$$

In this case we are given the input voltage but we will still take V^+ at the receiving end as our reference phasor. In order to find the value of V^+ we will go to the sending end where the voltge is known. The phasor diagram at the receiving end is shown in Figure 3-20.

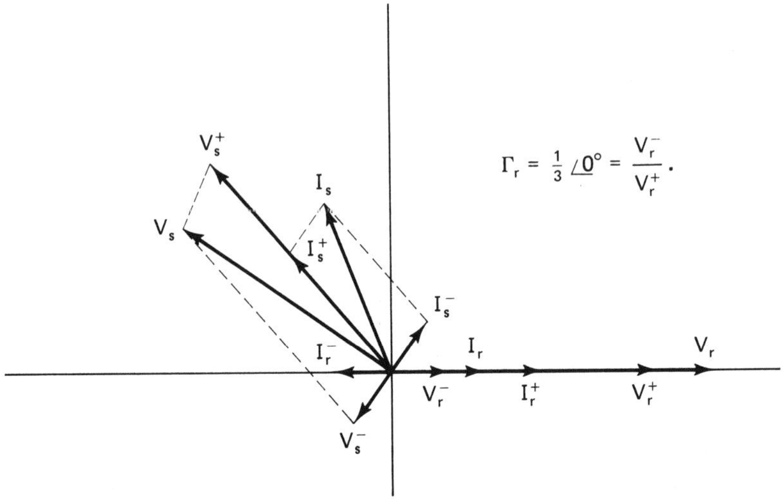

Figure 3-20. Combined phasor diagrams for the sending and receiving ends.

$$V_s = |V^+| \angle 114.59° + |V^-| \angle -114.59° = |V^+| \angle 114.59° + 1/3\, |V^+| \angle -114.59°$$
$$= .822\, |V^+| \angle 132.47°.$$

$$|V_s| = 100 = |V^+ (.822)|. \qquad |V^+| = \left|\frac{100}{.822}\right| = 121.68 \text{ volts.}$$

$$|V^-| = 1/3\, |V^+| = \frac{121.68}{3} = 40.56 \text{ volts.}$$

$$|I^+| = \left|\frac{V^+}{Z_o}\right| = \left|\frac{121.68}{50}\right| = 2.43 \text{ amperes.}$$

$$|I^-| = \left|\frac{V^-}{Z_o}\right| = \left|\frac{40.56}{50}\right| = 0.81 \text{ amperes} = 1/3\, |I^+|.$$

$$I_s = |I^+| \angle 114.59° - |I^-| \angle -114.59° = 2.43 \angle 114.59° - 0.81 \angle -114.59° =$$

$$I_s = 3.02 \angle 102.88° \text{ amps.}$$

$$V_s = 100 \angle 132.47°.$$

$$Z_s = \frac{V_s}{I_s} = \frac{100 \angle 132.47°}{3.02 \angle 102.88°} = 33.04 \angle 29.58° = 28.73 + j16.31 \text{ ohms.}$$

$$V_r = |V^+| \angle 0° + V^- \; 0° = 121.68 + 40.56 = 162.24 \angle 0° \text{ volts.}$$

$$I_r = |I^+| \angle 0° + |I^-| \angle 180° = 2.43 - 0.81 = 1.62 \angle 0° \text{ amperes.}$$

$$P_s = V_sI_s\cos\theta_s = (100)(3.02)\cos 29.58° = 263.24 \text{ watts.}$$

$$P_r = V_rI_r\cos\theta_r = (162.24)(1.62)\cos 0° = 263.24 \text{ watts.}$$

$$Q_s = V_sI_s\sin\theta_s = (100)(3.02)\sin 29.58° = 179.08 \text{ vars.}$$

$$Q_r = V_rI_r\sin\theta_r = (162.24)(1.62)\sin 0° = 0 \text{ vars.}$$

$P_s = P_r$, since we have assumed a lossless line. In the phasor diagram we have indicated the positions of the phasors at each end by the addition of the subscripts, s and r. This is unnecessary after one becomes used to the technique.

3.10 Example of the Pure Resistance Load, $Z_r < Z_o$

When we looked at the case where Z_r and Z_o were both real and Z_r was greater than Z_o, it turned out that Γ_r had an angle of zero degrees and we had a voltage maximum at the load. When Z_r is less than Z_o and both are real it turns out that we have a voltage minimum at the load, and the angle of the reflection coefficient, Γ_r, is 180 degrees.

Let us assume that $Z_r = 50$ ohms and $Z_o = 100$ ohms.

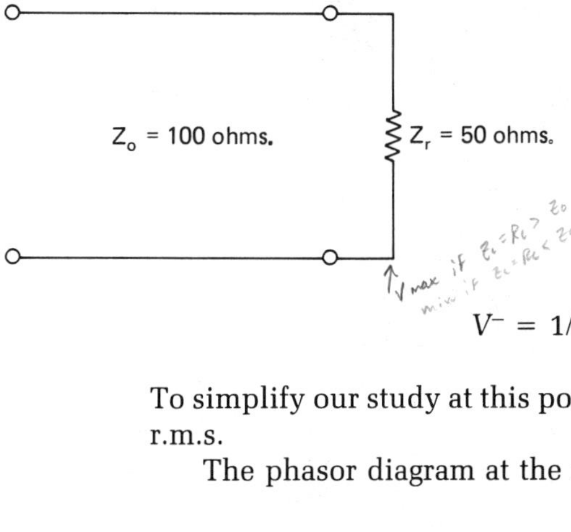

$$\Gamma_r = \frac{Z_r - Z_o}{Z_r + Z_o} = \frac{50 - 100}{50 + 100} = \frac{-50}{150} = -\frac{1}{3} = \frac{1}{3}\angle 180° = \frac{V_r^-}{V_r^+} \text{ or}$$

$$V^- = 1/3\,|V^+|\,\angle 180°.$$

To simplify our study at this point let us assume that $V_r = 100$ volts, r.m.s.

The phasor diagram at the receiving end becomes:

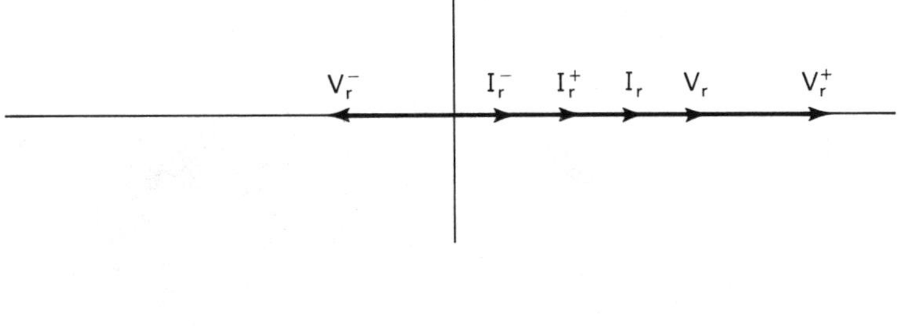

$$V_r = V_r^+ + V_r^- = |V_r^+| \angle 0^\circ + |V_r^- \angle 180^\circ = |V^+| - 1/3\,|V^+| = 2/3\,|V^+|.$$

$$2/3\,|V^+| = 100. \qquad |V^+| = 3/2\,(100) = 150 \text{ volts.}$$

$$|V^-| = 1/3\,|V^+| = 1/3(150) = 50 \text{ volts.}$$

$$|I^+| = \left|\frac{V^+}{Z_o}\right| = \left|\frac{150}{100}\right| = 1.5 \text{ amperes.}$$

$$|I^-| = \left|\frac{V^-}{Z_o}\right| = \left|\frac{50}{100}\right| = 0.5 \text{ amperes} =$$

$$1/3\,|I^+| = \frac{1.5}{3} = 0.5 \text{ amperes.}$$

Associated with the voltage minimum at the load we have a current maximum which gives a minimum value of impedance at the load of 50 ohms. Suppose we now move down the line so that we are 1/8 wavelength from the load, as shown in Figure 3-21. At this point on the line we find that the impedance looking towards the load is equivalent to a resistor in series with an inductance. The current, I, in the phasor diagram is lagging the voltage, V. This inductive effect will last until we move down the line 1/4 wavelength where we will find a voltage maximum and a current minimum as shown in Figure 3-22. Since this is the maximum value of the voltage and the minimum value of the current, 200 ohms is the maximum magnitude of the impedance we can have.

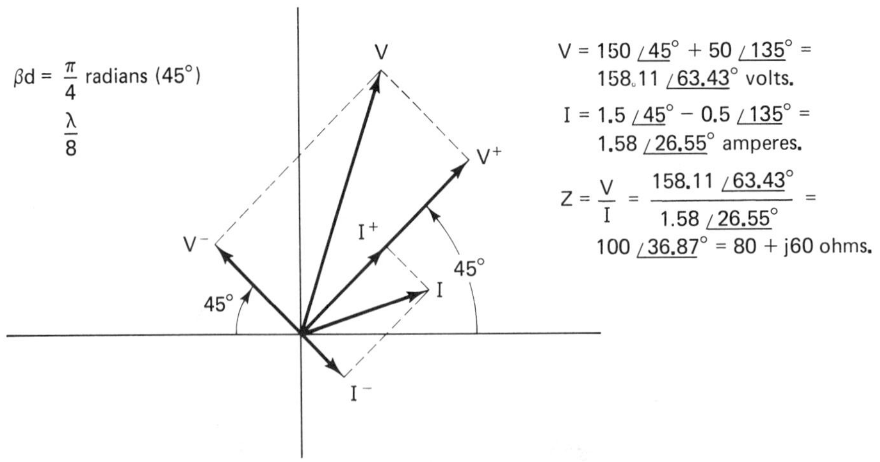

Figure 3-21. Phasor diagram for 1/8 wavelength from the load.

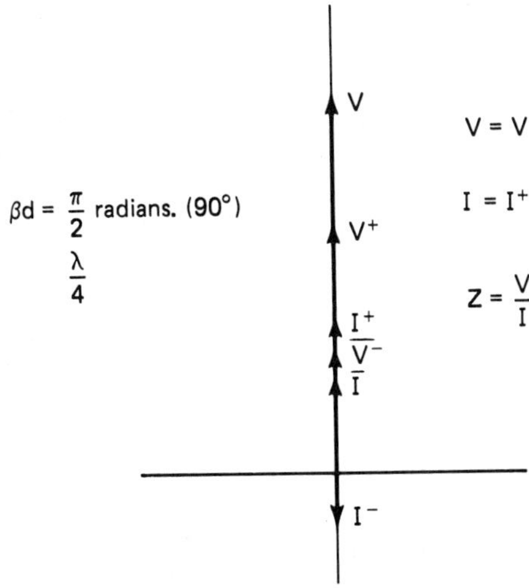

$$V = V^+ + V^- = 150\,\angle 90^\circ + 50\,\angle 90^\circ = 200\,\angle 90^\circ = 0 + j200 \text{ volts.}$$

$$I = I^+ + I^- = 1.5\,\angle 90^\circ + 0.5\,\angle -90^\circ = 1.0\,\angle 90^\circ = 0 + j1 \text{ amperes.}$$

$$Z = \frac{V}{I} = \frac{200\,\angle 90^\circ}{1\,\angle 90^\circ} = 200\,\angle 0^\circ = 200 + j0 \text{ ohms.}$$

Figure 3-22. Phasor diagram for 1/4 wavelength from the load.

At $3\lambda/8$ (135°) we would have the phasor diagram, shown in Figure 3-23. Thus we find that after being inductive for the first 1/4 wavelength the impedance becomes capacitive after passing the point of a voltage maximum. When we refer to the impedance as being inductive or capacitive we are not talking about pure inductances

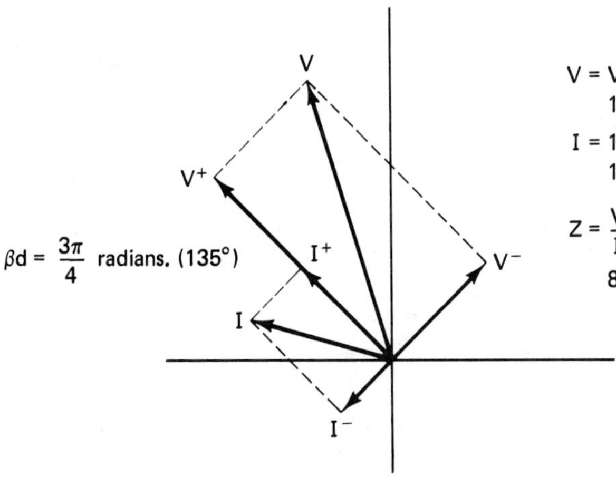

$$V = V^+ + V^- = 150\,\angle 135^\circ + 50\,\angle 45^\circ = 158.11\,\angle 116.56^\circ \text{ volts.}$$

$$I = 1.5\,\angle 135^\circ - 0.5\,\angle 45^\circ = 1.58\,\angle 153.43^\circ \text{ amperes.}$$

$$Z = \frac{V}{I} = \frac{158.11\,\angle 116.56^\circ}{1.58\,\angle 153.43^\circ} = 100\,\angle -36.87^\circ = 80 - j60 \text{ ohms.}$$

Figure 3-23. Phasor diagram for 3/8 wavelength from the load.

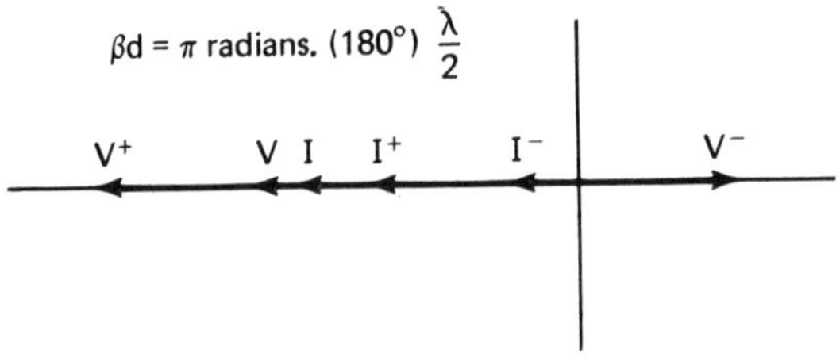

$V = 150 \angle 180^\circ + 50 \angle 0^\circ =$
$100 \angle 180^\circ$ volts.

$I = 1.5 \angle 180^\circ + 0.5 \angle 180^\circ =$
$2 \angle 180^\circ$ amperes.

$$Z = \frac{V}{I} = \frac{100 \angle 180^\circ}{2 \angle 180^\circ} = 50 \angle 0^\circ =$$
$50 + j0$ ohms.

Figure 3-24. Phasor diagram for 1/2 wavelength from the load.

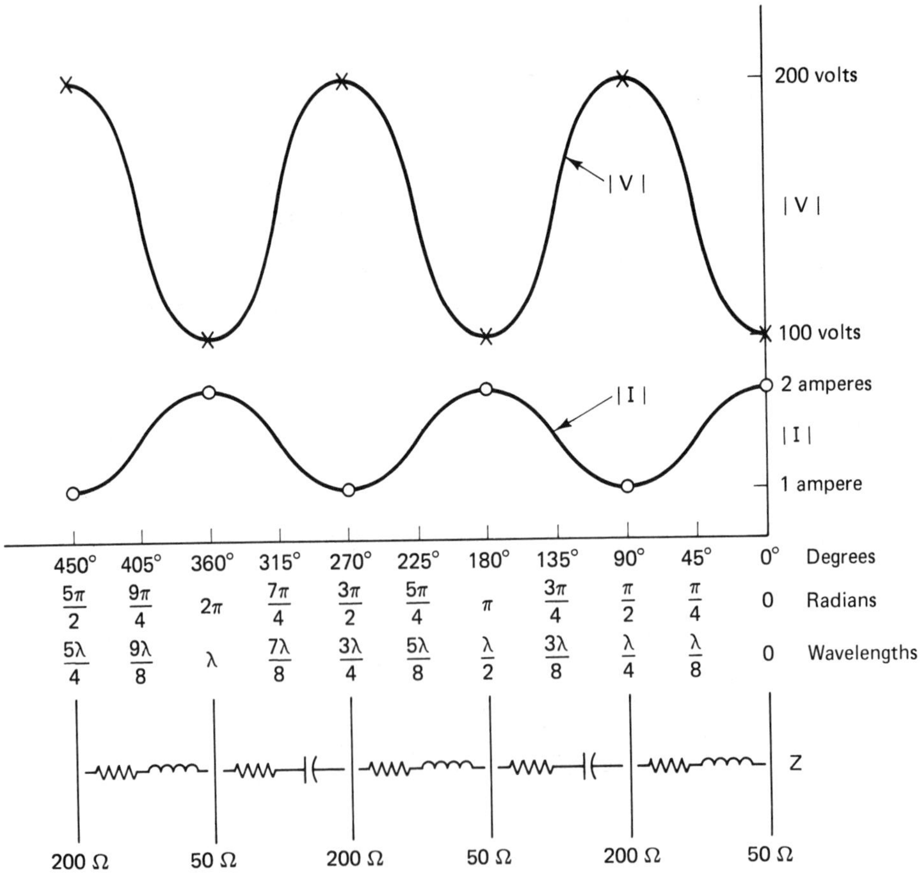

Figure 3-25. Patterns on line for resistive load.

or capacitors but the case where they are in series with some value of resistance. By this time we should be able to guess what happens at 1/2 wavelength from the load. This is shown in Figure 3-24. Now the conditions at the load are repeated, except that the voltage and current are 180 degrees out of phase with the voltage and current at the load. The line's behavior is depicted in Figure 3-25.

Either Figure 3-25 or our equations show that the VSWR is equal to 2.0. In this case the magnitude of the impedance varies between 50 and 200 ohms and, as before, there are an infinite number of impedances possible. Figure 3-25 should be compared with Figure 3-19. It might appear that the shape of the voltage and current magnitude curves is sinusoidal, but for these cases it is not true. We will see later that the voltage curve is broadened at the maximum value and narrowed at the minimum value.

3.11 Example of the Pure Resistance Load, $Z_r = Z_o$

While this case has been discussed, it is worthwhile to look at it from the phasor diagram basis. Let us assume that $Z_r = Z_o = 50$ ohms and that $V_r = 100$ volts, r.m.s.

$$\Gamma_r = \frac{Z_r - Z_o}{Z_r + Z_o} = \frac{50 - 50}{50 + 50} = \frac{0}{100} = 0 = \frac{V^-}{V^+}. \qquad V^- = 0.$$

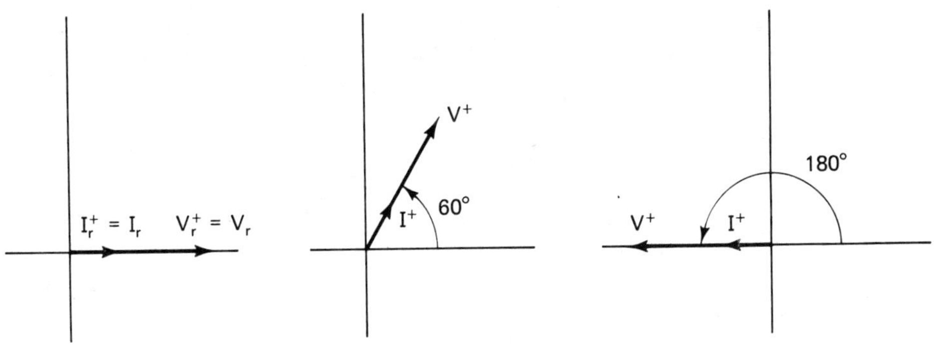

$V^+ = V_r = 100$ volts.

$$I^+ = I_r = \frac{V^+}{Z_o} = \frac{100}{50} = 2 \text{ amperes.}$$

Figure 3-26. Phasor diagrams for the case when $Z_r = Z_o$.

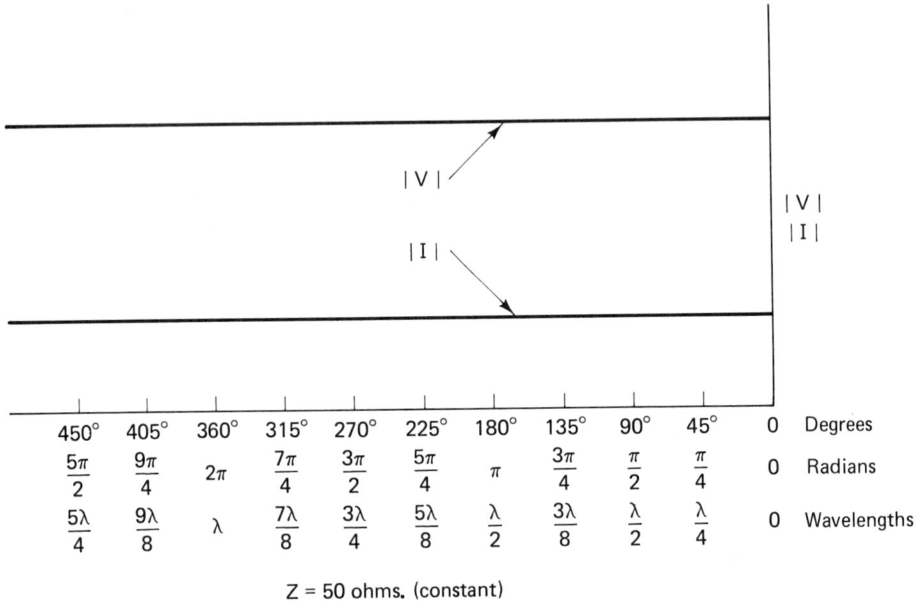

Figure 3-27. Pattern on line for $Z_r = Z_o$.

At the receiving end V^- and I^- are both zero and we just have the two phasors, V^+ and I^+, rotating counterclockwise as we move from the receiving end to the sending end. The magnitudes of the voltage and current remain constant on the line and, since they are always in phase, the impedance stays constant at the value of the characteristic impedance, Z_o. This is shown in Figure 3-26.

In Figure 3-27 we see the patterns on the line.

Looking at Figure 3-27 we see why the line is referred to as being "flat." There are no voltage or current maximums or minimums. The VSWR is equal to one and the impedance looking towards the load is constant at the value of the load impedance which is equal to the characteristic impedance. Since V^- and I^- are zero we have no reflected wave on the line.

3.12 A Rather Special Result When Z_r and Z_o Are Both Real

We have studied in detail the case where both the load impedance and the characteristic impedance are pure resistances. They can be equal or one can be larger than the other. In the latter case we can find the VSWR in terms of the two resistances.

$$\Gamma_r = \frac{Z_r - Z_o}{Z_r + Z_o}.$$

If $Z_r = R_r$ and $Z_o = R_o$ and to emphasize that they are pure resistances we can rewrite the equation in the following form:

$$\Gamma_r = \frac{\dfrac{R_r}{R_o} - 1}{\dfrac{R_r}{R_o} + 1}. \tag{3-16}$$

(3-15) $$\text{VSWR} = \frac{1 + |\Gamma_r|}{1 - |\Gamma_r|}.$$

If $\dfrac{R_r}{R_o} > 1$, $|\Gamma_r|$ is positive and $|\Gamma_r| = \Gamma_r$.

So,

$$\text{VSWR} = \frac{1 + \Gamma_r}{1 - \Gamma_r} = \frac{1 + \dfrac{\dfrac{R_r}{R_o} - 1}{\dfrac{R_r}{R_o} + 1}}{1 - \dfrac{\dfrac{R_r}{R_o} - 1}{\dfrac{R_r}{R_o} + 1}} = \frac{\dfrac{R_r}{R_o} + 1 + \dfrac{R_r}{R_o} - 1}{\dfrac{R_r}{Z_o} + 1 - \dfrac{R_r}{Z_o} + 1} = \frac{2\dfrac{R_r}{R_o}}{2} =$$

$$\frac{R_r}{R_o} = \text{VSWR}, \quad \frac{R_r}{R_o} > 1. \tag{3–17}$$

If the load resistance is larger than the characteristic resistance the VSWR is simply the larger value over the smaller value. If $Z_r = 100$ ohms and $Z_o = 50$ ohms the VSWR is equal to $100/50 = 2.0$.

If $\dfrac{R_r}{R_o} < 1$, $|\Gamma_r|$ is negative and $|\Gamma_r| = -\Gamma_r$.

So

$$\text{VSWR} = \frac{1 + |\Gamma_r|}{1 - |\Gamma_r|} = \frac{1 - \Gamma_r}{1 + \Gamma_r} = \frac{1 - \dfrac{\dfrac{R_r}{R_o} - 1}{\dfrac{R_r}{R_o} + 1}}{1 + \dfrac{\dfrac{R_r}{R_o} - 1}{\dfrac{R_r}{R_o} + 1}} =$$

$$\frac{\dfrac{R_r}{R_o} + 1 - \dfrac{R_r}{R_o} + 1}{\dfrac{R_r}{R_o} + 1 + \dfrac{R_r}{R_o} - 1} = \frac{2}{2\dfrac{R_r}{R_o}} = \frac{R_o}{R_r} = \text{VSWR.} \quad \frac{R_r}{R_o} < 1. \tag{3–18}$$

Again we should note that we simply take the larger of the two values and divide by the smaller number to obtain the VSWR.

Summary

$$\text{VSWR} = \frac{R_r}{R_o}. \quad \text{If } \frac{R_r}{R_o} > 1. \tag{3-17}$$

$$\text{VSWR} = \frac{R_o}{R_r}. \quad \text{If } \frac{R_r}{R_o} < 1. \tag{3-18}$$

$$\text{VSWR} = 1. \quad \text{If } R_r = R_o. \tag{3-19}$$

3.13 Example of the General Load Impedance on a Line With a Real Characteristic Impedance

The last case we will consider under the assumption that the line is lossless will be one where the load is a complex impedance. We will assume that $Z_o = 50$ ohms and that the line is terminated in a load impedance of $40 + j30$ ohms. The line will be 20 meters long and β will be 1.309 radians per meter. To make it as general as possible we will assume the input voltage is 100 volts, r.m.s. For the first time we will obtain an angle for the reflection coefficient other than 0° or 180° (See Figure 3-28).

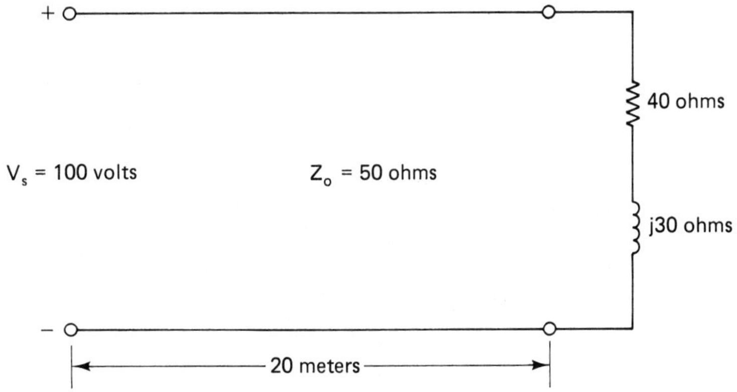

Figure 3-28. Line with complex load impedance.

$$\Gamma_r = \frac{Z_r - Z_o}{Z_r + Z_o} = \frac{40 + j30 - 50}{40 + j30 + 50} = \frac{-10 + j30}{90 + j30}$$

$$= 0.333 \angle 90° = \frac{V^-}{V^+}.$$

$V^- = 0.333\, V^+ \angle 90°$ at the receiving end. Notice that the 90 degree angle associated with V^- is purely accidental due to the numbers selected for the example. It could just as well have been 68.2 or 129.11 degrees. Our phasor diagram at the receiving end is now different from those we had before and is given in Figure 3-29.

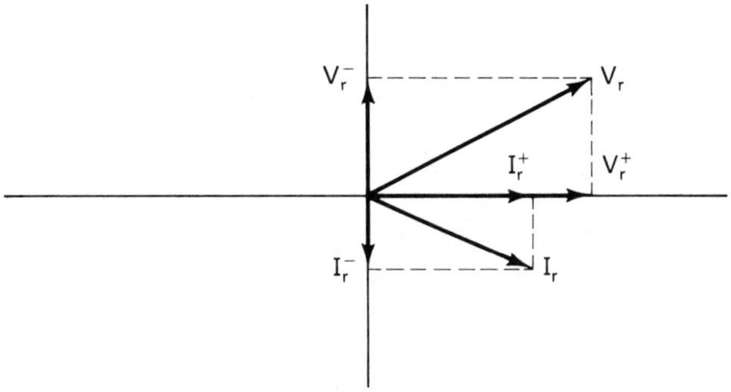

Figure 3-29. Receiving end phasor diagram.

$$V_r = V^+ + j0.333\ V^+ = V^+\ (1 + j.333) = 1.054\ V^+ \underline{/18.43^\circ} \text{ volts.}$$

$$I_r = \frac{V^+}{50} - \frac{j.333V^+}{50} = \frac{V^+(1 - j.333)}{50} = 0.021V^+ \underline{/18.43^\circ} \text{ amperes.}$$

$\beta\ell = (1.309)(20) = 26.18$ radians (1500°). So each phasor goes through $1500/360 = 4.166$ revolutions to arrive at the sending end. Since four revolutions take the phasors back to their starting point it is only the 0.166 revolution that concerns us. $0.166\ (360) = 60$ degrees so each phasor rotates around four times plus 60 degrees. The total shift for each phasor is then 60 degrees with respect to its position at the receiving end. At the sending end our phasor diagram is shown in Figure 3-30.

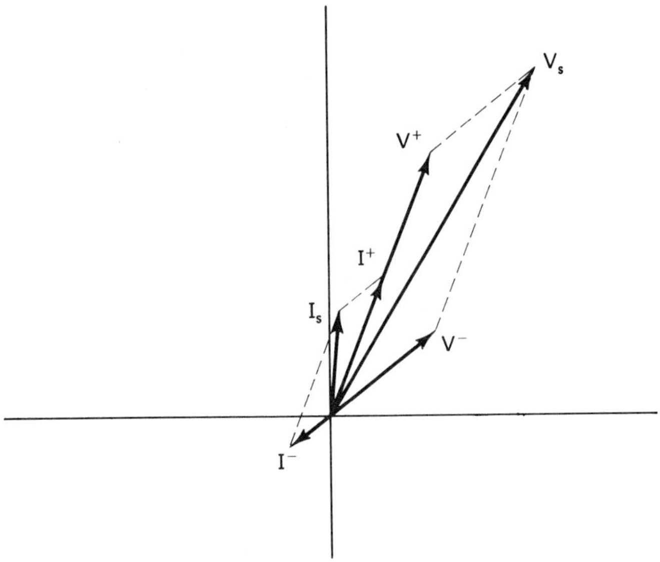

Figure 3-30. Sending end phasor diagram.

V^+ has rotated a net 60 degrees in the counterclockwise direction and V^- has a net angle of 30 degrees (90° − 60°).

$$V_s = V^+ \underline{/60^\circ} + 0.333V^+ \underline{/30^\circ} = 1.29V^+ \underline{/52.63^\circ} \text{ volts}$$
$$= 100 \underline{/52.63} \text{ volts.}$$

$$\therefore |V^+| = \left|\frac{100}{1.29}\right| = 76.96 \text{ volts.}$$

$$|V^-| = 0.333(76.96) = 25.65 \text{ volts. } I_s = \frac{V^+}{50}\angle 60^\circ - \frac{0.333V^+}{50}\angle 30^\circ =$$

$$\frac{76.96}{50}\angle 60^\circ - \frac{25.65}{50}\angle 30^\circ = 1.124\angle 73.18^\circ \text{ amperes.}$$

$$Z_s = \frac{V_s}{I_s} = \frac{100\angle 52.63^\circ}{1.124\angle 73.18^\circ} = 88.94\angle 20.55^\circ = 83.27 - j31.23 \text{ ohms.}$$

$$V_r = 1.054\,|V^+|\angle 18.43^\circ = 1.054(76.96)\angle 18.43^\circ = 81.12\angle 18.43^\circ \text{ volts.}$$

$$I_r = 0.021\,|V^+|\angle -18.43^\circ = 0.021(76.96)\angle -18.43^\circ = 1.62\angle -18.43^\circ \text{ amperes.}$$

$$P_s = V_sI_s\cos\theta_s = 100(1.124)\cos 20.55^\circ = 105.25 \text{ watts.}$$

$$P_r = V_rI_r\cos\theta_r = 81.12(1.62)\cos 36.87^\circ = 105.25 \text{ watts.}$$

$$Q_s = V_sI_s\sin\theta_s = 100(1.124)\sin 20.55^\circ = 39.46 \text{ vars}$$

$$Q_r = V_rI_r\sin\theta_r = (81.12)(1.62)\sin 36.87^\circ = 78.85 \text{ vars.}$$

The phasor diagram will allow us to find the points of voltage and current maximums and minimums which are not located at the load or 1/4 wavelength from the load as in previous examples. When we look at the phasor diagram at the receiving end and realize that V^+ is rotating counterclockwise and V^- is rotating clockwise, it is apparent that after 45 degrees in this particular case we will have a voltage maximum (and a current minimum). If they rotate another 90 degrees we will have a voltage minimum and a current maximum. We can also find the values of the maximums and minimums, the VSWR and the impedance at these points without trying to remember any equations. The phasor diagrams are shown in Figure 3-31.

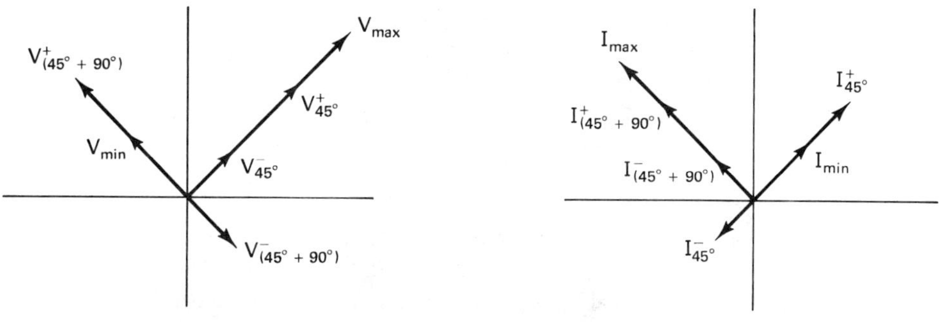

Figure 3-31. Phasor diagrams for voltage and current maximums and minimums.

Under these conditions, the distance to a voltage maximum and the corresponding current minimum turns out to be one-half of the angle of the voltage reflection coefficient. (In our case 90 degrees divided by 2 equals 45 degrees.) This will always be true if the angle of the reflection coefficient is positive. Moving from the load towards the sending end will result in a voltage maximum appearing first and then 90 degrees later a voltage minimum. If one sketches the situation with the angle of the voltage reflection coefficient as negative it is apparent that the opposite happens; a voltage minimum first and then 90 degrees later a voltage maximum. In this particular case we can continue to find out more information from the phasor diagram. The voltage maximum is at 45 degrees.

$$45^\circ \times \frac{\lambda}{360^\circ} = 0.125\lambda.$$

$$\text{(1-52)}\ \lambda = \frac{2\pi}{\beta} = \frac{2\pi}{1.309} = 4.8 \text{ meters.}$$

So the first maximum voltage is (0.125)(4.8) = 0.6 meters from the load. This is equivalent to 1/8 wavelength or $\pi/4$ radians.

$$V_{max} = |V^+| + |V^-| = 76.96 + 25.65 = 102.61 \text{ volts.}$$

$$V_{min} = |V^+| - |V^-| = 76.96 - 25.65 = 51.31 \text{ volts.}$$

$$\text{VSWR} = \frac{V_{max}}{V_{min}} = \frac{102.61}{51.31} = 2.$$

$$|I^+| = \left|\frac{V^+}{Z_o}\right| = \frac{76.96}{50} = 1.54 \text{ amperes.}$$

$$|I^-| = \left|\frac{V^-}{Z_o}\right| = \frac{25.65}{50} = 0.51 \text{ amperes.}$$

$$I_{max} = |I^+| + |I^-| = 1.54 + 0.51 = 2.05 \text{ amperes.}$$

$$I_{min} = |I^+| - |I^-| = 1.54 - 0.51 = 1.03 \text{ amperes.}$$

At points of voltage maximums and voltage minimums the currents are in phase with the voltages, and the impedance looking towards the load is purely resistive. This will always be true if the angle of the characteristic impedance is zero degrees. A voltage maximum and a current minimum gives a maximum value of the impedance.

$$Z_{max} = R_{max} = \frac{V_{max}}{I_{min}}. \tag{3-20}$$

In our case:

$$R_{max} = \frac{102.61}{1.03} = 100 \text{ ohms.}$$

$$Z_{min} = R_{min} = \frac{V_{min}}{I_{max}}. \tag{3-21}$$

In our case:

$$R_{min} = \frac{51.31}{2.05} = 25 \text{ ohms.}$$

At a voltage maximum we can write the following equation:

$$Z_{max} = R_{max} = \frac{V_{max}}{I_{min}} = \frac{|V^+| + |V^-|}{\left|\frac{V^+}{Z_o}\right| - \left|\frac{V^-}{Z_o}\right|} = Z_o\left[\frac{|V^+| + |V^-|}{|V^+| - |V^-|}\right] =$$

$$Z_o\frac{V_{max}}{V_{min}} = Z_o\,(\text{VSWR}) = R_{max}. \tag{3-22}$$

In our case:

$$R_{max} = (50)(2) = 100 \text{ ohms.}$$

At a voltage minimum we can write the following equation:

$$Z_{min} = R_{min} = \frac{V_{min}}{I_{max}} = \frac{|V^+| - |V^-|}{\left|\frac{V^+}{Z_o}\right| + \left|\frac{V^-}{Z_o}\right|} = Z_o\left[\frac{|V^+| - |V^-|}{|V^+| + |V^-|}\right] =$$

$$Z_o\frac{V_{min}}{V_{max}} = Z_o\frac{1}{\frac{V_{max}}{V_{min}}} = \frac{Z_o}{\text{VSWR}} = R_{min}. \tag{3-23}$$

In our case:

$$R_{min} = \frac{50}{2} = 25 \text{ ohms.}$$

Once again we can sketch pretty rapidly the picture of what is happening on this line. From the load to the first voltage maximum the impedance looking towards the load appears as a resistance in series with an inductor. In our example the first voltage maximum is at 45 degrees or 1/8 wavelength. Our results are given in Figure 3-32. Unlike the short-circuit, open-circuit or the pure resistance cases we no longer have a voltage maximum or minimum at the load. Notice some possibilities that we have for this case. We could take

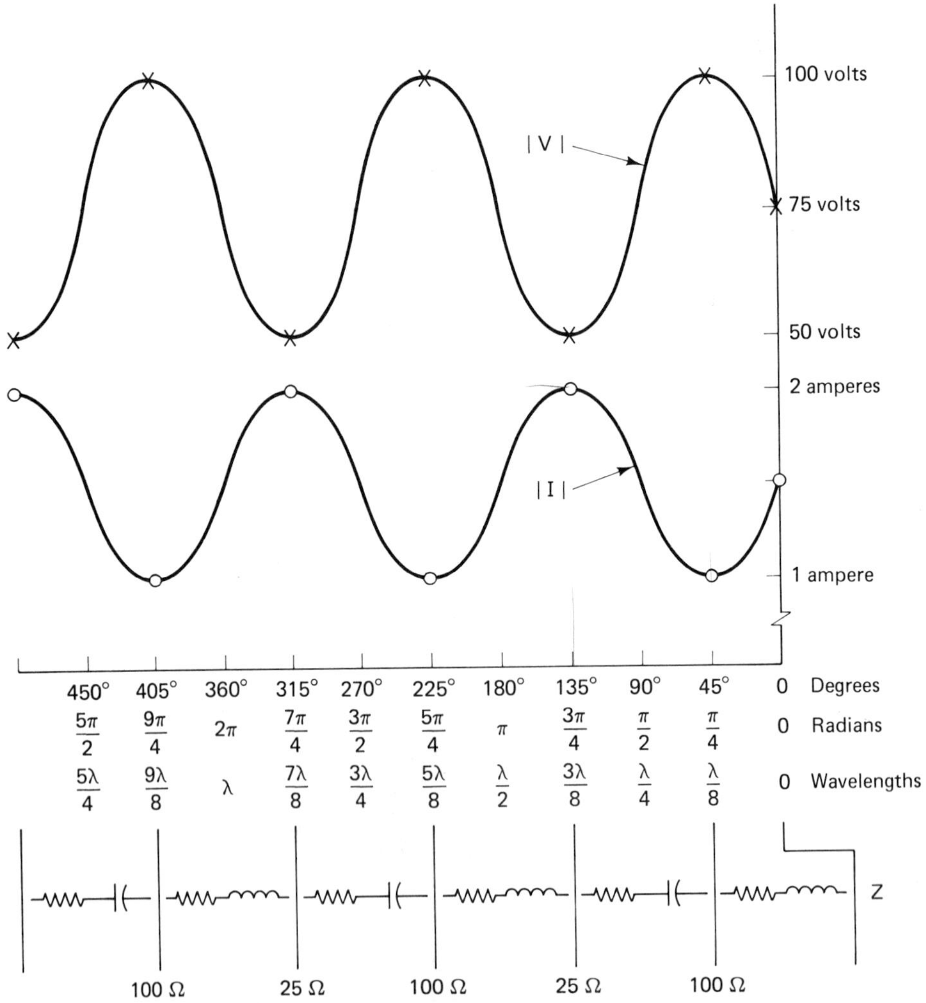

Figure 3-32. Patterns on line with complex impedance load.

a load of 40 + j30 ohms and connect it to a line with a characteristic impedance of 50 ohms; chop it off at 3/8 wavelength when the input impedance is 25 ohms and connect it to a section of line with a characteristic impedance of 25 ohms and have the 25-ohm line matched. The 50-ohm line acts as a transformer, transforming a load impedance of 40 + j30 ohms into a load impedance of 25 ohms. If the length of the line was 1/8 wavelength we could transform our load impedance into a load of 100 ohms real.

As we look at Figure 3-32 we see that once again everything repeats itself every half wavelength on the line. The input impedance of the line would be 40 + j30 ohms if the line is exactly one half wavelength long or any multiple of half wavelengths (1, 1.5, 2, 2.5, 3 wavelengths, etc.).

3.14 Measurement of the Magnitude of the Reflection Coefficient

One of the problems associated with any physical phenomenon is to make measurements of the quantities involved in the process. As we go through this book we will try to point out, from time to time, those concepts which are used in the field and in the laboratory. How do we measure the voltage reflection coefficient? Mathematically it is the ratio of two phasors at the receiving end and all we can do is measure the total receiving-end voltage. From our previous discussions it might appear to the reader at this point that the angle of the voltage reflection coefficient might be determined by the distance to a voltage maximum or minimum from the load. For instance we have just had a case where the angle of the reflection coefficient was 90 degrees and the distance to the voltage maximum was 45 degrees from the load or half of the reflection coefficient angle. And it turns out that this is a way of doing it, as we shall see later. The VSWR is normally a very easy thing to measure and from it we can determine the magnitude of the voltage reflection coefficient. We will start with equation (3-15).

$$\text{VSWR} = \frac{1 + |\Gamma_r|}{1 - |\Gamma_r|}.$$

Solving for $|\Gamma_r|$ we have:

$$|\Gamma_r| = \frac{\text{VSWR} - 1}{\text{VSWR} + 1}. \tag{3-24}$$

Equation (3-24) can be used to find the magnitude of the voltage reflection coefficient by using the VSWR measured on the line.

3.15 Distance to a Voltage Minimum for a Lossless Line

By using phasor diagrams we can obtain a great deal of new useful information about a transmission line; even more than we have up to this point. Let us take Figure 3-33 as our starting point.

We will let d_{min} be the distance from the load to the first voltage minimum.

$$\Gamma_r = \frac{Z_r - Z_o}{Z_r + Z_o} = \frac{V^-}{V^+} = |\Gamma_r| \angle\phi. \text{ (}\phi\text{ is in radians.)}$$

We will start out by assuming that ϕ is positive (See Figure 3-34). When ϕ is positive we will have a voltage maximum first as V^+ rotates counterclockwise and V^- rotates clockwise. It will occur when $\beta d = \phi/2$ radians as shown in Figure 3-35.

As indicated in Figure 3-36 we will obtain a voltage minimum when each phasor rotates another $\pi/2$ radians.

V_{min} occurs at $\alpha = \frac{\pi}{2} + \frac{\phi}{2}$ radians $= \beta d_{min}$. Since $\beta = \frac{2\pi}{\lambda}$,

$$\frac{2\pi}{\lambda} d_{min} = \frac{\pi}{2} + \frac{\phi}{2} = \frac{\pi}{2}\left[1 + \frac{\phi}{\pi}\right]. \qquad \frac{d_{min}}{\lambda} = \frac{1}{2\pi}\frac{\pi}{2}\left[1 + \frac{\phi}{\pi}\right] = 1/4\left[1 + \frac{\phi}{\pi}\right].$$

We will also have a voltage minimum every half wavelength down the line or $n/2$ wavelengths where $n = 1, 2, 3, \ldots$ (a positive integer).

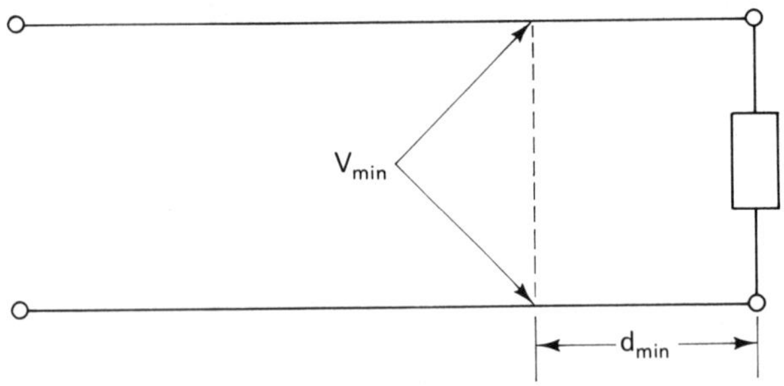

Figure 3-33.

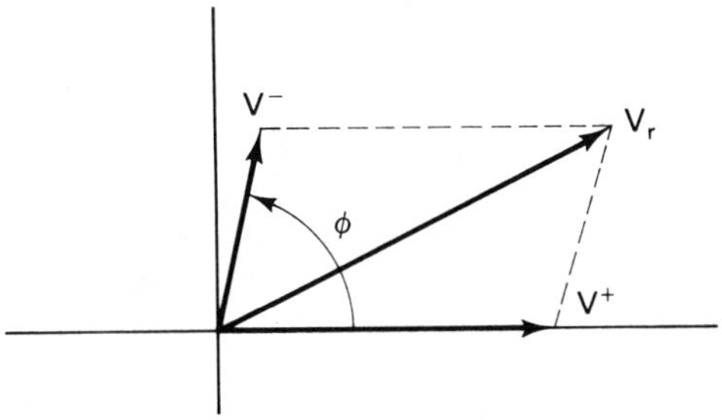

Figure 3-34. Conditions at receiving end with ϕ positive.

$$\frac{d_{min}}{\lambda} = 1/4\left[1 + \frac{\phi}{\pi}\right] + \frac{n}{2}.$$

If we were at a voltage minimum somewhere down the line we could write,

$$\frac{d_{min}}{\lambda} = 1/4\left[1 + \frac{\phi}{\pi}\right] - \frac{n}{2}.$$

In general we can write,

$$\frac{d_{min}}{\lambda} = 1/4\left[1 + \frac{\phi}{\pi}\right] \pm \frac{n}{2}. \qquad n = 0, 1, 2, 3, \text{etc.} \qquad (3\text{-}25)$$

If we assume that ϕ is negative we will have a voltage minimum first as shown in Figure 3-37.

We will have a voltage minimum when each phasor rotates α degrees. So

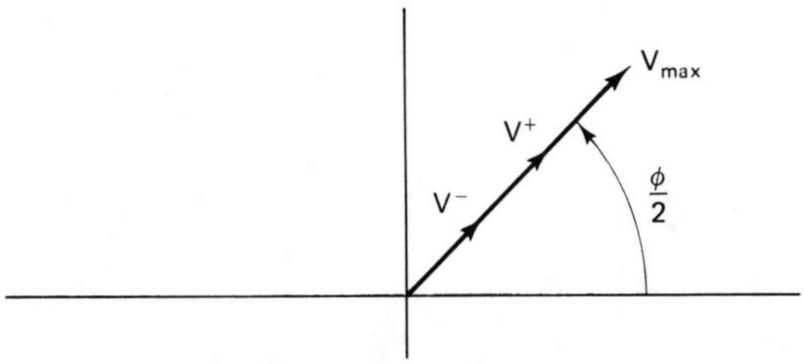

Figure 3-35. Phasor diagram for V_{max} when V^+ and V^- are in phase.

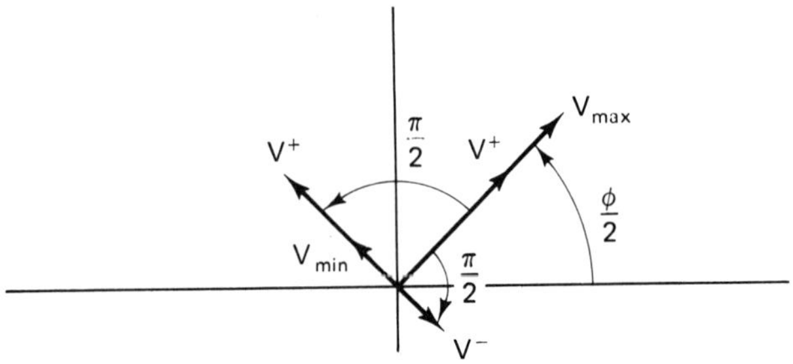

Figure 3-36. Phasor diagram for V_{min} when V^+ and V^- have traveled 90 degrees from V_{max} position.

$$2\alpha + \phi = \pi \qquad \alpha = \frac{\pi}{2} - \frac{\phi}{2} = \beta d_{min} = \frac{2\pi}{\lambda} d_{min}.$$

$$\frac{2\pi}{\lambda} d_{min} = \frac{\pi}{2} - \frac{\phi}{2} = \frac{\pi}{2}\left[1 - \frac{\phi}{\pi}\right] \text{ or } \frac{d_{min}}{\lambda} = \frac{1}{2\pi}\frac{\pi}{2}\left[1 - \frac{\phi}{\pi}\right] = \frac{1}{4}\left[1 - \frac{\phi}{4}\right].$$

This will also happen every half wavelength so we can write as a general equation:

$$\frac{d_{min}}{\lambda} = \frac{1}{4}\left[1 - \frac{\phi}{\pi}\right] \pm \frac{n}{2}. \qquad n = 0, 1, 2, 3, \text{ etc.} \qquad (3\text{-}26)$$

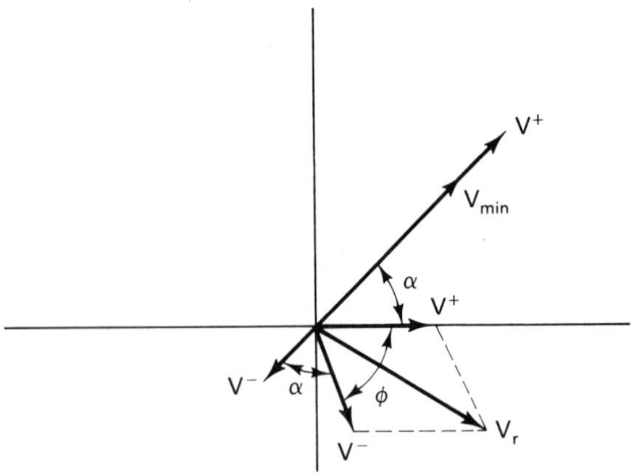

Figure 3-37. Phasor relationship with ϕ negative.

If we carry the + or − sign with ϕ we can write a general equation for either case.

$$\frac{d_{min}}{\lambda} = \frac{1}{4}\left[1 + \frac{\phi}{\pi}\right] \pm \frac{n}{2}. \tag{3-27}$$

In equation (3-27) we will let ϕ be + or − depending on its value from the reflection coefficient. Electrical engineers usually prefer to use degrees in angles rather than radians.

$$\frac{\phi^\circ}{180^\circ} = \frac{\phi\ (\text{rad})}{\pi\ (\text{rad})}. \qquad \phi_{rad} = \frac{\phi^\circ \pi}{180^\circ}.$$

$$\frac{d_{min}}{\lambda} = \frac{1}{4}\left[1 + \frac{\phi^\circ \pi}{180\pi}\right] \pm \frac{n}{2} = \frac{1}{4}\left[1 + \frac{\phi^\circ}{180}\right] \pm \frac{n}{2}. \tag{3-28}$$

We can solve equation (3-27) for ϕ.

$$4\frac{d_{min}}{\lambda} = 1 + \frac{\phi}{\pi} \pm 2n. \qquad \frac{\phi}{\pi} = 4\frac{d_{min}}{\lambda} - 1 \pm 2n.$$

$$\phi(\text{radians}) = 4\pi\frac{d_{min}}{\lambda} - \pi \pm 2n\pi. \tag{3-29}$$

This will give us the angle for Γ_r and we can find the magnitude from equation (3-24).

Let us look at several examples.

Example 3-A

Determine the fraction of a wavelength from the load to a voltage minimum if Z_r = 200 ohms and Z_o = 100 ohms.

Solution

$$\Gamma_r = \frac{Z_r - Z_o}{Z_r + Z_o} = \frac{200 - 100}{200 + 100} = \frac{100}{300} = 1/3\ \underline{/0^\circ}.$$

$\phi = 0^\circ$ or 0 radians

$$\frac{d_{min}}{\lambda} = \frac{1}{4}\left[1 + \frac{\phi}{\pi}\right] \pm \frac{n}{2} = \frac{1}{4}[1 + 0] \pm \frac{n}{2} = \frac{1}{4} \pm \frac{n}{2}.$$

With n = 0,

$$d_{min} = \frac{\lambda}{4} = 0.25\ \lambda.$$

So the first voltage minimum is a quarter wavelength from the load, which is equivalent to ninety electrical degrees. The phasor diagram is given as:

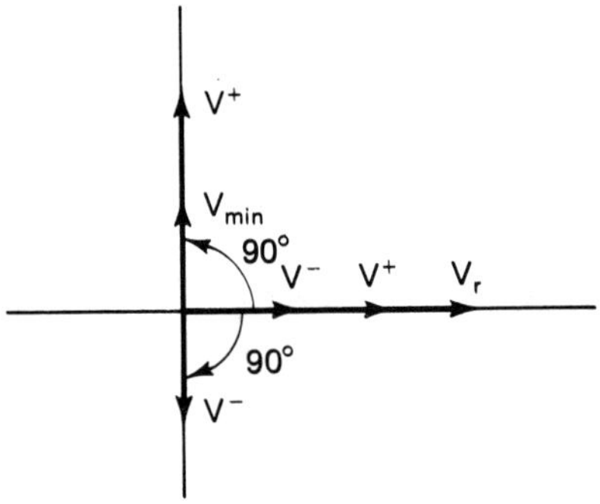

Example 3-B

Repeat example 3-A if Z_r = 50 ohms and Z_o = 100 ohms.

Solution

$$\Gamma_r = \frac{Z_r - Z_o}{Z_r + Z_o} = \frac{50 - 100}{50 + 100} = -\frac{50}{150} = -\frac{1}{3} = \frac{1}{3}\angle 180^\circ.$$

ϕ = 180° or π radians.

$$\frac{d_{min}}{\lambda} = \frac{1}{4}\left[1 + \frac{\pi}{\pi}\right] \pm \frac{n}{2} = \frac{1}{4}[1 + 1] \pm \frac{n}{2}$$
$$= \frac{2}{4} \pm \frac{n}{2} = \frac{1}{2} \pm \frac{n}{2}.$$

With n = 0,

$$\frac{d_{min}}{\lambda} = \frac{1}{2}. \qquad \text{With } n = -1. \qquad \frac{d_{min}}{\lambda} = 0.$$

Here the first voltage minimum occurs at the load. We can see this from the phasor diagram.

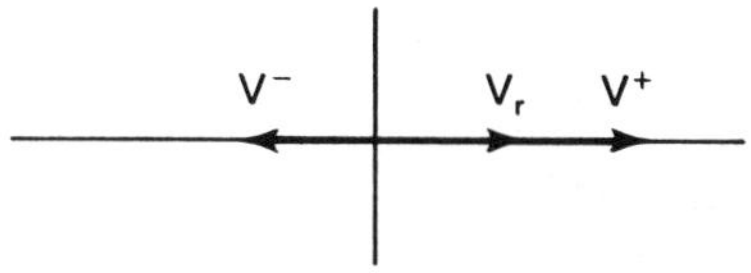

Example 3-C

Repeat example 3-A if $Z_r = 60 + j80$ and $Z_o = 100$ ohms.

Solution

$$\Gamma_r = \frac{Z_r - Z_o}{Z_r + Z_o} = \frac{60 + j80 - 100}{60 + j80 + 100} = \frac{-40 + j80}{160 + j80}$$

$$= 0.5 \angle 90°. \qquad \phi = 90° \text{ or } \frac{\pi}{2} \text{ radians.}$$

$$\frac{d_{min}}{\lambda} = \frac{1}{4}\left[1 + \frac{\phi}{180°}\right] \pm \frac{n}{2} = \frac{1}{4}\left[1 + \frac{90}{180}\right] \pm \frac{n}{2}$$

$$= \frac{3}{8} \pm \frac{n}{2}. \qquad \text{With } n = 0, \frac{d_{min}}{\lambda} = \frac{3}{8}.$$

Now the first voltage minimum is three-eighths wavelength from the load. The phasor diagram becomes:

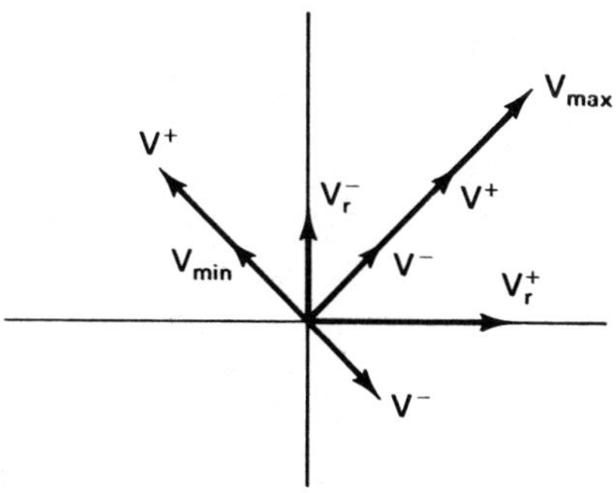

While equations are nice, nobody cares to remember more of them than necessary. The student should notice here that he or she could have obtained the same answers by simply sketching the phasor diagram, remembering only the equation for the voltage reflection coefficient.

Example 3-D

A lossless transmission line has a characteristic impedance of 100 ohms. It is terminated with a load impedance of 80 − *j*60 ohms. It is operated at a frequency so that the wavelength is 2 meters and the line is 5 meters long. Determine the following: (a) input impedance; (b) the impedance looking towards the load 0.25 meters from the load; (c) what the reflection coefficient is 0.25 meters from the load.

Solution

$$\Gamma_r = \frac{Z_r - Z_o}{Z_r + Z_o} = \frac{80 - j60 - 100}{80 - j60 + 100} = .333\,\underline{/-90} = 0 - j0.333.$$

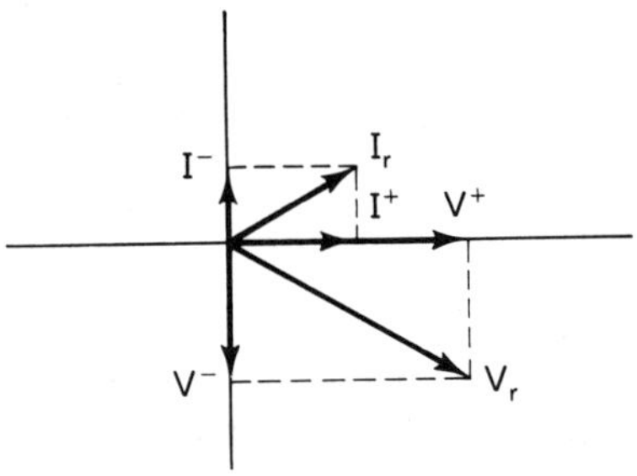

$$\frac{\ell}{\lambda} = 5/2 = 2.5.$$

$$\ell = 2.5\lambda.$$

$$.5(360^\circ) = 180^\circ.$$

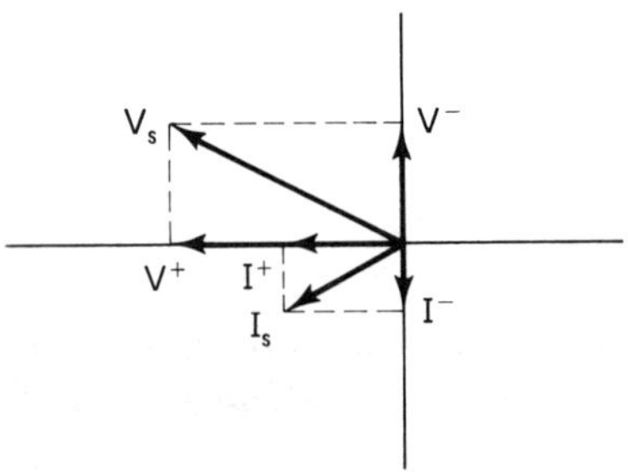

$$V_s = V^+ (-1 + j.333) = V^+ 1.054\underline{/161.565^\circ} \text{ volts.}$$

$$I_s = \frac{V^+}{Z_o}(-1 - j.333). \qquad = \frac{V^+}{Z_o} 1.054\underline{/-161.565^\circ} \text{ volts.}$$

$$Z_s = \frac{V^+ 1.054\underline{/161.566^\circ}}{\dfrac{V^+}{Z_o} 1.054\underline{/-161.565^\circ}} = Z_o \underline{/323.13^\circ} = Z_o \underline{/-36.870^\circ}$$

$$= 100 \underline{/-36.870^\circ} = 80 - j60 \text{ ohms.} \qquad \text{(a)}$$

Notice that we obtain the right answer since everything should repeat itself every one-half wavelength.

$$\frac{.25}{2} = .125. \qquad \ell = .125\lambda. \qquad .125(360) = 45^\circ.$$

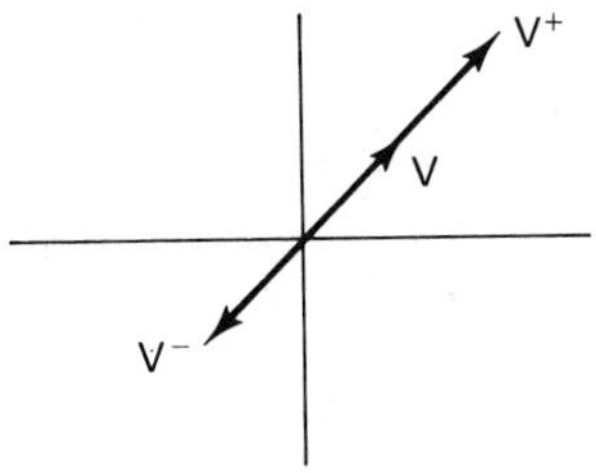

$$Z = \frac{V}{I} = \frac{V^+(1 - .333)\underline{/45^\circ}}{\dfrac{V^+}{Z_o}(1 - .333)\underline{/45^\circ}}. = \frac{(100)(0.666)}{1.333} = 50 \text{ ohms.} \qquad \text{(b)}$$

$$\Gamma = \frac{Z - Z_o}{Z + Z_o} = \frac{50 - 100}{50 + 100} = \frac{-50}{150} = -1/3 = .333 \underline{/180^\circ}. \qquad \text{(c)}$$

The student should note that for part (c) we are replacing the load and 0.25 meters of line by the impedance at that point. This does not change anything to the left of this point. We can always cut off a piece of a transmission line and replace it by the impedance at that point without affecting the rest of the line. We can now calculate the voltage reflection coefficient using that value of impedance. Notice the magnitude remains the same but the angle is changed.

Example 3-E

Measurements are made on a lossless 50-ohm slotted line in the laboratory. The first voltage maximum is located 22.5 cm from the load and the second voltage maximum is located 67.5 cm from the load. Its value is 100 volts. A voltage minimum is located 45 cm from the load and is 10 volts. Assuming the phase velocity is the speed of light, determine: (a) the VSWR, (b) the frequency, (c) the load impedance in complex form, (d) the voltage reflection coefficient, (e) the input impedance to the slotted line if it is 142 cm long.

Solution

The distance between voltage maximums is one-half wavelength.

$$67.5 - 22.5 = 45 \text{ cm} = \frac{\lambda}{2}$$

$$\lambda = 90 \text{ cm}$$

$$\lambda = .9 \text{ meter.}$$

$$\text{VSWR} = \frac{V_{max}}{V_{min}} = \frac{100}{10} = 10. \qquad \text{(a)}$$

$$f = \frac{c}{\lambda} = \frac{3 \times 10^8}{.9} = 333.3 \times 10^6 \text{ hertz.} \qquad \text{(b)}$$

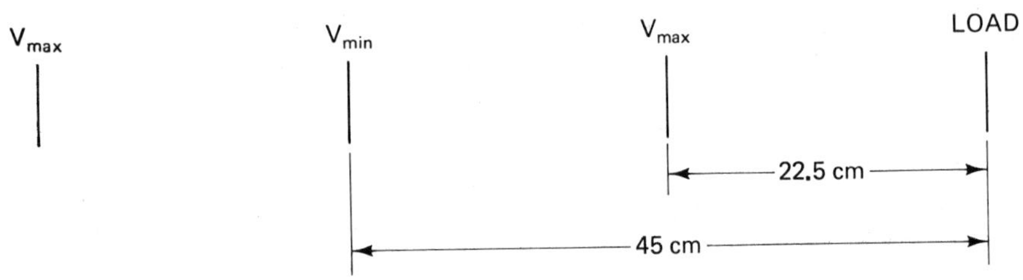

Note: V_{min} at load.

$$R_{min} = Z_r = \frac{Z_o}{\text{VSWR}} = \frac{50}{10} = 5 \text{ ohms.} \qquad \text{(c)}$$

$$\frac{V_{max}}{V_{min}} = \frac{|V^+| + |V^-|}{|V^+| - |V^-|} = \frac{100}{10} = 10$$

$$|V^+| + |V^-| = 10\,|V^+| - 10\,|V^-|$$

$$11\,|V^-| = 9\,|V^+|. \qquad \frac{|V^-|}{V^+} = \frac{9}{11} = .818.$$

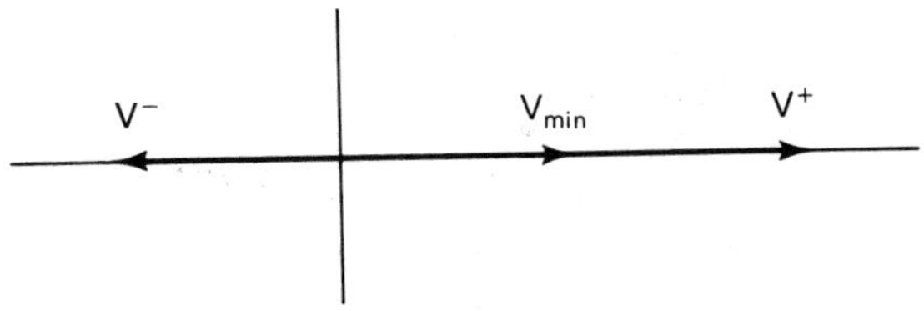

$\Gamma_r = .818\ \angle 180°$ from phasor diagram; or

$$\Gamma_r = \frac{Z_r - Z_o}{Z_r + Z_o} = \frac{5 - 50}{5 + 50} = \frac{-45}{55} = .818\ \angle 180°. \qquad \text{(d)}$$

$$\frac{\ell}{\lambda} = \frac{142}{90} = 1.578. \qquad \ell = 1.578\lambda. \qquad .578(360) = 208.08°.$$

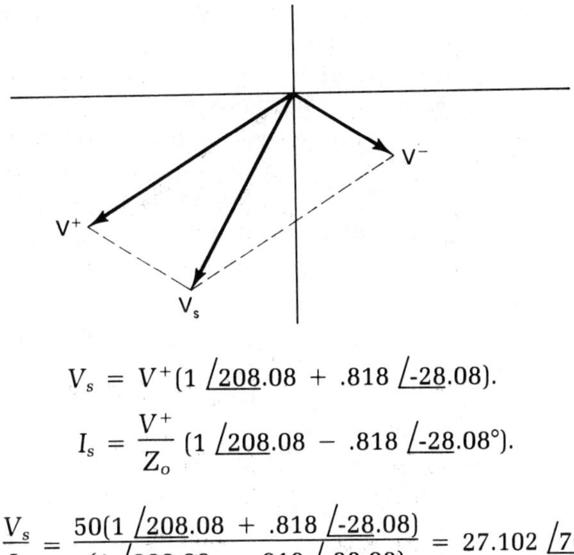

$$V_s = V^+(1\ \angle 208.08 + .818\ \angle -28.08).$$

$$I_s = \frac{V^+}{Z_o}(1\ \angle 208.08 - .818\ \angle -28.08°).$$

$$Z_s = \frac{V_s}{I_s} = \frac{50(1\ \angle 208.08 + .818\ \angle -28.08)}{(1\ \angle 208.08 - .818\ \angle -28.08)} = 27.102\ \angle 76.31°$$

$$= 6.412 + j26.33 \text{ ohms.} \qquad \text{(e)}$$

3.16 Chapter Summary

1. The voltage reflection coefficient is given as:

$$\Gamma_r = \frac{Z_r - Z_o}{Z_r + Z_o} = \frac{V^-}{V^+}.$$

This gives the relationship of the phasor value of the reflected component to the phasor value of the forward component at the load.

2. The forward component phasor, V^+, rotates in a counterclockwise direction as one moves from the load to the sending end whereas the reflected component, V^-, rotates in the clockwise direction as one moves to the sending end. At any point on the line the phasor value of the voltage may be determined by adding the two phasor components. The angle of rotation in radians is given as βd. 2π radians or 360 electrical degrees is equivalent to one wavelength.
3. At the same time as the voltage phasors rotate in opposite directions, the forward component phasor of the current, I^+, and the reflected component phasor of the current, I^-, rotate in the counter and clockwise directions respectively.
4. If the characteristic impedance, Z_o, is real, the forward component of the current is in phase with the forward component of the current and the reverse component of the current is 180 degrees out of phase with the reverse component of the voltage.
5. At any point on the line if one divides the total phasor voltage by the total phasor current the impedance at that point is obtained. This impedance is the impedance at that point on the line looking towards the load, and would be the input impedance to the line if the line were that long.

$$Z \text{ at point x or } d = \frac{V \text{ at point x or } d}{I \text{ at point x or } d}.$$

This impedance may be real, imaginary, complex, zero or infinite and varies continuously as one moves down the line from the load except in the one special case when the load is matched to the line ($Z_r = Z_o$).

6. If the load impedance is greater than zero but less than infinity and is not equal to the characteristic impedance of the line, there will be a standing wave on the line with a finite maximum value and a minimum value that does not go to zero. The ratio of the maximum value to the minimum value of the voltage is defined as the voltage standing-wave ratio.

$$\text{VSWR} = \frac{V_{max}}{V_{min}} = \frac{1 + |\Gamma_r|}{1 - |\Gamma_r|}.$$

If the load is matched to the line the VSWR equals one, which is the ideal case.

7. If the load is greater than zero, less than infinity and not equal to the characteristic impedance, the impedance looking toward the load will become real at some point on the line and a quarter wavelength down the line the impedance will again be real, providing that the characteristic impedance is real. In one case we will have a voltage maximum, a current minimum and a maximum value of resistance. In the other case we will have a voltage minimum, a current maximum and a minimum resistance. In between these two cases the impedance will be complex and will contain a resistive component and a reactive component which may be inductive or capacitive depending on the distance to the load.
8. By sketching the phasor diagram at any point d units from the load end, the student may determine what is happening at that point. The voltage, current and impedance are easily determined from the diagram.
9. As the two phasors rotate, a maximum will be obtained whenever they are in phase and a minimum will be obtained whenever they are 180 degrees out of phase. This will happen whether the characteristic impedance is real or complex.
10. Measuring the distance from the load to a voltage minimum or maximum allows one to determine the angle of the reflection coefficient. Measuring the VSWR then allows one to determine an unknown impedance from the equation.

$$Z_R = \frac{Z_o\,(1 + \Gamma_r)}{(1 - \Gamma_r)}.$$

11. The distance to a voltage minimum may be determined by the equation

$$\frac{d_{min}}{\lambda} = \frac{1}{4}\left[1 + \frac{\phi}{\pi}\right] \pm \frac{n}{2}$$

where ϕ is the angle of the voltage reflection coefficient in radians or it may be determined in a much easier manner by sketching the phasor diagram relationship and working it out.

3.17. Review and Study Questions.

1. What is the magnitude of the following quantities?
(a) e^{j377t} (b) $e^{j60°}$ (c) $100\ e^{-2}e^{j1000t}$
2. What is a phasor? What is a rotating phasor?

3. How do we make a phasor rotate?
4. Define the voltage reflection coefficient at the receiving end of the transmission line.
5. What is the voltage reflection coefficient the ratio of in terms of the traveling waves on a line?
6. Define the voltage standing wave ratio.
7. If the angle of the voltage reflection coefficient is positive do we have a voltage minimum or voltage maximum closest to the load?
8. As we move down the line how often do things repeat themselves on the line?
9. Everything in this chapter was done on the assumption that the line was lossless and therefore the characteristic impedance was real. How do you guess things would change if losses were taken into account?
10. If both Z_r and Z_o are real what is an easy expression for determining the VSWR on the line?
11. What is the angle of the voltage reflection coefficient if Z_o is real and: (a) $Z_r = Z_o$? (b) $Z_r = 2Z_o$? (c) $Z_r = 0.5\ Z_o$? (d) $Z_r = 0$? (e) Z_r is infinite?
12. Sketch the magnitudes of the voltage and current on a line if Z_o is real and: (a) $Z_r = Z_o$, (b) $Z_r = 2Z_o$, (c) $Z_r = 0.5Z_o$, (d) $Z_r = (0.5 + j0.5)\ Z_o$, (e) $Z_r = (0.5 - j0.5)\ Z_o$.
13. Draw the phasor diagrams for every 1/4 wavelength from the load for one wavelength if the load is $40 + j40$ ohms and Z_o is 75 ohms, real. Indicate where the impedance looking towards the load is real and where it is inductive or capacitive.
14. If $Z_r = 40 + j60$ and $Z_o = 75 + j0$ how far from the load is the first voltage minimum?
15. If the distance between two adjacent voltage minimums is 33.65 cm and the waves are traveling with the speed of light what is the frequency?
16. What are the limits on the magnitude of the voltage standing wave ratio?
17. What are the limits on the magnitude of the voltage reflection coefficient?
18. If the line is exactly two wavelengths long how do the input voltage, current and impedance compare with the load voltage, current and impedance if the line is lossless?

Problems

3-1. Using phasors, determine the output voltage of a 2.35 wavelength, open-circuited line, having a Z_o of 600 ohms, for an input voltage of 100 volts, r.m.s. Determine the current at the input to the line.

3-2. Using phasors, determine the output voltage of a 3.42 wavelength, open-circuited line, having a Z_o of 300 ohms, for an input voltage of 10 volts, r.m.s. Determine the input current to the line.

3-3. Using phasors, determine the output current of a 2.35 wavelength, short-circuited line, having a Z_o of 600 ohms, for an input voltage of 100 volts, r.m.s. Determine the current at the input to the line.

3-4. Using phasors, determine the output current of a 3.42 wavelength, short-circuited line, having a Z_o of 300 ohms, for an input voltage of 10 volts, r.m.s. Determine the current at the input to the line.

3-5. A lossless line has a characteristic impedance of 50 ohms. It is 3 meters long. It has an open-circuit load, and the wavelength is 1.5 meters. Determine, using phasors: (a) The input impedance of the line, (b) The impedance looking toward the load at 0.5 meters from the load, (c) The impedance looking toward the load at 1.5 meters from the load, (d) If the line is now short-circuited find the answers for (a), (b), and (c).

3-6. While in the laboratory working on a transmission line where the voltage and current waves travel with the speed of light, a student finds that the distance between a voltage minimum and a voltage maximum is 3.5 centimeters and that the maximum value of the voltage is 3.36 volts and the minimum value is 1.87. Determine (a) the VSWR, (b) the frequency.

3-7. A lossless transmission line is 7.2 wavelengths long, has a characteristic impedance of 100 ohms and a load impedance of $60 + j80$ ohms. Using phasors determine the following if the load voltage is 100 volts:
(a) Load current in phasor form.
(b) Sending-end voltage in phasor form.
(c) Sending-end impedance in phasor form.
NOTE: Use V^+ as your reference phasor.

3-8. In problem 3-7 there are certain points on the line where the impedance looking toward the load is purely resistive. Determine:
(a) The first point on the line as one moves away from the load. (Give the answer in terms of the wavelength. What is the magnitude of the impedance?)
(b) Find the next point on the line.

3-9. Repeat problem 3-7 if the input voltage is 100 volts, r.m.s.

3-10. A lossless line has a 200 ohm characteristic impedance and is terminated in a load of 200 + j300 ohms. Using phasors:
(a) Plot for 2 wavelengths the voltage and current standing waves.
(b) For 1.0 volt., r.m.s. input voltage, find the load voltage and current. Assume the line is 2.81 wavelengths long.
(c) Find the voltage standing wave ratio.
(d) Plot the impedance on the line, looking toward the load, for 2 wavelengths, showing separately the magnitude and the angle.

3-11. A lossless transmission line has a characteristic impedance of 50 ohms. If β is 2.35 radians per meter and the line 3.2 meters long and using a phasor solution:
(a) Determine the input impedance if the load impedance is 150 + j50 ohms.
(b) What is the load voltage if the input voltage is 50 volts?
(c) Determine the sending-end current under the conditions of (b).
(d) Determine the receiving-end current under the conditions of (b).
(e) Determine the load power under the conditions of (b).
(f) Determine the sending-end power under the conditions of (b).

3-12. A lossless line has a Z_o of 100 ohms and is terminated in a load of 50 + j50 ohms. The input voltage is 10 volts, r.m.s. Using phasors:
(a) Plot for 2 wavelengths the voltage and current standing waves.
(b) Find the load voltage and load current, assuming the line is 3.58 wavelengths long.
(c) Find the voltage standing wave ratio.
(d) Plot the impedance on the line, looking toward the load, for 2 wavelengths, showing separately the magnitude and the angle.

3-13. A lossless transmission line with a characteristic impedance of 50 ohms has an input impedance of 50 − j45 ohms. It is 357.65 centimeters long and the wavelength is 50 centimeters. Using phasors:
(a) Determine the load impedance.
(b) Magnitude of the reflection coefficient at the receiving end.
(c) Angle of the reflection coefficient at the receiving end.
(d) VSWR.

(e) How far in centimeters is the load from the first voltage maximum on the line?

(f) What is the value of the impedance at the voltage maximum found in (e)?

3-14. A line has a characteristic impedance of 50 ohms. It is assumed to be a lossless line. The VSWR is found to be 1.39 and the first voltage minimum is 5 centimeters from the load and the next minimum is 30 centimeters from the load. Using phasor diagrams determine:

(a) The load impedance.

(b) The location of the first voltage maximum from the load.

3-15. An air dielectric transmission line operating at a frequency of 200 megahertz has a characteristic impedance of $73 + j0$ ohms. It is terminated in a 30 ohm resistor shunted by a capacitance of 15 picofarads. The line losses are assumed to be negligible. Using phasors:

(a) What is the VSWR on the line?

(b) How far is the nearest voltage minimum in the standing wave pattern from the load teminals in meters?

(c) If the capacitance is disconnected, what is the VSWR?

(d) If the capacitance is disconnected, how far is the nearest voltage minimum in the standing wave pattern from the load terminals?

(e) If the capacitor is put back and the resistor removed what is the VSWR?

3-16. In the laboratory the following measurements were made on a lossless 50 ohm, coaxial slotted line.

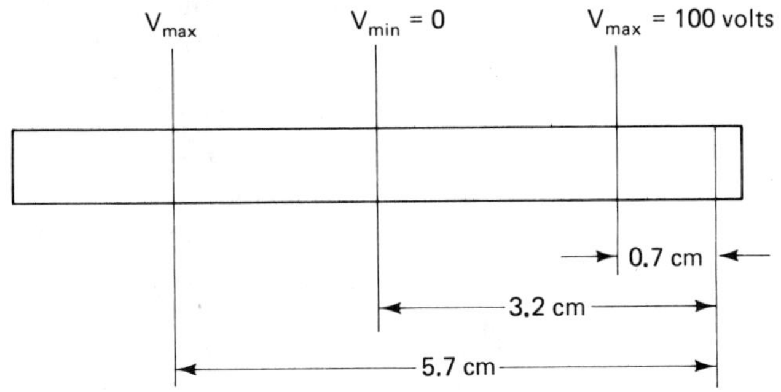

If $v_p = 3 \times 10^8$ meters/second determine (a) the frequency, (b) the VSWR, (c) the load impedance in polar form.

3-17. A section of lossless line is terminated in an open circuit. It is fed from a generator with an internal emf of 100 volts, r.m.s. and an internal impedance of 10 ohms, real. If the line has a characteristic impedance of 100 ohms and is 0.70 wavelengths long find the output voltage using phasors.

3-18. A lossless line with a characteristic impedance of 100 ohms is terminated in a short circuit. The line is 7.125 wavelengths long. Using phasors determine the following if the input voltage is 100 volts, r.m.s. (use V^+ as reference):
(a) The input current.
(b) The load voltage.
(c) The load current.
(d) The input impedance.
(e) The power to the load.

3-19. A lossless line has a load impedance of 97.622 + j65.108 ohms and a characteristic impedance of 50 ohms. The line is 10 meters long and β is 1.9635 radians per meter. The input voltage to the line is 50 volts, r.m.s. Using phasors determine:
(a) The voltage reflection coefficient.
(b) The distance from the load (in meters) to the first voltage minimum.
(c) The value of the impedance, looking toward the load, at the first voltage maximum.
(d) The receiving end current in phasor form.
(e) The input impedance to the line.
(f) The VSWR.

3-20. Measurements were made on a lossless 50 ohm line in the laboratory. A voltage minimum is located 22.5 cm from the load and the next is located 67.5 cm from the load. Its value is 20 volts. A voltage maximum of 80 volts is located 45 cm from the load. Assuming that v_p is the velocity of light, determine (a) the VSWR, (b) the frequency, (c) the load impedance in polar form, (d) the voltage reflection coefficient.

3-21. A lossless line with a characteristic impedance of 50 ohms is terminated in an admittance of 0.01 + j0 mhos.
(a) What impedance must be placed in parallel with the load in order to have an input impedance of 50 + j0 ohms?
(b) If the impedance in (a) is placed in parallel with the load and a voltage of $100\angle 0^\circ$ r.m.s. is applied to the input terminals what is:
(1) The power to the combined load?
(2) The voltage across the combined load?
(3) The input current to the line?
(4) The load current?

3-22. A lossless line with a characteristic impedance of 300 ohms is terminated in an open circuit. The line is 6.375 wavelengths long. If the input voltage is 100 volts, r.m.s., determine (a) the input current, (b) the load voltage, (c) the load current, (d) the input impedance.

3-23. A lossless line has a load impedance of $80 - j60$ ohms and a Z_o of 100 ohms. The line is 15 meters long and β is 2.0 radians per meter. The input voltage to the line is 100 volts, r.m.s. Determine:

(a) The distance from the load (in meters) to the first voltage minimum.

(b) The value of the impedance, looking towards the load, at the first voltage maximum.

(c) The receiving-end current in phasor form.

(d) The receiving-end voltage in phasor form.

(e) The input impedance to the line.

(f) The VSWR.

(g) The power to the load.

3-24. A lossless transmission line has a characteristic impedance of 50 ohms. It is operated at a frequency so that the wavelength is 2 meters with a standing wave ratio (VSWR) of 4.265. The first voltage maximum is 146.04 volts, r.m.s., and occurs at 0.73 meters from the load. Determine (a) the load voltage, (b) the load impedance in phasor form.

3-25. A lossless transmission line has a Z_o of 100 ohms. It is terminated in a load impedance of $80 - j60$ ohms. It is operated at a frequency so that the wavelength is 2 meters and the line is 5 meters long. Determine (a) the input impedance, (b) the impedance looking towards the load 0.25 meters from the load, (c) what would be the voltage reflection coefficient 0.25 meters from the load.

4

Lossless Lines and Smith Charts

4.1 Introduction

Before the days of digital computers and electronic calculators, engineers used slide rules and mechanical calculators to perform their calculations for design and analysis. Slide rules were limited when it came to accuracy and mechanical calculators could perform only the basic functions of addition, subtraction, multiplication, and division. Even after electric motors were added to the calculators they were still slow and the work was tedious and boring. Many hours and sometimes days were spent finding answers that today might take minutes. To reduce the work required, engineers developed all sorts of aids in their work such as tables, charts, and graphs; these reduced the time and effort needed in their work. The literature is full of these techniques and the Smith chart is one which is still worth looking at since it allows us to see rapidly what is happening on the line and makes certain types of calculations quite simple and quick while usually giving reasonable engineering accuracy.

An engineer who is familiar with the Smith chart can visualize what happens to the impedance looking towards the load as one moves from the load to the sending end. He or she can rapidly come up with answers which would take more time if equations or even the rotating phasor technique was used. As an aid to the engineer working with high frequency lines, the Smith chart becomes another tool which not only helps the solution but gives a physical interpretation of what is happening on the line. It is easy to use, and after a few examples the student can become quite proficient.

The basic idea behind the Smith chart involves the concept that one can trace in the complex plane the locus of constant resistance and constant reactance. The locus of both parameters turns out to be circles that intersect at right angles to each other so that the intersection gives a value of impedance with both a real and an imaginary term. Another circle, drawn by the engineer, gives a great deal of information as one moves around it. One can find the VSWR, the voltage reflection coefficient, the impedance anywhere on the line, the input impedance, the values of the minimum and maximum resistance and where they are located with respect to the load, where to place stub tuners to bring the VSWR down close to a value of unity and much other information of practical value. While the Smith chart may be used in the lossy case its primary beneficial use is for the assumed "lossless" case and so the information obtained will be quite similar to what we learned in Chapter 3. When making measurements in the laboratory or on an actual line, the Smith chart will prove to be of considerable value in making quick interpretations and coming up with some easy answers.

4.2 The Smith Chart

The Smith chart is particularly useful in the cases we have been considering in the first three chapters; namely the lossless and the high frequency cases. The mathematical derivation of the Smith chart is covered very adequately in the literature so we will leave most of the mathematics for your instructor so that he can impress you with the fact that he still remembers some algebra after a number of years. We will, instead, concentrate on its use and what we can see about the line from its use. The basic circuit involved is given in Figure 4-1.

We have defined the voltage reflection coefficient at the receiving end as Γ_r. Let us assume that somewhere on the line we replace

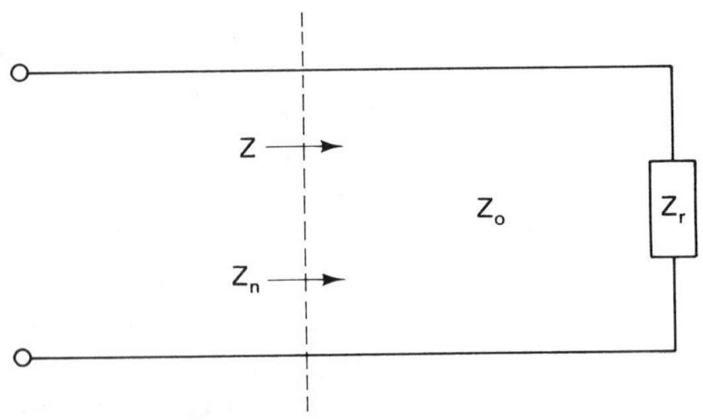

Figure 4-1. Smith chart transmission line.

everything to the right by the impedance, Z, at that point. The sending end wouldn't know the difference and the line to the left of this point would act exactly as it did before we made the change. Let us define the voltage reflection coefficient at this point by equation (4-1).

$$\Gamma_z = \frac{Z - Z_o}{Z + Z_o} = \frac{\dfrac{Z}{Z_o} - 1}{\dfrac{Z}{Z_o} + 1} = \frac{V^-}{V^+} \text{ (at z)}. \tag{4-1}$$

V^- in equation (4-1) would represent the phasor value of the reflected component of the voltage at point z and V^+ the incident component. It is convenient at this point to define a normalized impedance which is simply any impedance divided by some base impedance. In electrical power courses we call this the per-unit value of the impedance but we will call it the normalized impedance since this is a more common term when using the Smith chart. In our case we will use the characteristic impedance, Z_o, of the line as our base or reference impedance. In equation (4-1) Z/Z_o is a normalized impedance.

$$Z_n = \text{(normalized impedance)} = \frac{Z}{Z_o}$$

$$= \frac{R + jX}{Z_o} = \frac{R}{Z_o} + \frac{jX}{Z_o} = r_n + jx_n. \tag{4-2}$$

We will now assume that Z_o is real. Substituting into (4-1) gives:

$$\Gamma_z = \frac{r_n + jx_n - 1}{r_n + jx_n + 1}$$

which in general is a complex number so we will set it equal to the complex number, $U + jV$.

$$\Gamma_z = \frac{r_n + jx_n - 1}{r_n + jx_n + 1} = U + jV.$$

Cross multiplying gives:

$$r_nU + jUx_n + U + jr_nV + jV - x_nV = r_n + jx_n - 1.$$

$$Ur_n + U - Vx_n + j(Ux_n + Vr_n + V) = r_n - 1 + jx_n.$$

By equating the real and imaginary terms we obtain two equations:

$$r_n(U - 1) - x_nV = -(U + 1). \tag{4-3}$$
$$r_nV + x_n(U - 1) = -V. \tag{4-4}$$

By multiplying (4-3) by $(U - 1)$ and (4-4) by V we obtain two new equations which can be added to eliminate x_n. This equation is (4-5).

$$U^2 - \frac{2Ur_n}{r_n + 1} + V^2 = \frac{1 - r_n}{1 + r_n}. \tag{4-5}$$

After completing the square in U and rearranging, we obtain (4-6) which we recognize from our work in analytic geometry as the equation of a circle.

$$\left[U - \frac{r_n}{r_n + 1}\right]^2 + V^2 = \frac{1}{(r_n + 1)^2}. \tag{4-6}$$
$$(x - h)^2 + (y - k)^2 = r^2$$

is the equation of a circle with its center displaced h units on the x axis and k units on the y axis. In our case we have the result that the center is at

$$\left[\frac{r_n}{r_n + 1}, 0\right]$$

and the radius is

$$\frac{1}{r_n + 1}.$$

We will look at four cases of special values for the normalized resistance, r_n.

r_n	Center at	Radius
0	0, 0	1
0.5	0.333, 0	0.666
1.0	0.5, 0	0.5
2.0	0.666, 0	0.333

While there are several forms of the Smith chart we will look at one which gives the normalized values for both the impedance and the admittance. The exact center of this chart is the point 0,0 and the chart goes one unit to the right or left or up and down from this point. Since k equals zero the centers of our circles for all values of the normalized resistance will always be on the real or x axis. The basic coordinate system is shown in Figure 4-2. The circles for $r_n = 0$, 0.5, 1.0, and 2.0 are shown in Figure 4-3. In Figure 4-3 we notice that the normalized resistance, $r_n = 0$ gives the outside circle of the

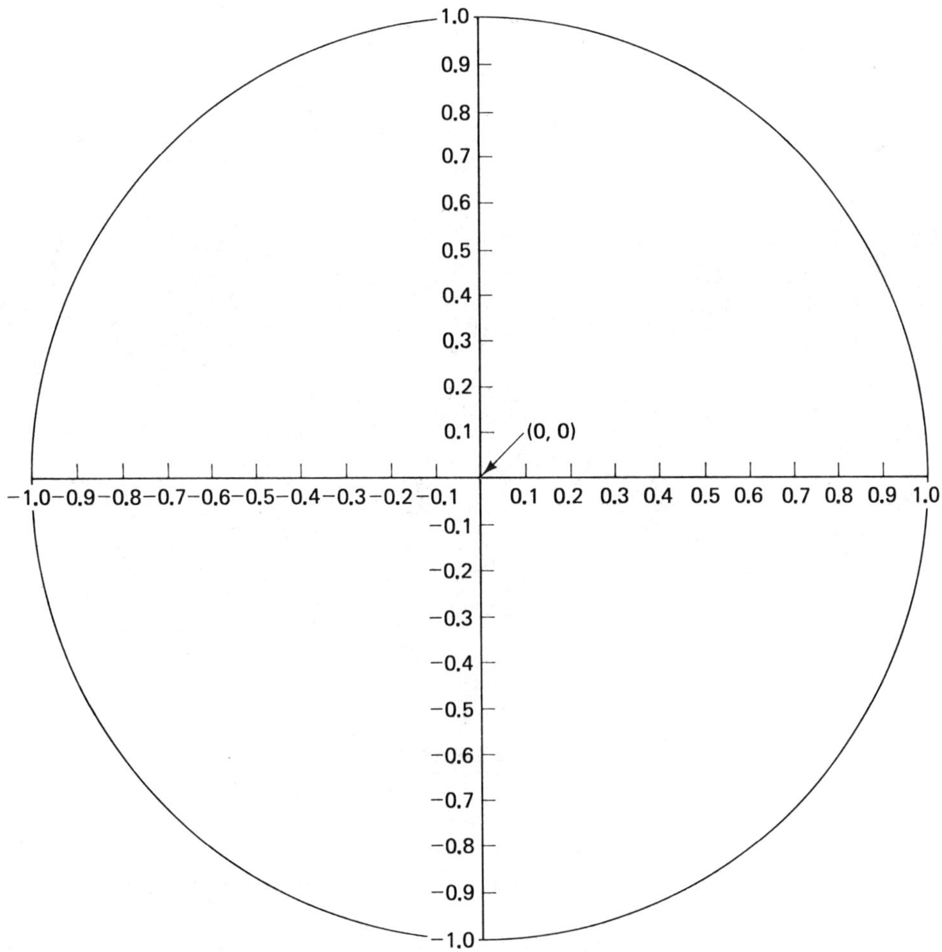

Figure 4-2. Basic coordinate system for Smith chart.

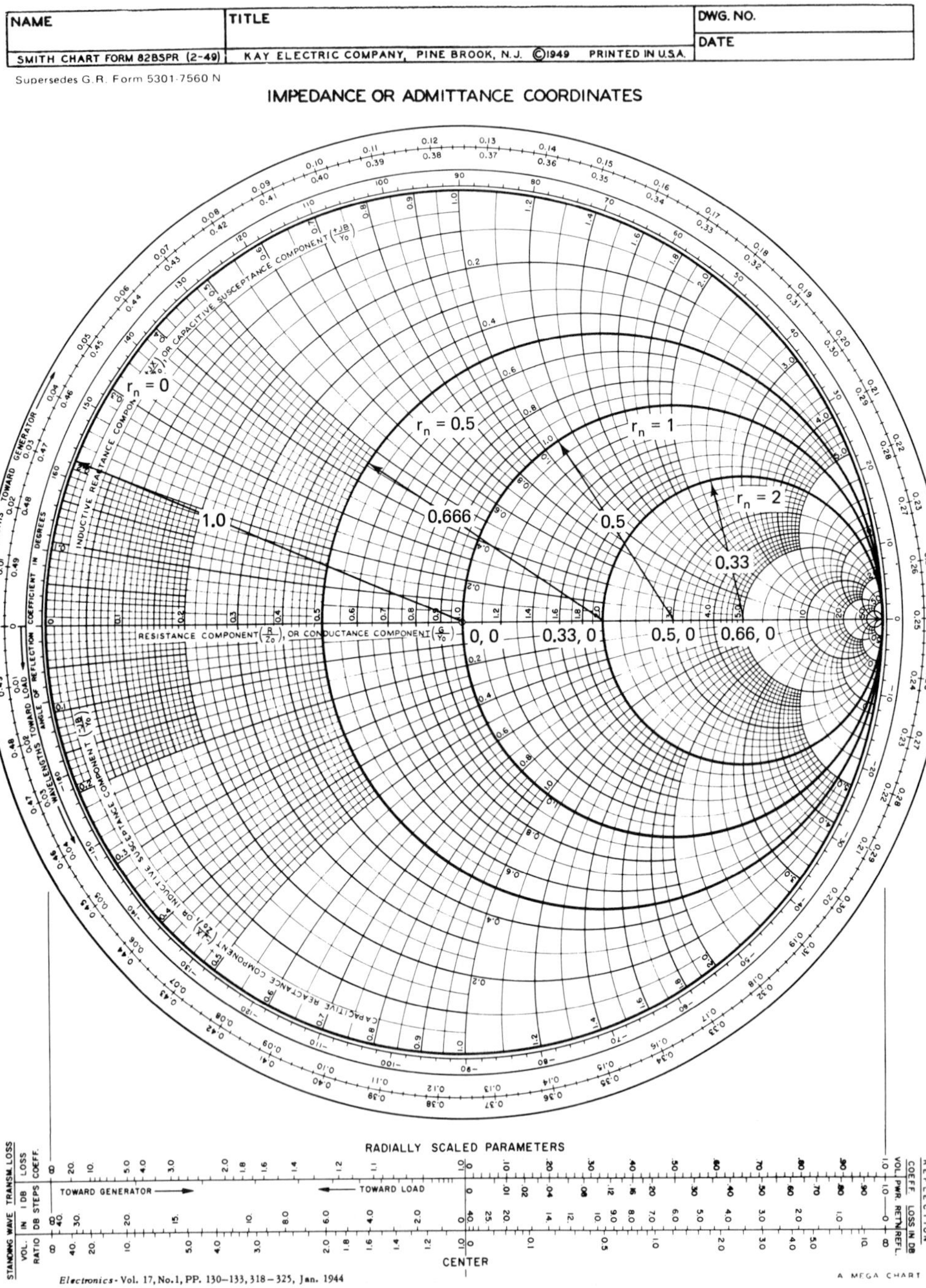

Figure 4-3. r_n circles.

Smith chart, with its center at the center of the chart. As the resistance increases from zero the circles become smaller and the centers move to the right on the positive real axis. When r_n approaches infinity the center approaches (1,0) and the radius approaches zero. Also notice that when r_n is between 50 and infinity there is very little area to work in on the Smith chart. In a practical sense this causes no difficulties since in most cases r_n would be much closer to 1 than to infinity or even to 50.

If we take equations (4-3) and (4-4) and eliminate r_n we obtain another equation for a circle.

$$(U - 1)^2 + \left(V - \frac{1}{x_n}\right)^2 = \frac{1}{x_n^2}. \tag{4-7}$$

This is the equation of a circle with a center on the right-hand edge with a value that is always one. The other coordinate of the center can be plus or minus since x and x_n in equation (4-2) can be either positive or negative. We have a center at $(1, \frac{1}{x_n})$ and a radius of $\frac{1}{x_n}$. As before we can set up a limited table for specialized values of x_n.

x_n	*Center at*	*Radius*
0	1, ∝	∞
0.2	1, 5	5
−.2	1, −5	5
1.0	1, 1	1
−1.0	1, −1	1
2.0	1, 0.5	0.5
−2.0	1, −0.5	0.5
∞	1, 0	0

One interesting circle is for $x_n = 0$. The radius is infinite with the center at $(1, \pm\infty)$. On the Smith chart it turns out to be the real axis. When $x_n = \infty$ the radius is 0 and the center is at (1,0). Figure 4-4 shows the circles for $x_n = 0, \pm 1$ and ± 2. The Smith chart, therefore, consists of two different sets of circles, one for r_n and one set for x_n. Each normalized impedance is represented by one fixed point on the Smith chart which is at the intersection of an r_n and x_n circle. This is indicated by Figure 4-5. Instead of starting with a normalized impedance we could have used a normalized admittance which would have resulted in two sets of circles; one for g_n and one for b_n, the normalized values of the conductance and the susceptance.

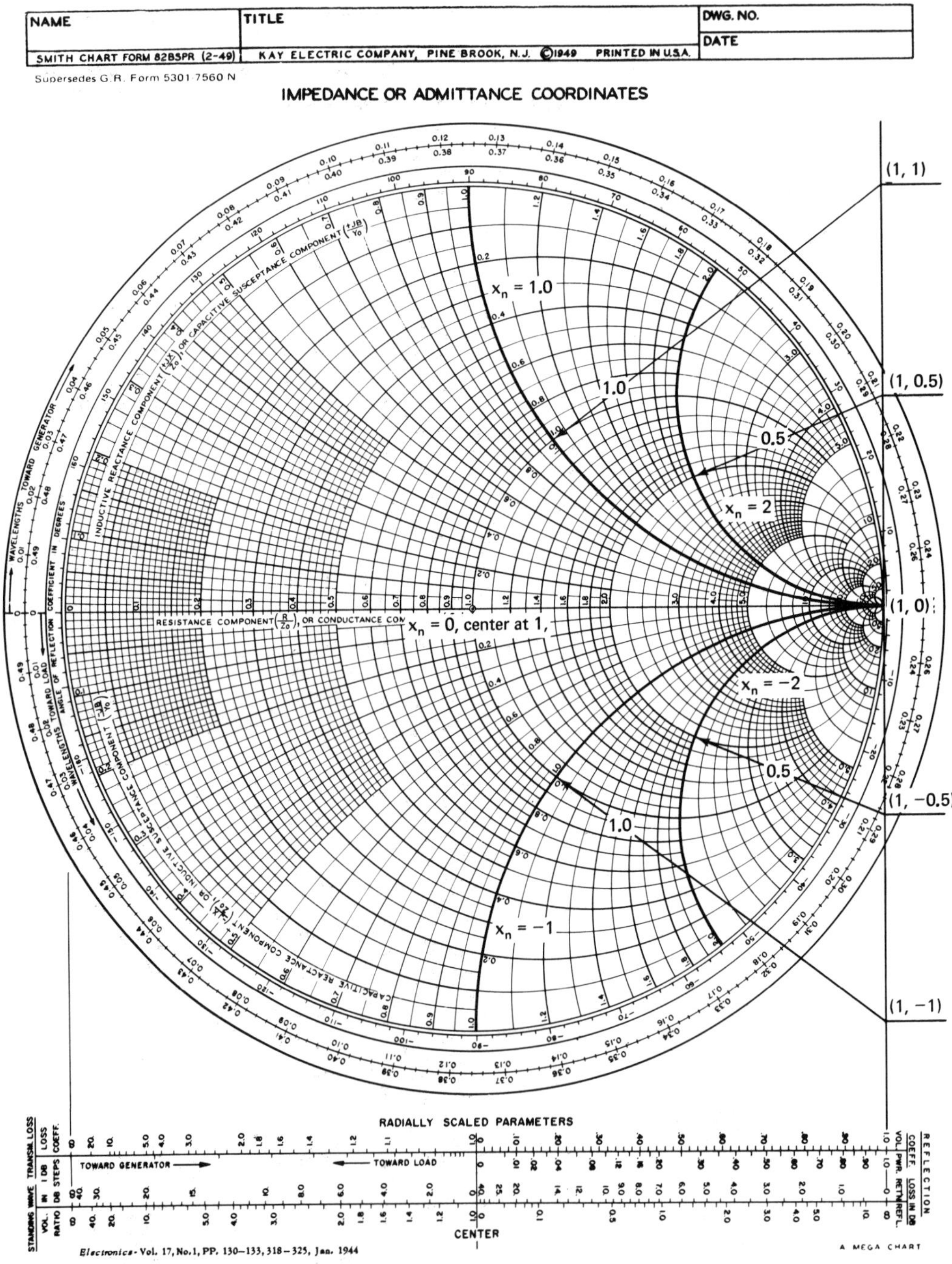

Figure 4-4. x_n circles.

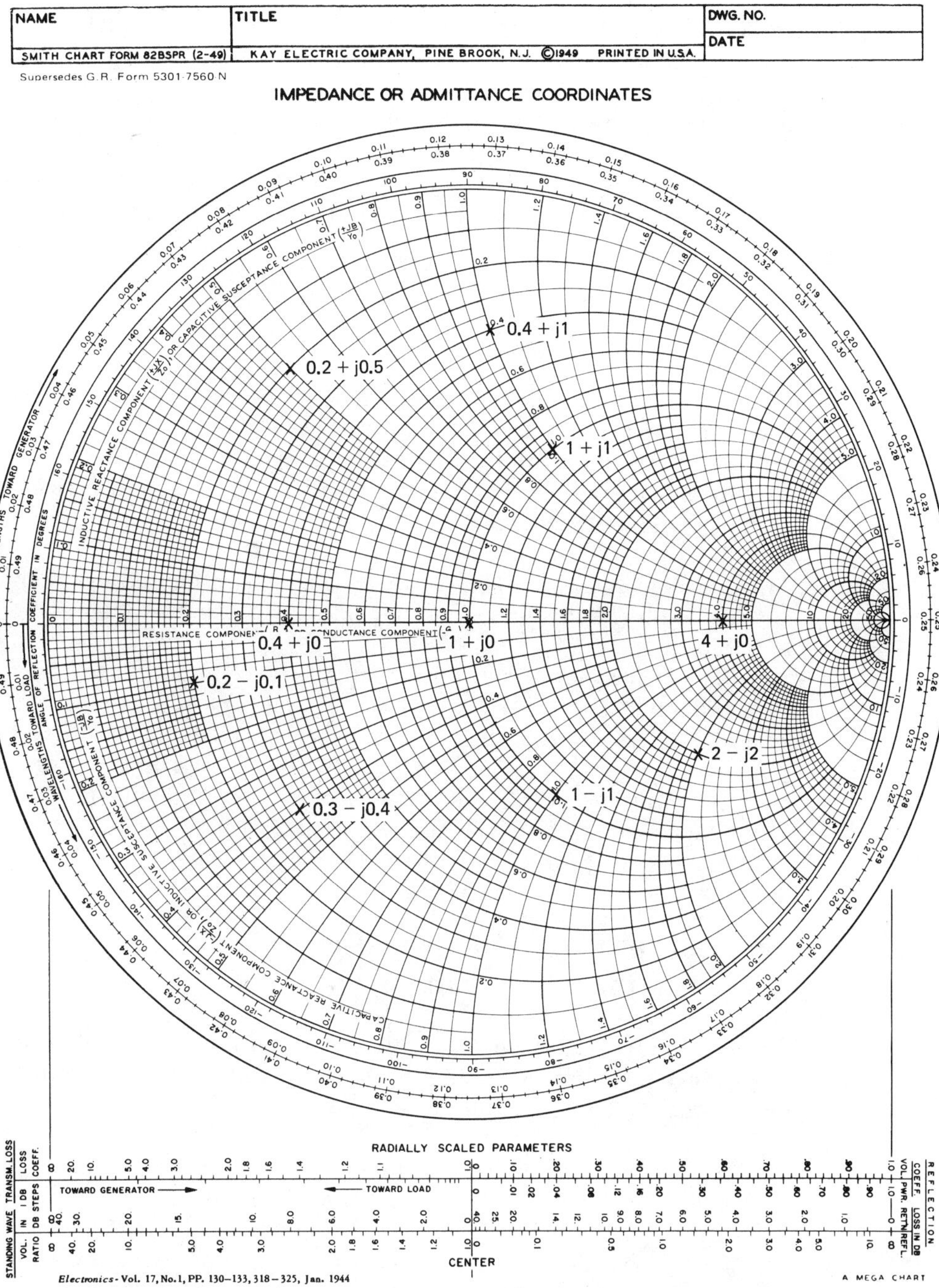

Figure 4-5. Normalized impedance locations.

Figure 4-5 would then show selected values of the normalized admittance. Here it should be pointed out that if we are using the Smith chart for impedance, the upper half represents an impedance with a resistor in series with an inductance and the bottom half represents a resistor in series with a capacitance. If we use the Smith chart for admittances, the top half is for a resistor and capacitor in parallel and the bottom half is for a resistor and inductance in parallel.

4.3 Changing an Impedance to an Admittance and Vice Versa

The Smith chart may be used to change an impedance to an admittance or we can go in the opposite direction. This means we can find the reciprocal of a complex number by using the Smith chart. Two examples are shown in Figure 4-6. Suppose $z_n = 1 + j1$ and we wish to find y_n. The procedure is as follows:

(a) Through the point $1 + j1$ draw a circle with the radius determined by the distance from the point $1 + j1$ to the center of the Smith chart.
(b) Draw a straight line through the point $1 + j1$ and the center of the Smith chart.
(c) The intersection of the circle and the straight line, 180 degrees from the $1 + j1$ point gives y_n. In this case it is $0.5 - j0.5$. Suppose we take $0.2 + j0.9$ as a second case. Following the same procedure we find that the admittance is $0.24 - j1.06$. If the impedance were $0.24 - j1.6$ the admittance would be $0.2 + j0.9$.

4.4 Example of Finding the Voltage Reflection Coefficient on the Smith Chart

The Smith chart can be used to find the magnitude and the angle of the voltage reflection coefficient for a load impedance. This is shown in Figure 4-7. The procedure is as follows:

(a) If we have a normalized value of the load impedance, such as $0.98 + j0.7$, we draw a circle through the point $0.98 + j0.7$ with the radius as the distance from the point to the center of the Smith chart. At the same time we draw an arc with the same radius on the top line at the lower right hand side of the Smith chart on the scale marked Reflection Coeff.-Vol. Where the arc

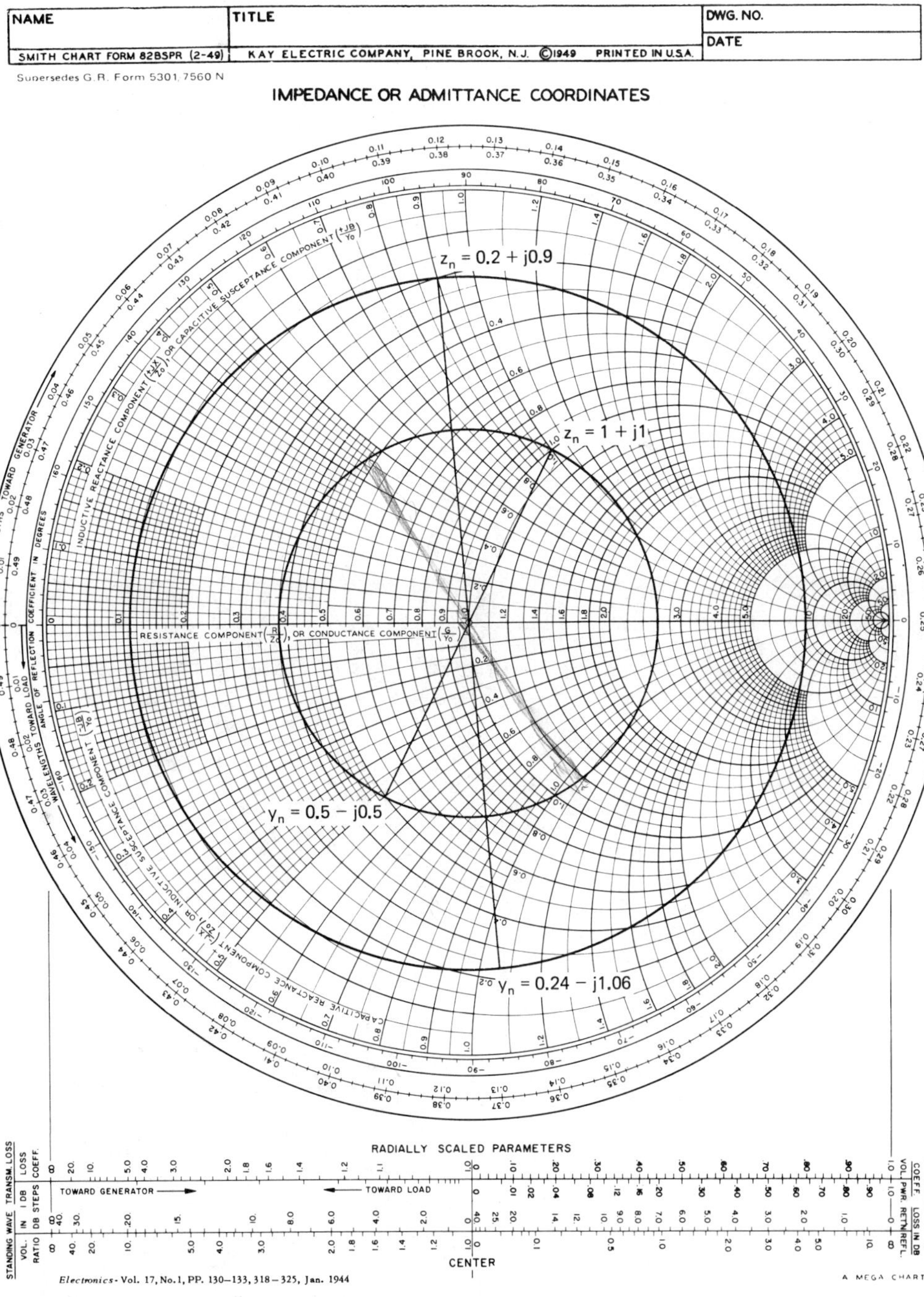

Figure 4-6. Finding the reciprocal of a complex number.

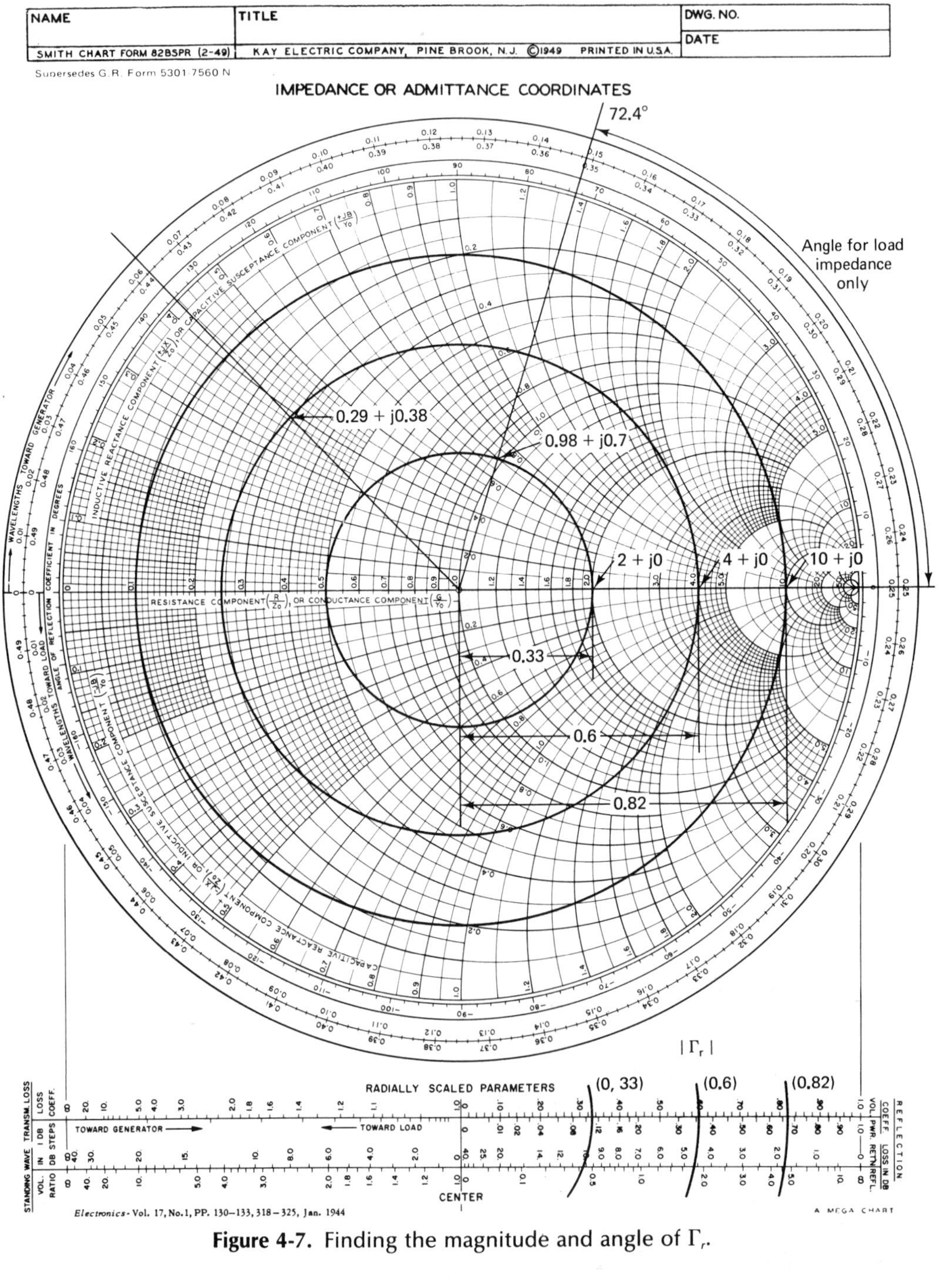

Figure 4-7. Finding the magnitude and angle of Γ_r.

crosses the scale we can read the magnitude of the reflection coefficient.

(b) Through the point and the center of the Smith chart we draw a straight line extending out beyond the outer circle.

(c) The angle of the voltage reflection coefficient is given at the point where the straight line cuts the inner scale of the three outer scales on the chart.

In this example we read 0.33 for the magnitude and 72.4 degrees for the angle. Other examples are given in Figure 4-7. For instance, a normalized impedance of $0.29 + j0.38$ gives $0.6 \angle 135°$ for Γ_r. An impedance of $4 + j0$ gives the same magnitude but the angle would be 0 degrees.

Other things of interest to us are evident by studying the chart. If the load impedance is a pure resistance with a normalized value greater than one, the angle is always zero degrees but if the normalized value is less than one the angle is always 180 degrees. If the normalized impedance has a positive x_n term, the angle is always positive and if the x_n term is negative the angle is always negative. All normalized impedances that lie on any given circle have the same magnitude for the voltage reflection coefficient but each has a different angle. There are an infinite number of impedances along each circle with the same magnitude. If the load admittance is given we must first convert it to a load impedance as discussed in Section 4-3 and then find the reflection coefficient. Since we are determining values graphically we would not expect the accuracy to be as good as if we used equations but if the person is reasonably careful the results are good enough in most practical situations.

4.5 Example of Finding the Voltage Standing Wave Ratio (VSWR) on the Smith Chart

The voltage standing wave ratio is found very easily from the Smith chart. If we look again at Figure 4-7 the procedure is as follows:

(a) Through the impedance (or admittance) at the load end we draw a circle with a radius determined by the distance to the center of the Smith chart.

(b) Where the circle crosses the positive real axis gives us the value of the VSWR.

The examples in Figure 4-7 give VSWR's of 2, 4, and 10 for the three circles shown. This means that all normalized impedances that fall on the inner circle of Figure 4-7 have a Γ_r with a magnitude of 0.33 and a VSWR of 2. If we recall equation (3-15) we can see that this must be true.

$$\text{VSWR} = \frac{1 + \Gamma_r}{1 - \Gamma_r} \tag{3-15}$$

4.6 Example of Finding Impedances on the Line Using the Smith Chart

The Smith chart will give us the impedance looking towards the load anywhere on the line. In fact, knowing the impedance at any point on the line will allow us to determine the impedance at any other point on the line. This is best illustrated by an example. Assume a line is 1.2 wavelengths long with a load impedance of $90 + j0$ ohms and that the characteristic impedance is 50 ohms. The procedure would be as follows:

(a) Determine the normalized value of the load impedance.

$$z_n = \frac{Z_r}{Z_o} = \frac{90 + j0}{50} = 1.8 + j0$$

(b) Drawing the usual circle through the point 1.8 and placing the arc on the lower right-hand magnitude scale, we know that,

$$\Gamma_r = 0.285 \angle 0°,\ \text{VSWR} = 1.8.$$

(c) The Smith chart has two scales on the outside of the chart. One is marked "wavelengths toward generator" and the other "wavelengths toward load." One revolution around the Smith chart is equivalent to one half wavelength. Our previous work showed us that everything repeats itself every half wavelength in the lossless case. As we start at the load ($1.8 + j0$) and move towards the generator (clockwise direction) each point on the circle gives the impedance at that point. The straight line drawn from the center of the chart through the load impedance gives not only the angle of Γ_r but a point on the wavelength scale. In this case

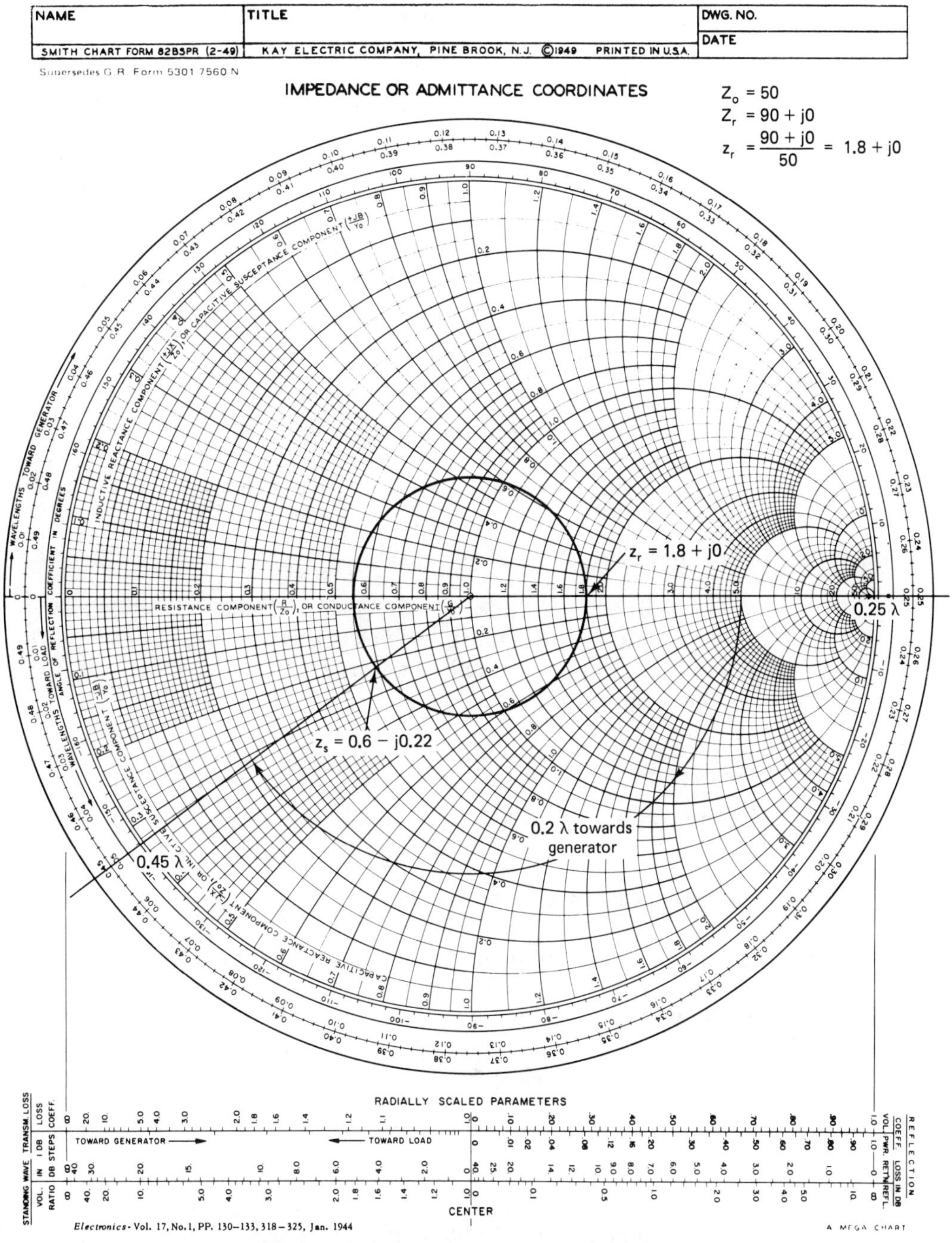

Figure 4-8. Finding input impedance.

we read 0.25 wavelength on the outer scale. As we move towards the sending end (toward the generator) we find that for the first quarter wavelength (from 0.25 to 0.50 wavelength) the impedance has a negative x_n term and is equivalent to a resistance in series with a capacitor. For the next quarter wavelength (0 to 0.25 wavelength) we have a positive x_n term so the impedance is equivalent to a resistor in series with an inductance. We are now back to our starting point after going 0.5 wavelength from the load. The normalized impedance at this point is once again $1.8 + j0$ and the actual impedance is $90 + j0$ ohms. This is the same result that we obtained when we discussed the lossless line using the phasor solution. Our next step is to move 1.2 wavelengths from the load to reach the sending end. Two revolutions take us back to the starting point so it is only the 0.2 wavelength that need concern us. So we move 0.2 wavelength towards the generator. In our case we have 0.25λ (at load) $+ 0.2\lambda = 0.45$ wavelength on the outer scale. We now draw a straight line through the center of the chart and the 0.45 wavelength mark on the "wavelengths toward the generator" scale.

(d) The point where this straight line intersects the circle is the input impedance of the line which in this case is $z_s = 0.6 - j0.22$. This normalized value is then used to find the actual value of the input impedance.

$$Z_s = z_s Z_o = (0.6 - j0.22)(50) = 30 - j11 \text{ ohms.}$$

This procedure is illustrated in Figure 4-8.

4.7 Another Example of Using the Smith Chart

A lossless line is terminated in a load of $0 + j150$, is 0.7 wavelength long and has a characteristic impedance of 50 ohms. Determine: Γ_r, VSWR, and the input impedance.

Solution (See Figure 4-9)

$$z_n = \frac{Z_r}{Z_o} = \frac{j150}{50} = j3.$$

Since the outer circle was obtained when r_n was equal to zero a pure reactance is always located on the outer circle of the Smith chart. We do not have to draw a circle since it is already on the chart.

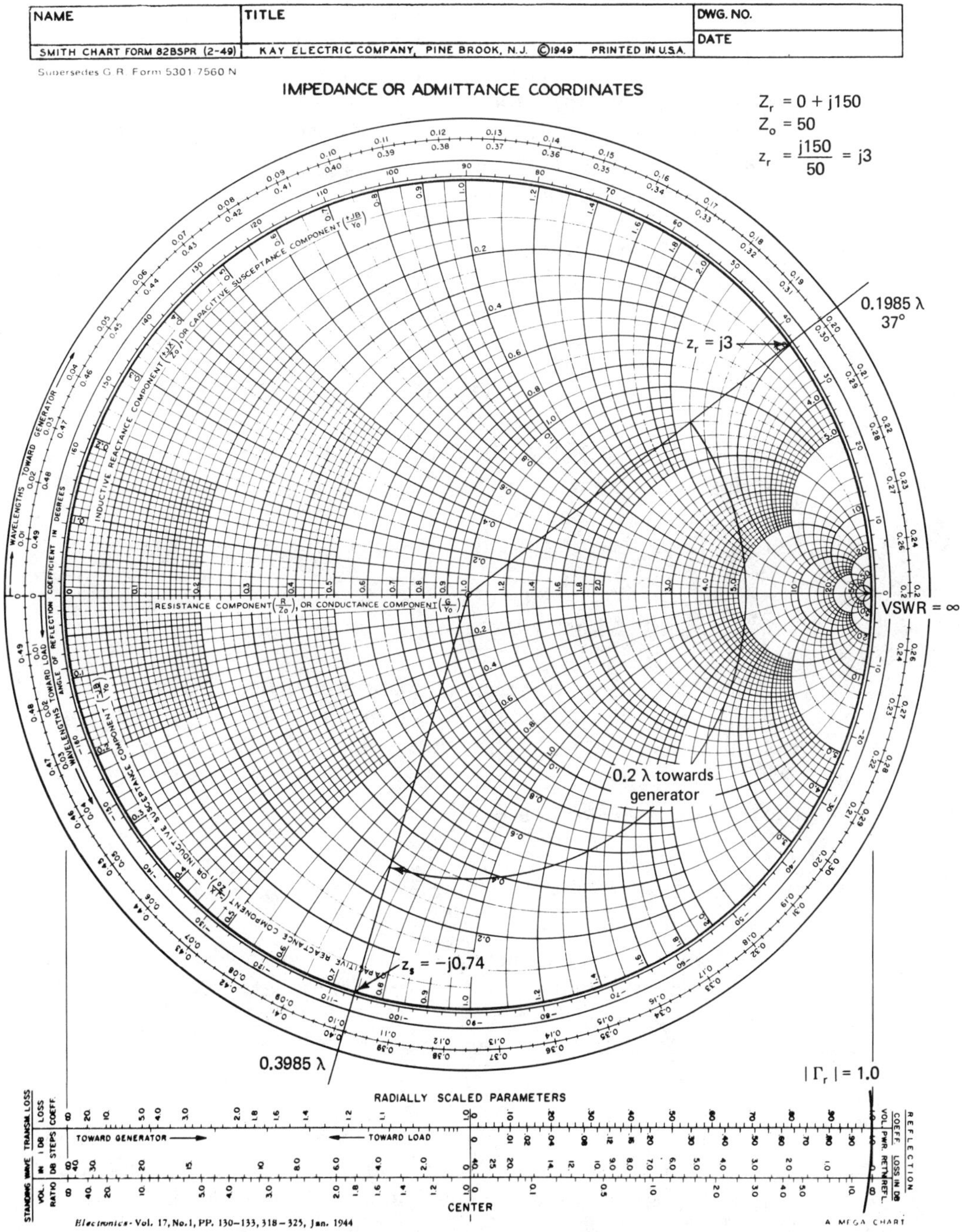

Figure 4-9. Finding input impedance.

Drawing a straight line through the center and the $j3$ point indicates that the angle of Γ_r is 37 degrees. We also notice that the magnitude of $\Gamma_r = 1$ and the VSWR is infinite. The outer scale gives a value of 0.1985 wavelength at the load. We must move 0.7 wavelength towards the generator. A half wavelength would bring us back to the starting point at $j3$ so we have to move an additional 0.2 wavelength to arrive at the sending end.

Since $0.1985\lambda + 0.2\lambda = 0.3985$ wavelength we move around on the outer scale till we find 0.3985 wavelength and draw a straight line through this point and the center of the chart. The intersection of this line and the circle gives the input impedance as $0 - j0.74$ in its normalized form. From this value we can find the actual value.

$$Z_s = z_s Z_o = (0 - j0.74)(50) = 0 - j37 \text{ ohms.}$$

Notice that the Smith chart does not give us as complete a solution as we had when we used phasors. If we knew the input voltage and found the input impedance we could find the input current but there is no direct way to find the load voltage and current.

4.8 Another Smith Chart Example

A lossless line with a characteristic impedance of 50 ohms is terminated in a load, $Z_r = 90 - j25$ ohms. The line is 2.11 wavelengths long. Determine, using the Smith chart, Γ_r, VSWR, and Z_s.

Solution (See Figure 4-10)

$$z_n = \frac{90 - j25}{50} = 1.8 - j0.5.$$

Note that 2.11 wavelengths means 4 revolutions plus 0.11 wavelength from the load. Using the same techniques as we just discussed we find that

$$\Gamma_r = 0.335 \angle -21.95^\circ, \text{ VSWR} = 2$$

and the wavelength scale at the load reads 0.28 wavelength; $0.28\lambda + 0.11\lambda = 0.39$ wavelength so after drawing the straight line and finding the point of intersection with the circle our answer is:

$$z_s = 0.72 - j0.53 \text{ and } Z_s = (0.72 - j0.53)(50) = 36 - j26.5 \text{ ohms.}$$

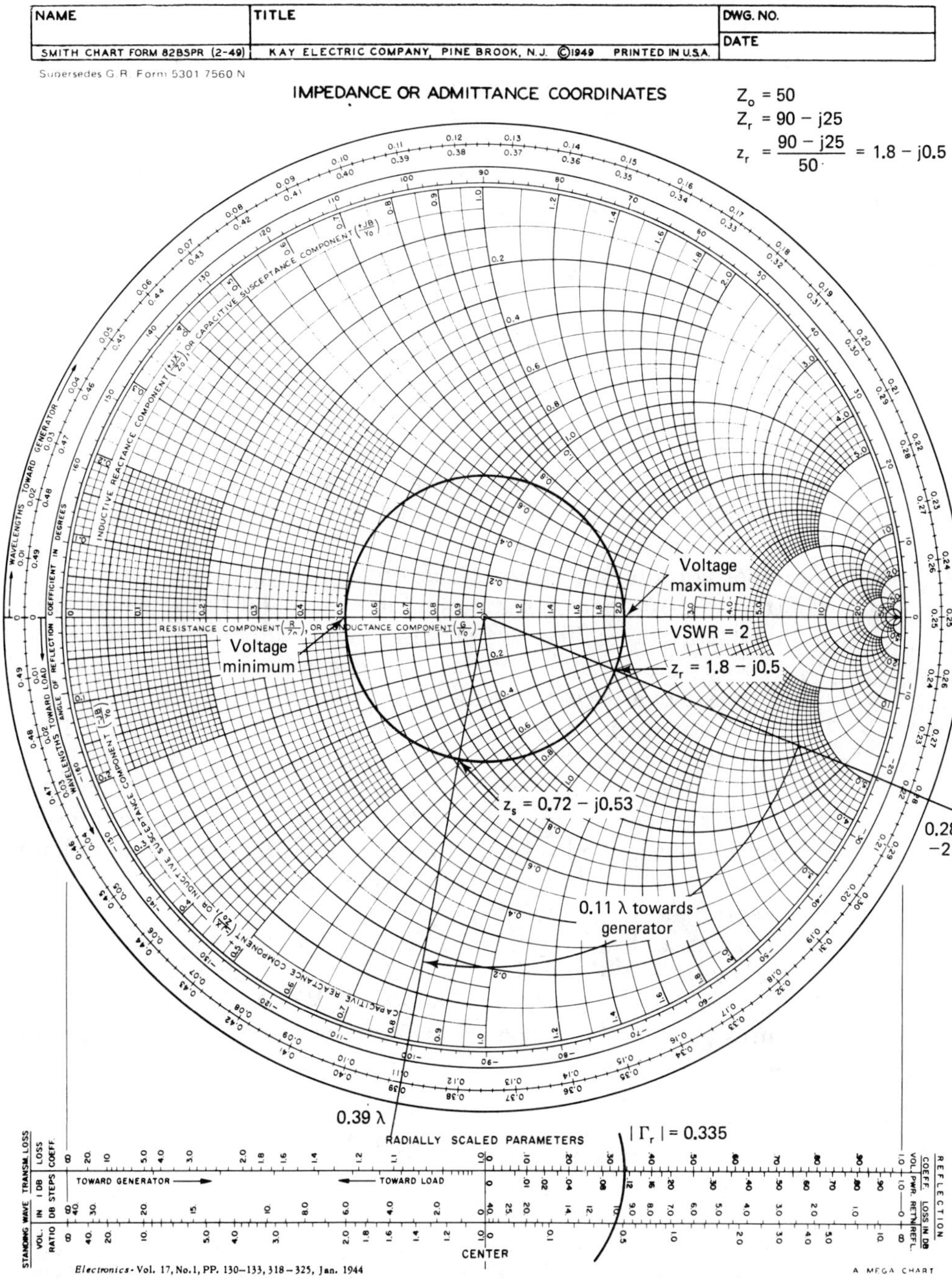

Figure 4-10. Finding input impedance.

4.9 Further Comments on the Smith Chart

We have now looked at enough examples to recognize that as we go around the impedance circle, moving in either direction, there are two points on the circle where the normalized impedance is real. The maximum value is on the positive real axis and has the same value as the VSWR. We know from the phasor solution that we obtain a maximum real resistance when we have a voltage maximum and a current minimum. So a voltage maximum is located at the resistance maximum and the location may be determined with respect to any other point on the line in terms of wavelengths by using the Smith chart. For instance, in the last example (Figure 4-10) we would have to travel from 0.28 wavelength to 0.5 wavelength (where the scale starts over at 0 wavelengths) or $0.5 - 0.28 = 0.22$ wavelength plus 0.25 wavelength to reach a voltage maximum. The minimum resistance is located on the negative real axis and has a normalized value less than one. This is a point of voltage minimum and current maximum so we can locate the voltage minimum with respect to any other point on the line in terms of wavelength. In our last example (Figure 4-10) the voltage minimum is located 0.22 wavelength from the load. The Smith chart tells us that as we proceed from the load towards the generator that we reach a voltage minimum first and then one quarter wavelength later we have a voltage maximum.

Notice that if the load is a pure resistance with a value larger than Z_o, then we have a voltage maximum at the load and if it is smaller than Z_o we have a voltage minimum at the load. If we know the wavelength in meters the chart will allow us to determine the distance from the load to a voltage maximum or minimum in terms of meters. It should now be obvious that if we know the VSWR from measurement, then we can draw a circle without knowing the load impedance. Knowing the distance from a voltage minimum to the load will then allow us to determine the unknown load impedance in terms of the VSWR and the distance to a voltage minimum. It turns out that this becomes a rather useful technique in the laboratory.

4.10 Voltage Standing Wave Ratio or Standing Wave Ratio?

Engineers often use the terms voltage standing wave ratio and standing wave ratio interchangeably and sometimes they are confused in the literature. But we will use the following definition for the standing wave ratio (SWR).

$$\text{Standing Wave Ratio} = \text{SWR} = 20 \log_{10} \frac{V_{max}}{V_{min}}. \quad (4\text{-}8)$$

Since

$$\text{VSWR} = \frac{V_{max}}{V_{min}}$$

it follows that

$$\text{SWR} = 20 \log_{10} \text{VSWR}.$$

Thus, we could set up a table showing the relationship between the two terms.

VSWR	*SWR*
1.00	0.0
1.26	2.0
2.0	6.0
3.16	10.0
4.0	12.0
10.0	20.0
40.0	32.0
100.00	40.0
1,000.00	60.0

The Smith chart can be used to determine these relationships. They are given on the lower scale on the lower left-hand corner of the chart. The heading is Standing Wave and beneath are two columns—one labeled Vol. Ratio which is the same as VSWR and the other says DB which is the SWR. The reader must be careful in the laboratory since it is not always clear whether the instrument is measuring VSWR or SWR. Always check the instruction manual when in doubt.

4.11 Example of Going the Other Way—From the Sending End to the Receiving End

The Smith chart can be used quite easily in going the other way—that is, knowing the input impedance and the length of the line, what is the load impedance, the VSWR and Γ_r? Let us look at the following circuit, while looking at Figure 4-11.

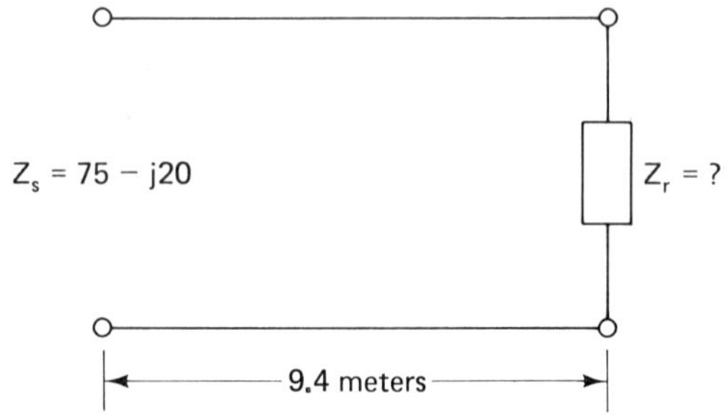

$$Z_o = 50 \text{ ohms}, \lambda = 1.2 \text{ meters}$$

Find: Z_r, VSWR and Γ_r. Since the line is 9.4 meters long and the wavelength is 1.2 meters the line contains:

$$\frac{9.4 \text{ meters}}{1.2 \text{ meters/wavelength}} = 7.833 \text{ wavelengths.}$$

Seven and one half wavelengths would be 15 revolutions and would get us back to the starting point so we are interested only in $7.833 - 7.5 = 0.333$ wavelength.

$$z_s = \frac{Z_s}{Z_o} = \frac{75 - j20}{50} = 1.5 - j0.4.$$

A circle is drawn with the distance from $1.5 - j0.4$ to the center as the radius. A straight line is drawn from the center through the point $1.5 - j0.4$ and we read 0.2085 wavelength on the inner of the two outer scales (wavelengths toward load). We must now go around the circle 0.333 wavelength from this point.

$$0.2085\lambda + 0.333\lambda = 0.5415 \text{ wavelength.}$$

Our scale only goes to 0.5 wavelength so we must proceed $0.5415 - 0.5 = 0.0415$ wavelength from the 0.5 (or 0) common point on the scale. This takes us to the 0.0415 point on the scale. We draw a straight line through this point and the center of the circle. Where the straight line intersects the circle we find the normalized load impedance to be $0.63 - j0.16$ so the actual load impedance is:

$$Z_r = z_n Z_o = (0.63 - j0.16)(50) = 31.5 - j8.0 \text{ ohms.}$$

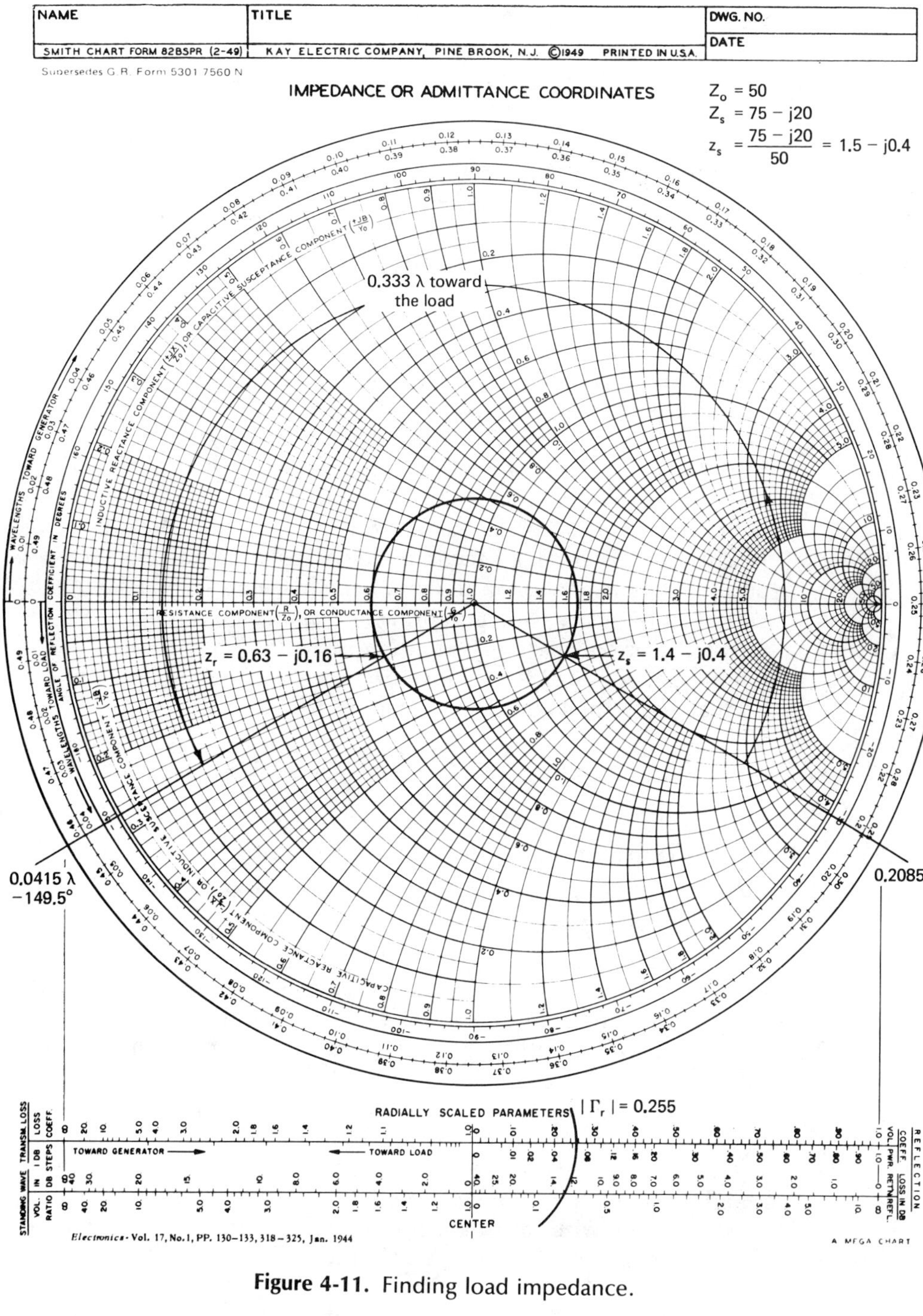

Figure 4-11. Finding load impedance.

We also find from the circle and the other construction:

$$\Gamma_r = 0.255 \angle -149.5^\circ. \qquad \text{VSWR} = 1.68.$$

In this case we move in the counterclockwise direction from the sending end to the load end.

4.12 Determining Z_r from Laboratory Measurements

The Smith chart is very handy in determining an unknown load impedance from measurements in the laboratory. Suppose we have determined from measurement that the VSWR = 1.5 and that the first voltage minimum is 0.37 wavelength from the load. As shown in Figure 4-12 we draw a circle with the radius determined by the VSWR (1.5 on the positive real axis.) The voltage minimum is the point on the circle of minimum real resistance or where the circle intersects the negative real axis. Now we must proceed 0.37 wavelength towards the load (counterclockwise direction). In this case our scale starts at zero on the wavelength scale so we move to 0.37 wavelength on the scale and draw a straight line through 0.37 and the center of the chart. Where it intersects the circle is the value of the normalized load impedance. Here we read $Z_n = 0.95 + j0.4$ in normalized form so our actual load impedance may be calculated.

$$Z_r = z_n Z_o = (0.95 + j0.4)(50) = 47.5 + j20 \text{ ohms}.$$

4.13 Further Comments on the Use of the Smith Chart

Once we have the circle drawn for impedances on the line we automatically have the circle for the admittances since for each impedance we have the admittance 180 degrees around the circle. All we have to remember is that the voltage reflection coefficient is determined for the load impedance and not for the load admittance. If we find or have the load admittance we would have to go 180 degrees around the circle to find the angle of Γ_r.

We have two points on the Smith chart that deserve special attention; $(0 + j0)$ on the left side and $(\infty + j\infty)$ on the right side. If we are dealing with impedances $0 + j0$ is a point of a short circuit and $\infty + j\infty$ is a point of an open circuit. In both cases the VSWR is

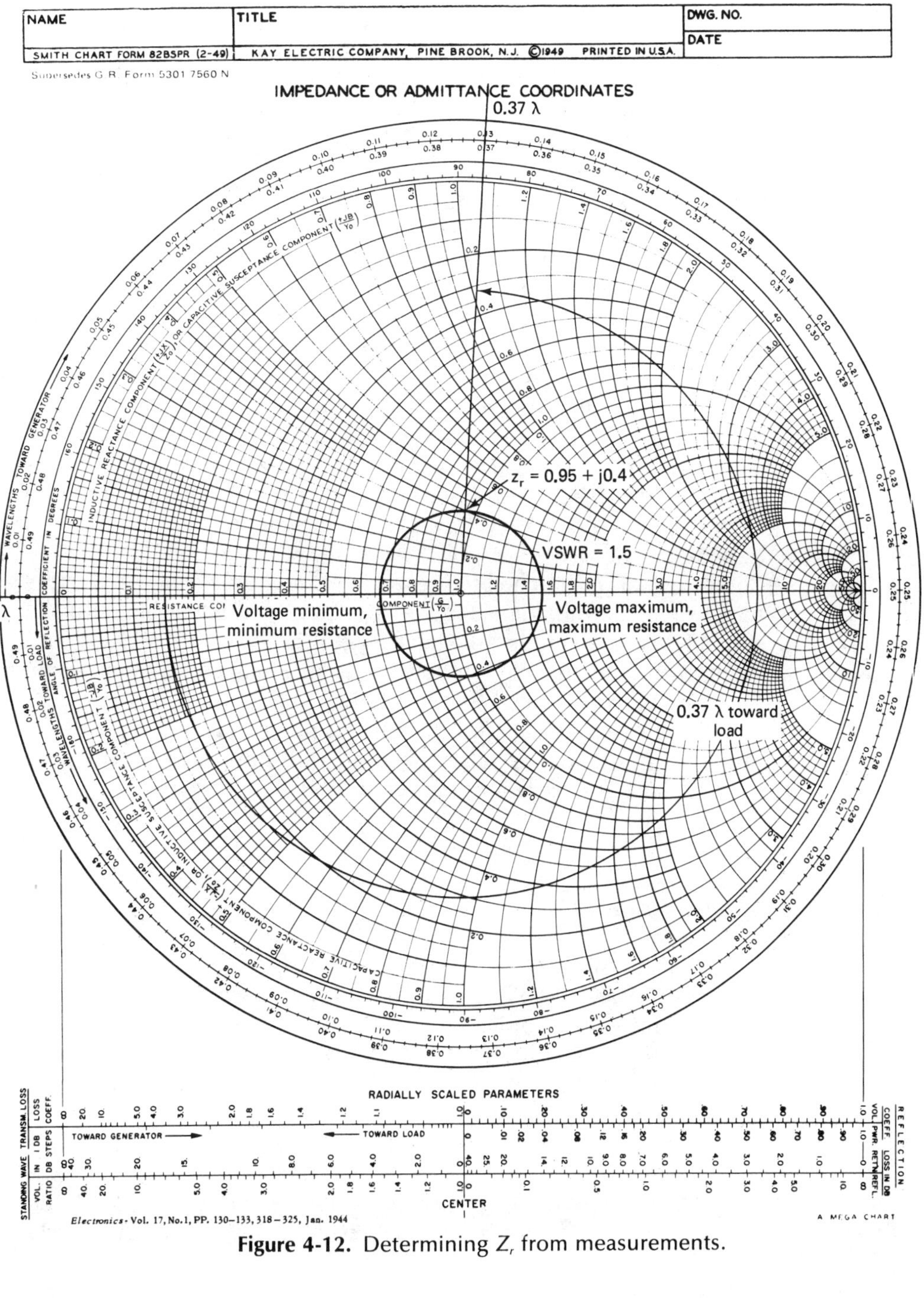

Figure 4-12. Determining Z_r from measurements.

infinite and the magnitude of Γ_r is one. The angle of Γ_r is zero degrees for the open circuit and 180 degrees for the short circuit. The circle has a radius of one with the center in the center of the chart, which makes the circle the $r_n = 0$ circle. If we start at the short circuit all of our impedances around the circle are purely reactive starting out with a $+j$ or inductive reactance until we reach 1/4 wavelength where the impedance becomes infinite. As we proceed in the clockwise direction around the outer circle the impedance becomes capacitive with a $-j$ value until we go another 1/4 wavelength and return to our starting point of a short circuit. These are the same results we had previously but are apparent immediately on the Smith chart and simply visualizing the form of the Smith chart will help us remember the situation on the line.

When dealing with admittances the opposite is true. The admittance $0 + j0$ now becomes our open circuit and $\infty + j\infty$ becomes our short circuit. The use of the chart will show how the admittances act as we proceed from a short- or open-circuit load. The reader should recall that an $a + jb$ admittance is equivalent to a $c - jd$ impedance.

If the load is matched to the characteristic impedance of the line the normalized impedance is $1 + j0$. Thus our circle has a radius of zero. (There is no circle—we have a point in the center of the chart.) The Smith chart is not of much use to us since as we move down the line the impedance stays constant at Z_o.

We can also use the Smith chart for other practical engineering problems but it will be more useful to wait until later in this book to show how the techniques relate to specific problems.

4.14 Chapter Summary

1. The Smith chart is a useful tool for the engineer working on transmission lines that can be approximated as lossless lines. Its use requires only a compass, a straightedge, and a pencil.
2. The locus of constant resistance is a circle on the chart as is the locus of constant reactance. The constant resistance and constant reactance circles intersect at right angles to each other. Each impedance or admittance is represented by one point on the chart.
3. The engineer draws one important circle on the chart with the center of the chart as the center of the circle and this circle

gives a great deal of information about the line as one moves around its perimeter. This circle may be drawn by knowing any of the following values: a) the normalized value of the load impedance, b) the normalized value of the input impedance, c) the VSWR, d) the magnitude of the voltage reflection coefficient, e) the normalized value of the load admittance, f) the normalized value of the input admittance.

4. Once the circle is drawn the engineer may find a number of unknown facts about his transmission line, such as: a) VSWR, b) the magnitude of the voltage reflection coefficient and its angle, c) the values of the impedance anywhere along the line, d) the input or load impedance, as the case may be if the length of the line is known, e) the locations of voltage maximums and voltage minimums, f) the values of the maximum and minimum resistances that exist on the line.
5. Knowing the normalized values of the impedance one can determine the magnitude and angle of the voltage reflection coefficient quite easily.
6. The Smith chart does not allow us to calculate directly the voltages and currents on the line.
7. One of the most useful techniques is to enter the Smith chart with the VSWR and the distance from a voltage minimum or maximum to the load. This will allow us to find all the pertinent information along with the value of the unknown load impedance.
8. The Smith chart can also be used for other things such as changing an admittance to an impedance or vice versa.
9. One of the more important uses of the Smith chart is to visualize the chart and then think of a certain engineering case. By remembering the form of the chart, one can easily determine if a voltage maximum or minimum will come up first as one moves away from the load or if the impedance will be inductive or capacitive and in what locations.
10. The chart is useful only if the impedances involved on the line are in a range of 1 to 5 or 6 times the characteristic impedance, unless one is concerned with the open- or short-circuit case.
11. The left-hand edge of the chart on the real axis represents an impedance or admittance of $0 + j0$. This represents a short for the impedance and an open for an admittance.
12. The right-hand edge of the chart on the real axis represents an impedance or admittance of $\infty + j\infty$. This represents an open for an impedance and a short for an admittance.

4.15 Review and Study Questions

1. Define the normalized impedance and the normalized admittance.
2. Sketch the circles for the normalized values of resistance of 0, 0.2, 1, and 4.
3. Sketch the circles for the normalized values of reactance of 0, ± 0.4, ± 1, and ± 4.
4. Locate the points of a short-circuit impedance, a short-circuit admittance, an open-circuit impedance, and an open-circuit admittance.
5. How is the normalized value of admittance found if given the normalized value of the impedance?
6. Given the normalized value of the load impedance, how does one determine the magnitude and the angle of the voltage reflection coefficient?
7. Given the normalized value of the load admittance, how does one determine the magnitude and angle of the voltage reflection coefficient?
8. Given the load impedance and the length of the line, how does one determine the input impedance and the distance to the first voltage minimum and voltage maximum from the load?
9. Given the load impedance, how does one determine the VSWR and the SWR?
10. Given the input impedance of the line and the length of the line, how does one determine the load impedance, the voltage reflection coefficient, the VSWR and the SWR?
11. If the load is inductive, do you have a voltage maximum or voltage minimum closest to the load?
12. If the load is capacitive, do you have a voltage minimum or voltage maximum closest to the load?
13. If the VSWR is 4.0, what is the minimum value of the impedance looking towards the load if the characteristic impedance is 100 ohms?
14. Under what conditions will the angle of the voltage reflection coefficient be 180 degrees? 0 degrees?
15. If the line is long enough will the admittance looking towards the load ever go through the point $1 \pm jB$ where B could be any value of susceptance?
16. If one is given the VSWR and the distance in wavelengths to the nearest voltage minimum explain how one may determine an unknown load impedance.

17. From your analytic geometry studies give the following: (a) the equation of a straight line going through the origin, (b) the equation of a circle with its center at 0,0, (c) the equation of a circle with its center at -1, $+2$.
18. If you have a short circuit on a line and move towards the sending end is the impedance going to be inductive or capacitive just as you leave the short?
19. How is the SWR related to the VSWR?
20. How is a decibel defined?
21. What is the minimum value of the VSWR that appears on the Smith chart? the maximum value?
22. How many wavelengths is once around the circle on a Smith chart?
23. If you move from the sending end to the load end do you move counterclockwise or clockwise?
24. Could you determine the load impedance if you knew the VSWR and the angle of the voltage reflection coefficient?

Problems

4-1. Using the Smith chart find the admittances for the following normalized values of impedance:
(a) $2 + j2$, (b) $0.5 - j1$, (c) $3 + j3$, (d) $1 + j0$.

4-2. Using the Smith chart find the impedances for the following normalized values of admittance:
(a) $0.4 - j0.8$, (b) $0.8 + j1$, (c) $0 - j4$, (d) $2 + j1$.

4-3. A lossless transmission line has the following values of normalized load impedance. (Z_r/Z_o) For each one find the magnitude and angle of the voltage reflection coefficient using the Smith chart.
(a) $2 + j0$, (b) $1/2$, (c) $1/4$, (d) $2 \angle 30°$, (e) $j/2$, (f) $2 - j1$, (g) $1.5 \angle -30°$.

4-4. Using the Smith chart, determine the input impedance of a 2.35 wavelength, open-circuited line having a Z_o of 600 ohms. Assume the line is lossless. What would be the input current if the input voltage is 100 volts, r.m.s.?

4-5. A lossless line having a 200 ohm characteristic impedance is terminated in $200 + j300$ ohms. Assume the line is 2.81 wavelengths long. Using the Smith chart:
(a) Find the input impedance of the line.
(b) Find the VSWR.

(c) Find the magnitude and angle of the voltage reflection coefficient.

4-6. A lossless line has a characteristic impedance of 50 ohms. If β is 2.35 radians per meter and the line is 3.2 meters long, determine the following using the Smith chart, when $Z_r = 150 + j50$ ohms:
(a) Input impedance.
(b) VSWR.
(c) Magnitude and angle of the voltage reflection coefficient.

4-7. A line having a Z_o of 100 ohms is found to have a VSWR of 3.0. The first voltage minimum occurs 0.4 meter from the load and the next one occurs 2.5 meters from the load. Using the Smith chart determine:
(a) the load impedance.
(b) the wavelength.

4-8. A lossless transmission line with a characteristic impedance. of 50 ohms, real, has an input impedance of $50 - j45$ ohms. It is 357.65 centimeters long and the wavelength is 50 centimeters. Find, using the Smith chart:
(a) Load impedance.
(b) Magnitude of the reflection coefficient at the receiving end.
(c) Angle of the reflection coefficient at the receiving end.
(d) VSWR.
(e) How far, in centimeters, the load is from the first voltage minimum on the line.
(f) What is the value of the impedance at the voltage minimum found in (e)?

4-9. A lossless transmission line has a characteristic impedance of 50 ohms, real. The load is $100 + j65$ ohms. Find, using the Smith chart:
(a) VSWR.
(b) Magnitude and angle of the voltage reflection coefficient.

4-10. In the laboratory, a student determined the following information while using a 50-ohm coaxial slotted line: a voltage maximum of 100 volts, r.m.s. was located 0.7 cm from the load; a voltage minimum of 0 volts was located 3.2 cm from the load; the next voltage maximum was 5.7 cm from the load. Determine, using the Smith chart:
(a) Load impedance in rectangular form.
(b) Frequency assuming v_p is the velocity of light.
(c) VSWR.

4-11. A 50-ohm transmission line is found to have a voltage standing wave ratio of 1.39. The first voltage minimum is 5 cm from the load and the next is 30 cm from the load. Using the Smith chart determine the load impedance.

4-12. Determine the voltage reflection coefficient in polar form for the following normalized values of admittance:
(a) $1 + j1$.
(b) $0.4 - j0.6$.
(c) $2 + j0.5$.

4-13. The normalized admittances $1.299 + j0.75$, $1.061 + j1.061$, and $0.75 + j1.299$ all have the normalized magnitude of 1.5. Which one of these admittances produces the lowest VSWR when connected as a terminal load admittance on a lossless line?

4-14. What is the input impedance of a lossless line which is terminated in an open circuit if the line is 9.0 cm long and the operating frequency is 1 gigahertz? The characteristic impedance is 100 ohms. Use the Smith chart.

4-15. Repeat problem 4-14 if the line is terminated in a short circuit.

4-16. The impedance of an antenna is measured to be $40 + j70$ ohms. It is connected to a transmitter through a lossless line 2.10 wavelengths long. If Z_o is 50 ohms what is the load presented to the transmitter? Use the Smith chart.

4-17. A slotted-line measurement on a lossless 75-ohm transmission line gives a standing wave ratio of 2.1, with the first voltage minimum 0.175 wavelength from the load. What is the load impedance?

4-18. A lossless transmission line 1.25 wavelengths long and having a Z_o of 50 ohms is terminated by the input terminals of a second lossless line 1.25 wavelengths long having a Z_o of 75 ohms. The second line is terminated in a pure resistance of 100 ohms. Determine, using the Smith chart, the input impedance of the first line.

4-19. A lossless line has a Z_o of 50 ohms and is 0.8 meters long. $\beta = 5$ radians/meter. If $Z_r = 100$ ohms, determine, using the Smith chart:
(a) The input impedance of the line.
(b) Γ_r in polar form.
(c) VSWR.
(c) The location of the first voltage maximum in wavelengths from the load.

4-20. A transmission line is lossless, and is terminated in a short circuit. What is the impedance looking towards the load:
(a) A quarter wavelength from the load if Z_o is 50 ohms?
(b) A half wavelength from the load if Z_o is 100 ohms?
(c) One-eighth wavelength from the load if Z_o is 50 ohms?

4-21. If a line is lossless and has a Z_o of 50 ohms, what is the VSWR if the load impedance is:
(a) $50 + j0$.
(b) $100 + j0$.
(c) $25 + j0$.
(d) $30 + j40$.
(e) $40 - j20$.

4-22. The VSWR on a lossless transmission line is 1.75. If Z_o is 50 ohms determine:
(a) The value of the impedance at the voltage maximum.
(b) The value of the impedance at a voltage minimum.

4-23. A lossless line with a Z_o of 100 ohms is terminated in a load of $75 + j90$ ohms. The line is 5.78 wavelengths long. Using the Smith chart determine:
(a) The voltage reflection coefficient.
(b) The input impedance.
(c) The location of the first voltage minimum from the load. (Express the answer in wavelengths.)

4-24. A lossless transmission line has a characteristic impedance of 100 ohms. It is operated at a frequency so that the wavelength is 3 meters with a VSWR of 3.0. The first voltage minimum is 0.73 meters from the load and has a magnitude of 160 volts, r.m.s. Determine the load impedance, using the Smith chart.

4-25. A lossless transmission line has a characteristic impedance of 50 ohms. It is terminated with a load impedance of 100 ohms and is operated at a frequency so that the wavelength is 3 meters and the line is 6.2 meters long. Determine, using the Smith chart:
(a) The input impedance.
(b) The impedance looking toward the load 0.25 meters from the load.

4-26. A lossless transmission line is 8.4 wavelengths long, has a Z_o of 50 ohms and a load impedance of $40 + j30$ ohms. The impedance on the line looking towards the load is purely resistive at certain points on the line. Actually there are two purely resistive values.

(a) Find the higher of the two.
(b) How far from the load does the first value in (a) occur? (Express in wavelengths.)
(c) Find the lower of the two values and how far does it occur from the load?

4-27. A lossless transmission line whose characteristic impedance is 50 ohms has a VSWR of 1.75. What is the value of the impedance at the voltage maximum and at the voltage minimum?

4-28. Measurements were made on a lossless 50 ohm slotted line in the laboratory. The first voltage maximum is located 22.5 cm from the load. The second voltage maximum is located 67.5 cm from the load. Its value is 100 volts, r.m.s. A voltage minimum of 10 volts, r.m.s. is located 45 cm from the load. Assuming that v_p is the speed of light determine:
(a) The VSWR.
(b) The frequency.
(c) The load impedance in complex form.
(d) The voltage reflection coefficient.
(e) The input impedance to the line if it is 142 cm long.

4-29. A lossless line has a load impedance of $80 - j60$ ohms and a characteristic impedance of 100 ohms. The line is 15 meters long and β is 2.0 radians per meter. Determine:
(a) The distance from the load to the first voltage minimum.
(b) The value of the impedance, looking toward the load, at the first voltage maximum.
(c) The VSWR.
(d) The input impedance of the line.

4-30. A lossless line with a Z_o of 50 ohms is terminated in an admittance of $0.01 + j0$ mhos.
(a) What impedance must be placed in parallel with the load in order to have an input impedance of the line equal to 50 ohms, real?
(b) What is the VSWR of the line with and without the impedance in (a) in parallel?

5

Using Equations for the Lossless Case

5.1 Introduction

Up to this point in our studies we have been looking at the specialized case of the lossless line by using the phasor concept and in the last chapter we introduced the Smith chart. The objective was to give the reader a fundamental understanding of what is happening physically on the transmission line and to become familiar with a process which is easy to remember and quick to obtain usable solutions to practical engineering problems. Traveling waves, standing waves, impedance values, voltage, current and power values, reflection, absorption, and other items are all important in describing the physical situation to the engineer.

The lossless line has been basic for our preliminary work since it is easy to consider, and represents a close approximation to a large number of practical cases. Even for the 60-hertz lossy case, the lossless approximation is used to consider transients on the line due to switching or lightning strokes. In the past 30 years engineering education has become more and more mathematical and this has allowed the engineer to solve more sophisticated engineering problems. However in the process some engineering graduates have been able to use the mathematics but have been unable to relate to the physical realities of the real world with little understanding of what is happening on the transmission line. When the author worked as a transmission line engineer for a public utility company many years ago it was quite a shock to find out that not all of the problems in the real world could be solved with just the use of mathematics.

Although the physical concepts are important it is also true that equations are vital to our understanding of what is happening on the line. Modern computers and even relatively inexpensive calculators can now solve many complicated transmission line problems with speed and accuracy. So while we know a great deal about the lossless line already, we will reexamine this special case by looking at the appropriate equations. Since the lossless line is just a special case of the lossy line with α equal to zero we are really approaching the general case as an extension of the lossless case. We will see that we can arrange our equations in slightly different forms and that the problem is often to select the equation that is the most directly applicable to our given situation.

In this chapter we will also discuss how we can remove the standing wave on a transmission line by the use of stub tuners. At high frequencies the stub tuner becomes a practical way to improve the performance of the line. Consideration at this time will help give additional insight into the way lossless lines operate.

We have been assuming that the standing waves on a line are sinusoidal in shape but this is not necessarily true. In the vast majority of cases the voltage maximums are broadened out over the pure sine wave and the voltage minimums are sharpened. In the laboratory or on an actual line this results in the engineer normally making measurements at the voltage minimums since it is easier to find the point of the actual minimum.

5.2 General Equations

Our starting point will be to repeat some equations we had before for the case when $\alpha = 0$ and Z_O is real.

$$V = V^+e^{j\beta d} + V^-e^{-j\beta d}. \tag{2-3}$$

$$I = \frac{V^+e^{j\beta d}}{Z_o} - \frac{V^-e^{-j\beta d}}{Z_o}. \tag{2-4}$$

$$\Gamma_r = \frac{Z_r - Z_o}{Z_r + Z_o} = \frac{V^-}{V^+}. \tag{3-9}$$

$$V^- = \Gamma_r V^+. \tag{3-10}$$

So

$$V = V^+e^{j\beta d} + \Gamma_r V^+e^{-j\beta d} = V^+(e^{j\beta d} + \Gamma_r e^{-j\beta d}) =$$

$$V^+(\cos\beta d + j\sin\beta d + \Gamma_r\cos\beta d - j\Gamma_r\sin\beta d) =$$

$$V = V^+[(1 + \Gamma_r)\cos\beta d + j(1 - \Gamma_r)\sin\beta d]. \tag{5-1}$$

When $d = 0$, $V = V_r$ so:

$$V_r = V^+[(1 + \Gamma_r) + j(0)] \text{ and } V^+ = \frac{V_r}{1 + \Gamma_r}. \tag{5-2}$$

$$V = \frac{V_r}{1 + \Gamma_r}[(1 + \Gamma_r)\cos\beta d + j(1 - \Gamma_r)\sin\beta d] =$$

$$V_r\left[\cos\beta d + j\frac{(1 - \Gamma_r)}{(1 + \Gamma_r)}\sin\beta d\right].$$

From equation (3-8) we can write (5-3).

$$\frac{Z_o}{Z_r} = \frac{1 - \Gamma_r}{1 + \Gamma_r}. \tag{5-3}$$

Making the substitution gives:

$$V = V_r\left[\cos\beta d + j\frac{Z_o}{Z_r}\sin\beta d\right] = V_r\cos\beta d + j\frac{V_r Z_o}{Z_r}\sin\beta d =$$

$$V = V_r\cos\beta d + jI_r Z_o\sin\beta d. \tag{5-4}$$

From equation (2-4) we obtain:

$$I = \frac{V^+}{Z_o}[(1 - \Gamma_r)\cos\beta d + j(1 + \Gamma_r)\sin\beta d].$$

For the moment let us use the V^+ from the voltage equation.

$$I = \frac{V_r}{Z_o(1+\Gamma_r)}[(1 - \Gamma_r)\cos\beta d + j(1 + \Gamma_r)\sin\beta d] =$$

$$\frac{V_r}{Z_o}\left[\frac{(1 - \Gamma_r)}{(1 + \Gamma_r)}\cos\beta d + j\sin\beta d\right] = \frac{V_r}{Z_o}\left[\frac{Z_o}{Z_r}\cos\beta d + j\sin\beta d\right] =$$

$$\frac{V_r}{Z_r}\cos\beta d + j\frac{V_r}{Z_o}\sin\beta d = I_r\cos\beta d + j\frac{V_r}{Z_o}\sin\beta d. \tag{5-5}$$

$$Z = \frac{V}{I} = \frac{V_r\cos\beta d + jI_r\sin\beta d}{I_r\cos\beta d + j\frac{V_r}{Z_o}\sin\beta d} = \frac{I_r Z_r\cos\beta d + jI_r Z_o\sin\beta d}{I_r\cos\beta d + j\frac{I_r Z_r}{Z_o}\sin\beta d} =$$

$$Z = \frac{Z_o(Z_r\cos\beta d + jZ_o\sin\beta d)}{(Z_o\cos\beta d + jZ_r\sin\beta d)}. \tag{5-6}$$

Suppose instead of using V^+ from the voltage equation we use V^+ from the current equation.

$$I = \frac{V^+}{Z_o}[(1 - \Gamma_r)\cos\beta d + j(1 + \Gamma_r)\sin\beta d].$$

When $d = 0, I = I_r$.

$$I_r = \frac{V^+}{Z_o}[(1 - \Gamma_r) + j(0)] \text{ so } V^+ = \frac{I_r Z_o}{(1 - \Gamma_r)}. \tag{5-7}$$

Therefore:

$$I = \frac{I_r Z_o}{Z_o(1 - \Gamma_r)}[(1 - \Gamma_r)\cos\beta d + j(1 + \Gamma_r)\sin\beta d] =$$

$$I_r\left[\cos\beta d + j\frac{(1 + \Gamma_r)}{(1 - \Gamma_r)}\sin\beta d\right] = I_r\left[\cos\beta d + j\frac{Z_r}{Z_o}\sin\beta d\right] =$$

$$I_r\cos\beta d + j\frac{I_r Z_r}{Z_o}\sin\beta d = I_r\cos\beta d + j\frac{V_r}{Z_o}\sin\beta d.$$

This is the same equation as (5-5).

Summary

$$\begin{cases} V = V_r\cos\beta d + jI_r Z_o\sin\beta d. & (5\text{-}4) \\ I = I_r\cos\beta d + j\dfrac{V_r}{Z_o}\sin\beta d. & (5\text{-}5) \\ Z = \dfrac{Z_o(Z_r\cos\beta d + jZ_o\sin\beta d)}{(Z_o\cos\beta d + jZ_r\sin\beta d)}. & (5\text{-}6) \end{cases}$$

When $d = \ell$, $V = V_s$, $I = I_s$, $Z = Z_s$.

$$\begin{cases} V_s = V_r\cos\beta\ell + jI_r Z_o\sin\beta\ell. & (5\text{-}8) \\ I_s = I_r\cos\beta\ell + j\dfrac{V_r}{Z_o}\sin\beta\ell. & (5\text{-}9) \\ Z_s = Z_o\dfrac{(Z_r\cos\beta\ell + jZ_o\sin\beta\ell)}{(Z_o\cos\beta\ell + jZ_r\sin\beta\ell)}. & (5\text{-}10) \end{cases}$$

If we rearrange (5-9), we have the following results.

$$V_s = V_r\cos\beta\ell + jI_rZ_o\sin\beta\ell.$$

$$I_s = \frac{jV_r}{Z_o}\sin\beta\ell + I_r\cos\beta\ell.$$

And using determinants we have the following results.

$$V_r = \frac{\begin{vmatrix} V_s & jZ_o\sin\beta\ell \\ I_s & \cos\beta\ell \end{vmatrix}}{\begin{vmatrix} \cos\beta\ell & jZ_o\sin\beta\ell \\ \dfrac{j\sin\beta\ell}{Z_o} & \cos\beta\ell \end{vmatrix}} = \frac{V_s\cos\beta\ell - jI_sZ_o\sin\beta\ell}{\cos^2\beta\ell + \sin^2\beta\ell} =$$

$$V_r = V_s\cos\beta\ell - jI_sZ_o\sin\beta\ell. \qquad (5\text{-}11)$$

And solving for I_r gives equation (5-12).

$$I_r = I_s\cos\beta\ell - j\frac{V_s}{Z_o}\sin\beta\ell \qquad (5\text{-}12)$$

Now let us rearrange equation (5-6).

$$Z = Z_o\frac{(Z_r\cos\beta d + jZ_o\sin\beta d)}{(Z_o\cos\beta d + jZ_r\sin\beta d)}. \qquad Y = \frac{1}{Z} = \frac{(Z_o\cos\beta d + jZ_r\sin\beta d)}{Z_o(Z_r\cos\beta d + jZ_o\sin\beta d)}.$$

$$Y = Y_o\frac{\left(\frac{1}{Y_o}\cos\beta d + j\frac{1}{Y_r}\sin\beta d\right)}{\left(\frac{1}{Y_r}\cos\beta d + j\frac{1}{Y_o}\sin\beta d\right)} \times \frac{Y_oY_r}{Y_oY_r} =$$

$$Y = Y_o\frac{(Y_r\cos\beta d + jY_o\sin\beta d)}{(Y_o\cos\beta d + jY_r\sin\beta d)}. \qquad (5\text{-}13)$$

And as before, when $d = \ell$, $Y = Y_s$.

$$Y_s = Y_o\frac{(Y_r\cos\beta\ell + jY_o\sin\beta\ell)}{(Y_o\cos\beta\ell + jY_r\sin\beta\ell)}. \qquad (5\text{-}14)$$

The equations derived are simple and easy to use if one has a computer or calculator handy. They are not the type of equations that one would wish to memorize since they are readily available in transmission line books and in handbooks. We will now use them to solve a practical type problem.

5.3. Example Using Equations, Phasors, and the Smith Chart

Let us assume that we have a lossless line that is 3.0194 meters long, has a characteristic impedance of 50 ohms, a β of 5 radians per meter, a load impedance of 30 + *j*40 ohms and an input voltage of 100 volts, r.m.s. Since the equations for the voltage and current contain an unknown voltage or current we will first solve for the input impedance, Z_s, so that we can find I_s.

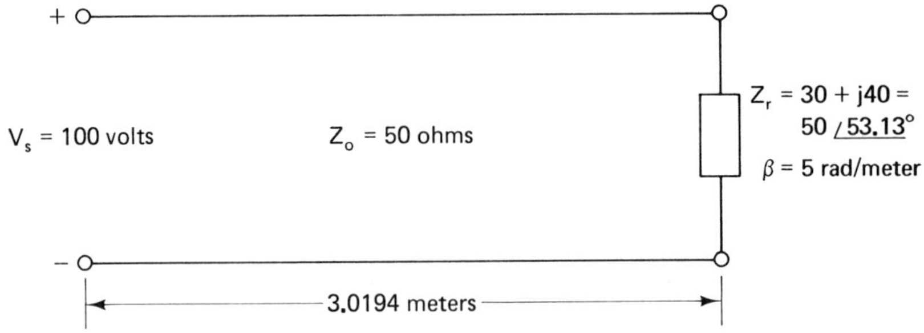

$$Z_s = Z_o\frac{(Z_r\cos\beta\ell + jZ_o\sin\beta\ell)}{(Z_o\cos\beta\ell + jZ_r\sin\beta\ell)} = 50\left[\frac{(50)\underline{/53.13°}(\cos 865°) + j(50)(\sin 865°)}{(50)(\cos 865°) + j(50\underline{/53.13°})(\sin 865°)}\right] =$$

$$18.82\underline{/24.514°} = Z_s = 17.12 + j7.81 \text{ ohms.}$$

$$I_s = \frac{V_s}{Z_s} = \frac{100\underline{/0°}}{18.82\underline{/-24.514°}} = 5.31\underline{/-24.514°} \text{ amperes.}$$

$$V_r = V_s\cos\beta\ell - jI_sZ_o\sin\beta\ell =$$

$$(100\underline{/0°})(\cos 865°) - (1\underline{/90°})(5.31\underline{/-24.514°})(\sin 865°) =$$

$$200.63\underline{/-136.32°} \text{ volts.}$$

$$I_r = \frac{V_r}{Z_r} = \frac{200.63\underline{/-136.32°}}{50\underline{/53.13°}} =$$

$$4.01\underline{/-189.45°} = 4.01\underline{/170.55°} \text{ amperes.}$$

$$P_s = V_sI_s\cos\theta_s = (100)(5.31)\sin 24.514° = 483.14 \text{ watts.}$$

$$P_r = V_rI_r\cos\theta_r = (200.63)(4.01)\cos 53.13° = 483.14 \text{ watts.}$$

$$Q_s = V_s I_s \sin\theta_s = (100)(5.31)\sin 24.514° = 220.32 \text{ vars.}$$
$$Q_r = V_r I_r \sin\theta_r = (200.63)(4.01)\sin 53.13° = 643.62 \text{ vars.}$$

A good check of our numerical work is to look at P_s and P_r which must be equal if the line is lossless. If we use the phasor solution all we need to remember is the equation for Γ_r.

$$\Gamma_r = \frac{Z_r - Z_o}{Z_r + Z_o} = \frac{30 + j40 - 50}{30 + j40 + 50} = \frac{-20 + j40}{80 + j40}$$
$$= \frac{44.72 \angle 116.57°}{89.44 \angle 26.57°} = 0.5 \angle 90° = \frac{V^-}{V^+}.$$

$V^- = 0.5V^+ \angle 90°$ at the load end. $\dfrac{865°}{360°} = 2.402$ revolutions of the phasors but we are only interested in $(.402)(360°) = 145°$. So V^+, as well as the other phasors, goes around two times plus 145 degrees.

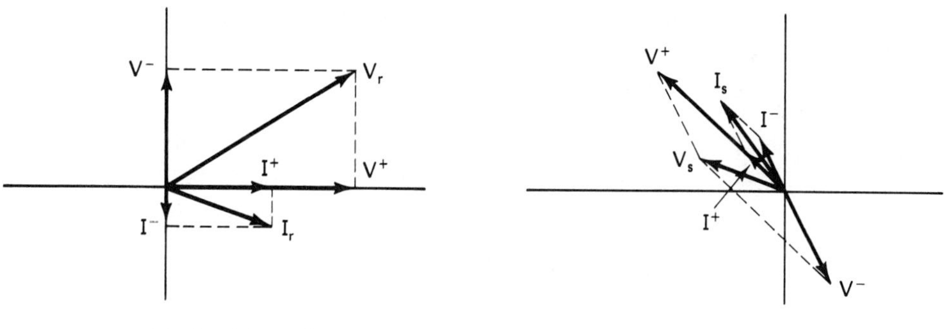

(a) Receiving end (b) Sending end

Figure 5-1. Phasor diagrams for example problem.

From the phasor diagrams we have the following information.

$$V_r = V^+ + V^- = V^+ + j0.5V^+ = V^+(1.118 \angle 26.565°).$$
$$I_r = \frac{V^+}{50} - \frac{V^-}{50} = \frac{V^+}{50} - \frac{j0.5V^+}{50} = \frac{V^+}{50}(1.118 \angle -26.565°).$$
$$V_s = V^+ \angle 145° + 0.5\, V^+ \angle -55° = V^+ (.557 \angle 162.87°).$$

$$|V_s| = |.557V^+| = 100. \qquad |V^+| = \frac{100}{.557} = 179.516 \text{ volts.}$$

$$|V^-| = 0.5|V^+| = 0.5(179.516) = 89.76 \text{ volts.}$$

$$|I^+| = \left|\frac{V^+}{50}\right| = \frac{179.516}{50} = 3.59 \text{ amperes.}$$

$$|I^-| = 0.5|I^+| = 0.5(3.59) = 1.795 \text{ amperes.}$$

So now we may summarize our results.

$$V_s = 100 \angle 162.87^\circ \text{ volts.} \qquad I_s = 3.59 \angle 145^\circ - 1.795 \angle -55^\circ = 5.31 \angle 138.36^\circ \text{ amperes.}$$

$$V_r = (179.516)(1.118 \angle 26.565^\circ) = 200.7 \angle 26.565^\circ \text{ volts.}$$

$$I_r = \frac{179.516}{50}(1.118 \angle -26.565^\circ) = 4.01 \angle -26.565^\circ \text{ amperes.}$$

$$P_s = V_s I_s \cos\theta_s = (100)(5.31)\cos 24.51^\circ = 483.1 \text{ watts.}$$

$$P_r = V_r I_r \cos\theta_r = (200.7)(4.01)\cos 53.13^\circ = 483.1 \text{ watts.}$$

$$Q_s = V_s I_s \sin\theta_s = (100)(5.31)\sin 24.51^\circ = 220.32 \text{ vars.}$$

$$Q_r = V_r I_r \sin\theta_r = (200.7)(4.01)\sin 53.13^\circ = 643.62 \text{ vars.}$$

If we compare the answers for the two different techniques we find that the magnitudes are the same but the angles are different since we used V_s as the reference phasor when using equations and V^+ as the reference when solving by the phasor method.

Figure 5-2 gives the solution for the input impedance, Z_s, using the Smith chart.

$$z_n = \frac{Z_r}{Z_o} = \frac{30 + j40}{50} = 0.6 + j0.8.$$

Drawing our circle and straight line gives a wavelength at the load of 0.125λ. We have to proceed towards the generator a net of 0.402λ or $0.402\lambda + 0.125\lambda = 0.527\lambda$. Since the scale only goes to 0.5λ we have to proceed $0.527\lambda - 0.5\lambda = 0.027\lambda$ beyond the starting point of the scale. Drawing in the straight line gives the normalized value of the impedance at the sending end of $0.345 + j0.16$ so Z_s, the sending end impedance, is $(0.345 + j0.16)(50) = 17.25 + j8$ which compares with the equation's solution value of $17.12 + j7.81$ ohms. The Smith chart shows that $\Gamma_r = 0.5 \angle 90^\circ$ and the additional information that the VSWR = 3.0.

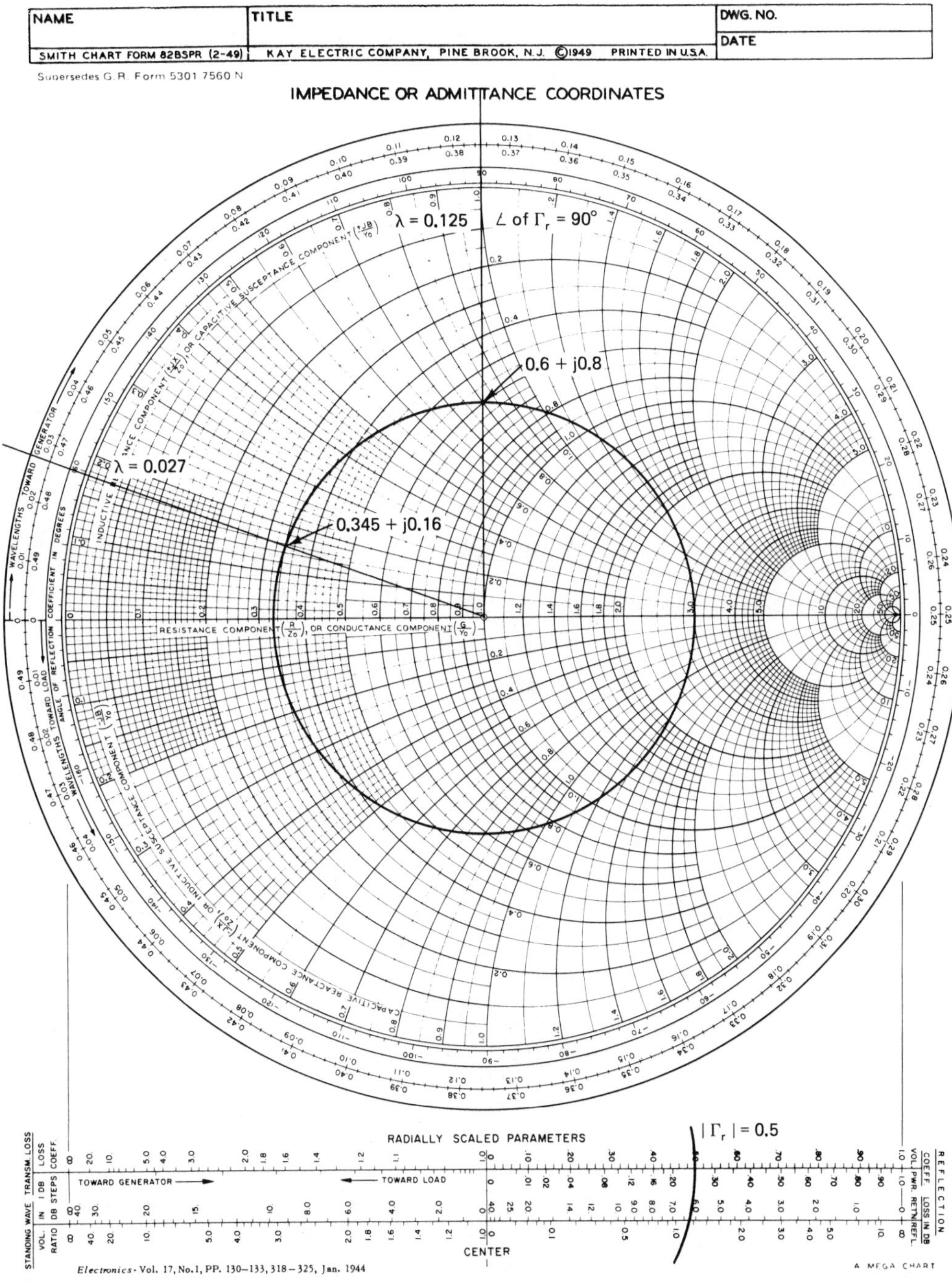

Figure 5-2. Input impedance for example problem.

Both the phasor diagrams and the Smith chart allow us to rapidly find where the voltage maximums and minimums are located with respect to the load (and also with respect to the sending end). The phasor diagram indicates that in this special case we will obtain a voltage maximum when V^+ and V^- each rotate 45 degrees towards each other from their receiving-end positions.

$$V_{max} \text{ when } \beta d = \frac{\pi}{4} \text{ radians} = 5d.$$

$$d = \frac{\pi}{(4)(5)} = 0.157 \text{ meters.}$$

$$V_{max} = V^+ + V^- = 179.516 + 89.76 = 269.28 \text{ volts.}$$

The Smith chart indicates a voltage maximum after we go 0.125 wavelengths since the load is at 0.125 wavelength and the voltage maximum is at 0.25 wavelength. When the phasors rotate another 90 degrees each from the point of a voltage maximum we obtain a voltage minimum at 90 + 45 = 135 degrees.

$$V_{min} \text{ when } \beta d = \frac{3\pi}{4} \text{ radians} = 5d. \qquad d = \frac{3\pi}{(4)(5)} = 0.471 \text{ meters.}$$

5.4 Exponential Form of Solution

Sometimes it is convenient just to leave our equations in exponential form for computational purposes. One of the intermediate equations we had in deriving equation (5-1) was:

$$V = V^+(e^{j\beta d} + \Gamma_r e^{-j\beta d}). \tag{5-15}$$

When deriving the current expression we had:

$$I = \frac{V^+}{Z_o}(e^{j\beta d} - \Gamma_r e^{-j\beta d}). \tag{5-16}$$

When $d = 0$, $V = V_r$ and $I = I_r$.

$$V_r = V^+(1 + \Gamma_r). \tag{5-17}$$

$$I_r = \frac{V^+}{Z_o}(1 - \Gamma_r). \tag{5-18}$$

If we use the V^+ from the voltage equation we obtain:

$$V = \frac{V_r}{1 + \Gamma_r}(e^{j\beta d} + \Gamma_r e^{-j\beta d}). \tag{5-19}$$

$$I = \frac{V_r}{Z_o(1 + \Gamma_r)}(e^{j\beta d} - \Gamma_r e^{-j\beta d}). \tag{5-20}$$

We can solve for the impedance, Z.

$$Z = \frac{V}{I} = \frac{\dfrac{V_r}{1 + \Gamma_r}(e^{j\beta d} + \Gamma_r e^{-j\beta d})}{\dfrac{V_r}{Z_o(1 + \Gamma_r)}(e^{j\beta d} - \Gamma_r e^{-j\beta d})} = Z_o\frac{(e^{j\beta d} + \Gamma_r e^{-j\beta d})}{(e^{j\beta d} - \Gamma_r e^{-j\beta d})}. \tag{5-21}$$

If instead we use the value of V^+ from the current equation (5-18) we will obtain:

$$V = \frac{I_r Z_o}{(1 - \Gamma_r)}(e^{j\beta d} + \Gamma_r e^{-j\beta d}). \tag{5-22}$$

$$I = \frac{I_r}{(1 - \Gamma_r)}(e^{j\beta d} - \Gamma_r e^{-j\beta d}). \tag{5-23}$$

When $d = \ell$, $V = V_s$, $I = I_s$, and $Z = Z_s$.

$$V_s = \frac{V_r}{(1 + \Gamma_r)}(e^{j\beta\ell} + \Gamma_r e^{-j\beta\ell}). \tag{5-24}$$

$$I_s = \frac{V_r}{Z_o(1 + \Gamma_r)}(e^{j\beta\ell} - \Gamma_r e^{-j\beta\ell}). \tag{5-25}$$

$$V_s = \frac{I_r Z_o}{(1 - \Gamma_r)}(e^{j\beta\ell} + \Gamma_r e^{-j\beta\ell}). \tag{5-26}$$

$$I_s = \frac{I_r}{(1 - \Gamma_r)}(e^{j\beta\ell} - \Gamma_r e^{-j\beta\ell}). \tag{5-27}$$

Whether we use the trigonometric or exponential form we must be careful to include the angle for those quantities which are complex. Forgetting the angle makes the work easier but gives the wrong answer.

$$Z_s = Z_o \frac{(e^{j\beta\ell} + \Gamma_r e^{-j\beta\ell})}{(e^{j\beta\ell} - \Gamma_r e^{-j\beta\ell})}. \tag{5-28}$$

5.5 Example of an Exponential Solution

Let us rework the example in Section 5.3 using exponential equations. The pertinent information given in Section 5.3 was as follows: $\ell = 3.0194$ meters; $Z_o = 50$ ohms; $\beta = 5$ radians/meter; $Z_r = 30 + j40$ ohms; $V_s = 100$ volts, r.m.s.

$$\beta\ell = (5)(3.0194) = 15.097 \text{ radians } (865°).$$

$$\Gamma_r = \frac{Z_r - Z_o}{Z_r + Z_o} = \frac{30 + j40 - 50}{30 + j40 + 50} = 0.5 \angle 90°.$$

$$\frac{865°}{360°} = 2.4027 \text{ wavelengths}.$$

$$0.4027(360) = 145°.$$

So for $\beta\ell$ we can use either 865° or 145°.

$$V_r = \frac{V_s (1 + \Gamma_r)}{(e^{j\beta\ell} + \Gamma_r e^{-j\beta\ell})} = \frac{(100 \angle 0°)(1 + j0.5)}{1 \angle 45° + (.5 \angle 90°)(1 \angle -145°)} = 200.7 \angle -136.312° \text{ volts}.$$

$$I_r = \frac{V_r}{Z_r} = \frac{200.7 \angle -136.31°}{30 + j40} = 4.014 \angle -189.44° \text{ amperes}.$$

$$Z_s = Z_o \frac{(e^{j\beta\ell} + \Gamma_r e^{-j\beta\ell})}{(e^{j\beta\ell} - \Gamma_r e^{-j\beta\ell})} = 50 \frac{1 \angle 145° + (.5 \angle 90°)(1 \angle -145°)}{1 \angle 145° - (.5 \angle 90°)(1 \angle -145°)} =$$

$$Z_s = 18.82 \angle 24.514° = 17.125 + j7.809 \text{ ohms}.$$

$$I_s = \frac{V_s}{Z_s} = \frac{100 \angle 0°}{18.82 \angle 24.514°} = 5.313 \angle -24.574° \text{ amperes}.$$

There is not much choice between the two techniques when using equations. You should use the one you are more comfortable with and which seems the easiest for you. The phasor solution, as before, offers the possibility of more information, such as the location of the voltage maximums and voltage minimums without resorting to equations.

5.6 Other Forms for the Impedance and the Admittance Equations

Before leaving the lossless case we can convert the equations for Z and Y to other forms which are often found in the literature. Let us start with equation (5-6).

$$Z = Z_o\frac{(Z_r\cos\beta d + jZ_o\sin\beta d)}{(Z_o\cos\beta d + jZ_r\sin\beta d)} = Z_o\frac{\left[Z_r + jZ_o\dfrac{\sin\beta d}{\cos\beta d}\right]}{\left[Z_o + jZ_r\dfrac{\sin\beta d}{\cos\beta d}\right]} =$$

$$Z = Z_o\frac{(Z_r + jZ_o\tan\beta d)}{(Z_o + jZ_r\tan\beta d)}. \qquad (5\text{-}29)$$

This can also be written in a slightly different form.

$$\frac{Z}{Z_o} = \frac{\dfrac{Z_r}{Z_o} + j\tan\beta d}{1 + \dfrac{jZ_r}{Z_o}\tan\beta d}. \qquad (5\text{-}30)$$

Here Z/Z_o and Z_r/Z_o are normalized or per-unit values of Z and Z_r.

If we start with equation (5-14), and go through the same type of conversion techniques, we will end up with some new forms for the admittance.

$$Y = Y_o\frac{(Y_r\cos\beta d + jY_o\sin\beta d)}{(Y_o\cos\beta d + jY_r\sin\beta d)} = Y_o\frac{(Y_r + jY_o\tan\beta d)}{(Y_o + jY_r\tan\beta d)}. \qquad (5\text{-}31)$$

$$\frac{Y}{Y_o} = \frac{\dfrac{Y_r}{Y_o} + j\tan\beta d}{1 + \dfrac{jY_r}{Y_o}\tan\beta d}. \qquad (5\text{-}32)$$

Now Y/Y_o and Y_r/Y_o are normalized or per-unit values of Y and Y_r. We can also write when $d = \ell$.

$$Z_s = Z_o\frac{(Z_r + jZ_o\tan\beta\ell)}{(Z_o + jZ_r\tan\beta\ell)}. \qquad (5\text{-}33)$$

$$Y_s = Y_o\frac{(Y_r + jY_o\tan\beta\ell)}{(Y_o + jY_r\tan\beta\ell)}. \qquad (5\text{-}34)$$

Let us look at another example of calculations for the lossless line case.

Example

Given a lossless transmission line with the following parameters: $Z_o = 50$ ohms; $Z_r = 37.5 + j25$ ohms; $\beta = 1.5$ radians/meter; length of line = 4.89 meters; $V_s = 56$ volts, r.m.s.

Find:

(a) V_r, (b) I_r, (c) I_s, (d) Power to the load, (e) Input power factor.

Solution

$$\beta\ell = (1.5)(4.89) = 7.335 \text{ radians.}$$

$$\frac{7.335 \text{ radians}}{2\pi \text{ radians/wavelength}} = 1.167 \text{ wavelengths.}$$

In terms of our calculations, 1.167 wavelengths is the same as 0.167 wavelengths since everything repeats itself every half-wavelength.

$$0.167(360°) = 60.26°.$$

$$Z_s = Z_o\frac{(Z_r\cos\beta\ell + jZ_o\sin\beta\ell)}{(Z_o\cos\beta\ell + jZ_r\sin\beta\ell)} =$$

$$50\left[\frac{(37.5 + j25)\cos 60.26° + j(50)\sin 60.26°}{50\cos 60.26° + j(37.5 + j25)\sin 60.26°}\right] =$$

$$89.33\angle{-12.98°} = 87.04 - j20 \text{ ohms.}$$

$$I_s = \frac{V_s}{Z_s} = \frac{56\angle 0°}{89.33\angle{-12.98°}} = 0.626\angle 12.98° \text{ amperes.}$$

$V_r = V_s\cos\beta\ell - jI_sZ_o\sin\beta\ell = (56\,\underline{/0^\circ})\cos 60.26^\circ - j(.626\,\underline{/12.98^\circ})(50)\sin 60.26^\circ = 43.07\,\underline{/-37.87^\circ}$ volts.

$$I_r = \frac{V_r}{Z_r} = \frac{43.07\,\underline{/-37.87^\circ}}{45.07\,\underline{/33.69^\circ}} = 0.95\,\underline{/-76.06^\circ} \text{ amperes.}$$

$P_r = V_rI_r\cos\theta_r = (43.07)(.95)\cos 33.69^\circ = 34.04$ watts.

Power factor = P.F. = $\cos 12.98^\circ = 0.97$.

5.7 Single-Stub Matching

We will find out later that when we have standing waves on the line we have additional losses and, of course, we do not have maximum power transfer to the load. One way to improve the situation at high frequencies is to use a short-circuited stub tuner in parallel with the line at a particular point on the line. This particular point is where the real component of the impedance, looking toward the load, is equal to the characteristic impedance of the line. If the stub tuner can cancel out the reactance at this point, the line will be matched at the stub tuner location to the characteristic impedance of the line. Since admittances in parallel add, it is more convenient to deal with admittances than impedances when we are talking about stub tuner matching. Let us repeat equation (5-32).

$$\frac{Y}{Y_o} = \frac{\dfrac{Y_r}{Y_o} + j\tan\beta d}{1 + \dfrac{jY_r}{Y_o}\tan\beta d} \tag{5-35}$$

Let

$$\frac{Y_r}{Y_o} = G + jB$$

$$\frac{Y}{Y_o} = \frac{G + jB + j\tan\beta d}{1 + j(G + jB)\tan\beta d} = \frac{G + j(B + \tan\beta d)}{1 - B\tan\beta d + jG\tan\beta d}.$$

$$\frac{Y}{Y_o} = \frac{G + j(B + \tan\beta d)}{(1 - B\tan\beta d) + jG\tan\beta d} \times \frac{(1 - B\tan\beta d) - jG\tan\beta d}{(1 - B\tan\beta d) - jG\tan\beta d} =$$

$$\frac{G(1 - B\tan\beta d) + (G\tan\beta d)(B + \tan\beta d)}{(1 - B\tan\beta d)^2 + (G\tan\beta d)^2} + j\frac{(\quad)}{(\quad)}.$$

The real term is the term we are interested in.

$$\text{The real term} = \frac{G - GB\tan\beta d + GB\tan\beta d + G\tan^2\beta d}{(1 - B\tan\beta d)^2 + (G\tan\beta d)^2} = \frac{G(1 + \tan^2\beta d)}{(1 - B\tan\beta d)^2 + (G\tan\beta d)^2}.$$

Now when $\frac{Y}{Y_o} = G' + jB' = \text{real} + j(\text{imaginary}) = 1 + jB'$ we can place in parallel the admittance, $Y/Y_o = -jB'$.

At the location of the stub tuner the total normalized admittance will be:

$$\frac{Y'}{Y_o} = 1 + jB' - jB' = 1 + j0 \text{ and } Y' = Y_o \text{ or } Z' = Z_o.$$

When we set the real term equal to unity we can find the point on the line where we wish to place the stub tuner in parallel with the line.

$$\frac{G(1 + \tan^2\beta d)}{(1 - B\tan\beta d)^2 + (G\tan\beta d)^2} = 1.$$

The algebra that follows is very simple.

$$\tan^2\beta d - \frac{2B}{B^2 + G^2 - G}\tan\beta d + \frac{1 - G}{B^2 + G^2 - G} = 0.$$

$$\tan\beta d = \frac{\frac{+2B}{B^2 + G^2 - G} \pm \sqrt{\frac{4B^2}{(B^2 + G^2 - G)^2} - \frac{4(1 - G)}{(B^2 + G^2 - G)}}}{2}.$$

But $G^2 + B^2 = |Y|^2$, so

$$\tan\beta d = \frac{\frac{2B}{|Y|^2 - G} \pm \sqrt{\frac{4B^2}{(|Y|^2 - G)^2} - \frac{4(1 - G)(|Y|^2 - G)}{(|Y|^2 - G)^2}}}{2}.$$

$$\tan\beta d = \frac{2B}{2(|Y|^2 - G)} \pm \frac{2}{2}\sqrt{\frac{B^2}{(|Y|^2 - G)^2} - \frac{(1 - G)(|Y|^2 - G)}{(|Y|^2 - G)^2}}.$$

$$\tan\beta d = \frac{B \pm \sqrt{B^2 - |Y|^2 + G + G|Y|^2 - G^2}}{|Y|^2 - G}.$$

We will add and subtract G^2 under the square root sign.

$$\tan\beta d = \frac{B \pm \sqrt{B^2 - |Y|^2 + G + G|Y|^2 - G^2 + G^2 - G^2}}{|Y|^2 - G}.$$

$$\tan\beta d = \frac{B \pm \sqrt{G^2 + B^2 - |Y|^2 + G + G|Y|^2 - 2G^2}}{|Y|^2 - G}.$$

$$\tan\beta d = \frac{B \pm \sqrt{|Y|^2 - |Y|^2 + G + G|Y|^2 - 2G^2}}{|Y|^2 - G}.$$

$$\tan\beta d = \frac{B \pm \sqrt{G + G|Y|^2 - 2G^2}}{|Y|^2 - G} = \frac{B \pm \sqrt{G(1 + |Y|^2 - 2G^2)}}{|Y|^2 - G}.$$

$$\beta d = \tan^{-1}\frac{B \pm \sqrt{G(1 + |Y|^2 - 2G)}}{|Y|^2 - G}.$$

$$d = \frac{1}{\beta}\tan^{-1}\frac{B \pm \sqrt{G(1 + |Y|^2 - 2G)}}{|Y|^2 - G} \pm \frac{n\lambda}{2}. \qquad (5\text{-}36)$$

$n = 0, 1, 2,$ etc.

The $n\lambda/2$ term, with n equal to a positive integer, may be added since the condition where the real term is equal to 1 will occur every half wavelength down the line.

Equation (5-36) gives the location of the stub tuner. Now we would like to find out the magnitude of the susceptance term to cancel out the j term.

$$\frac{Y}{Y_o} = 1 + jA = \frac{G + jB + j\tan\beta d}{1 + j(G + jB)\tan\beta d}.$$

$$(1 + jA)(1 - B\tan\beta d + jG\tan\beta d) = G + j(B + \tan\beta d).$$

$$1 - B\tan\beta d + jG\tan\beta d + jA - jAB\tan\beta d - AG\tan\beta d = G + j(B + \tan\beta d).$$

$$1 - B\tan\beta d - AG\tan\beta d + j(G\tan\beta d + A - AB\tan\beta d) = G + j(B + \tan\beta d).$$

Equating the real and the j terms gives:

$$1 - B\tan\beta d - AG\tan\beta d = G. \qquad (5\text{-}37)$$

$$G\tan\beta d + A - AB\tan\beta d = B + \tan\beta d. \qquad (5\text{-}38)$$

From equation (5-37) we have:

$$1 - \tan\beta d(B + AG) = G \text{ or } \tan\beta d = \frac{1 - G}{B + AG} \qquad (5\text{-}39)$$

From equation (5-38) we have:

$$(\tan\beta d)(G - AB - 1) = B - A \text{ or}$$

$$\tan\beta d = \frac{B - A}{G - AB - 1}. \qquad (5\text{-}40)$$

Equating equations (5-39) and (5-40) we obtain:

$$\frac{B - A}{G - AB - 1} = \frac{1 - G}{B + AG}.$$

$$(B - A)(B + AG) = (1 - G)(G - AB - 1).$$

$$B^2 + BAG - AB - A^2G = G - AB - 1 - G^2 + GAB + G.$$

$$B^2 + G^2 + 1 - 2G = A^2G \text{ or } |Y|^2 + 1 - 2G = A^2G.$$

$$A^2 = \frac{|Y|^2 + 1 - 2G}{G}. \qquad A = \pm\sqrt{\frac{|Y|^2 + 1 - 2G}{G}}. \qquad (5\text{-}41)$$

Actually, once we know the distance to the point where the stub tuner will be put in parallel with the line, it is usually more convenient to substitute back into equation (5-35) to find the j term to be cancelled. This also serves as a check on our work since the real term should come out to be equal to one.

To find the length of the stub tuner we go back to equation (2-14).

$$Z_s = jZ_o\tan\beta\ell. \qquad \frac{Z_s}{Z_o} = j\tan\beta\ell \text{ or } \frac{Y_o}{Y_s} = j\tan\beta\ell.$$

$$\frac{Y_s}{Y_o} = \frac{1}{j\tan\beta\ell} = -j\cot\beta\ell = \frac{-j}{\tan\beta\ell}.$$

$$\frac{|Y_s|}{|Y_o|} = \cot\beta\ell. \qquad \beta\ell = \cot^{-1}\frac{|Y_s|}{|Y_o|}$$

$$\ell = \frac{1}{\beta}\cot^{-1}\frac{|Y_s|}{|Y_o|} \tag{5-42}$$

Let us look at an example of using a stub tuner on a line.

Example

We will assume a load impedance of 100 + j50 ohms and a characteristic impedance of 50 ohms.

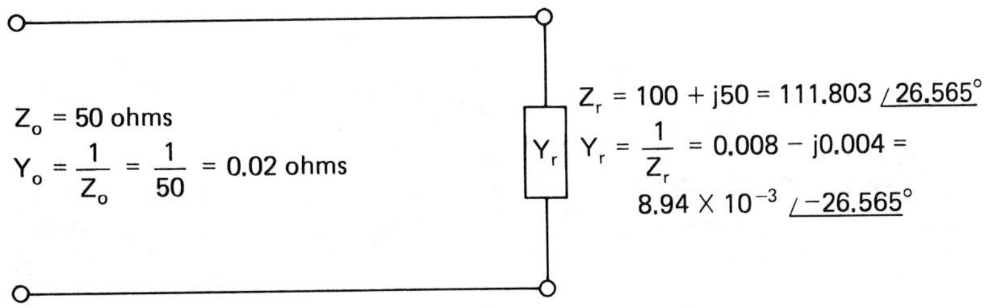

$$\frac{Y_r}{Y_o} = \frac{0.008 - j0.004}{0.02} = 0.4 - j0.2$$

$$= G + jB. \qquad G = 0.4,\ B = -0.2.$$

$$\beta d = \tan^{-1}\frac{B \pm \sqrt{G(1 + |Y|^2 - 2G)}}{|Y|^2 - G}.$$

$$Y = 0.4 - j0.2 = 0.447 \angle -26.565°$$

$$Y^2 = 0.447^2 = 0.2$$

$$\beta d = \tan^{-1}\frac{-0.2 \pm \sqrt{0.4(1 + 0.2 - 2(0.4))}}{0.2 - 0.4} = \tan^{-1}\frac{-0.2 \pm \sqrt{0.4(0.4)}}{-0.2} =$$

$$\tan^{-1}\frac{0.2 \pm 0.4}{-0.2}. \qquad \beta d = \tan^{-1}\frac{-0.2 - 0.4}{-0.2} = \tan^{-1} 3.0.$$

$$\beta d = 71.57° \text{ or } -108.43°.$$

Or when we take the + sign we have:

$$\beta d = \tan^{-1} \frac{-0.2 + 0.4}{-0.2} = \tan^{-1} -1.0. \qquad \beta d = -45° \text{ or } 135°.$$

Picking only the positive values of βd (why?) gives 71.57° and 135° which would indicate that there are two positions where the real part of the normalized admittance is equal to unity. Let us take the angle, 71.57°, first.

$$\frac{Y}{Y_o} = \frac{\frac{Y_r}{Y_o} + j\tan\beta d}{1 + \frac{jY_r}{Y_o}\tan\beta d} = \frac{0.4 - j0.2 + j\tan 71.57°}{1 + j(0.4 - j0.2)\tan 71.57°} =$$

$$\frac{0.4 - j0.2 + j3}{1 + j(0.4 - j0.2)3} = \frac{0.4 + j2.8}{1.6 + j1.2} = 1 + j1.$$

In order to cancel out the j term we will need an input admittance of short-circuited stub equal to $-j1$.

$$-j1 = \frac{-j}{\tan\beta\ell} = \frac{Y_s}{Y_o} \text{ or } 1 = \frac{1}{\tan\beta\ell}. \qquad \tan\beta\ell = 1.0.$$

Therefore:

$$\beta\ell = \frac{\pi}{4}(45°), \frac{5\pi}{4}(225°), \text{ etc.}$$

This means that if we pick a position for the stub tuner that is 71.57° or 71.57/360 = 0.199 wavelength from the load that the length of the short-circuited stub tuner should be 45/360 = 0.125 wavelength.

If we pick the 135° position, we would have:

$$\frac{Y}{Y_o} = \frac{\frac{Y_r}{Y_o} + j\tan\beta d}{1 + \frac{jY_r}{Y_o}\tan\beta d} = \frac{0.4 - j0.2 + j\tan 135°}{1 + j(0.4 - j0.2)\tan 135°} =$$

$$\frac{0.4 - j0.2 - j1}{1 + j(0.4 - j0.2)(-1)} = \frac{0.4 - j1.2}{0.8 - j0.4} = 1 - j1.$$

In order to cancel out the j term we will now need an input admittance of the short-circuited stub equal to $+j1$.

$$\frac{Y_s}{Y_o} = +j1 = \frac{-j}{\tan\beta\ell}. \qquad 1 = \frac{-1}{\tan\beta\ell}. \qquad \tan\beta\ell = -1.$$

So $\beta\ell$ is now equivalent to 135°, 315°, etc.

It turns out that there are two primary points where we can place a stub. And, of course, we can place them every half wavelength from the two primary points. It is easier to express our answers in terms of wavelength.

Position 1: $\beta d = 71.57°$ or 0.198λ. Stub length 45° or 0.125λ.

Position 2: $\beta d = 135°$ or 0.375λ. Stub length 135° or 0.375λ.

Let us go back and use equation (5-41) to find the value of the j term when the real term is equal to one.

$$A = \pm\sqrt{\frac{Y^2 + 1 - 2G}{G}} = \pm\sqrt{\frac{0.2 + 1 - 2(0.4)}{0.4}} =$$

$$\pm\sqrt{\frac{0.4}{0.4}} = \pm 1.0.$$

The trouble with using equation (5-41) is that we don't know which value of A goes with which position. We could end up with the wrong stub length for position 1 or position 2. If we use the Smith chart to solve the same problem we have no such difficulty. Let us solve the same problem using the Smith chart.

5.8 Single-Stub Tuning Using the Smith Chart

Let us repeat the example given in Section 5.7 using the Smith chart. We will enter the Smith chart at the normalized admittance $0.4 - j0.2$ and draw the usual circle through this point with the center of the chart as the center of the circle. The construction will be shown on Figure 5-3. We read, after drawing the usual straight line through the center and the point $0.4 - j0.2$, a wavelength of 0.463 on the outer scale (wavelengths toward generator). We proceed in a clockwise direction around the circle until we reach the point $1 + jB$, where B is to be determined. In our case, it just happens to turn out that B also equals 1. We now draw a straight line through the center and this point $(1 + j1)$ and read on the wavelength scale that the wavelength, λ, is 0.16. Thus we have moved from the load towards the generator a total of $0.50 - 0.463 = 0.037 + 0.16 = 0.197$ wavelength. The equation solution gave an answer of 0.198 wavelength.

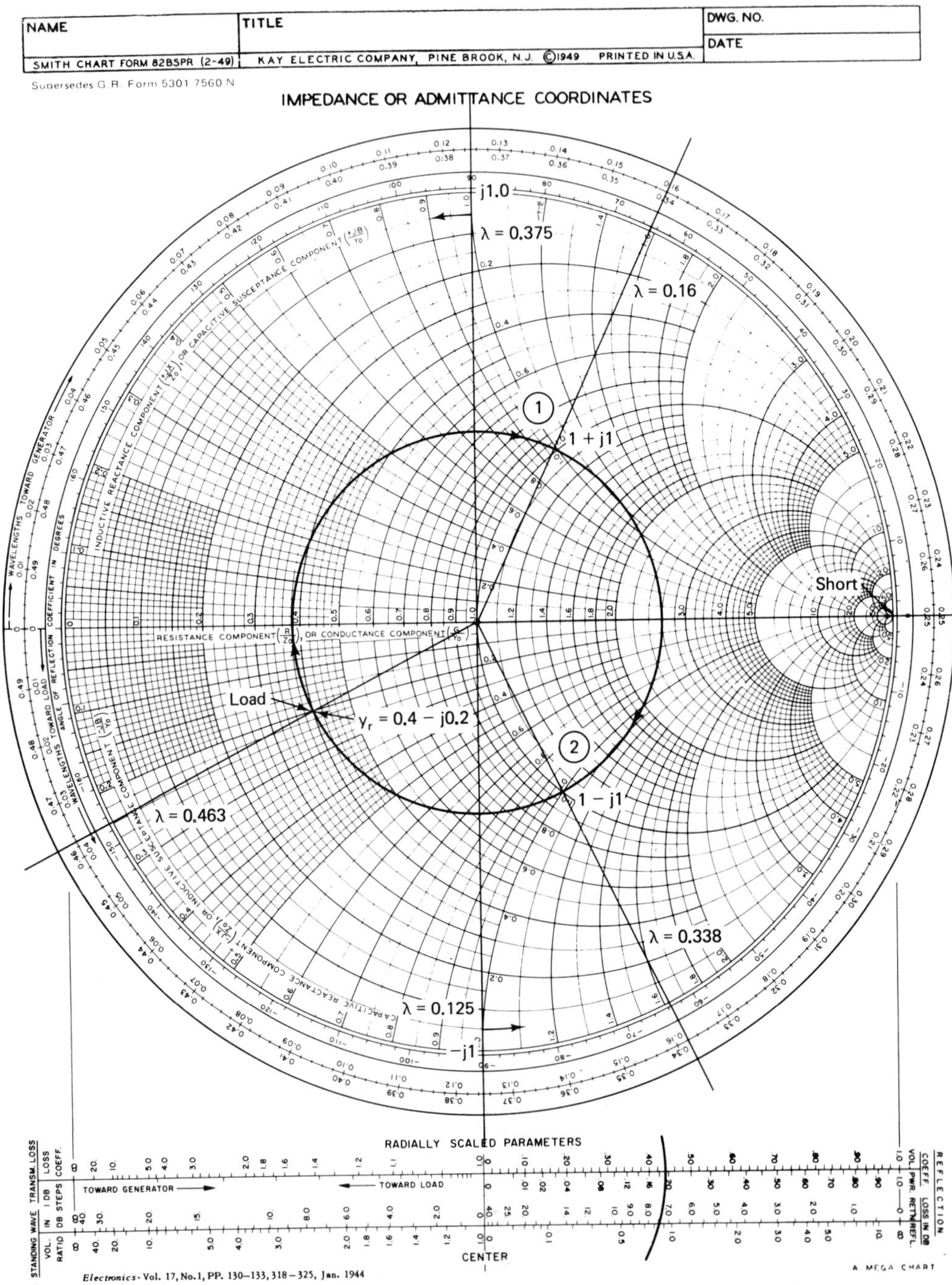

Figure 5-3. Stub tuning construction.

If we continue around the circle from the $1 + jB$ point we find that the circle hits the $1 - jB$ point. In other words, our circle hits the $g = 1$ circle twice. Again in our example the point is $1 - j1$ where we read a wavelength of 0.338 after we draw a straight line through the center and the point $1 - j1$. Now we have moved from the load a total distance of $0.5 - 0.463 = 0.037 + 0.338 = 0.375$ wavelength. The equation solution gave an answer of 0.375 wavelength also.

Now we can use the Smith chart to determine the length of the short-circuit stub tuner. For position 1 we had an admittance looking towards the load of $1 + j1$ in normalized form. To cancel the B term we need an input admittance of $0 - j1$ for the stub tuner. So we go to $0 - j1$ on the Smith chart and draw a straight line through it and the center of the chart and read on the inner scale(wavelength toward load) 0.125 wavelength. Now we proceed to the load, which is a short circuit and represented on the Smith chart as $Y = \infty + j\infty$. Here the reading on the wavelength scale is 0.25 wavelength so we have proceeded from the input of the tuner to the short a distance of $0.25 - 0.125 = 0.125$ wavelength. This is the same answer we obtained by using equations.

For position 2 we need an input admittance of $0 + j1$ so we go to the $0 + j1$ position on the Smith chart, draw our straight line and read that the wavelength is 0.375. Again we proceed in the counterclockwise direction to the short-circuit load a total distance of $0.50 - 0.375 = 0.125 + 0.25 = 0.375$ wavelength, and the stub tuner would have to be 0.375 wavelength long if we placed it in the second position. This is the same answer we obtained by using equations. However when we use the Smith chart there is no question as to where the different stub lengths should be placed. The results of our calculations or our use of the Smith chart are shown in Figure 5-4.

We can place the stub tuner at either of two positions or we could place them one half wavelength down the line from either of these two positions. After we have placed a stub tuner at either of the two positions, the admittance that the line sees just to the left of the tuner is normalized form would be:

$$\frac{Y}{Y_o} = 1 + j0.$$

So,

$$Y = Y_o \text{ and } Z = Z_o.$$

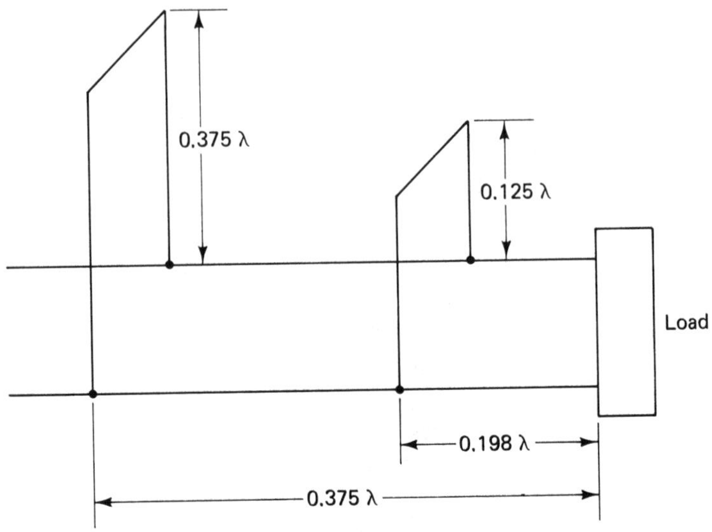

Figure 5-4. Stub tuner matching positions.

The impedance looking towards the stub tuner becomes the characteristic impedance. We have eliminated the standing waves from the point of the stub tuner to the sending end of the line. Our VSWR is now 1.0 for this portion of the line.

Single-stub tuners are easy to use and the calculations or the use of the Smith chart are straightforward. They do have one disadvantage and that is that their location is fixed with respect to the distance from the load. The location may not be realizable from a physical standpoint. We may find that we need to place the tuner in the middle of a wall. Double-stub tuners do not have this restriction. A double-stub tuner is simply two short-circuited stub tuners separated by a fixed distance which is a small fraction of a wavelength (approximately 3/8 λ). They can be placed anywhere, which makes them convenient to use with coaxial cables since it is difficult to move a single-stub tuner around on a coaxial cable. Double-stub tuners can be analyzed by use of the Smith chart, but often it is easier in the practical case to set the lengths of the two stub tuners by trial and error techniques.

5.9 Standing Wave Patterns

Before leaving the lossless case let us look in detail at the patterns we obtain when we plot the magnitude of the voltage on a transmission line. Our previous sketches tended to imply that the patterns are sinusoidal in form, but this is not necessarily true. Let us look first at the short-circuit load situation (See Figure 5-5).

$$\Gamma_r = \frac{Z_r - Z_o}{Z_r + Z_o} = \frac{0 - Z_o}{0 + Z_o} = -1 = \frac{V^-}{V^+}.$$

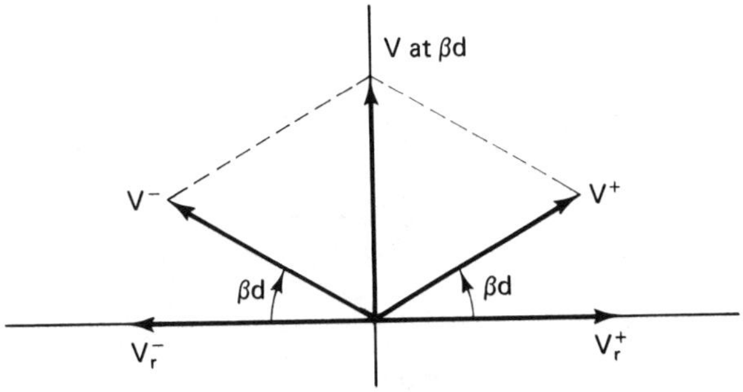

Figure 5-5. Short-circuit load.

We can write the expression for the voltage at any point on the line in terms of the angle, βd, radians, that we have moved from the load.

$$V = V^+ \underline{/\beta d} + V^- \underline{/\pi\text{-}\beta d}.$$

Let $\beta d = \theta$ degrees and convert to degrees.

$$V = V^+ \underline{/\theta} + V^- \underline{/180^\circ - \theta} = V^+ [1 \underline{/\theta} + \underline{/180^\circ - \theta}] =$$

$$V^+ [\cos\theta + j\sin\theta + \cos 180^\circ \cos\theta + \sin 180^\circ \sin\theta + j(\sin 180^\circ \cos\theta - \cos 180^\circ \sin\theta)] =$$

$$V^+ [\cos\theta + j\sin\theta - \cos\theta + 0 + j(0 - (-1)\sin\theta)] =$$

$$V^+ (2j\sin\theta).$$

$$|V| = |V^+ 2j\sin\theta| = |2V^+ \sin\theta| = |2V^+ \sin\beta d|. \qquad (5\text{-}43)$$

When we plot the magnitude of the voltage as a function of the distance from the load we do, indeed, have a sinusoidal distribution.

If we repeat our work for an open-circuit load we would end up with equation (5-44).

$$|V| = |2V^+ \cos\theta|. \qquad (5\text{-}44)$$

In both of these cases the magnitude of V^- equals the magnitude of V^+ and we obtain a sinusoidal (or cosinusoidal) distribution of the magnitude of the voltage on the line. However let us see what happens when V^- and V^+ do not have the same magnitude. Let us assume a condition as shown in Figure 5-6.

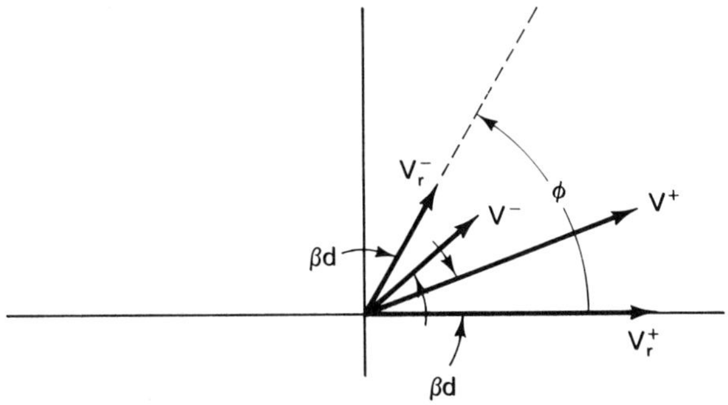

Figure 5-6. General case.

Let $V^- = V^+ \Gamma_r \angle\phi$ so that as we proceed down the line we can write:

$$V = V^+ \angle\beta d + |\Gamma_r| V^+ \angle\theta d - \beta d.$$

$$V = V^+ (\cos\beta d + j\sin\beta d) + |\Gamma_r|\, V^+ [\cos(\phi - \beta d) + j\sin(\phi - \beta d)].$$

$$V = V^+ [\cos\beta d[1 + |\Gamma_r| \cos\phi] + |\Gamma_r| \sin\phi\sin\beta d] + jV^+ [\sin\beta d(1 - |\Gamma_r|\cos\phi) + |\Gamma_r| \sin\phi\cos\beta d]. \quad (5\text{-}45)$$

There is no way to break this equation down into a simple form of a sine or cosine wave. It will be easier to use a numerical example, so let us assume that:

$$\Gamma_r = 0.5 \angle 30^\circ.$$

$$V = V^+[\cos\beta d(1 + 0.5\cos 30^\circ) + |0.5| \sin 30^\circ \sin\beta d] + jV^+ [\sin\beta d(1 - 0.5\cos 30^\circ) + |0.5| \sin 30^\circ \cos\beta d]$$

$$V = V^+[\cos\beta d(1.433) + 0.25\sin\beta d] + jV^+ [\sin\beta d(0.567) + 0.25 \cos\beta d] = V^+ [1.433 \cos\beta d + 0.25 \sin\beta d] + jV^+ [0.567 \sin\beta d + 0.25 \cos\beta d].$$

When we let $V^+ = 10$ volts we obtain the curve indicted in Figure 5-7. To compare this with a sinusoidal wave shape, we can develop an equation which has the same maximum and minimum values as $V = 10 + 5\cos 2(15 - \theta)$, where the angle is in degrees.

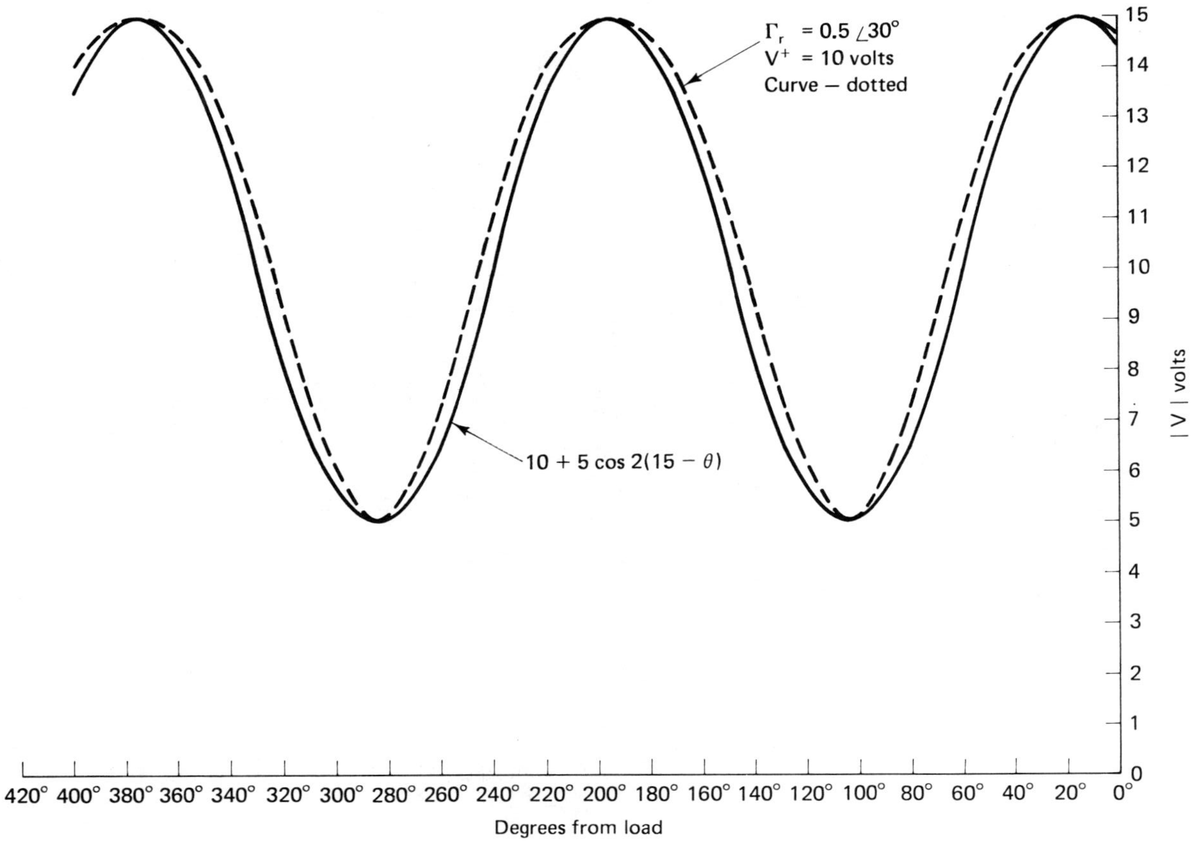

Figure 5-7. Comparison of standing wave patterns.

This is also plotted in Figure 5-7. The pattern from the line is seen to be close to a sine wave in form but it is broadened in the region of the maximum value and narrowed in the region of the minimum value. The minimum is sharper than the maximum and it is for this reason that measurements made in the laboratory are usually made at the voltage minimums since they are more sharply defined and a more exact position can be determined as the minimum.

5.10. Chapter Summary

1. The equations for the lossless line operating in the steady-state, a.c. case are easy to use and need not be memorized, since they are always available.
2. A summary of the lossless equations is as follows:

(5-4) $$V = V_r\cos\beta d + jI_rZ_o\sin\beta d.$$

(5-5) $$I = I_r\cos\beta d + \frac{jV_r}{Z_o}\sin\beta d.$$

(5-6) $$Z = Z_o\frac{(Z_r\cos\beta d + jZ_o\sin\beta d)}{(Z_o\cos\beta d + jZ_r\sin\beta d)}.$$

(5-8) $$V_s = V_r\cos\beta\ell + jI_rZ_o\sin\beta\ell.$$

(5-9) $$I_s = I_r\cos\beta\ell + \frac{jV_r}{Z_o}\sin\beta\ell.$$

(5-10) $$Z_s = Z_o\frac{(Z_r\cos\beta\ell + jZ_o\sin\beta\ell)}{(Z_o\cos\beta\ell + jZ_r\sin\beta\ell)}.$$

(5-11) $$V_r = V_s\cos\beta\ell - jI_sZ_o\sin\beta\ell.$$

(5-12) $$I_r = I_s\cos\beta\ell - \frac{jV_s}{Z_o}\sin\beta\ell.$$

(5-14) $$Y_s = Y_o\frac{(Y_r\cos\beta\ell + jY_o\sin\beta\ell)}{(Y_o\cos\beta\ell + jY_r\sin\beta\ell)}.$$

(3-9) $$\Gamma_r = \frac{Z_r - Z_o}{Z_r + Z_o} = \frac{V^-}{V^+}.$$

(5-19) $$V = \frac{V_r}{1 + \Gamma_r}(e^{j\beta d} + \Gamma_r e^{-j\beta d}).$$

(5-20) $$I = \frac{V_r}{Z_o(1 + \Gamma_r)}(e^{j\beta d} - \Gamma_r e^{-j\beta d}).$$

(5-21) $$Z = Z_o\frac{(e^{j\beta d} + \Gamma_r e^{-j\beta d})}{(e^{j\beta d} - \Gamma_r e^{-j\beta d})}.$$

3. For a single stub short-circuited tuner with

$$\frac{Y_r}{Y_o} = G + jB \text{ and } |Y|^2 = G^2 + B^2,$$

the distance (d) to the stub location is:

(5-36) $$d = \frac{1}{\beta}\tan^{-1}\frac{B \pm \sqrt{G(1 + |Y|^2 - 2G)}}{|Y|^2 - G} \pm \frac{n\lambda}{2}$$

At the position d,

(5-41) $$\frac{Y}{Y_o} = 1 + jA \text{ where } A = \pm\sqrt{\frac{|Y|^2 + 1 - 2G}{G}}.$$

4. A short-circuited, stub tuner may be placed on the line at a particular location and, if it has the proper input admittance, the voltage standing wave ratio may be improved to unity on that part of the line which is left of the tuner since from that point on the load will match the characteristic impedance of the line. While determining the location of the stub tuner and its length may be done with equations it is most easily obtained by using a Smith chart.
5. Whenever the load on the line is greater than zero ohms and less than infinite ohms, the pattern of the standing wave is not sinusoidal but instead the maximum is broadened and the minimum is narrowed. Thus, if one is finding the location of a voltage minimum or voltage maximum it is easier to determine the location of the minimum. If one is finding the wavelength on the line, the accuracy will be better if the measurement for one half wavelength is obtained by measuring the distance between two voltage minimums if at all possible.

5.11 Review and Study Questions

1. Why are the equations for a lossless line of concern to the power line engineer?
2. What transmission line parameter is zero for a lossless line?
3. How do we change the general equations for a lossless line to the equations relating the sending and receiving end quantities?
4. What do we know about the relationships between the following quantities on a lossless line?
   (a) The sending and receiving end power.
   (b) The sending and receiving end reactive power.
5. What quantities on a line repeat themselves:
   (a) Every half wavelength?
   (b) Every wavelength?
6. What is the purpose of a short-circuited stub tuner?
7. If a stub tuner is properly used on a lossless transmission line what is the effect on the VSWR
   (a) Between the load and the tuner?
   (b) Between the tuner and the sending end?
8. What is the input impedance of a lossless line if a stub tuner is properly applied to the line?
9. Sketch the shape of the standing wave on a lossless transmission line terminated in a load
   (a) Equal to the characteristic impedance.
   (b) With zero impedance.
   (c) With infinite impedance.
   (d) That is not zero, infinite or equal to Z_o.
   How does each pattern compare to a sinusoidal type of waveshape?
10. What is the sending end impedance when one uses equation (5-10) if
    (a) The load is a short circuit?
    (b) The load is an open circuit?
11. Express e^{jx} in terms of sines and cosines.
12. Express the sine of x in terms of exponentials.
13. How do the magnitudes of your answers change if you change from a phasor solution to a solution using equations?
14. In placing a stub tuner on a line which procedure would you rather use?
    (a) Equations.
    (b) Smith charts.

15. The input power always equals the load power on a lossless line. Explain why it is possible for the input reactive power to be greater than the load reactive power.
16. What happens to the characteristic impedance of a lossless line if the frequency is doubled?
17. Explain how you would solve the lossless case if you were given the magnitude of the input voltage and the receiving end current along with the length of the line and its parameters.

Problems

5-1. A lossless line has a Z_o of 600 ohms and is terminated in an open circuit. It is 2.35 wavelengths long and has an input voltage of 100 volts, r.m.s. Using V_s as the reference determine, using equations:
(a) Input current.
(b) Receiving end voltage.
(c) Input power.
(d) Load current.

5-2. A lossless line is 7.2 wavelengths long, has a characteristic impedance of 100 ohms and a load impedance of $60 + j80$ ohms. If the load voltage is 100 volts, r.m.s., determine, using equations:
(a) Load current.
(b) Sending end voltage.
(c) Sending end current.
(d) Power to the load.

5-3. Repeat problem 5-2 if the input voltage is 100 volts, r.m.s. Take V_s as the reference.

5-4. A lossless line is 3.2 wavelengths long and has a characteristic impedance of 50 ohms. If β is 2.35 radians per meter and the load impedance is $150 + j50$ ohms, determine, using equations, the following if the input voltage is 50 volts r.m.s. and is taken as the reference
(a) Load voltage.
(b) Sending end current.
(c) Receiving end current.
(d) Power to the load.

5-5. A lossless line has a Z_o of 100 ohms and is terminated in a load of $50 + j50$ ohms. The input voltage is 10 volts, r.m.s. Determine, using equations and V_s as the reference:

(a) Load voltage and load current if the line is 3.58 wavelengths long.
(b) VSWR.
(c) Input current.
(d) Power to the load.

5-6. A lossless transmission line with a characteristic impedance of 50 ohms has an input impedance of $50 - j45$ ohms. It is 357.65 cm long and the wavelength is 50 cm. Determine using equations:
(a) Load impedance.
(b) VSWR.
(c) Load voltage if the input voltage is used as the reference and has a magnitude of 100 volts, r.m.s.

5-7. A lossless line has a load impedance of $97.662 + j65.108$ ohms and a characteristic impedance of 50 ohms. The line is 10 meters long and β is 1.9635 radians per meter. The input voltage is 50 volts, r.m.s. Using V_s as the reference determine, using equations:
(a) Input current.
(b) Input impedance.
(c) Load voltage.
(d) Load current.
(e) Input power to the line.
(f) Power to the load.

5-8. A lossless line is 4.82 wavelengths long, has a characteristic impedance of 50 ohms, and is terminated in an admittance of $0.01 + j0$ mhos. If the input voltage is 100 volts, r.m.s, and is taken as the reference, determine:
(a) Load current.
(b) Load voltage.
(c) Input impedance.
(d) Input current.
(e) Input power.

5-9. A lossless line has a load impedance of $80 - j60$ ohms and a Z_o of 100 ohms. The line is 15 meters long and β is 2.0 radians per meter. The input voltage to the line is 100 volts, r.m.s. Using V_s as the reference determine, using equations:
(a) Receiving end voltage.
(b) Receiving end current.
(c) VSWR.
(d) Load power.
(e) Input current.

5-10. A lossless transmission line has a Z_o of 100 ohms. It is terminated in a load impedance of $80 - j60$ ohms. It is operated at a frequency so that the wavelength is 2 meters and the line is 5 meters long. Determine, using equations:
(a) Input impedance.
(b) Load current if the input voltage is used as the reference and has a magnitude of 50 volts, r.m.s.

5-11. Find the input impedance for a lossless line that is 15.2 meters long if the wavelength is 2.6 meters and the load impedance has the values given below, when Z_o is 50 ohms:
(a) $30 - j40$
(b) $40 + j30$
(c) $50 + j0$
(d) $100 + j0$ 5-12.

5-12. A lossless line has Z_o of 600 ohms, is 2.35 wavelengths long, and is terminated in an open circuit. Using the input voltage as the reference with a value of 100 volts, r.m.s., determine:
(a) Input current.
(b) Load voltage.
(c) Input impedance.

5-13. A lossless line has a 200-ohm characteristic impedance and is terminated in a load of $200 + j300$ ohms. If the line is 2.81 wavelengths long and has an input voltage of 100 volts, r.m.s., determine, using V_s as the reference:
(a) Load voltage.
(b) Load current.
(c) VSWR.
(d) Input current.
(e) Power to the load.
(f) Input admittance

5-14. A lossless line has the following parameters: $Z_o = 50$ ohms, $\beta = 2.35$ radians per meter, $Z_r = 150 + j50$ ohms, $\ell = 3.2$ meters, $V_s = 75$ volts, r.m.s.
Determine, using V_s as the reference:
(a) Load current.
(b) Load voltage.
(c) Load power.
(d) Input current.

5-15. The line of problem 5-14 has a load voltage of 50 volts, r.m.s. Determine, using V_r as the reference:
(a) Input voltage.
(b) Input current.

(c) Load current.
(d) Input power.

5-16. A lossless line with a characteristic impedance of 50 ohms has an input impedance of 50 − $j45$ ohms and a load voltage of 45 volts, r.m.s. It is 357.65 cm long and the wavelength is 50 cm. Using equations, determine the following with V_r as the reference:
(a) Load impedance.
(b) Input voltage.
(c) Input current.
(d) Load current.

5-17. A lossless line has a Z_o of 50 ohms and is terminated in a load of 40 − $j30$ ohms.
(a) As one moves away from the load, determine, using equations, the first point on the line where the normalized admittance has the form:

$$Y_{\text{normalized}} = 1 \pm jB.$$

Express your answer in wavelengths.
(b) Find the second point on the line where the normalized admittance is $1 \pm jB$.
(c) Find the values of B at the two points.
(d) Find the lengths of the stub tuners with short-circuit terminations for the two cases.

5-18. A lossless, 50-ohm line is terminated in an impedance 30 − $j15$. An engineer wishes to use a single-stub (short circuited) tuner to improve the VSWR on this line to approximately 1.0. Determine the distance in wavelengths from the load to the first two positions where the tuners might be placed. Determine the lengths of the tuners at the two different positions. Note: Use the Smith chart.

5-19. A short-circuited stub has a characteristic impedance of 50 ohms and may be considered lossless. Determine the input admittance of the stub if the stub is:
(a) 1 wavelength long.
(b) 1/4 wavelength long.
(c) 0.6 wavelength long.

5-20. The impedance of an antenna is measured to be 50 + $j100$ ohms. It is connected to a transmitter through a lossless line 2.3 wavelengths long. What is the impedance presented to the transmitter if the line has a characteristic impedance of 50 ohms? Design a single-stub (short-circuited) tuner to match this impedance to the line. Give the length of the stub and its distance from the antenna in terms of the wavelength.

5-21. A lossless line has a characteristic impedance of 50 ohms. The load is $100 + j65$ ohms. Determine:

(a) VSWR.

(b) A single short-circuited stub is to be used to correct the VSWR to approximately 1.0. Find the location of the first stub from the load to do this and the length of the stub. Express your answers in wavelengths.

5-22. Repeat problem 5-21 if the load is $200 + j0$ ohms.

6

The Lossy Case

6.1 Introduction

Losses are a fact of life and the engineer spends much of his life trying to minimize them in order to maximize the efficiency of a device or a system. Fortunately, electrical engineers are often working with equipment which inherently has very high efficiencies when comparing them with other common devices. Electrical transformers may have efficiencies close to 100% whereas the engineer who designs them will drive to work in a device which has an efficiency that is about one-fourth of that amount. Losses may or may not be important, depending on what the engineer is trying to do. All transmission lines have some losses and it becomes an engineering judgement as to whether we have to consider them in a particular case or if they can be ignored. The type of line, frequency, length of line, and what is being done on the line all have a bearing on the decision. As we go through the rest of the text we will see what considerations are involved in making this decision.

Up to this point in our work we have been considering a special case in assuming that the transmission line was lossless. The lossless case is really the special case of the more general situation where the line is lossy. One way to look at transmission lines would be to take the lossy case first and to modify the results for the lossless case. However, in order to obtain a clearer picture of the physical situation on the line it is easier to look at the lossless case first. Then we can modify our understanding by adding the lossy concept to our thinking. Obviously if we are moving large amounts of expensive electrical energy over long distances the problem is different than if we start out with a minute amount and move it only a few feet. The power engineer has developed special procedures for handling transmission lines and so we will delay our consideration of those techniques until later. In this chapter we will look at the equations for the lossy case which can be used in either the exponential form or in terms of hyperbolic functions and, once again, approach an intuitive concept by considering the counter-rotating phasor solution.

6.2 Lossy Line Equations

Our starting point will be some equations from Chapter 3.

(3-5) $$V = V^+e^{\gamma d} + V^-e^{-\gamma d}$$

and

(3-6) $$I = \frac{V^+e^{\gamma d}}{Z_o} - \frac{V^-e^{-\gamma d}}{Z_o}$$

where

$$\gamma = \alpha + j\beta = \sqrt{(R + j\omega L)(G + j\omega C)}$$

and

(3-3) $$V = Ae^{-\gamma \ell}$$

and

(3-4) $$V^- = Be^{\gamma \ell},$$

both constants for a given line. Since $V^- = \Gamma_r V^+$ we obtain:

$$V = V^+(e^{\gamma d} + \Gamma_r e^{-\gamma d}). \quad (6\text{-}1)$$

$$I = \frac{V^+}{Z_o}(e^{\gamma d} - \Gamma_r e^{-\gamma d}). \quad (6\text{-}2)$$

Since

(1-35) $$Z_o = \sqrt{\frac{(R + j\omega L)}{(G + j\omega C)}},$$

both Z_o and γ are complex in the general case.

At $d = 0$, $V = V_r$, $I = I_r$, so:

$$V_r = V^+(1 + \Gamma_r). \tag{6-3}$$

$$I_r = \frac{V^+}{Z_o}(1 - \Gamma_r). \tag{6-4}$$

Using V^+ from (6-3) gives us the following equations.

$$V = \frac{V_r}{(1 + \Gamma_r)}(e^{\gamma d} + \Gamma_r e^{-\gamma d}). \tag{6-5}$$

$$I = \frac{V_r}{Z_o(1 + \Gamma_r)}(e^{\gamma d} - \Gamma_r e^{-\gamma d}). \tag{6-6}$$

$$Z = \frac{V}{I} = \frac{Z_o(e^{\gamma d} + \Gamma_r e^{-\gamma d})}{(e^{\gamma d} - \Gamma_r e^{\gamma d})}. \tag{6-7}$$

When $d = \ell$, $V = V_s$, $I = I_s$, $Z = Z_s$ and we have, as before.

$$V_s = \frac{V_r}{(1 + \Gamma_r)}(e^{\gamma \ell} + \Gamma_r e^{-\gamma \ell}). \tag{6-8}$$

$$I_s = \frac{V_r}{Z_o(1 + \Gamma_r)}(e^{\gamma \ell} - \Gamma_r e^{-\gamma \ell}). \tag{6-9}$$

$$Z_s = Z_o\frac{(e^{\gamma \ell} + \Gamma_r e^{-\gamma \ell})}{(e^{\gamma \ell} - \Gamma_r e^{-\gamma \ell})}. \tag{6-10}$$

If instead of taking V^+ from (6-3), we take it from (6-4), our equations would become:

$$V = \frac{I_r Z_o(e^{\gamma d} + \Gamma_r e^{-\gamma d})}{(1 - \Gamma_r)}. \tag{6-11}$$

$$I = \frac{I_r(e^{\gamma d} - \Gamma_r e^{-\gamma d})}{(1 - \Gamma_r)}. \tag{6-12}$$

$$V_s = \frac{I_r Z_o(e^{\gamma \ell} + \Gamma_r e^{-\gamma \ell})}{(1 - \Gamma_r)}. \tag{6-13}$$

$$I_s = \frac{I_r(e^{\gamma \ell} - \Gamma_r e^{-\gamma \ell})}{(1 - \Gamma_r)}. \tag{6-14}$$

From mathematics, we have the identities:

$$\sinh x = \frac{e^x - e^{-x}}{2}. \tag{6-15}$$

$$\cosh x = \frac{e^x + e^{-x}}{2}. \tag{6-16}$$

$$e^x = \cosh x + \sinh x. \tag{6-17}$$

$$e^{-x} = \cosh x - \sinh x. \tag{6-18}$$

$$\cosh^2 x - \sinh^2 x = 1. \tag{6-19}$$

When we substitute into equation (6-7) we obtain:

$$Z = Z_o \frac{(e^{\gamma d} + \Gamma_r e^{-\gamma d})}{(e^{\gamma d} - \Gamma_r e^{-\gamma d})}.$$

$$Z = Z_o \left[\frac{(\cosh\gamma d + \sinh\gamma d + \dfrac{(Z_r - Z_o)}{(Z_r + Z_o)}(\cosh\gamma d - \sinh\gamma d)}{(\cosh\gamma d + \sinh\gamma d) - \dfrac{(Z_r - Z_o)}{(Z_r + Z_o)}(\cosh\gamma d - \sinh\gamma d)} \right]$$

$$Z = Z_o \left[\frac{Z_r \cosh\gamma d + Z_o \sinh\gamma d}{Z_o \cosh\gamma d + Z_r \sinh\gamma d} \right]. \tag{6-20}$$

$$V = \frac{V_r(e^{\gamma d} + \Gamma_r e^{-\gamma d})}{(1 + \Gamma_r)} =$$

$$\frac{V_r}{(1 + \Gamma_r)} [(\cosh\gamma d + \sinh\gamma d) + \Gamma_r(\cosh\gamma d - \sinh\gamma d)] =$$

$$\frac{V_r}{(1 + \Gamma_r)} [(1 + \Gamma_r)\cosh\gamma d + (1 - \Gamma_r)\sinh\gamma d] =$$

$$V_r \cosh\gamma d + V_r \frac{(1 - \Gamma_r)}{(1 + \Gamma_r)} \sinh\gamma d.$$

Using (5-3)

$$\frac{Z_o}{Z_r} = \frac{1 - \Gamma_r}{1 + \Gamma_r}$$

gives:

$$V = V_r \cosh\gamma d + \frac{V_r Z_o}{Z_r} \sinh\gamma d = V_r \cosh\gamma d + I_r Z_o \sinh\gamma d. \tag{6-21}$$

Continuing with a like process for the current gives:

$$I = I_r\cosh\gamma d + \frac{V_r}{Z_o}\sinh\gamma d. \qquad (6\text{-}22)$$

In general, γd is a complex number so we will have the hyperbolic sine and cosine of complex arguments. From mathematics we have:

$$\sinh\gamma d = \sinh(\alpha d + j\beta d) = \sinh\alpha d\cos\beta d + j\cosh\alpha d\sin\beta d. \qquad (6\text{-}23)$$

$$\cosh\gamma d = \cosh(\alpha d + j\beta d) = \cosh\alpha d\sin\beta d + j\sinh\alpha d\sin\beta d. \qquad (6\text{-}24)$$

When we substitute, the equations look complicated but are easy to calculate if we have a hand calculator with the hyperbolic functions or ready access to a digital computer.

$$V = V_r\cosh\alpha d\cos\beta d + I_rZ_o\sinh\alpha d\cos\beta d + j(V_r\sinh\alpha d\sin\beta d + I_rZ_o\cosh\alpha d\sin\beta d). \qquad (6\text{-}25)$$

From the definitions we notice that $\cosh 0 = 1$ and the $\sinh 0 = 0$, so if $\alpha = 0$, we obtain $V = V_r\cos\beta d + jI_rZ_o\sin\beta d$, which is the same as we obtained in equation (5-4) for the lossless case.

For the current, after substitution in equation (6-26), we would have

$$I = I_r\cosh\alpha d\cos\beta d + \frac{V_r}{Z_o}\sinh\alpha d\sin\beta d + j\left[I_r\sinh\alpha d\sin\beta d + \frac{V_r}{Z_o}\cosh\alpha d\sin\beta d\right]. \qquad (6\text{-}26)$$

When $d = \ell$, $V = V_s$, $I = I_s$, and $Z = Z_s$, so equations (6-21) and (6-22) become:

$$V_s = V_r\cosh\gamma\ell + I_rZ_o\sinh\gamma\ell. \qquad (6\text{-}27)$$

$$I_s = I_r\cosh\gamma\ell + \frac{V_r}{Z_o}\sinh\gamma\ell. \qquad (6\text{-}28)$$

$$Z_s = Z_o\left[\frac{Z_r\cosh\gamma\ell + Z_o\sinh\gamma\ell}{Z_o\cosh\gamma\ell + Z_r\sinh\gamma\ell}\right]. \qquad (6\text{-}29)$$

By rearranging the form of equation (6-28) we can write:

$$V_s = V_r\cosh\gamma\ell + I_rZ_o\sinh\gamma\ell. \tag{6-27}$$

$$I_s = \frac{V_r}{Z_o}\sinh\gamma\ell + I_r\cosh\gamma\ell. \tag{6-28}$$

Using determinants and equation (6-19) we find the equation for V_r.

$$V_r = V_s\cosh\gamma\ell - I_sZ_o\sinh\gamma\ell. \tag{6-30}$$

Solving for the current gives:

$$I_r = I_s\cosh\gamma\ell - \frac{V_s}{Z_o}\sinh\gamma\ell. \tag{6-31}$$

Summarizing, we have:

$$V_s = V_r\cosh\gamma\ell + I_rZ_o\sinh\gamma\ell. \tag{6-27}$$

$$I_s = I_r\cosh\gamma\ell + \frac{V_r}{Z_o}\sinh\gamma\ell. \tag{6-28}$$

$$V_r = V_s\cosh\gamma\ell - I_sZ_o\sinh\gamma\ell. \tag{6-30}$$

$$I_r = I_s\cosh\gamma\ell - \frac{V_s}{Z_o}\sinh\gamma\ell. \tag{6-31}$$

$$Z_s = Z_o\left[\frac{Z_r\cosh\gamma\ell + Z_o\sinh\gamma\ell}{Z_o\cosh\gamma\ell + Z_r\sinh\gamma\ell}\right]. \tag{6-29}$$

$$\cosh\gamma\ell = \cosh\alpha\ell\cos\beta\ell + j\sinh\alpha\ell\sin\beta\ell. \tag{6-24}$$

$$\sinh\gamma\ell = \sinh\alpha\ell\cos\beta\ell + j\cosh\alpha\ell\sin\beta\ell. \tag{6-23}$$

These are not the type of equations we would like to memorize for future reference. However the phasor solution is not much more difficult than it was in the lossless case and is fairly easy to remember.

6.3 Phasor Solution—Lossy Case

The lossy case can be handled with a phasor solution with only a slight modification of what we did in Chapter 3. We start with the same equations we used at the beginning of this chapter.

$$V = V^+e^{\gamma d} + V^-e^{\gamma d}. \tag{3-5}$$

$$I = \frac{V^+e^{\gamma d}}{Z_o} - \frac{V^-e^{\gamma d}}{Z_o}. \tag{3-6}$$

Since $\gamma = \alpha + j\beta$, we have:

$$V = V^+e^{\alpha d}e^{j\beta d} + V^-e^{-\alpha d}e^{-j\beta d}. \tag{6-32}$$

$$I = \frac{V^+}{Z_o}e^{\alpha d}e^{j\beta d} - \frac{V^-}{Z_o}e^{-\alpha d}e^{-j\beta d}. \tag{6-33}$$

As we noted in the lossless case, the term $e^{j\beta d}$ rotates the phasor, V^+, in the counterclockwise direction but does not change its magnitude. The term $e^{\alpha d}$ changes only the magnitude and increases the magnitude of V^+ as d increases (as we move from the load end to the sending end.) The term $e^{-\alpha d}$ decreases the magnitude of V^- as d increases and, of course, the term $e^{-j\beta d}$ has a magnitude of unity and rotates the phasor, V^-, in the clockwise direction as before. So one major difference between the lossless and lossy cases is the change in magnitude of both V^+ and V^- as we move down the line. The other major change is that I^+ is no longer in phase with V^+ and I^- is no longer 180 degrees out of phase with V^- when Z_o has an angle.

It seems obvious that the phasor V^+ becomes larger in a spiraling manner as V^+ rotates towards the sending end. V^- does the opposite; it spirals inward to smaller values and, if the line is long enough, it would approach zero as d increases in value. The action of each is sketched in Figure 6-1.

How fast V^+ gets larger and V^- gets smaller depends on the value of α. If the attenuation is small, the effect is such that we approach the lossless case as shown in Figure 6-2. As we move down the line, the voltage maximums and minimums are changing but not very

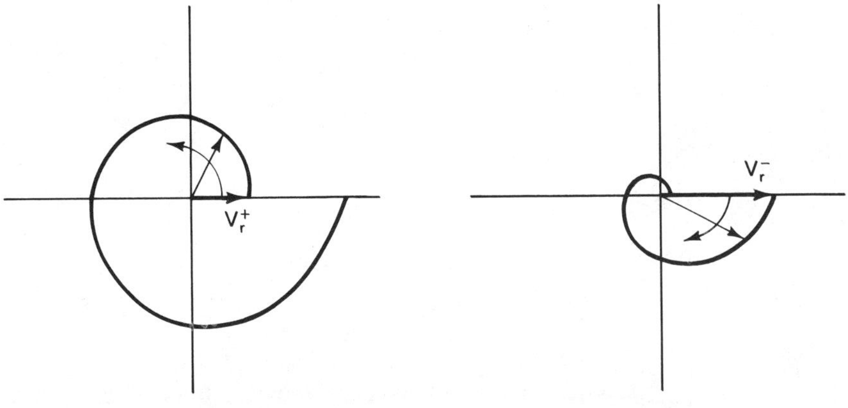

Figure 6-1. Rotating phasors for lossy case.

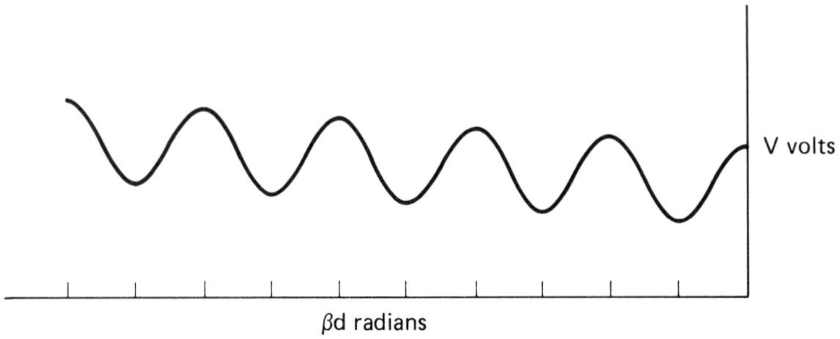

Figure 6-2. Standing wave pattern when attenuation is small.

rapidly. In such a case we often still find it convenient to talk about the VSWR, at least over a limited region of the line. However, if the attenuation is large so that V^+ increases at a rapid rate and V^- approaches zero very quickly, then the VSWR has no meaning since we end up with the pattern shown in Figure 6-3.

When V^- goes to zero we have V^+ increasing exponentially and this becomes the magnitude of the voltage on the line:

$$V = V^+ e^{\alpha d} e^{j\beta d} + 0.$$

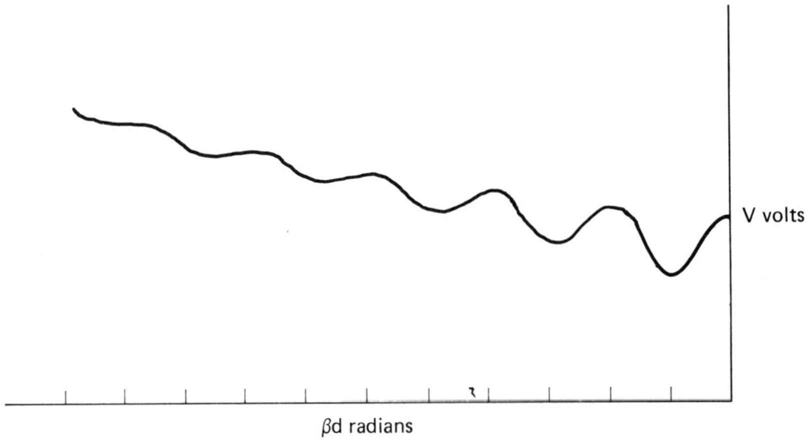

Figure 6-3. Standing wave pattern when attenuation is higher.

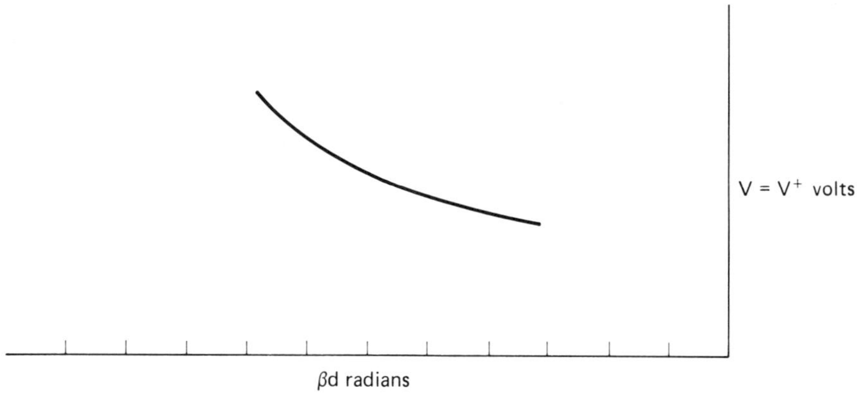

Figure 6-4. Pattern of voltage on line after V^- becomes zero.

When this happens, the pattern of Figure 6-3 would continue and would look as it does in Figure 6-4.

The easiest way to illustrate the lossy case is to give an example.

6.4 Example of the Solution of a Lossy Line

A transmission line is lossy and has the following parameters: $R = 19.9$ ohms/meter, $L = 9.989 \times 10^{-3}$ henrys/meter, $C = 1 \times 10^{-6}$ farads/meter, $G = 0$. $Z_r = 300 + j0$ ohms. $V_s = 100$ volts, r.m.s. The line is 2.9088 meters long. $\omega = 30 \times 10^3$ radians/second.

Determine the following quantities: α, β, γ, Z_o, I_s, V_r, I_r, P_s, P_r, Z_s, Q_r, and Q_s.

Solution

We will do it first by using the phasor method and then by using equations.

$$\omega L = (30 \times 10^3)(9.989 \times 10^{-3}) = 299.67 \text{ ohms.}$$

$$\omega C = (30 \times 10^3)(1 \times 10^{-6}) = 0.03 \text{ mhos.}$$

$$Z_o = \sqrt{\frac{(R + j\omega L)}{(G + j\omega C)}} = \sqrt{\frac{19.9 + j299.67}{0 + j0.03}} = 100.05 \angle -1.89^\circ = 100 - j3.316 \text{ ohms.}$$

Notice that the characteristic impedance is almost real.

$$\gamma = \sqrt{(R + j\omega L)(G + j\omega C)} = \sqrt{(19.9 + j299.67)(0 + j0.03)} =$$
$$3.00 \angle 88.1^\circ = 0.0995 + j3.00 = \gamma$$

so

$$\alpha = 0.0995 \text{ nepers/meter}$$

and

$$\beta = 3 \text{ radians/meter.}$$

$$v_p = \frac{\omega}{\beta} = \frac{30 \times 10^3}{3} = 10 \times 10^3 \text{ meters/second.}$$

Notice that v_p, the phase velocity, is considerably less than the velocity of light.

$$\beta\ell = (3)(2.9088) = 8.726 \text{ radians.}$$

This is equivalent to 500 degrees.

$$\frac{500^\circ}{360^\circ} = 1.388 \text{ revolutions.}$$

The important part is the 0.388: (0.388)(360°) = 140°. So, our phasors go around one time plus 140 degrees.

$$\Gamma_r = \frac{Z_r - Z_o}{Z_r + Z_o} = \frac{300 - (100 - j3.316)}{300 + (100 - j3.316)} = \frac{200 + j3.316}{400 - j3.316} =$$
$$\frac{200.02 \angle 0.95^\circ}{400.01 \angle -0.475^\circ} = 0.500 \angle 1.424^\circ = 0.499 + j0.012.$$

The voltage reflection coefficient is approximately 0.5 at an angle of 0 degrees. On the phasor diagram we will show the angle larger than it is so we can see it.

It would be very difficult to place the current phasors on the diagram since there is a small angle associated with Z_o. The current phasor, I^+, is no longer in phase with the phasor, V^+, and I^- is no longer 180 degrees out of phase with V^- as was true in the case of the lossless case. When we go to the sending end, the phasor V^+ not

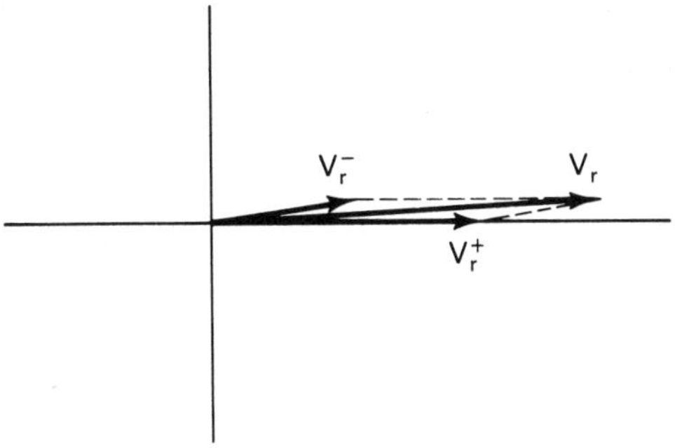

$$V_r = V^+ \angle 0^\circ + V^- \angle 1.425^\circ = V^+ \angle 0^\circ + 0.5V^+ \angle 1.425^\circ = 1.499V^+ \angle 0.475^\circ$$

only rotates a net of 140 degrees but its magnitude must be corrected because we now have attenuation and the attenuation factor must be taken into account. V^- must also have its magnitude adjusted for the sending end.

$$e^{\alpha d} = e^{(0.0995)(2.9088)} = e^{0.2894} = 1.335.$$

$$e^{-\alpha d} = e^{-0.2894} = 0.7486.$$

$$V_s = V^+(1)(1.335) \angle 140^\circ + V^+(0.5)(0.7486) \angle -138.57^\circ$$

Here we have to be careful to place V^- 140° in the clockwise direction from the angle at the load end (1.425° − 140° = −138.575°). $V_s = 1.439\ V^+ \angle 154.898^\circ$. Since the magnitude of V_s = 100 volts we can find the magnitude of V^+.

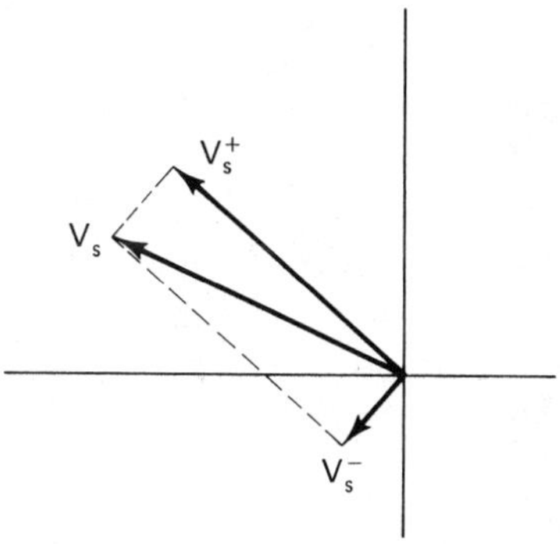

$$|V^+| = \frac{100}{1.439} = 69.449 \text{ volts.} \qquad |V^-| = 0.5\,|V^+| = 34.728 \text{ volts.}$$

These are, of course, the magnitudes at the load end.

$$I^+(\text{at receiving end}) = \frac{V^+}{Z_o} = \frac{69.449\angle 0^\circ}{100.05\angle -1.89^\circ} = 0.694\angle 1.89^\circ \text{ amperes.}$$

$$I^-(\text{at receiving end}) = -\frac{V^-}{Z_o} = \frac{-34.728\angle 1.425^\circ}{100.05\angle -1.89^\circ} = -0.347\angle 3.324^\circ \text{ amperes.}$$

$$I_r = I^+ + I^- = 0.694\angle 1.89^\circ - 0.347\angle 3.324^\circ = 0.347\angle 0.475^\circ \text{ amperes.}$$

$$V_r = (1.499)(69.449)\angle -.475^\circ = 104.14\angle 0.475^\circ \text{ volts.}$$

It is always good practice to check our numerical work whenever possible.

$$Z_r = \frac{V_r}{I_r} = \frac{104.17\angle 0.475^\circ}{0.347\angle 0.475^\circ} = 299.99\angle 0^\circ \simeq 300.0\angle 0^\circ \text{ ohms.}$$

V_s as a phasor may be written as $V_s = 100\angle 154.89^\circ$ volts. I_s is slightly more difficult since we have an angle for Z_o. As the angle is negative, the currents will lead their respective voltage phasors. We also have to take into account the change in magnitude at the sending end.

$$I_s = (0.694)(1.335)\angle 140^\circ + 1.89^\circ - (0.347)(0.7486)\angle -138.57^\circ + 1.89^\circ = 0.927\angle 125.76^\circ \text{ amperes.}$$

$$Z_s = \frac{V_s}{I_s} = \frac{100\angle 154.89^\circ}{0.9247\angle 125.76^\circ} = 108.137\angle 29.13^\circ = 94.45 + j52.64 \text{ ohms.}$$

$$P_s = V_sI_s\cos\theta_s = (100)(0.9247)\cos(154.89^\circ - 125.76^\circ) = 80.77 \text{ watts.}$$

$$P_r = V_rI_r\cos\theta_r = (104.17)(0.347)\cos 0^\circ = 36.17 \text{ watts.}$$

$$Q_s = V_sI_s\sin\theta_s = (100)(0.9247)\sin(154.89^\circ - 125.76^\circ) = 44.87 \text{ vars.}$$

$$Q_r = V_rI_r\sin\theta_r = (104.17)(0.347)\sin 0^\circ = 0 \text{ vars.}$$

The sending end power is no longer equal to the receiving end power and the difference is the power losses in the line. Line losses = 80.77 − 36.17 = 44.60 watts. The efficiency of a transmission line can be defined as:

$$\text{Transmission Line Efficiency} = \frac{P_r}{P_s}(100)\ \% = \eta. \qquad (6\text{-}34)$$

In our example we would have:

$$\text{Efficiency} = \frac{36.17}{80.77}(100) = 44.78\%.$$

This is not a very high value for efficiency, but we are not dealing with a large amount of power. This value would be completely unacceptable for the power line of a utility company.

Even though this line is lossy it is almost identical to the lossless case as far as the calculations are made with only a small amount of additional complications for us. The attenuation constant causes a modification of the magnitudes of the phasors and any angle associated with the characteristic impedance, Z_o, causes a phase angle difference between the voltage and current phasors. Some important results are apparent with just a little thought. First of all, a minimum current does not occur at the same point on the line as a maximum voltage because of the angle of Z_o. Also, a maximum current does not occur at the same point as a minimum voltage. In this example, a point of maximum voltage and a point of minimum current would occur very close to each other since the angle of Z_o is almost zero. Another difference between the lossy and lossless cases is also evident. The load impedance does not repeat itself one half wavelength down the line because of the magnitude changes as it does in the lossless case.

Let us now work the same problem using equations. In order to solve the equations we need to know the hyperbolic sine and cosine of a complex argument.

$\alpha\ell = (0.0995)(2.9088) = 0.2894.\ \beta\ell = (3)(2.9088) = 8.726.$

$$\sinh(0.2894 + j8.726) = \sinh 0.2894\cos 8.726 + j\cosh 0.2894\sin 8.726 = 0.767\,\underline{/108.543^\circ} = -0.224 + j0.67.$$

Notice that 8.726 is in radians and not in degrees (8.726 radians is equivalent to 500°).

$$\cosh(0.2894 + j8.726) = \cosh 0.2894\cos 8.726 + j\sinh 0.2894\sin 8.726 = 0.8201\,\underline{/166.698^\circ} = -0.7981 + j0.1887.$$

$$Z_s = Z_o\left[\frac{Z_r\cosh\gamma\ell + Z_o\sinh\gamma\ell}{Z_o\cosh\gamma\ell + Z_r\sinh\gamma\ell}\right]. \tag{6-29}$$

$$Z_s = 100.054\underline{/-1.899^\circ}$$

$$\left[\frac{(300)(0.8201\,\underline{/166.698^\circ}) + (100.054\,\underline{/-1.899^\circ})(0.7067\,\underline{/108.54^\circ})}{(100.054\,\underline{/-1.899^\circ})(0.8201\,\underline{/166.698^\circ}) + (300)(0.7067\,\underline{/108.54^\circ})}\right] =$$

$$108.106\,\underline{/29.13^\circ} = 94.429 + j52.631 \text{ ohms}.$$

$$I_s = \frac{V_s}{Z_s} = \frac{100\,\underline{/0^\circ}}{108.106\,\underline{/29.13^\circ}} = 0.925\,\underline{/-29.13^\circ} \text{ amperes}.$$

$$V_r = V_s\cosh\gamma\ell - I_sZ_o\sinh\gamma\ell. \tag{6-30}$$

$$V_r = (100\,\underline{/0^\circ})(0.820\,\underline{/166.698^\circ}) - (0.925\,\underline{/-29.13^\circ})(100.054\,\underline{/-1.899^\circ})$$
$$(0.7067\,\underline{/108.54^\circ}) = 104.18\,\underline{/-154.41^\circ})\text{ volts.}$$

$$I_r = \frac{V_r}{Z_r} = \frac{104.18\,\underline{/-154.41^\circ}}{300\,\underline{/0^\circ}} = 0.347\,\underline{/-154.41^\circ}\text{ amperes.}$$

$$P_s = V_sI_s\cos\theta_s = (100)(0.925)\cos 29.13^\circ = 80.8\text{ watts.}$$

$$P_r = V_rI_r\cos\theta_r = (104.18)(0.347)\cos 0^\circ = 36.15\text{ watts.}$$

$$Q_s = V_sI_s\sin\theta_s = (100)(0.925)\sin 29.13^\circ = 45.03\text{ vars.}$$

$$Q_r = V_rI_r\sin\theta_r = (104.18)(0.347)\sin 0^\circ = 0\text{ vars.}$$

If we compare our answers using equations with our answers using the phasor technique, we find that the magnitudes are very close. The angles are different since we used V^+ as our reference for the phasor method and V_s as our reference when we used equations.

Comparison	Phasor Solution	Equation Solution
V_s	$100\,\underline{/154.89^\circ}$	$100\,\underline{/0^\circ}$
I_s	$0.924\,\underline{/125.76^\circ}$	$0.925\,\underline{/-29.13^\circ}$
V_r	$104.17\,\underline{/0.475^\circ}$	$104.18\,\underline{/-154.41^\circ}$
I_r	$0.347\,\underline{/0.475^\circ}$	$0.347\,\underline{/-154.41^\circ}$
Z_s	$108.137\,\underline{/29.13^\circ}$	$108.106\,\underline{/29.13^\circ}$
P_r	80.77	80.8
P_s	36.17	36.15
Q_s	44.87	45.03
Q_r	0.0	0.0

The student should recognize that this example illustrates a case where the line is relatively short, since it is only a little over one wavelength long and the attenuation factor does not change the values of V^+ and V^- very much in that distance. If, however, the line were 10 times as long we would have the following results: $\ell = 10(2.9088) = 29.088$ meters; $e^{\alpha d} = e^{0.0995(29.088)} = e^{2.89} = 18.07$; $e^{-\alpha d} = e^{-2.89} = 0.0555$. Now the forward phasor would be about 18 times as big at the sending end as it is at the receiving end and the reverse phasor would be almost zero at the sending end in comparing it with its value at the receiving end. For all practical purposes V^- could be neglected when determining the value of V^+ at the sending end. The length of the line and the attenuation constant must both be considered in making a judgment when considering the lossy line.

6.5 Example of a Lossy Short-Circuited Line

Let us assume that the line used in the example in Section 6.4 is terminated in a short circuit and we wish to find the same quantities.

Solution

From Section 6.4 we have:

$$\sinh\gamma\ell = 0.7067\,\underline{/108.543^\circ}. \qquad \cosh\gamma\ell = 0.8201\,\underline{/166.698^\circ}.$$

With $Z_r = 0$, equation (6-29) becomes:

$$Z_s = Z_o\frac{\sinh\gamma\ell}{\cosh\gamma\ell}. \qquad Z_s = 100.054\,\underline{/-1.899^\circ}\left[\frac{0.7067\,\underline{/108.543^\circ}}{0.8201\,\underline{/166.698^\circ}}\right] =$$

$$86.188\,\underline{/-60.11^\circ}\text{ ohms.}$$

$$I_s = \frac{V_s}{Z_s} = \frac{100\,\underline{/0^\circ}}{86.188\,\underline{/-60.11^\circ}} = 1.16\,\underline{/60.11^\circ}\text{ amperes.}$$

For kicks, let us calculate V_r.

$$V_r = V_s\cosh\gamma\ell - I_sZ_o\sinh\gamma\ell.$$

$$V_r = (100\,\underline{/0^\circ})(0.8201\,\underline{/166.698^\circ})$$

$$-\ (1.16\,\underline{/60.11^\circ})(100.054\,\underline{/-1.89^\circ})(0.7067\,\underline{/108.54^\circ} =$$

$$82.01\,\underline{/166.69^\circ} - 81.99\,\underline{/166.69^\circ} =$$

$$-79.81 + j18.87 + 79.79 - j18.86 \simeq 0. \tag{6-30}$$

We can not use the equation $I_r = V_r/Z_r$ because both V_r and Z_r are zero and the equation becomes indeterminate. The easiest equation to use is (6-28) since with V_r equal to zero it becomes:

$$I_r = \frac{I_s}{\cosh\gamma\ell} = \frac{1.16\,\underline{/60.11^\circ}}{0.8201\,\underline{/166.69^\circ}} = 1.414\,\underline{/-106.58^\circ}\text{ amperes.}$$

$$P_r = V_rI_r\cos\theta_r = (0)(1.414)\cos\theta_r = 0\text{ watts.}$$

$$P_s = V_sI_s\cos\theta_s = (100)(1.160)\cos 60.11^\circ = 57.80\text{ watts.}$$

$$Q_r = V_rI_r\sin\theta_r = (0)(1.414)\sin\theta_r = 0\text{ vars.}$$

$$Q_s = V_sI_s\sin\theta_s = (100)(1.16)\sin 60.11^\circ = 100.57\text{ vars.}$$

In this example we see some changes from the results we had in the lossless case. The input impedance now contains a real term and the power absorbed by this real term becomes the power loss in the line since the load is not absorbing any real power. We notice immediately that this lossy section of line would not be of much help if we were trying to use it as a stub tuner for matching another line.

6.6 Chapter Summary

1. All lines are lossy and the engineer must make the educated determination as to when the losses must be considered and when they may be neglected.
2. Lossy lines may be considered by the phasor technique, which is relatively easy to remember, or by the use of equations which are mathematically fairly easy to use but hard to remember.
3. The exponential form of the equations could be used to find solutions but it is normally easier to use them to develop the hyperbolic forms and then use the hyperbolic equations for finding solutions.
4. If attenuation is high the reverse phasor, V^-, goes toward zero very rapidly and the revolving phasor technique becomes just one phasor, V^+, increasing exponentially as it moves towards the sending end.
5. Unless the line is lossless or has very low attenuation, the concept of talking about the VSWR is meaningless.
6. All lossy lines will have an equivalent input impedance which will have a resistive term. This resistive term will absorb the same amount of real power as the line and the load impedance. If the load impedance takes no real power, the resistive component of the input impedance takes the same real power as the total losses in the line.
7. If the line is lossy, I^+, is not in phase with V^+ and I^- is not 180 degrees out of phase with V^-. This shows that V_{max} and I_{min} do not occur at the same point on the line and V_{min} and I_{max} also do not occur at the same point. Thus we do not have any points on the line where the minimum or maximum impedance is purely resistive.
8. When using the phasor technique we only have to remember the equation for Γ_r and the forms $e^{\alpha d}$ and $e^{-\alpha d}$, and remember which goes with V^+ and which goes with V^-.
9. If α becomes zero the lossy equations become equal to the lossless equations found in Chapter 5.

10. The important equations from this chapter are given below.

$$V_s = V_r\cosh\gamma\ell + I_rZ_o\sinh\gamma\ell. \tag{6-27}$$

$$I_s = I_r\cosh\gamma\ell + \frac{V_r}{Z_o}\sinh\gamma\ell. \tag{6-28}$$

$$V_r = V_s\cosh\gamma\ell - I_sZ_o\sinh\gamma\ell. \tag{6-30}$$

$$I_r = I_s\cosh\gamma\ell - \frac{V_s}{Z_o}\sinh\gamma\ell \tag{6-31}$$

$$Z_s = Z_o\left[\frac{Z_r\cosh\gamma\ell + Z_o\sinh\gamma\ell}{Z_o\cosh\gamma\ell + Z_r\sinh\gamma\ell}\right]. \tag{6-29}$$

$$\cosh\gamma\ell = \cosh\alpha\ell\cos\beta\ell + j\sinh\alpha\ell\sin\beta\ell. \tag{6-24}$$

$$\sinh\gamma\ell = \sinh\alpha\ell\cos\beta\ell + j\cosh\alpha\ell\sin\beta\ell. \tag{6-23}$$

6.7 Review and Study Questions

1. Give the expression for the cosine of a complex number. How does it differ from the hyperbolic cosine of a complex number?
2. Is a d. c. transmission line a lossy or a lossless line? How are the equations in this chapter affected when the d. c. case is considered?
3. How is the phasor, V^+, modified along a transmission line if the line is lossy? How is the phasor, V^-, modified?
4. If $\alpha = \beta$ would you consider the line lossy, very lossy, or lossless? Why?
5. If a line is lossy and you are given its parameters, describe how you would find the voltage at the load end and the input current by using equations if you were given the input voltage and the load was an open circuit.
6. Why do you suppose the phasor technique is not too useful for the power transmission line engineer?
7. Can the input impedance of an ordinary transmission line connected to loads with real resistance, inductance and capacitance ever contain a negative real term? Explain.
8. Sketch the standing wave on a transmission line with a short-circuited load if: (a) it is lossless, (b) it has low attenuation per wavelength, (c) it has high attenuation per wavelength.
9. Express the hyperbolic sine and cosine of a complex argument in terms of exponentials.

10. Write equations (6-27), (6-28), (6-29), (6-30), and (6-31) out in terms of sines, cosines and hyperbolic sines and cosines of real arguments.
11. Can you prove that $\cosh^2 x - \sinh^2 x = 1$? If so, do it.
12. What is a simplified expression for the input impedance of a lossy line in terms of Z_o if the line is: (a) short circuited? (b) open circuited?
13. Give three different forms for the equation for the efficiency of a transmission line, in terms of input power, load power and losses in the line.
14. Why might it be difficult to draw the phasor diagrams of the current along with the voltage at different points along the line?
15. Explain how the magnitudes of V^+ and V^- on a lossy line change as one moves (a) from the sending end to the receiving end, (b) from the receiving end to the sending end.
16. Electrical engineers are used to discussing attenuation in terms of decibels. Can you find a relationship between decibels and nepers in any of your texts?

Problems

6-1. Compute the values of the cos x for the following values of x: (a) 1.0, (b) $0.6 \angle -30°$, (c) $1.5 \angle 45°$, (d) j2.

6-2. Compute the values of the cosh x and the sinh x for the following values of x: (a) 2.0 (b) $0.6 \angle -30°$ (c) $1 + j1$ (d) j2

6-3. A certain transmission line has the following parameters at 10 kilohertz: $\alpha = 0.28$ nepers per mile. $\beta = 0.59$ radians per mile. $Z_o = 150 - j70$ ohms. If the line is 5 miles long and terminated in an impedance of $100 + j100$ ohms determine the following if the input voltage is 5 volts, r.m.s.
(a) The voltage reflection coefficient.
(b) The input impedance.
(c) The input current.
(d) The input power.
(e) The load voltage.
(f) The load current.
(g) The load power and the efficiency of the line.

6-4. If you worked problem 6-3 by equations resolve the problem by using phasors. If you used phasors in problem 6-3 resolve it by using equations.

6-5. A transmission line is 100 miles long and has the following parameters per mile: $R = 0.86$ ohms; $L = 3.978 \times 10^{-3}$ henrys; $G = 0$ mhos. $C = 4.244 \times 10^{-8}$ farads. It is operating at 60 hertz and the load impedance is $250 + j250$ ohms. Determine the following if the sending-end voltage is 1,000 volts, r.m.s.

(a) The receiving-end voltage.
(b) The receiving-end current.
(c) The sending-end current.
(d) The sending-end power factor.
(e) The sending-end power.
(f) The receiving-end power and the efficiency of the line.

6-6. Assume the line in problem 6-5 has the receiving end open circuited when 1,000 volts, r.m.s. is applied to the sending end. Determine:

(a) The receiving-end voltage.
(b) The sending-end current.
(c) The sending-end power factor.
(d) The sending-end power.

6-7. Assume the line in problem 6-5 has a load at the receiving end of 2,500 kw at 0.8 power factor lagging when the receiving end voltage is 40 kv. Determine:

(a) The sending-end voltage.
(b) The sending-end current.
(c) The sending-end power.
(d) The efficiency of the line.

6-8. A line is 20 kilometers long and is operating at a frequency of 1 kilohertz when the input voltage is 50 volts, r.m.s. It has the following parameters: $R = 6.30$ ohms per kilometer; $L = 2.1 \times 10^{-3}$ henrys per kilometer; $G = 6.8 \times 10^{-8}$ mhos per kilometer; $C = 0.0056 \times 10^{-6}$ farads per kilometer.

Determine: (a) α, β, and Z_o; (b) V_r, I_r, I_s, P_r, and P_s if the load is $200 + j0$ ohms.

6-9. Let us assume that the line in problem 6-8 is 10 kilometers long. If $Z_r = 300 + j300$ ohms and $V_s = 100$ volts, r.m.s., determine:

(a) V_r.
(b) I_r.
(c) The receiving-end power.
(d) The sending-end power.
(e) The sending-end power factor.
(f) The efficiency of the line.
(g) The vars the source must provide.

6-10. Suppose the line in problem 6-8 is 20 kilometers long and the load impedance equals the characteristic impedance of the line. If the input voltage is 1,000 volts, r.m.s., determine:

(a) I_s.
(b) V_r.
(c) I_r.
(d) The sending-end power.
(e) The receiving-end power and the efficiency of the line.

6-11. The voltage at the receiving end for the line in problem 6-8 is 200 volts, r.m.s. If the load is $200 + j200$ ohms sketch the magnitude of the voltage for one wavelength from the load. Assume the line is several wavelengths long.

6-12. A single-phase power transmission line is 100 miles long and operating at 60 hertz. The constants per mile (for both wires) are as follows: $R = 0.86$ ohms; $L = 3.978 \times 10^{-3}$ henrys; $G = 0$; $C = 4.244 \times 10^{-8}$ farads. With the receiving end open circuited, 35,000 volts is applied at the sending end. Determine:

(a) The receiving-end voltage.
(b) The receiving-end current.
(c) The sending-end current.
(d) If a load of 1,500 kw at 0.85 power factor, lagging, is placed at the receiving end and the receiving-end voltage is 60 kv, determine the sending-end current.
(e) What is the efficiency of the line for the conditions of part (d)?

6-13. Repeat problem 6-12 if the line is only 30 miles long.

6-14. Repeat parts (a), (b), and (c) for problem 6-12 if the load is a short circuit.

7

Power Lines

7.1 Introduction

Until the energy crisis in 1974, the use of electrical energy in the United States doubled approximately every nine years. This growth required huge investments in generating stations, substations, transmission lines and distribution systems and a great deal of effort on the part of many engineers who were responsible for the design, construction and installation of these systems. It is reasonable to predict now that in the coming years the use of electricity will continue to expand as electrical heat and electrically operated transportation units reduce our dependence on gas and oil. Increased efficiencies and reduced costs are goals that engineers will have for decades in the future. Utility engineers, along with design and manufacturing engineers, have been and will continue to be important to the development of increased capacity in our power systems.

Higher efficiency on a transmission line demands higher voltages along with the reduced currents to lower the line losses due to the line resistance. The use of direct-current transmission lines, after a period of many years, has now been made possible by the use of new techniques which allow the conversion from alternating to direct current and back again with excellent efficiency. The difficulty of securing right-of-way for the placement of transmission lines has presented new challenges to the utility engineer who must now upgrade present lines or design new lines that may not take the most direct path from the generating station to the consumers.

In the early years, transmission line engineers were mainly concerned with operation at relatively low frequencies. The first use of transmission line theory was involved with the design and construction of telegraph, telephone and power lines. Recent advances in microwave and optical fiber communications circuits have reduced the importance and use of two-wire communications circuits but the importance of the power engineer continues to grow. In this chapter we will look at some of the work that was the concern of and still is the concern of the power system engineer. They have, over the years, developed equations and techniques which were suitable to their needs and made their work easier and quicker without the aid of modern calculating devices. With the development of the digital computer their work is now not only faster but they are able to handle complicated problems that were impossible to solve under the old techniques. However, in our process of trying to secure a feel for the physical operation of a line, we will look at some of the older ways of handling the power line.

Since all transmission of large amounts of electrical power is always done with either d. c. lines or three-phase lines, students should refresh their memories by reviewing their circuits books. The square root of three causes a great deal of trouble for many students whose three-phase circuit studies were brief and hurried.

In this chapter, we shall introduce some new concepts such as the total series impedance and shunt admittance of the line. These fictional ideas simplify the calculations but we are not able to go to an actual line and say that there exists a total series impedance that we can point to since it really does not exist in a true physical way.

In many of your previous courses you have replaced physical objects by equivalent circuits. A simple equivalent circuit for a d. c. generator consists of an internal voltage source in series with a fixed resistance. The equivalent circuit of a transistor is more complicated but makes the calculations easier once the circuit constants are related to the physical constants of the actual transistor. It is also possible to replace our transmission line by an equivalent circuit and in this chapter we shall use some old friends as equivalent circuits of our power transmission line.

7.2 The Basic Equations

While theoretically we could handle a power line using phasors it is not done for several practical reasons. First, the wavelength at power frequencies is so long that even a relatively long line would involve only a rotation of a few degrees for the phasors from the load end to the sending end. Second, the equivalent load impedance is constantly changing and so the voltage reflection coefficient becomes a variable and of no use to us. Furthermore, no attempt is made to match the load to the characteristic impedance of the line since we have no real control over the changing load and we are more interested in maximum efficiency than in maximum power transfer. Third, the phasor technique is really useful only for lines operating at much higher frequencies. It was used in previous chapters when discussing either a theoretically lossless line or a line which may be approximated as a lossless line but losses must nearly always be taken into account when discussing power lines.

Our basic starting point will be equations developed in Chapter 6.

(6-27) $$V_s = V_r \cosh\gamma\ell + I_r Z_o \sinh\gamma\ell.$$

(6-28) $$I_s = I_r \cosh\gamma\ell + \frac{V_r}{Z_o}\sinh\gamma\ell.$$

(6-30) $$V_r = V_s \cosh\gamma\ell - I_s Z_o \sinh\gamma\ell.$$

(6-31) $$I_r = I_s \cosh\gamma\ell - \frac{V_s}{Z_o}\sinh\gamma\ell.$$

In Chapter 1 we had the equation for Z_o:

(1-35) $$Z_o = \sqrt{\frac{(R + j\omega L)}{(G + j\omega C)}}.$$

Let us multiply the numerator and denominator by the length of the line, ℓ.

$$Z_o = \sqrt{\frac{(R + j\omega L)\ell}{(G + j\omega C)\ell}}.$$

We will now define two new quantities.

The total series impedance of the line $= Z = (R + j\omega L)\ell$. (7-1)

The total shunt admittance of the line $= Y = (G + j\omega C)\ell$. (7-2)

Notice that there is no place on the actual line that we can point to a physical Z or Y. They are simply found by taking the impedance and admittance per unit length and multiplying by the length of the line. So we have:

$$Z_o = \sqrt{\frac{(R + j\omega L)\ell}{(G + j\omega C)\ell}} = \sqrt{\frac{Z}{Y}}. \tag{7-3}$$

In a like fashion we can go back to equation (1-30),

$$\gamma = \sqrt{(R + j\omega L)(G + j\omega C)}$$

and multiply both sides of the equation by ℓ.

$$\gamma\ell = \sqrt{(R + j\omega L)\ell(G + j\omega C)\ell} = \sqrt{ZY}. \tag{7-4}$$

Now we can substitute into our equations.

$$V_s = V_r\cosh\sqrt{ZY} + I_r\sqrt{\frac{Z}{Y}}\sinh\sqrt{ZY} = V_r\cosh\sqrt{ZY} + I_r\sqrt{\frac{Z^2}{ZY}}\sinh\sqrt{ZY} = V_r\cosh\sqrt{ZY} + I_rZ\frac{\sinh\sqrt{ZY}}{\sqrt{ZY}} \tag{7-5}$$

Let us define another quantity.

$$\sqrt{ZY} = \Theta. \tag{7-6}$$

Now equation (7-5) becomes equation (7-7).

$$V_s = V_r\cosh\Theta + I_rZ\frac{\sinh\Theta}{\Theta}. \tag{7-7}$$

We can do the same thing for the current expression.

$$I_s = I_r\cosh\gamma\ell + \frac{V_r}{Z_o}\sinh\gamma\ell = I_r\cosh\sqrt{ZY} + \frac{V_r}{\sqrt{\frac{Z}{Y}}}\sinh\sqrt{ZY} = I_r\cosh\sqrt{ZY} + \frac{V_r}{\sqrt{\frac{ZY}{Y^2}}}\sinh\sqrt{ZY} = I_r\cosh\sqrt{ZY} + \frac{V_rY}{\sqrt{ZY}}\sinh\sqrt{ZY} =$$

$$I_r\cosh\sqrt{ZY} + V_rY\frac{\sinh\sqrt{ZY}}{\sqrt{ZY}} = I_r\cosh\Theta + V_rY\frac{\sinh\Theta}{\Theta}. \tag{7-8}$$

The two other equations become:

$$V_r = V_s\cosh\Theta - I_sZ\frac{\sinh\Theta}{\Theta}. \tag{7-9}$$

$$I_r = I_s\cosh\Theta - V_sY\frac{\sinh\Theta}{\Theta}. \tag{7-10}$$

Many years ago, L. F. Woodruff published the curves for the $\cosh\Theta$ and for the $\sinh\Theta/\Theta$ in his text on transmission lines (Woodruff, 1948). They are reproduced here with the kind permission of the John Wiley publishing company. (NOTE: Specific references to material published in other texts will be given by the author's name and the year of publication in parentheses and refers to the Reference Section in the back of this text.)

These curves are useful since they illustrate that these complex numbers have a certain range of magnitudes and angles for our work with practical transmission lines. The accuracy in reading these charts is quite good when comparing the values with those obtained using computational methods.

The equations given relate the quantities at each end of the line since these are usually the items of most importance. However, we can also have equations for the voltage and current anywhere on the line. Let us start with equation (6-21).

$$V = V_r\cosh\gamma d + I_rZ_o\sinh\gamma d.$$

Using (7-9) and (7-10) we obtain:

$$V = (V_s\cosh\gamma\ell - I_sZ_o\sinh\gamma\ell)\cosh\gamma d + (I_s\cosh\gamma\ell - \frac{V_s}{Z_o}\sinh\gamma\ell)Z_o\sinh\gamma d =$$

$$V_s(\cosh\gamma\ell\cosh\gamma d - \sinh\gamma\ell\sinh\gamma d) + I_sZ_o(\cosh\gamma\ell\sinh\gamma d - \sinh\gamma\ell\cosh\gamma d).$$

We can now use two identities:

$$\sinh(U - V) = \cosh V\sinh U - \sinh V\cosh U.$$
$$\cosh(U - V) = \cosh U\cosh V - \sinh U\sinh V.$$

$$\begin{aligned} V = {} & V_s\cosh(\gamma\ell - \gamma d) + I_sZ_o\sinh(\gamma d - \gamma\ell) = \\ & V_s\cosh\gamma(\ell - d) + I_sZ_o\sinh\text{-}\gamma(\ell - d) = \\ & V_s\cosh\gamma(\ell - d) - I_sZ_o\sinh\gamma(\ell - d). \end{aligned}$$

but $x + d = \ell$, so $\ell - d = x$. Therefore,

$$V = V_s\cosh\gamma x - I_sZ_o\sinh\gamma x. \tag{7-11}$$

By a similar process we find that:

$$I = I_s\cosh\gamma x - \frac{V_s}{Z_o}\sinh\gamma x. \tag{7-12}$$

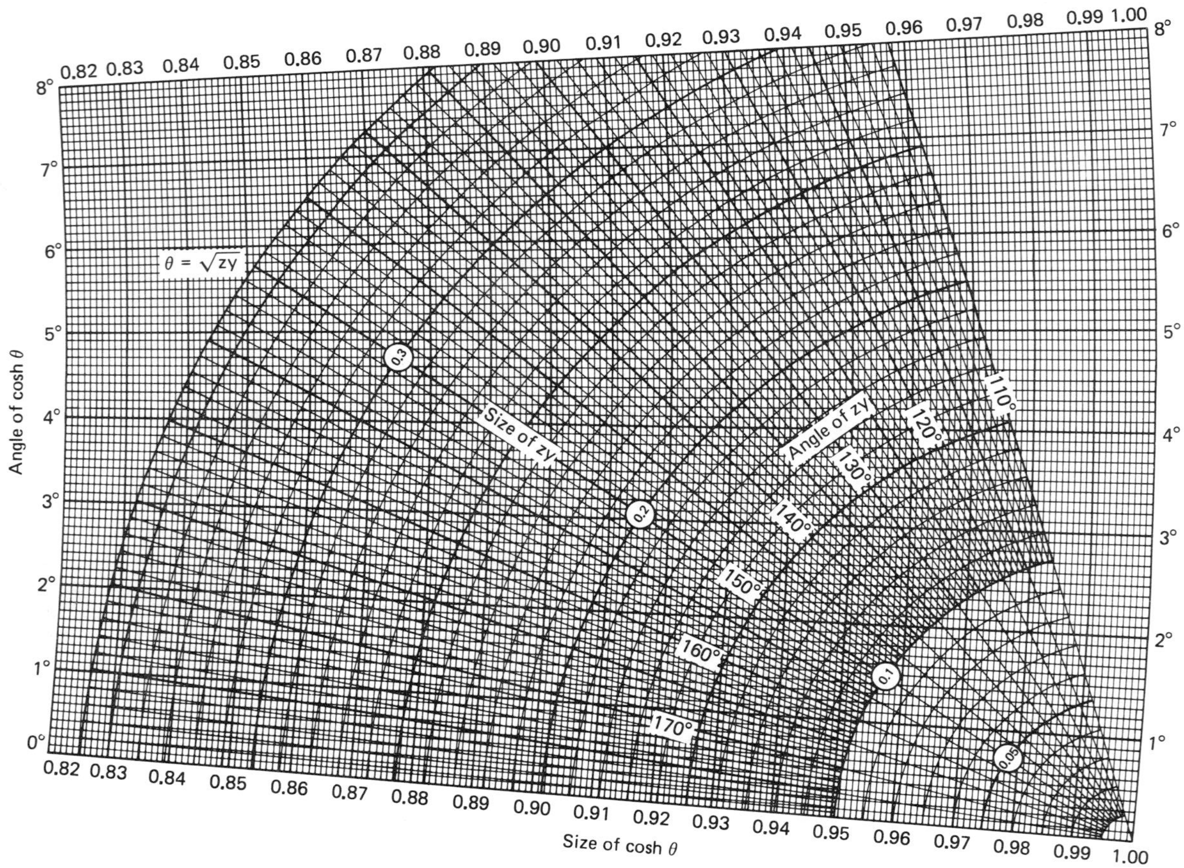

Figure 7-1. Hyperbolic cosine of Θ.

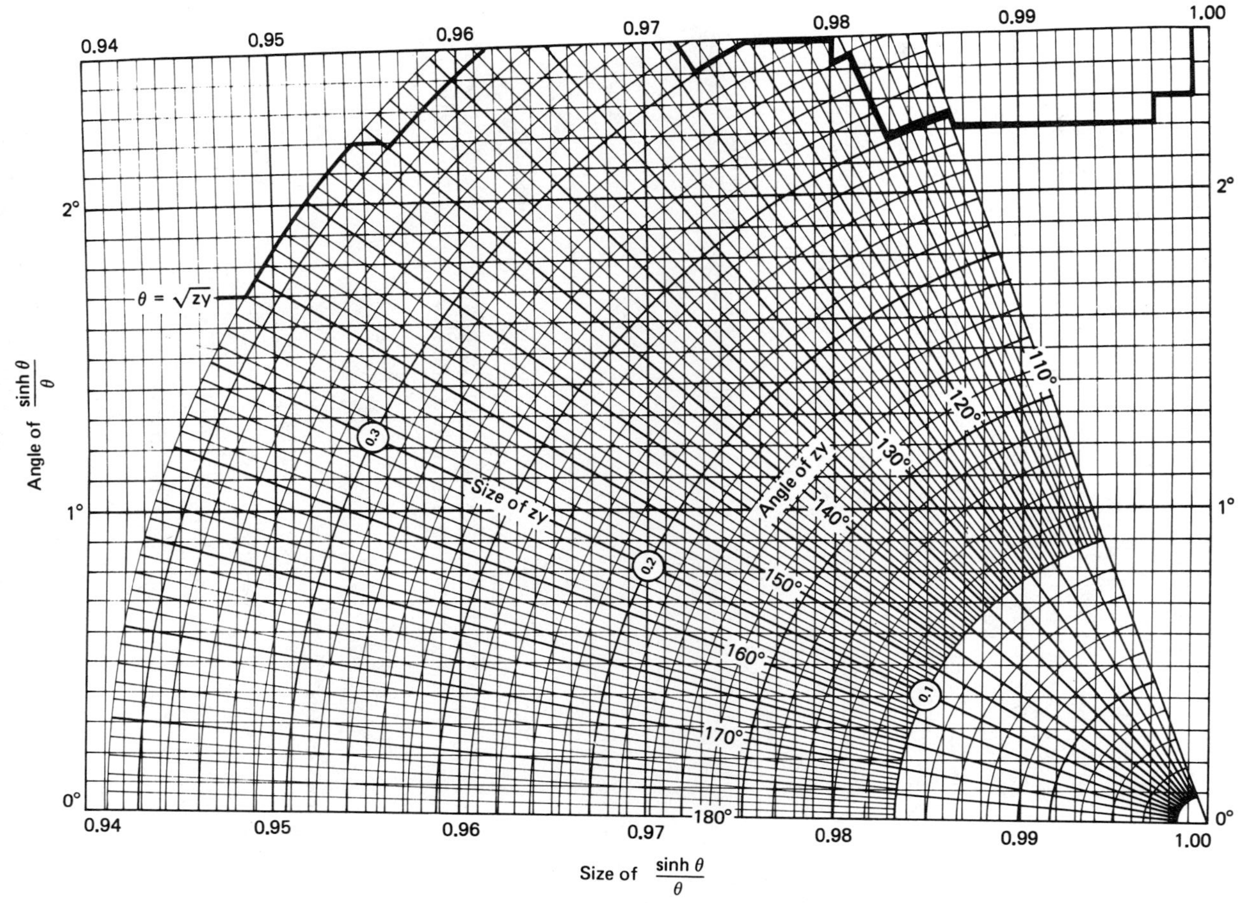

Figure 7-2. Hyperbolic sine of Θ divided by Θ.

7.3 Example of a 60-Hertz Transmission Line Problem

The previous equations may be used for either single-phase or three-phase lines but since most transmission lines are three-phase let us look at a typical problem. When solving three-phase transmission line problems we always work on a per-phase basis and assume a wye system. That is, our voltages will be the line-to-neutral voltages and our currents will be line currents. Line parameters are determined on a per-phase basis which is equivalent to a per-conductor basis. We will assume that we have found the line parameters by the proper methods to be discussed later in the text.

$\ell = 100$ miles. $z = R + j\omega L = 0.647 + j2.415$ ohms per mile.
$y = G + j\omega C = 0 + j1 \times 10^{-5}$ mhos per mile.

The line is supplying 100,000 kilowatts to a load with a power factor of 0.9 lagging and the receiving-end voltage is 230 kilovolts, line-to-line.

Find:

(a) $\cosh \Theta$; (b) $\dfrac{\sinh \Theta}{\Theta}$; (c) I_r; (d) V_s; (e) I_s; (f) Sending-end power factor; (g) P_s; (h) Efficiency of the line; (i) The values of V_r and I_s with the load removed and the voltage, V_s, kept at the value found in (d).

Solution

$$z = 0.647 + j2.425 = 2.5 \underline{/75^\circ}.$$
$$y = 0 + j1 \times 10^{-5} = 1 \times 10^{-5} \underline{/90^\circ}.$$
$$Z = z\ell = (2.5 \underline{/75^\circ})(100) = 250 \underline{/75^\circ}.$$
$$Y = y\ell = (1 \times 10^{-5} \underline{/90^\circ})(100) = 1 \times 10^{-3} \underline{/90^\circ}.$$
$$ZY = (250 \underline{/75^\circ})(1 \times 10^{-3} \underline{/90^\circ}) = 0.25 \underline{/165^\circ}.$$

If we use our charts we can find our answers directed from ZY without taking the square root of ZY.

$$\cosh\Theta = 0.882 \angle 2^\circ. \qquad \frac{\sinh\Theta}{\Theta} = 0.9602 \angle 0.63^\circ$$

If we wish to calculate our values we need to take the square root of ZY.

$$\cosh\Theta = \cosh\sqrt{ZY} = \cosh\sqrt{0.25 \angle 165^\circ} = \cosh 0.5 \angle 82.5^\circ = \cosh(0.06526 + j0.4957) = \cosh 0.06526 \cos 0.4957 + j\sinh 0.06526 \sin 0.4957 = 0.8814 + j0.03106 = 0.882 \angle 2.018^\circ.$$

$$\sinh\Theta = \sinh\sqrt{ZY} = \sinh 0.5 \angle 82.5^\circ = \sinh(0.06526 + j0.4957) = \sinh 0.06526 \cos 0.4957 + j\cosh 0.06526 \sin 0.4957 = 0.05744 + j0.4766 = 0.4801 \angle 83.128^\circ$$

so

$$\frac{\sinh\Theta}{\Theta} = \frac{0.4801 \angle 83.128^\circ}{0.5 \angle 82.5^\circ} = 0.9602 \angle 0.628^\circ.$$

Notice that the values found from the charts are quite close to the values found from equations. Since we are working on a wye-circuit basis we need to find V_r as the line-to-neutral value and use this as our reference phasor.

$$V_r = \frac{230{,}000 \angle 0^\circ}{\sqrt{3}} = 132{,}790.56 \angle 0^\circ \text{ volts.}$$

Since

$$P_r = \sqrt{3}\, V_{\ell-\ell} I_\ell \cos\theta_r,$$

$$I_r = \frac{P_r}{\sqrt{3} V_{\ell-\ell} \cos\theta_r} = \frac{100{,}000{,}000}{\sqrt{3}(230{,}000)(0.9)} = 278.91 \text{ amperes.}$$

With a power factor of 0.9 lagging (25.84°) and with V_r as our reference, we can write:

$$I_r = 278.91 \angle -25.84^\circ.$$

$$V_s = V_r \cosh\Theta + I_r Z \frac{\sinh\Theta}{\Theta}.$$

$$V_s = (132{,}790.56\,\underline{/0^\circ})(0.882\,\underline{/2^\circ}) +$$
$$(278.91\,\underline{/-25.84^\circ})(250\,\underline{/75^\circ})(0.9602\,\underline{/0.63^\circ}) =$$
$$160{,}275.91 + j55{,}216.98 = 169{,}520.74\,\underline{/19.00^\circ} \text{ volts.}$$

The sending-end voltage needed to supply this load under the conditions stated would then have to be on a line-to-line basis:

$$V_{\ell-\ell} = \sqrt{3}\,V_{\ell-n} = \sqrt{3}\,(169{,}520.74) = 293{,}618.54 \text{ volts.}$$

$$I_s = I_r \cosh\Theta + V_r Y \frac{\sinh\Theta}{\Theta}.$$

$$I_s = (278.91\,\underline{/-25.84^\circ})(0.882\,\underline{/2^\circ}) +$$
$$(132{,}790.56\,\underline{/0^\circ})(1 \times 10^{-3}\,\underline{/90^\circ})(0.9602\,\underline{/0.63^\circ}) =$$
$$223.60 + j28.06 = 225.35\,\underline{/7.15^\circ} \text{ amperes.}$$

The angle between the sending-end voltage and the sending-end current is $(19.0^\circ - 7.15^\circ) = 11.85^\circ$. Therefore the sending-end power factor is calculated as:

$$\text{P. F.} = \cos\theta_s = \cos 11.85^\circ = 0.98 \text{ lagging.}$$

The sending-end power, P_s, can be found in the normal manner.

$$P_s = \sqrt{3}\,V_{\ell-\ell} I_\ell \cos\theta_s = \sqrt{3}\,(293{,}618.54)(225.35)\cos 11.85^\circ =$$
$$112{,}164{,}000 \text{ watts} = 112{,}164 \text{ kilowatts.}$$

The efficiency of the line may be easily found.

$$\eta = \frac{\text{output (load) power}}{\text{input power}}(100)\,\% = \frac{100{,}000\,(100)}{112{,}164.0} = 89.15\%.$$

The difference between the input and output represents the losses in the line. In our case since G was assumed to be zero the losses represent the I^2R losses in the series resistance of the line.

$$112{,}164{,}000 - 100{,}000{,}000 = 12{,}164{,}000 \text{ watts lost in the line.}$$

Suppose we keep the input voltage constant at the value we just found and remove the load so that I_r becomes zero.

$$V_s = V_r\cosh\Theta + I_r Z\frac{\sinh\Theta}{\Theta}.$$

$$169{,}520.74\ \underline{/19.0^\circ} = V_r(0.882\ \underline{/2^\circ}) + 0. \qquad V_r = \frac{169{,}520.74\ \underline{/19.0^\circ}}{0.882\ \underline{/2^\circ}} =$$

192,200.38 $\underline{/17.0^\circ}$ volts, line-to-neutral or 332,900.82 volts, line-to-line.

The engineer would have to realize that, on this line with a nominal voltage of 230 kV, the voltage would climb to 332 kV if the load were removed and would have to provide for an insulation level on the equipment at the terminal end to take care of this high voltage or insulation breakdown could lead to a failure in the system. If the load is removed, what is the sending-end current? One might guess that it would be zero since the load current would be zero. However we already know that this is not so and equation (7-8) confirms our previous results in earlier chapters.

$$I_s = I_r\cosh\Theta + V_r Y\frac{\sinh\Theta}{\Theta}.$$

$$I_s = 0 + (192{,}200.38\ \underline{/17.0^\circ})(1 \times 10^{-3}\ \underline{/90^\circ})(0.9602\ \underline{/0.628^\circ}) = 184.55\ \underline{/107.6^\circ}\ \text{amperes}.$$

Since $V_s = 169{,}520.74\ \underline{/19.0^\circ}$, the angle between the current, I_s, and V_s when the load current, I_r, is zero, is $(107.62^\circ - 19.0^\circ) = 88.62^\circ$ with the current leading. The input power can be found as three times the power per phase.

$$P_{s(\text{no load})} = (3)(169{,}520.74)(184.55)\cos 88.62^\circ = 2{,}260{,}333.98\ \text{watts}.$$

That we have an input current of 184.55 amperes and an input power of 2,260,333.98 watts with no load on the line should not be surprising when we realize that we have shunt capacitance for a path for the current and series resistance to give us a power loss on the line. The open-circuit line looks like a large capacitor and we can think of this current as the charging current for our capacitor.

It may also seem strange that the receiving-end voltage can be higher than the sending-end voltage but again we must remember that we are dealing with distributed inductance and capacitance and the possibility of a resonant condition much as we might have with a lumped parameter circuit.

7.4 Equivalent Circuits of Transmission Lines

If we can develop an equivalent circuit of a transmission line it would make our calculations easier since we would then have a circuit to solve and could forget about the equations we have been using. By constructing the circuit using lumped circuit elements we have a model to make studies on the transmission line for varying conditions. In fact, models of transmission lines were used for many years to solve practical transmission line problems and to predict how the line would react to certain normal or abnormal conditions. The first circuit we will use is the equivalent pye circuit shown in Figure 7-3.

Using circuit theory the following equations result for the circuit in Figure 7-3.

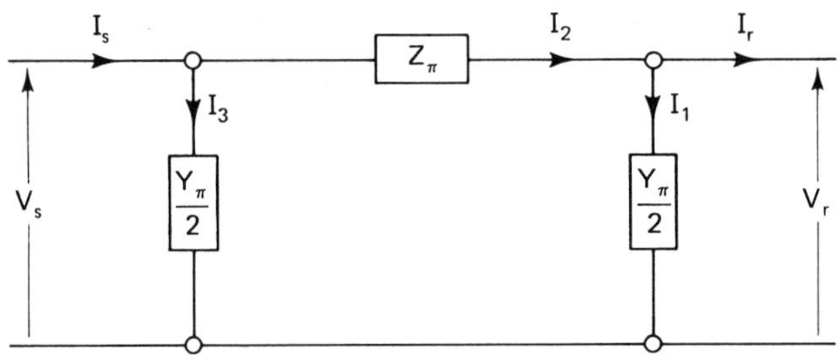

Figure 7-3. Equivalent pye circuit of a transmission line.

$$I_1 = \frac{V_r Y_\pi}{2}. \quad I_2 = I_r + I_1 = I_r + \frac{V_r Y_\pi}{2}. \quad V_s = V_r + I_2 Z_\pi = V_r + \left(I_r + \frac{V_r Y_\pi}{2}\right) Z_\pi. \quad V_s = V_r\left(1 + \frac{Y_\pi Z_\pi}{2}\right) + I_r Z_\pi. \tag{7-13}$$

$$I_3 = \frac{V_s Y_\pi}{2} = \left[V_r\left(1 + \frac{Y_\pi Z_\pi}{2}\right) + I_r Z_\pi\right]\frac{Y_\pi}{2} = V_r \frac{Y_\pi}{2}\left(1 + \frac{Y_\pi Z_\pi}{2}\right) + I_r \frac{Z_\pi Y_\pi}{2}.$$

$$I_s = I_2 + I_3 = I_r + \frac{V_r Y_\pi}{2} + \frac{V_r Y_\pi}{2}\left(1 + \frac{Y_\pi Z_\pi}{2}\right) + I_r \frac{Z_\pi Y_\pi}{2} = V_r Y_\pi\left(1 + \frac{Y_\pi Z_\pi}{4}\right) + I_r\left(1 + \frac{Z_\pi Y_\pi}{2}\right) = I_s. \tag{7-14}$$

If we compare equations we have:

$$V_s = V_r\cosh\gamma\ell + I_r Z_o \sinh\gamma\ell. \tag{6-27}$$

$$V_s = V_r\left(1 + \frac{Y_\pi Z_\pi}{2}\right) + I_r Z_\pi. \tag{7-13}$$

$$I_s = I_r\cosh\gamma\ell + \frac{V_r}{Z_o}\sinh\gamma\ell. \tag{6-28}$$

$$I_s = I_r\left(1 + \frac{Y_\pi Z_\pi}{2}\right) + V_r Y_\pi\left(1 + \frac{Z_\pi Y_\pi}{4}\right). \tag{7-14}$$

When comparing the equations we notice that (6-27) will be equivalent to (7-13) and (6-28) to (7-14) if the following relationships are true.

$$1 + \frac{Y_\pi Z_\pi}{2} = \cosh\gamma\ell. \tag{7-15}$$

$$Z_\pi = Z_o\sinh\gamma\ell. \tag{7-16}$$

$$Y_\pi\left(1 + \frac{Z_\pi Y_\pi}{4}\right) = \frac{1}{Z_o}\sinh\gamma\ell. \tag{7-17}$$

Equation (7-16) gives us our first result; namely, that for equivalence, $Z_\pi = Z_o\sinh\gamma\ell$. From equation (7-15) we have:

$$\frac{Y_\pi Z_\pi}{2} = \cosh\gamma\ell - 1, \text{ so } \frac{Y_\pi}{2} = \frac{\cosh\gamma\ell - 1}{Z_\pi} = \frac{\cosh\gamma\ell - 1}{Z_o\sinh\gamma\ell}.$$

But

$$\tanh\frac{U}{2} = \frac{\cosh U - 1}{\sinh U},$$

so:

$$\frac{Y_\pi}{2} = \frac{1}{Z_o}\tanh\frac{\gamma\ell}{2}. \tag{7-17}$$

But we really should check equation (7-17) to make certain that it is consistent with these results.

$$\frac{1}{Z_o}\sinh\gamma\ell = Y_\pi\left(1 + \frac{Y_\pi Z_\pi}{4}\right) = \frac{2}{Z_o}\frac{\tanh\gamma\ell}{2}\left[1 + \frac{1}{Z_o}\frac{\tanh\gamma\ell}{2}\frac{(Z_o\sinh\gamma\ell)}{2}\right] =$$

$$\frac{2}{Z_o}\left[\frac{\cosh\gamma\ell - 1}{\sinh\gamma\ell}\right]\left[1 + \frac{(\cosh\gamma\ell - 1)}{\sinh\gamma\ell}\frac{(\sinh\gamma\ell)}{2}\right] =$$

$$\frac{2}{Z_o}\left[\frac{\cosh\gamma\ell - 1}{\sinh\gamma\ell}\right]\left[\frac{2 + \cosh\gamma\ell - 1}{2}\right] =$$

$$\frac{1}{Z_o}\left[\frac{(\cosh\gamma\ell - 1)(\cosh\gamma\ell + 1)}{\sinh\gamma\ell}\right] = \frac{1}{Z_o}\frac{(\cosh^2\gamma\ell - 1)}{\sinh\gamma\ell}.$$

But $\cosh^2 U - \sinh^2 U = 1$ and $\cosh^2 U - 1 = \sinh^2 U$ so,

$$\frac{1}{Z_o}\sinh\gamma\ell = \frac{1}{Z_o}\frac{\sinh^2\gamma\ell}{\sinh\gamma\ell} = \frac{1}{Z_o}\sinh\gamma\ell.$$

Equation (7-17) is therefore consistent with our previous results. Our equivalent circuit is shown in Figure 7-4.

Since $\gamma\ell = \sqrt{ZY}$ and $Z_o = \sqrt{\frac{Z}{Y}}$ we can modify our expressions.

$$Z_\pi = Z_o\sinh\gamma\ell = \sqrt{\frac{Z}{Y}}\sinh\sqrt{ZY} = \sqrt{\frac{Z^2}{ZY}}\sinh\sqrt{ZY} =$$

$$Z\frac{\sinh\sqrt{ZY}}{\sqrt{ZY}} = Z_\pi = Z\frac{\sinh\Theta}{\Theta}. \qquad (7\text{-}18)$$

$$\frac{Y_\pi}{2} = \frac{1}{Z_o}\tanh\frac{\gamma\ell}{2} = \frac{1}{\sqrt{\frac{Z}{Y}}}\tanh\frac{\sqrt{ZY}}{2} = \sqrt{\frac{Y^2}{ZY}}\tanh\frac{\sqrt{ZY}}{2} =$$

$$\frac{Y}{\sqrt{ZY}}\tanh\frac{\sqrt{ZY}}{2} = \frac{Y}{2}\frac{\tanh\frac{\sqrt{ZY}}{2}}{\frac{\sqrt{ZY}}{2}} = \frac{Y}{2}\frac{\tanh\frac{\Theta}{2}}{\frac{\Theta}{2}} = \frac{Y_\pi}{2}. \qquad (7\text{-}19)$$

So we can redraw our circuit in Figure 7-4 either as Figure 7-5 or Figure 7-6. Woodruff has also furnished us with a set of curves for the $\frac{\tanh\Theta/2}{\Theta/2}$ and they are reproduced as Figure 7-7 (Woodruff 1948).

Now, suppose we had approximated our circuit for our transmission line by putting the total series impedance of the line Z, in

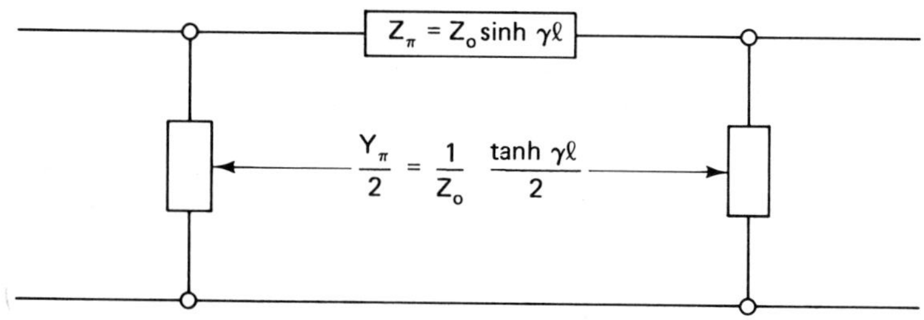

Figure 7-4. Equivalent pye circuit of a transmission line.

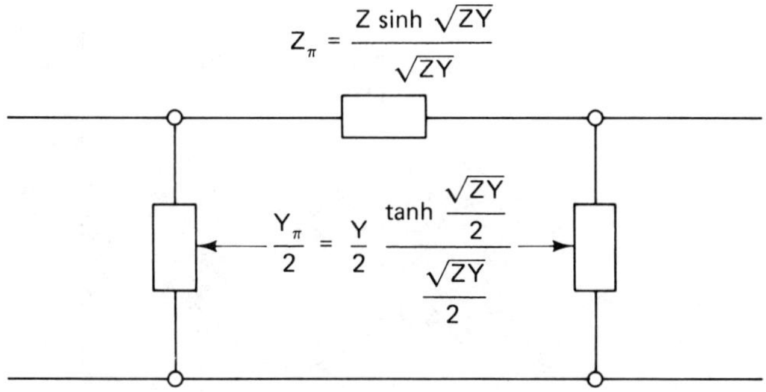

Figure 7-5. Equivalent pye circuit—exact equivalent circuit.

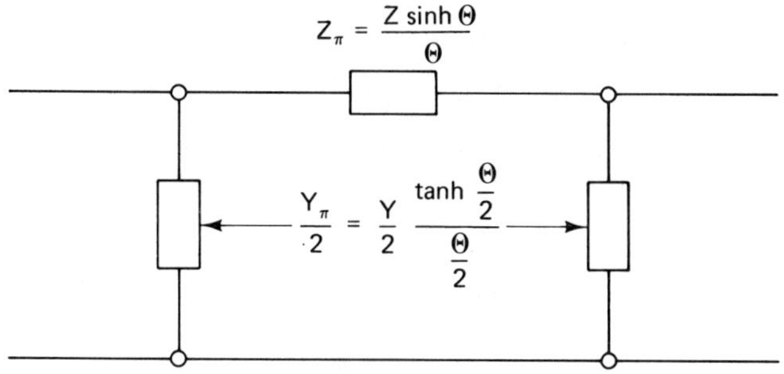

Figure 7-6. Equivalent pye circuit.

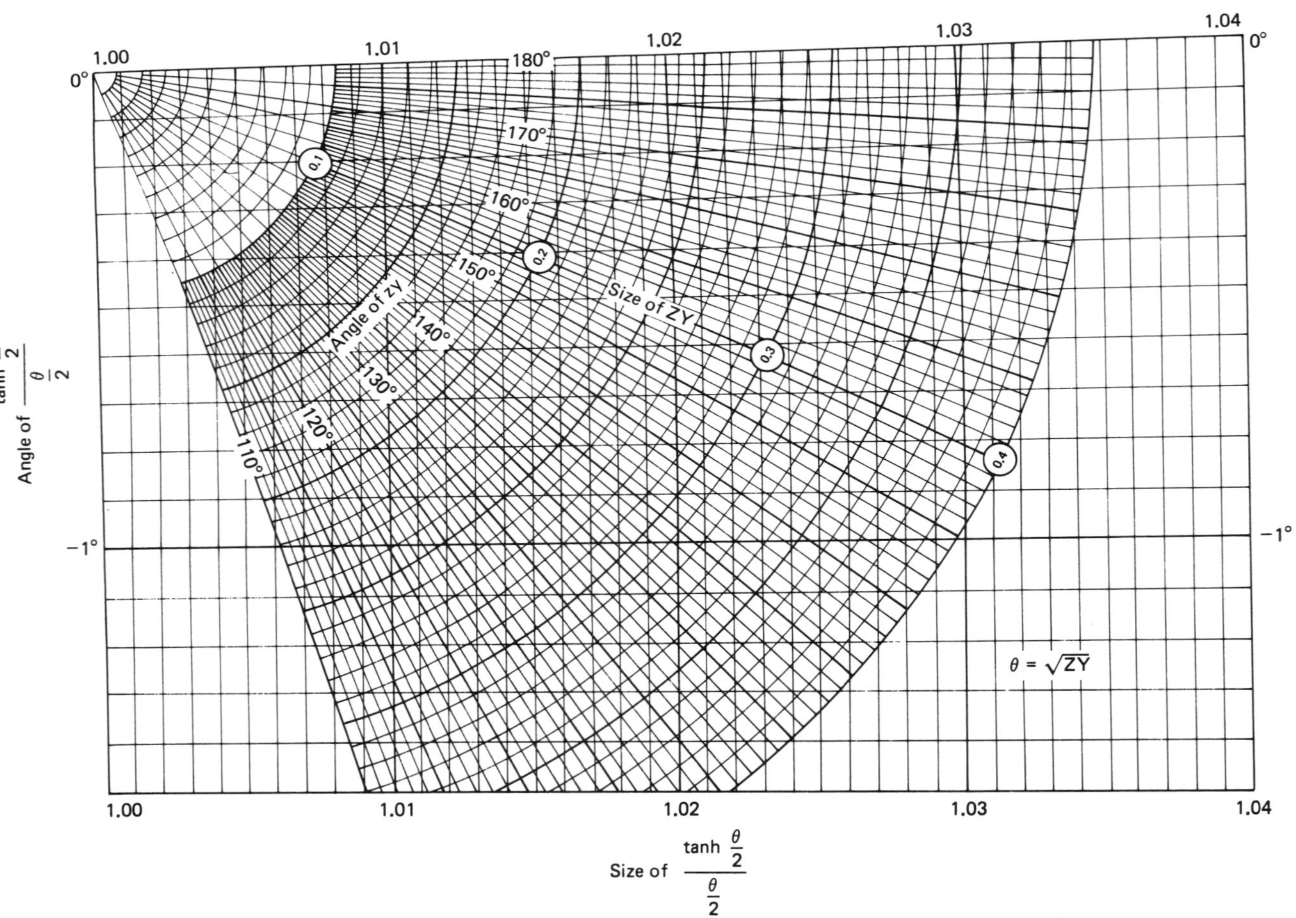

Figure 7-7. The hyperbolic tangent of θ/2 divided by θ/2.

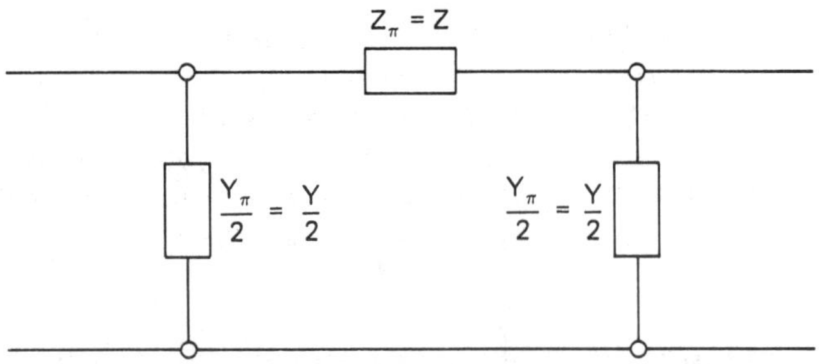

Figure 7-8. Approximate pye circuit.

the upper branch and one half of the total shunt admittance, $Y/2$, at each end. We would end up with Figure 7-8.

The difference between the circuit shown in Figure 7-8 and the circuit in Figure 7-5 are two multiplying factors. These factors convert the approximate circuit in Figure 7-8 to the exact equivalent circuit in Figure 7-5. If we look at these factors we can estimate when we can use the approximate circuit and when we need to use the exact circuit. Z_π in the approximate circuit is multiplied by the factor, $\dfrac{\sinh\sqrt{ZY}}{\sqrt{ZY}}$, to obtain the exact value. Since

$$\sinh x = x + \frac{x^3}{3!} + \frac{x^5}{5!} + \frac{x^7}{7!} + \ldots$$

$$\frac{\sinh\sqrt{ZY}}{\sqrt{ZY}} = \frac{1}{\sqrt{ZY}}\left[\sqrt{ZY} + \frac{(\sqrt{ZY})^3}{3!} + \frac{(\sqrt{ZY})^5}{5!} + \ldots\right] =$$

$$\frac{\sinh\sqrt{ZY}}{\sqrt{ZY}} = 1 + \frac{ZY}{3!} + \frac{(ZY)^2}{5!} + \frac{(ZY)^3}{7!} + \ldots \tag{7-20}$$

$$\tanh x = x - \frac{1}{3}x^3 + \frac{2}{15}x^5 + \ldots$$

so

$$\frac{\tanh\dfrac{\sqrt{ZY}}{2}}{\dfrac{\sqrt{ZY}}{2}} = \frac{2}{\sqrt{ZY}}\left[\frac{\sqrt{ZY}}{2} - \frac{1}{3}\frac{(\sqrt{ZY})^3}{8} + \frac{2}{15}\frac{(\sqrt{ZY})^5}{(32)} - \ldots\right] =$$

$$\frac{\tanh\dfrac{\sqrt{ZY}}{2}}{\dfrac{\sqrt{ZY}}{2}} = 1 - \frac{ZY}{12} + \frac{(ZY)^2}{120} - \ldots \tag{7-21}$$

Both multiplying factors approach 1 as ZY approaches zero, which is the same as saying that as the line becomes electrically shorter the exact circuit parameters approach the approximate circuit parameters. When the second and following terms become insignificant, the engineer may use the approximate circuit which is easy to find in terms of the line parameters.

7.5 Example of Finding the Correction Factors for the Approximate Equivalent Circuits

To obtain a feeling for the correction factors studied in Section 7.4, let us calculate them for a line with two different lengths, 25 and 100 miles. Given line data:

$$z = R + j\omega L = 0.694 + j3.939 = 4.0 \angle 80^\circ \text{ ohms per mile.}$$

$$y = G + j\omega C = 0 + j26 \times 10^{-6} = 26 \times 10^{-6} \angle 90^\circ \text{ mhos per mile.}$$

Solution

For a 25 mile length we would have the following results:

$$Z = z\ell = (4.0 \angle 80^\circ)(25) = 100 \angle 80^\circ = 17.364 + j98.48 \text{ ohms.}$$

$$Y = y\ell = (26 \times 10^{-6} \angle 90^\circ)(25) = 650 \times 10^{-6} \angle 90^\circ = 0 + j650 \times 10^{-6} \text{ mhos.}$$

$$ZY = (100 \angle 80^\circ)(650 \times 10^{-6} \angle 90^\circ) = 65 \times 10^{-3} \angle 170^\circ.$$

$$\sqrt{ZY} = \sqrt{65 \times 10^{-3} \angle 170^\circ} = 254.95 \times 10^{-3} \angle 85^\circ = 22.22 \times 10^{-3} + 253.98 \times 10^{-3}.$$

$$\sinh \sqrt{ZY} = \sinh 22.2 \times 10^{-3} \cos 253.98 \times 10^{-3} + j\cosh 22.2 \times 10^{-3} \sin 253.98 \times 10^{-3} = 253.2398 \times 10^{-3} \angle 85.108^\circ.$$

$$\frac{\sinh \sqrt{ZY}}{\sqrt{ZY}} = \frac{252.239 \times 10^{-3} \angle 85.108^\circ}{254.95 \times 10^{-3} \angle 85^\circ} = 0.98936 \angle 0.108^\circ = \frac{\sinh\Theta}{\Theta}.$$

If we use the series correction factor we need only the first two terms for reasonable accuracy.

$$\frac{\sinh \sqrt{ZY}}{\sqrt{ZY}} = 1 + \frac{ZY}{3!} = 1 + \frac{65 \times 10^{-3} \angle 170^\circ}{6} =$$
$$1 + 0.01083 \angle 170^\circ = 0.9893 \angle 0.108^\circ.$$

Since we are in the range we can also use our curves. Figure 7-2 gives a value of 0.9895 $\angle 0.11^\circ$ for the sin*h* function. For 25 miles, the correction factor is almost equal to one and is found much more easily by using the curves or the series expansion. The values found by any of the three methods are equal for all practical purposes.

The correction factor for the shunt admittance can also be found by three different means.

$$\frac{\sqrt{ZY}}{2} = \frac{254.95 \times 10^{-3} \angle 85^\circ}{2} = 127.475 \times 10^{-3} \angle 85^\circ =$$
$$11.11 \times 10^{-3} + j126.98 \times 10^{-3}.$$

$$\tanh \frac{\sqrt{ZY}}{2} = \tanh(11.11 \times 10^{-3} + j126.98 \times 10^{-3}).$$

We can use the identity

$$\tanh(U + jV) = \frac{\sinh 2U + j\sin 2V}{\cosh 2U + \cos 2V}.$$

$$\frac{\tanh \sqrt{ZY}}{2} = \frac{\sinh 22.22 \times 10^{-3} + j\sin 253.979 \times 10^{-3}}{\cosh 22.22 \times 10^{-3} + \cos 253.979 \times 10^{-3}} =$$
$$128.159 \times 10^{-3} \angle 84.945^\circ = 11.292 \times 10^{-3} + j127.66 \times 10^{-3}.$$

$$\frac{\tanh \frac{\sqrt{ZY}}{2}}{\frac{\sqrt{ZY}}{2}} = \frac{128.159 \times 10^{-3} \angle 84.945^\circ}{127.475 \times 10^{-3} \angle 85^\circ} = 1.00536 \angle -0.0543^\circ.$$

We will obtain almost the same answer with only two terms of the series expansion.

$$\frac{\tanh \frac{\sqrt{ZY}}{2}}{\frac{\sqrt{ZY}}{2}} = 1 - \frac{ZY}{12} = 1 - \frac{65 \times 10^{-3} \angle 170°}{12} =$$

$1 - 5.416 \times 10^{-3} \angle 170° = 1 + 5.334 \times 10^{-3} - j940.594 \times 10^{-6} =$
$1.0053 - j940.594 \times 10^{-6} = 1.0053 \angle -0.0536°$.

Figure 7-7 gives the correction factor for the hyperbolic tangent function as $1.0055 \angle -0.05°$. For 25 miles of line this correction factor is also almost equal to one and the factors found by any of the three methods are again almost equal. When we apply the correction factors to obtain the exact equivalent circuit we can compare the two circuits as shown in Figure 7-9. They are almost the same.

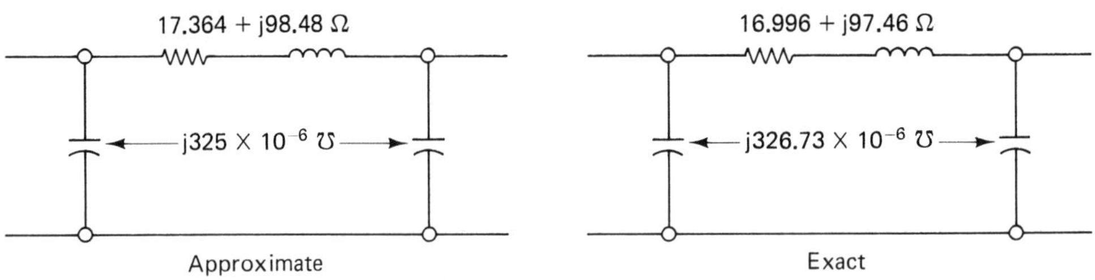

Figure 7-9. Approximate and exact equivalent circuits for 25-mile line.

Repeating the same type of calculations for the 100 mile line we obtain the following results:

$$Z = z\ell = (4 \angle 80°)(100) = 400 \angle 80° = 69.459 + j393.92 \text{ ohms.}$$

$$Y = y\ell = (26 \times 10^{-6})(100) = 2.6 \times 10^{-3} \angle 90° = 0 + j2.6 \times 10^{-3} \text{ mhos.}$$

$$ZY = (400 \angle 80°)(2.6 \times 10^{-3} \angle 90°) = 1.04 \angle 170°.$$

$$\frac{\sinh \sqrt{ZY}}{\sqrt{ZY}} = 838.018 \times 10^{-3} \angle 1.854° = 0.838 \angle 1.854°.$$

$$\frac{\tanh \frac{\sqrt{ZY}}{2}}{\frac{\sqrt{ZY}}{2}} = 1.0948 \angle -0.977°.$$

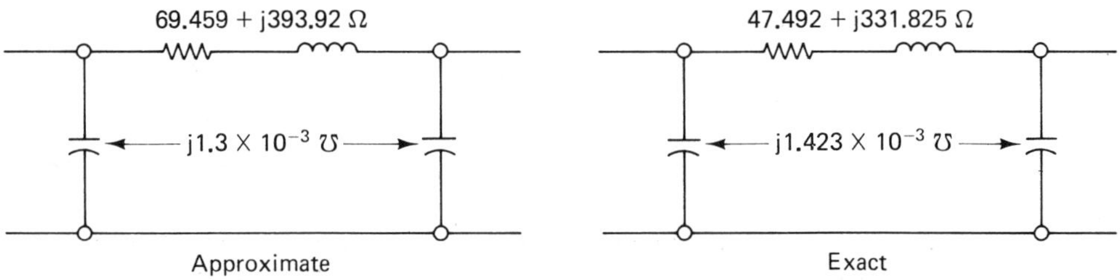

Figure 7-10. Approximate and exact equivalent circuits for 100-mile line.

The two circuits are shown in Figure 7-10.

As we compare the two equivalent circuits it is obvious that there is considerable difference between them for the longer line, and that the approximate circuit solution will give the engineer only an indication of the true picture of what is happening on the line.

7.6 Equivalent "T" Circuit of a Transmission Line

An equivalent "T" circuit may also be found for a transmission line as shown in Figure 7-11. By following the same procedures that we did for the equivalent pye circuit we find that the following relationships are true:

$$Y_T = Y\frac{\sinh\sqrt{ZY}}{\sqrt{ZY}} = Y\frac{\sinh\Theta}{\Theta}. \tag{7-22}$$

$$\frac{Z_T}{2} = \frac{Z}{2}\frac{\tanh\frac{\sqrt{ZY}}{2}}{\frac{\sqrt{ZY}}{2}} = \frac{Z}{2}\frac{\tanh\frac{\Theta}{2}}{\frac{\Theta}{2}} \tag{7-23}$$

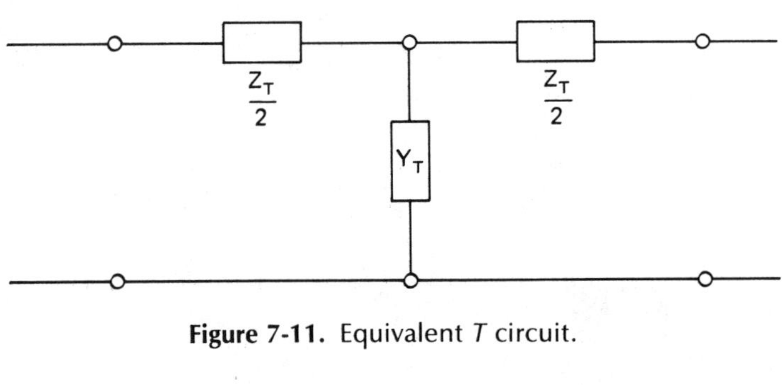

Figure 7-11. Equivalent *T* circuit.

7.7 D. C. Transmission Lines

A d. c. transmission line was put into operation in Sweden in 1954. It transmitted 20,000 kw at 100 kv, d. c., from the mainland to the island of Gotland located in the Baltic sea, 60 miles from the mainland. A single-conductor submarine cable was used as the transmission line with the sea as the return conductor. However, until recent years, long d. c. transmission lines were primarily concerned with telegraph engineering practice although in this case a constant d. c. current serves no purpose. The current must be interrupted so that the receiving end can detect useful information on the line. Since the rate at which dots and dashes can be generated by a human operator is limited, the d. c. telegraph line, in operation, becomes a low frequency line with non-sinusoidal signals being propagated from one end to the other. A straight d. c. line at low voltage becomes one with small currents and the currents through the shunt conductance may be of considerable importance. The telegraph lines were single wire with a ground return or they were constructed with two wires. In recent years, d. c. lines have become important for long-distance power transmission. They are usually a two-wire line with one wire positive with respect to ground and the other negative with respect to ground.

For those circuits where the operating speed is comparatively low, a fair degree of accuracy may be obtained by simply treating the d. c. line as a special case of the a. c. line with zero frequency. The currents on the line may be calculated with the key open or closed if we are considering the telegraph case. We neglect L and C and use only R and G. Under these conditions we would have:

$$Z_o = \sqrt{\frac{(R + j\omega L)}{(G + j\omega C)}} = \sqrt{\frac{R}{G}} = R_o. \tag{7-24}$$

$$\gamma = \sqrt{(R + j\omega L)(G + j\omega C)} = \sqrt{RG} = \alpha + j\beta. \tag{7-25}$$

$$\alpha = \sqrt{RG},\ \beta = 0.$$

We can also write:

$$\gamma\ell = \sqrt{(R\ell)\,(G\ell)}. \qquad \sqrt{RG}\,\ell = \alpha\ell = \Theta. \tag{7-26}$$

And as before:

$$V_s = V_r\cosh\Theta + I_rZ\frac{\sinh\Theta}{\Theta} = V_r\cosh\Theta + I_r(R\ell)\frac{\sinh\Theta}{\sqrt{(R\ell)(G\ell)}} =$$

$$V_r\cosh\Theta + I_r\sqrt{\frac{R^2\ell^2}{(R\ell)(G\ell)}}\sinh\Theta = V_r\cosh\Theta + I_r\sqrt{\frac{R\ell}{G\ell}}\sinh\Theta =$$

$$V_s = V_r\cosh\Theta + I_rR_o\sinh\Theta. \quad (7\text{-}27)$$

Also:

$$I_s = I_r\cosh\Theta + \frac{V_r}{R_o}\sinh\Theta. \quad (7\text{-}28)$$

$$V_r = V_s\cosh\Theta - I_sR_o\sinh\Theta. \quad (7\text{-}29)$$

$$I_r = I_s\cosh\Theta - \frac{V_s}{R_o}\sinh\Theta. \quad (7\text{-}30)$$

In this case we are dealing with the hyperbolic functions of real arguments and we can use the following equations:

$$\sinh\Theta = \Theta + \frac{\Theta^3}{3!} + \frac{\Theta^5}{5!} + \frac{\Theta^7}{7!} + \ldots$$

$$\cosh\Theta = 1 + \frac{\Theta^2}{2!} + \frac{\Theta^4}{4!} + \frac{\Theta^6}{6!} + \ldots$$

Or we can use the tables in the handbooks listed in the reference section.

7.8 Example of a Solution for a D. C. Line

Let us assume that we have a d. c. line with the following parameters: $R = 5$ ohms/mile; $G = 1 \times 10^{-5}$ mhos/mile; $V_s = 100$ volts; $R_{load} = 1{,}000$ ohms. Length of the line, ℓ, is 50 miles.

Find:

$$V_r, I_r, I_s, P_r, P_s.$$

Solution

$$R_o = \sqrt{\frac{R}{G}} = \sqrt{\frac{5}{1 \times 10^{-5}}} = \sqrt{5 \times 10^5} = 707.11 \text{ ohms.}$$

$$\Theta = \sqrt{(R\ell)(G\ell)} = \sqrt{RG}\,\ell = \sqrt{(5)(1 \times 10^{-5})}\,(50) = 0.353.$$

From tables:

$$\sinh\Theta = \sinh 0.353 = 0.360$$
$$\cosh\Theta = \cosh 0.353 = 1.0629$$

$$V_s = V_r\cosh\Theta + I_rR_o\sinh\Theta.$$

$$100 = V_r(1.0629) + \frac{V_r\,(707.11)(0.360)}{1{,}000} = 1.317\,V_r.$$

$$V_r = \frac{100}{1.317} = 75.90 \text{ volts. } I_r = \frac{V_r}{Z_r} = \frac{75.9}{1{,}000}\,75.9 \times 10^{-3} \text{ amperes.}$$

$$I_s = I_r\cosh\Theta + \frac{V_r}{R_o}\sinh\Theta =$$

$$(75.9 \times 10^{-3})(1.0629) + \frac{(75.9)(0.360)}{1{,}000} = 108.0 \times 10^{-3} \text{ amperes.}$$

$$P_r = V_rI_r = (75.90)(75.9 \times 10^{-3}) = 5.76 \text{ watts.}$$
$$P_s = V_sI_s = (100)(108 \times 10^{-3}) = 10.80 \text{ watts.}$$

7.9 Chapter Summary

1. While the hyperbolic equations in Chapter 6 may be used by the power engineer or by any engineer for all the lossy cases, the power engineer usually likes to modify his equations and use them in a slightly different form.
2. New quantities defined by the power engineer are:

$$Z_o = \sqrt{\frac{(R + j\omega L)\ell}{(G + j\omega C)\ell}} = \sqrt{\frac{Z}{Y}} = \sqrt{\frac{(R + j\omega L)}{(G + j\omega C)}}. \qquad (7\text{-}3)$$

The power engineer calls this the "Surge" impedance but it is the same as the characteristic impedance.

$$Z = (R + j\omega L)\ell, \text{ the total series impedance of the line.} \quad (7\text{-}1)$$

$$Y = (G + j\omega C)\ell, \text{ the total shunt admittance of the line.} \quad (7\text{-}2)$$

$$\gamma\ell = \sqrt{(R + j\omega L)\ell(G + j\omega C)\ell} = \sqrt{ZY} = \Theta. \quad (7\text{-}4)$$

3. Using these new quantities the lossy line equations become:

$$V_s = V_r\cosh\Theta + I_rZ\frac{\sinh\Theta}{\Theta}. \quad (7\text{-}7)$$

$$I_s = I_r\cosh\Theta + V_rY\frac{\sinh\Theta}{\Theta}. \quad (7\text{-}8)$$

$$V_r = V_s\cosh\Theta - I_sZ\frac{\sinh\Theta}{\Theta}. \quad (7\text{-}9)$$

$$I_r = I_s\cosh\Theta - V_sY\frac{\sinh\Theta}{\Theta}. \quad (7\text{-}10)$$

4. In terms of the distance from the sending end, x, our lossy equations become:

$$V = V_s\cosh\gamma x - I_sZ_o\sinh\gamma x. \quad (7\text{-}11)$$

$$I = I_s\cosh\gamma x - \frac{V_s}{Z_o}\sinh\gamma x. \quad (7\text{-}12)$$

V and I, as before, are the voltage and current anywhere on the line at point x.

5. Curves are available for the functions $\cosh\Theta = \cosh\sqrt{ZY}$ and $\frac{\sinh\Theta}{\Theta} = \frac{\sinh\sqrt{ZY}}{\sqrt{ZY}}$.
6. For three-phase transmission lines the calculations are made on a per-phase basis assuming a wye connected circuit. All voltages such as V_r, V_s, etc. are on a line-to-neutral basis. All currents such as I_r, I_s, etc. are line currents. The power per-circuit or per-phase is one third of the total three-phase power.

   $P_{total} = \sqrt{3}\, V_{\ell-\ell}I_\ell\cos\theta$, where θ is actually the angle between the phase line-to-neutral voltage and the phase line current and not the angle between the line-to-line voltage, $(V_{\ell-\ell})$ and the line current, (I_ℓ) used in the equation.

$$P_{phase} = \frac{P_{total}}{3} = V_pI_p\cos\theta_p$$

where V_p and I_p are the line-to-neutral and line current current respectively, and θ_p is the angle between them. As always, these equations are only true if the system load is balanced.

7. Any transmission line may be replaced by an equivalent circuit. Two possible choices are the pye and T circuits. Each of these may be developed into an exact or an approximate circuit. For shorter lines the approximate equivalent circuit is easy to find and is almost the same as the exact circuit. For longer lines the exact equivalent circuit should be used unless only very rough answers are needed. The exact circuit can be found from the approximate circuit by applying some correction factors which are fairly easy to calculate or which may be found in most cases by the use of curves. One of these correction factors is

$$\frac{\tanh\frac{\Theta}{2}}{\frac{\Theta}{2}} = \frac{\tanh\frac{\sqrt{ZY}}{2}}{\frac{\sqrt{ZY}}{2}}.$$

8. The correction factors are:

$$\frac{\sinh\Theta}{\Theta} = \frac{\sinh\sqrt{ZY}}{\sqrt{ZY}} = 1 + \frac{ZY}{3!} + \frac{(ZY)^2}{5!} + \frac{(ZY)^3}{7!} + \ldots \quad (7\text{-}20)$$

$$\frac{\tanh\Theta/2}{\Theta/2} = \frac{\tanh\sqrt{ZY}/2}{\sqrt{ZY}/2} = 1 - \frac{ZY}{12} + \frac{(ZY)^2}{120} - \ldots \quad (7\text{-}21)$$

9. When the correction factors are close to unity the approximate equivalent circuit is almost the same as the exact equivalent circuit and may be used with excellent accuracy.
10. D. C. lines are once again becoming important to the power engineer. The d. c. line in the steady state can be thought of as an a. c. line with zero frequency. In the transient state other techniques must be used.

7.10 Review and Study Questions

1. For three-phase circuits define the following quantities: (a) line-to-neutral voltage, (b) line-to-line voltage, (c) phase sequence ABC, (d) phase sequence ACB, (e) line current, (f) phase current, (g) per-phase power, (h) total power, (i) wye circuit, (j) delta

circuit, (k) methods of solutions for balanced circuits, (l) methods of solutions for unbalanced circuits.

2. For a balanced wye-connected three-phase circuit what is the relationship between the line-to-line voltages and the line-to-neutral voltages?
3. For a balanced delta-connected three-phase circuit what is the relationship between the line currents and the delta (phase) currents?
4. What is the surge impedance? What is the surge admittance?
5. How does the surge impedance compare with the characteristic impedance of the line?
6. Why does a transmission line have a sending-end current when the receiving end is open-circuited?
7. Is it possible for the receiving-end voltage to be higher than the sending-end voltage? If so, under what conditions might it happen?
8. Sketch two possible approximate equivalent circuits for a lossy transmission line.
9. Describe three possible ways of finding the correction factors used to convert the approximate equivalent circuit to the exact one.
10. Is it possible for the receiving-end current to be higher than the sending-end current? If so, under what conditions might it happen?
11. Can a transmission line be (a) electrically short and physically long? (b) electrically long and physically short? Explain your answers.
12. How can the losses on a transmission line be determined?
13. How can the efficiency of a transmission line be determined?
14. Give the expression for the $\sinh(x + jy)$ in exponential form.
15. Give the expression for the $\cosh(x + jy)$ in exponential form.
16. What advantages might a d. c. power transmission line have over an a. c. power transmission line? How do the field effects compare?

Problems

7-1. A three-phase transmission line has the following parameters on a per-phase basis: $R = 0.5$ ohm/mile; $\omega L = 2.0$ ohms/mile; $G = 0$; $\omega C = 0.9 \times 10^{-5}$ mhos/mile. The line is 80 miles long.

(a) Determine Z and Y.

(b) Determine $\cosh\Theta$ and $\dfrac{\sinh\Theta}{\Theta}$.

7-2. The line in problem 7-1 is delivering 80,000 kilowatts to a balanced load with a power factor of 0.85 lagging. The receiving-end voltage is 230 kilovolts, line-to-line. Determine the following:
(a) The sending-end voltage.
(b) The sending-end current.
(c) The sending-end power factor.
(d) The losses in the line. (total three-phase losses.)

7-3. The line in problem 7-1 is delivering 80,000 kilowatts to a balanced load with a power factor of 0.9 lagging. The sending-end voltage is 230 kilovolts, line-to-line. Determine the following:
(a) The receiving-end voltage.
(b) The receiving-end current.
(c) The sending-end power factor.
(d) The sending-end current.
(e) The sending-end power and the efficiency of the line.

7-4. Using the *Standard Handbook for Electrical Engineers* (which should be in your school library), determine the z and y per mile per phase for a 715,500 circular mil, ACSR conductor (26/7) if the equivalent spacing is 12 feet and the frequency is 60 hertz.

7-5. A single-phase power transmission line is 100 miles long and operates at 60 hertz. The constants per mile for both wires are given as: $R = 0.86$ ohms; $L = 3.978 \times 10^{-3}$ henrys; $G = 0$; $C = 4.244 \times 10^{-8}$ farads. The load is 2,500 kw at 0.8 power factor lagging when the receiving-end voltage is 40 kv. Determine:
(a) The sending-end voltage.
(b) The receiving-end current.
(c) The sending-end current.
(d) The sending-end power factor.
(e) The sending-end power.
(f) The losses in the line.

7-6. The line in problem 7-5 has 35,000 volts applied to the sending end with no load on the receiving end (open circuit) Determine:
(a) The receiving-end voltage.
(b) The sending-end current.
(c) The sending-end power.
(d) The sending-end power factor.

7-7. A three-phase power transmission line has the following parameters given by another engineer: $\cosh\Theta = 0.875 \angle 1.41°$,

$\dfrac{\sinh\Theta}{\Theta} = 0.9575 \angle 0.438°$, $Z = 200 \angle 80°$, $Y = 0.0013 \angle 90°$. The power at the receiving end is 120,000 kw, at a power factor of 0.85 lagging. The receiving-end voltage is 230 kilovolts, line-to-line. Determine:

(a) The sending-end voltage.
(b) The sending-end current.
(c) The sending-end power factor.
(d) The losses in the line.

7-8. Find the approximate and exact equivalent pye circuits for the line in problem 7-1. Then using the exact circuit resolve problem 7-2.

7-9. Derive equations (7-22) and (7-23).

7-10. Find the approximate and exact equivalent T circuits for the line in problem 7-1.

7-11. Find the load impedance per phase for the load in problem 7-2. How does this compare with the characteristic impedance of the line? What is the voltage reflection coefficient for this load?

7-12. A three-phase transmission line has a series impedance of $2.113 + j4.5315$ ohms per mile per phase and a shunt admittance of $0 + j20 \times 10^{-6}$ mhos per mile per phase. The line is 50 miles long. Determine for this line:

(a) $\cosh\Theta$.
(b) $\dfrac{\sinh\Theta}{\Theta}$.

7-13. For the line in problem 7-12 we will assume a load of 20,000 kilowatts at 44 kilovolts, line-to-line, and a power factor of 0.82 lagging. Determine the following:

(a) Sending-end current.
(b) Sending-end voltage.
(c) Efficiency.

7-14. The line in problem 7-12 is supplying a load of 20,000 kw at a power factor of 0.82 lagging, at 230 kv, line-to-line, instead of the 44 kv specified in problem 7-13. Determine the following:

(a) Sending-end current.
(b) Sending-end voltage.
(c) Compare the efficiency in this problem with the efficiency in problem 7-13.

7-15. The pye equivalent circuit is an example of a four-terminal (three node) circuit. When balanced, as in the following figure, the open-circuit impedance or the short-circuit impedance would be the same at either end. It can be shown that the characteristic impedance, Z_o, can be found by measuring the short-circuit and the open-circuit impedances.

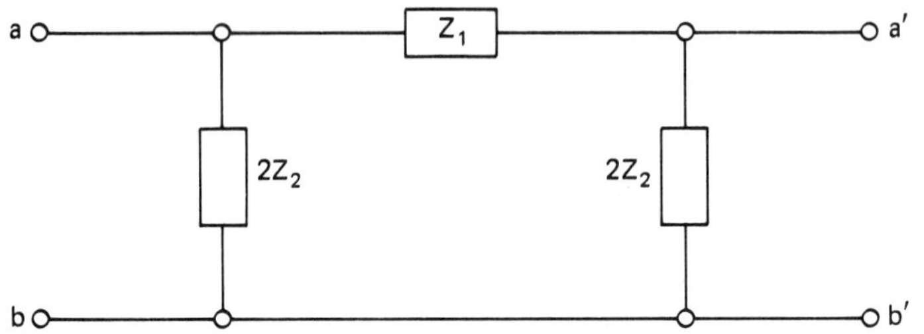

$$Z_o = \sqrt{Z_{sc} Z_{oc}}$$

For our circuit,

$$Z_o = \frac{\sqrt{Z_1 Z_2}}{T}.$$

Find T, in terms of Z_1 and Z_2.

8

Bounce Diagrams

8.1 Introduction

We have been discussing, for the most part, the sinusoidal, steady-state, response of transmission lines. This is an important part of our study since in many practical cases we have either a single-frequency sinusoidal wave transmitting power or a complex wave transmitting some sort of electrical signal. The complex wave often may be resolved into a number of sinusoidal-type waves with different frequencies and magnitudes. Using superposition the engineer can often treat the action of the transmission line on the different frequency components and achieve an understanding of how the output signal compares with the input signal. If signal reproduction at the receiving end is important, it is often possible to take steps to achieve this even without a complete analysis of the waveform.

However, many signals on a line are not of a form that is easy to analyze except by using special techniques. These signals often take the form of transient type waves which are either man-made or nature-made and become very important in design, analysis and, in many cases, equipment protection. These transients may be induced on power lines by faults on the system, by lightning strokes, or by sudden and large changes in loading. Transients on lines may be very important when one considers what happens as the pulses travel down the line. Modern electronic digital computers operate with pulses and the problem of communication between computers or between parts of the same computer involves an understanding of the transient operation of transmission lines. This analysis may be extremely important even though the distance the signal is traveling may be physically very short and is especially important as engineers try to design faster and faster operation into their digital circuits.

In this chapter we will consider only one special case of transients on a line. Although our study will be restricted it will present some fundamental ideas which again are more important than the special case we are discussing.

8.2 Fundamental Equations for the Instantaneous Values of Voltage and Current on the Line

If we assume a lossless line with $R = G = 0$, we have as before:

$$\gamma = \alpha + j\beta = \sqrt{(R + j\omega L)(G + j\omega C)} = 0 + j\omega\sqrt{LC},$$

$$\text{so } \alpha = 0 \text{ and } \beta = \omega\sqrt{LC}.$$

$$Z_o = \sqrt{\frac{(R + j\omega L)}{(G + j\omega C)}} = \sqrt{\frac{L}{C}}. \qquad v_p = \frac{\omega}{\beta} = \frac{\omega}{\omega\sqrt{LC}} = \frac{1}{\sqrt{LC}}.$$

So, v_p, the phase velocity, is independent of the frequency. Let us go back to some fundamental equations from Chapter 1.

$$\frac{\partial v}{\partial x} = -Ri - L\frac{\partial i}{\partial t}. \tag{1-10}$$

$$\frac{\partial i}{\partial x} = -Gv - C\frac{\partial v}{\partial t}. \tag{1-11}$$

$$\frac{\partial^2 v}{\partial x^2} = RGv + (RC + LG)\frac{\partial v}{\partial t} + LC\frac{\partial^2 v}{\partial t^2}. \tag{1-16}$$

$$\frac{\partial^2 i}{\partial x^2} = RGi + (RC + LG)\frac{\partial i}{\partial t} + LC\frac{\partial^2 i}{\partial t^2}. \tag{1-17}$$

When $R = G = 0$ for the lossless case we have:

$$\frac{\partial^2 v}{\partial x^2} = LC\frac{\partial^2 v}{\partial t^2} = \frac{1}{v_p^2}\frac{\partial^2 v}{\partial t^2}. \tag{1-18}$$

$$\frac{\partial^2 i}{\partial x^2} = LC\frac{\partial^2 i}{\partial t^2} = \frac{1}{v_p^2}\frac{\partial^2 i}{\partial t^2}. \tag{1-19}$$

As we saw in Chapter 1, the solution is any function of $(x - v_p t)$, or $(x + v_p t)$, such as $\cos(x - v_p t)$, $(x + v_p t)^3$, etc. Let us assume the following solutions:

$$v = m(x - v_pt) + n(x + v_pt). \tag{8-1}$$

$$i = Mm(x - v_pt) + Nn(x + v_pt). \tag{8-2}$$

M and N are constants and $m(x - v_pt)$ and $n(x + v_pt)$ represent any m and n functions of $(x - v_pt)$ and $(x + v_pt)$ respectively. Let us look at the forward wave only as a starting point.

$$\frac{v}{i} = \frac{m(x - v_pt)}{Mm(x - v_pt)} = \frac{1}{M}. \tag{8-3}$$

With $R = 0$, equation (1-10) becomes:

$$\frac{\partial v}{\partial x} = -L\frac{\partial i}{\partial t}. \tag{8-4}$$

$$\begin{aligned} v = m(x - v_pt), \text{ so } \frac{\partial v}{\partial x} &= \frac{\partial m(x - v_pt)}{\partial(x - v_pt)}\frac{\partial(x - v_pt)}{\partial x} = \\ &\frac{\partial m(x - v_pt)}{\partial(x - v_pt)}(1) = \frac{\partial v}{\partial x} = \frac{\partial m(x - v_pt)}{\partial(x - v_pt)}. \end{aligned} \tag{8-5}$$

$$\begin{aligned} i = Mm(x - v_pt), \text{ so } \frac{\partial i}{\partial t} &= \frac{M\partial m(x - v_pt)}{\partial(x - v_pt)}\frac{\partial(x - v_pt)}{\partial t} = \\ &\frac{M\partial m(x - v_pt)}{\partial(x - v_pt)}(-v_p) = \frac{\partial i}{\partial t}. \end{aligned} \tag{8-6}$$

Substituting (8-5) and (8-6) into equation (8-4) gives us:

$$\frac{\partial v}{\partial x} = -L\frac{\partial i}{\partial t}.$$

$$\frac{\partial m(x - v_pt)}{\partial(x - v_pt)} = -L\,M\frac{\partial m(x - v_pt)}{\partial(x - v_pt)}(-v_p)$$

or

$$\begin{aligned} 1 = MLv_p. \qquad M &= \frac{1}{Lv_p} = \frac{1}{L\frac{1}{\sqrt{LC}}} = \sqrt{\frac{LC}{L^2}} = \\ \sqrt{\frac{C}{L}} &= M = \frac{1}{\sqrt{\frac{L}{C}}} = \frac{1}{Z_o}. \end{aligned} \tag{8-7}$$

Going back to equation (8-3) gives us the result:

$$\frac{v}{i} = \frac{1}{M} = \frac{1}{\frac{1}{Z_o}} = Z_o. \tag{8-8}$$

We come up with the remarkable result that on any lossless line the voltage to the current ratio is Z_o. Since we have been discussing functions of time, we are saying that at any instant of time the ratio of the instantaneous voltage to the instantaneous current is the characteristic impedance, Z_o. This is the same relationship that we found between the voltage and current when they were sinusoidal functions of time—the steady-state case.

If we look at the reverse (backward) components we can do the same thing as we did for the forward components.

$$v = n(x + v_p t). \qquad i = Nn(x + v_p t).$$

$$\frac{\partial v}{\partial x} = \frac{\partial n(x + v_p t)}{\partial(x + v_p t)} \frac{\partial(x + v_p t)}{\partial x} = \frac{\partial n(x + v_p t)}{\partial(x + v_p t)}(1) = \frac{\partial n(x + v_p t)}{\partial(x + v_p t)}.$$

$$\frac{\partial i}{\partial t} = \frac{N\partial n(x + v_p t)}{\partial(x + v_p t)} \frac{\partial(x + v_p t)}{\partial t} = \frac{N\partial n(x + v_p t)}{\partial(x + v_p t)}(v_p).$$

Using equation (8-4) we have

$$\frac{\partial v}{\partial i} = -L\frac{\partial i}{\partial t}.$$

$$\frac{\partial n(x + v_p t)}{\partial(x + v_p t)} = -L\frac{N\partial n(x + v_p t)}{\partial(x + v_p t}(v_p),$$

or

$$1 = -LNv_p. \qquad N = -\frac{1}{Lv_p} = -\frac{1}{L\frac{1}{\sqrt{LC}}} =$$

$$-\frac{1}{\sqrt{\frac{L^2}{LC}}} = -\frac{1}{\sqrt{\frac{L}{C}}} = -\frac{1}{Z_o}. \tag{8-9}$$

If we look at only the reverse components of equations (8-1) and (8-2) we have:

$$\frac{v}{i} = \frac{n(x + v_p t)}{Nn(x + v_p t)} = \frac{1}{N} = \frac{1}{-\dfrac{1}{Z_o}} = -Z_o. \tag{8-10}$$

For the reverse wave the ratio of the instantaneous voltage to the instantaneous current is the negative of the characteristic impedance. This seems logical since we have defined v as a voltage rise going from the bottom wire to the top wire for both cases but the direction of the current is reversed for the reverse current as illustrated in Figure 8-1.

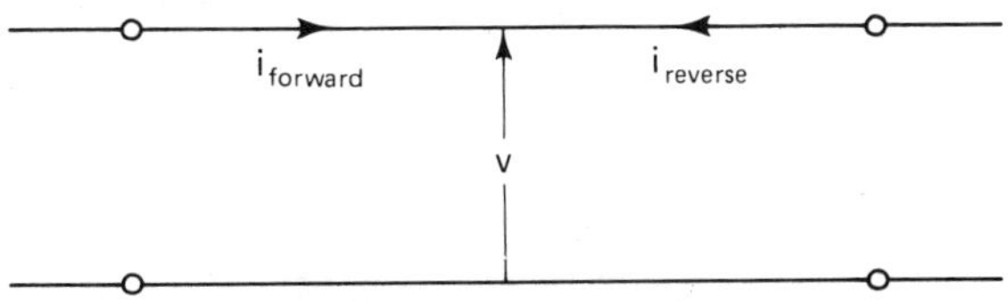

Figure 8-1. Forward and reverse currents.

So
$$\frac{v}{i}\text{(forward)} = Z_o. \tag{8-8}$$

And
$$\frac{v}{i}\text{(reverse)} = -Z_o. \tag{8-10}$$

8.3 A Study of the Special Transient D.C. Case

To illustrate our technique, we will assume that we are suddenly applying a d. c. voltage to a transmission line which has a characteristic impedance of Z_o, by closing a switch at $t = 0$ seconds. The source will have an internal resistance which is a real R_g ohms. The load will also be a pure resistance of R_r ohms. This situation is illustrated in Figure 8-2.

When we close the switch at $t = 0$ seconds, a voltage will appear at the input terminals of the line and a current will begin. We have

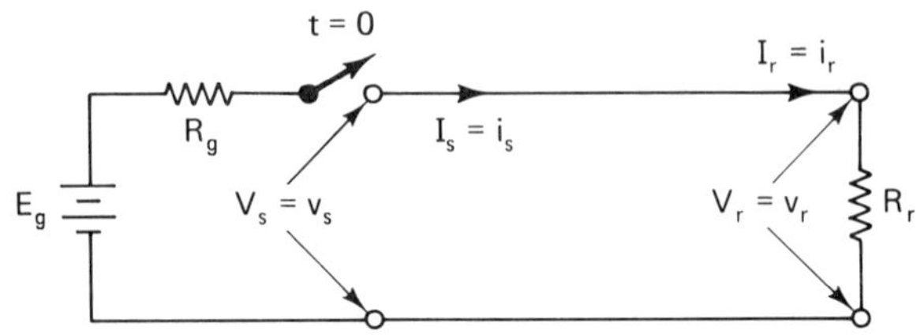

Figure 8-2. Our circuit for the Bounce Diagram study.

just seen that the instantaneous voltage divided by the instantaneous current is Z_o for the forward wave so, at the terminals at the instant the switch is closed, the input impedance (resistance) of the line will be Z_o (R_o). At $t = 0$ seconds the circuit diagram appears as shown in Figure 8-3.

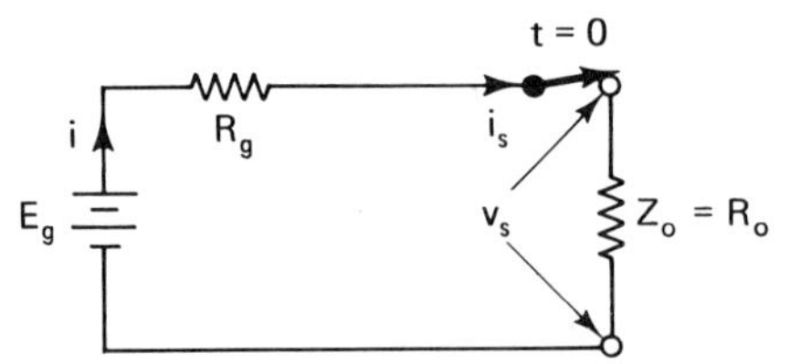

$$i = I = \frac{E_g}{R_g + Z_o}. \quad (8\text{-}11)$$

$$v_s = i_s Z_o = \frac{E_g Z_o}{R_g + Z_o}. \quad (8\text{-}12)$$

Figure 8-3. Circuit at $t = 0$ seconds.

At $t = 0$ seconds we have a d. c. value of voltage and current at the input terminals of the line. These instantaneous values of voltage and current do not appear everywhere along the line at that instant since they are traveling with a velocity of v_p meters per second and it takes time for the voltage and current to appear farther down the line. We have an initial pulse of voltage and an initial pulse of current with values determined by equations (8-11) and (8-12).

$$v_s = \frac{E_g Z_o}{R_g + Z_o}. \quad (8\text{-}13)$$

$$i_s = \frac{E_g}{R_g + Z_o}. \quad (8\text{-}14)$$

They are both traveling with a velocity of v_p. $v_p = 1/\sqrt{LC}$ meters/second. As time passes they would appear as shown in Figure 8-4.

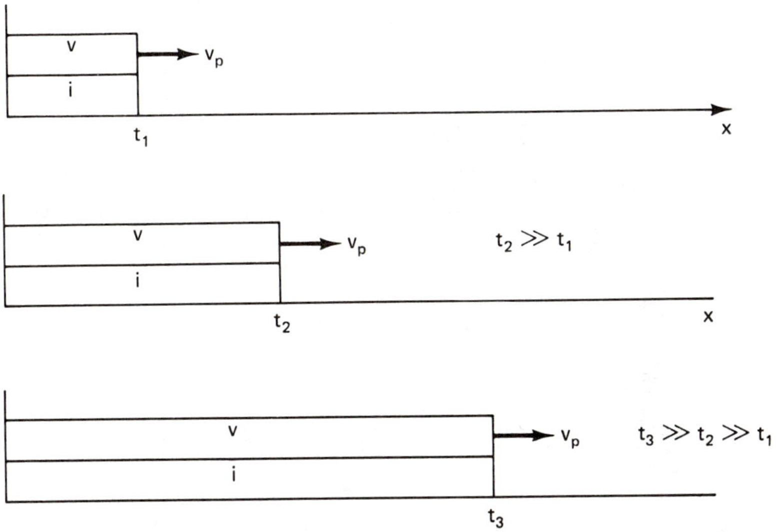

Figure 8-4. Instantaneous values of *v* and *i* for different values of time.

When the two pulses arrive at the load what happens? Let us first define some values of the voltages and currents on the line.

v^+ = voltage traveling towards the load (forward wave).
v^- = voltage traveling towards the generator (reverse wave).
i^+ = current traveling towards the load (forward wave).
i^- = current traveling towards the generator (reverse wave).

At the load we know that the following must be true.

$$v_r = v^+ + v^-. \qquad i_r = i^+ + i^- = \frac{v^+}{Z_o} - \frac{v^-}{Z_o},$$

since

$$i^- = i_{\text{reverse}} = -\frac{v^-}{Z_o}.$$

$$Z_r = R_r = \frac{v_r}{i_r} = \frac{v^+ + v^-}{\dfrac{v^+}{Z_o} - \dfrac{v^-}{Z_o}} = Z_o\frac{(v^+ + v^-)}{(v^+ - v^-)} = Z_o\frac{\left(1 + \dfrac{v^-}{v^+}\right)}{\left(1 - \dfrac{v^-}{v^+}\right)}.$$

Let us now define the voltage reflection coefficient at the receiving end as:

$$\Gamma_r = \frac{V^-}{V^+}. \tag{8-15}$$

Notice that this is a different reflection coefficient than the one defined in equation (3-7). Here we are defining it as the ratio of two instantaneous values whereas previously we defined it as the ratio of two phasor values. Our equation now becomes:

$$Z_r = R_r = Z_o\frac{(1 + \Gamma_r)}{(1 - \Gamma_r)}. \tag{8-16}$$

Solving for Γ_r gives the same form as we had for the sinusoidal case.

$$\Gamma_r = \frac{Z_r - Z_o}{Z_r + Z_o} = \frac{V^-}{V^+}. \tag{8-17}$$

All through the text, up to this point, we have been talking about the voltage reflection coefficient, stressing the voltage term. The reason was that now we are going to define for the first time a current reflection coefficient.

$$\Gamma_{r(i)} = \frac{i^-}{i^+} = -\frac{\dfrac{V^-}{Z_o}}{\dfrac{V^+}{Z_o}} = -\frac{V^-}{V^+} = -\Gamma_{r(v)}. \tag{8-18}$$

The current reflection coefficient turns out to be the negative of the voltage reflection coefficient. At this point it will be convenient to look at a specific example. Suppose we have the conditions shown in Figure 8-5.

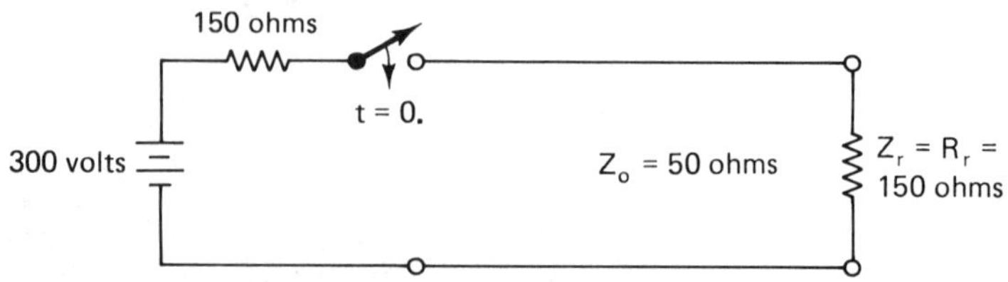

Figure 8-5. Line before switch is closed.

At $t = 0$ seconds we have:

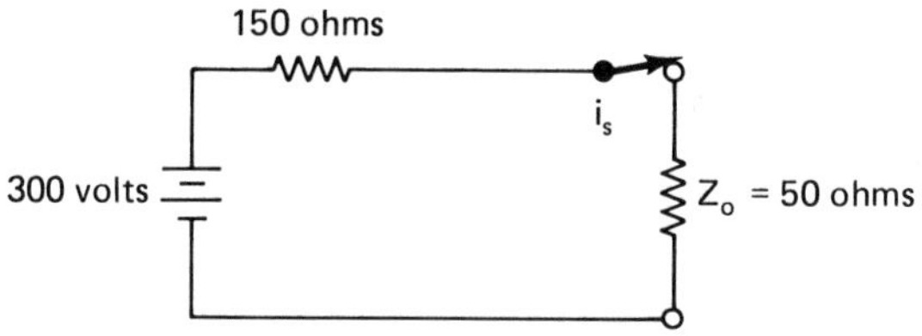

$$I_s = i_s = \frac{300}{150 + 50} = \frac{300}{200} = 1.5 \text{ amperes.}$$
$$V_s = v_s = i_s Z_o = (1.5)(50) = 75 \text{ volts.}$$

So the initial pulse of voltage with a value of 75 volts travels down the line towards the receiving end. The initial pulse of current with a value of 1.5 amperes travels along with the voltage at the same velocity, v_p, meters per second. At the receiving end:

$$\Gamma_{r(v)} = \frac{Z_r - Z_o}{Z_r + Z_o} = \frac{150 - 50}{150 + 50} = \frac{100}{200} = 0.5 = \frac{v^-}{v^+},$$

so

$$v = 0.5\ v^+.$$

As the initial voltage pulse of 75 volts hits the load end there is a reflected voltage pulse of $(0.5)(75) = 37.5$ volts which now travels towards the sending end at a velocity of v_p.

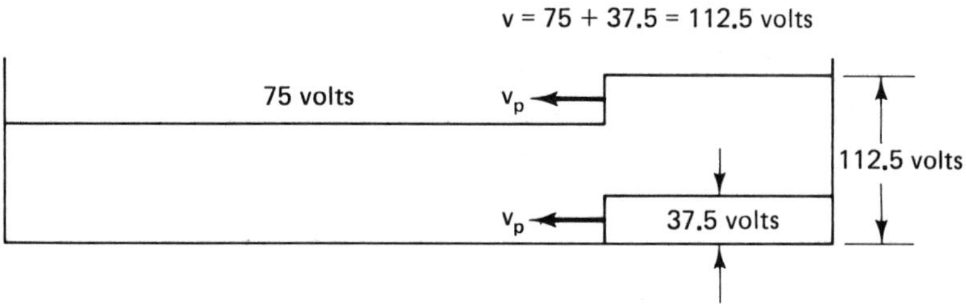

We can see that the voltage becomes 112.5 volts as the 37.5 volt pulse adds to the 75 volts that was present on the line after the original pulse reached the receiving end. The question can then be asked—What happens at the sending end as the 37.5 volt pulse arrives there? In Figure 8-6, we define three quantities at the sending end.

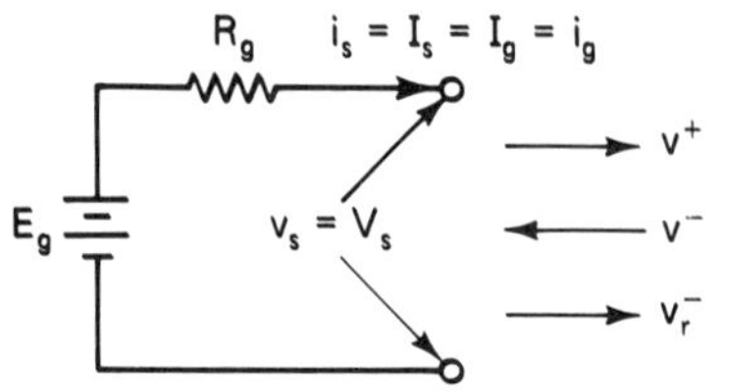

Figure 8-6. Conditions at the sending end.

v_r^- represents the reflected component of v^- as it hits the sending end. Circuit equations give us:

$$v_s = V_s = E_g - i_g R_g.$$

$$i_g = \frac{E_g - v_s}{R_g} = \frac{E_g}{R_g} - \frac{v_s}{R_g}.$$

$$V_s = v_s = v^+ + v^- + v_r^-.$$

But

$$i_s = \frac{v^+}{Z_o} + \frac{v_r^-}{Z_o} - \frac{v^-}{Z_o} = i_g.$$

So

$$\frac{E_g}{R_g} - \frac{v_s}{R_g} = \frac{v^+}{Z_o} + \frac{V_r^-}{Z_o} - \frac{v^-}{Z_o},$$

and

$$\frac{E_g}{R_g} - \frac{(v^+ + v^- + v_r^-)}{R_g} = \frac{v^+}{Z_o} + \frac{v_r^-}{Z_o} - \frac{v^-}{Z_o}.$$

$$\frac{E_g}{R_g} - \frac{v^+}{R_g} - \frac{v^-}{R_g} - \frac{v_r^-}{R_g} = \frac{v^+}{Z_o} + \frac{v_r^-}{Z_o} - \frac{v^-}{Z_o}$$

but

$$v^+ = \frac{E_g Z_o}{R_g + Z_o},$$

so we can substitute:

$$\frac{E_g}{R_g} - \frac{1}{R_g}\frac{E_gZ_o}{(R_g + Z_o)} - \frac{v^-}{R_g} - \frac{v_r^-}{R_g} = \frac{E_gZ_o}{Z_o(R_g + Z_o)} + \frac{v_r^-}{Z_o} - \frac{v^-}{Z_o}.$$

$$\frac{E_g(R_g + Z_o) - E_gZ_o}{R_g(R_g + Z_o)} - \frac{v^-}{R_g} - \frac{v_r^-}{R_g} = \frac{E_g}{(R_g + Z_o)} + \frac{v_r^-}{Z_o} - \frac{v^-}{Z_o}.$$

$$\frac{E_g}{R_g + Z_o} - \frac{v^-}{R_g} - \frac{v_r^-}{R_g} = \frac{E_g}{(R_g + Z_o)} + \frac{v_r^-}{Z_o} - \frac{v^-}{Z_o}.$$

So

$$v^-\left[\frac{1}{Z_o} - \frac{1}{R_g}\right] = v_r^-\left[\frac{1}{Z_o} + \frac{1}{R_g}\right] \text{ or } v^-\frac{(R_g - Z_o)}{R_gZ_o} = v_r^-\frac{(R_g + Z_o)}{R_gZ_o}$$

or

$$\frac{v_r^-}{v^-} = \frac{R_g - Z_o}{R_g + Z_o} = \Gamma_{s(\text{voltage})}. \tag{8-19}$$

The ratio of the reflected component to the incident component at the sending end turns out to have the same form as at the load end. Instead of R_r we have R_g. Also the current reflection coefficient is the negative of the voltage reflection coefficient.

$$\Gamma_{s(\text{current})} = -\Gamma_{s(\text{voltage})}. \tag{8-20}$$

In our example:

$$\Gamma_s = \frac{150 - 50}{150 + 50} = \frac{100}{200} = 0.5 = \frac{v_r^-}{v^-}.$$

So, as the initial pulse of voltage returns to the sending end with a value of 37.5 volts, we will have a reflected component:

$$v_r^- = 0.5\,v^- = (0.5)(37.5) = 18.75 \text{ volts}.$$

So as we follow the reflections back and forth we would obtain the following pictures.

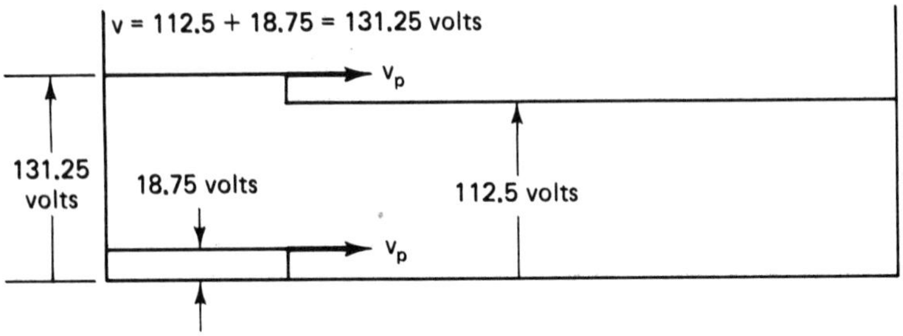

When the 18.75 volt pulse reaches the receiving end it is reflected as before: $(0.5)(18.75) = 9.375$ volts so we have $131.25 + 9.375 = 140.625$ volts.

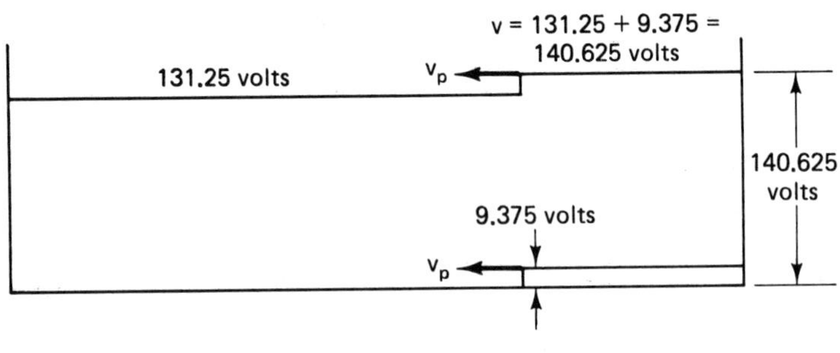

This process will continue until we reach a steady-state condition which would be what we would have calculated if we had used lumped parameter circuit theory.

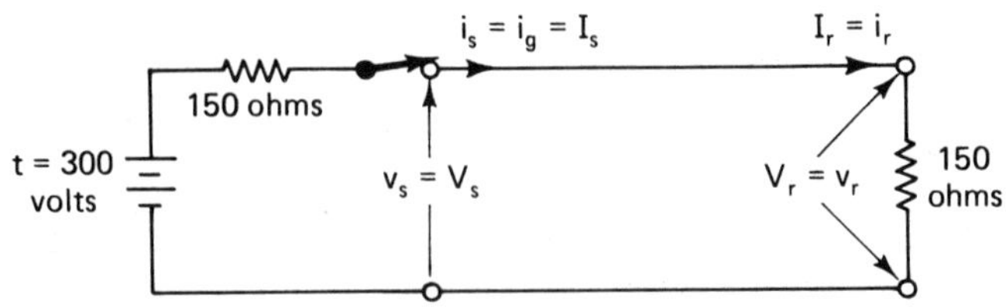

Figure 8-7. Circuit when $t = \infty$.

In Figure 8-7 we have the circuit for t = ∞.

$$i_s = i_r = I_s = I_r = \frac{300}{150 + 150} = \frac{300}{300} = 1 \text{ ampere.}$$

$$V_r = v_r = IR = (1)(150) = 150 \text{ volts} = V_s = v_s = E_g - IR_g = 300 - (1)(150) = 150 \text{ volts.}$$

We could also trace the current pulses back and forth until we reached the value of 1 ampere, but we are able to make the process easy and quick by using the Bounce Diagram.

8.4 Bounce Diagram

Instead of tracing the voltage and current pulses back and forth as we did in Section 8.3 it is much easier to construct what is called a "Bounce Diagram." We assume that it takes T seconds for a pulse to travel the length of the line in one direction. It then takes $2T$ seconds to go down and back. We draw two parallel vertical lines and a zigzag line representing the pulse starting at the sending end and going back and forth as shown in Figure 8-8. At each end we place the voltage reflection coefficient.

The bounce diagram does in a straightforward fashion what we previously did in tracing the reflections back and forth on the line. The voltage at the receiving end is zero until the initial pulse arrives. One-half of this pulse of 75 volts or 37.5 volts is reflected so the receiving-end voltage becomes 112.5 volts at T seconds and will remain at that value until the 37.5 volt pulse travels to the sending end and a new pulse of 18.75 volts arrives back at the receiving end. This takes $2T$ seconds to go down the line and back. Again, one-half of the 18.75-volt pulse is reflected so a 9.375-volt pulse travels back toward the sending end. The receiving-end voltage becomes 112.50 + 18.75 + 9.375 or 140.625 volts. All we have to do at either end is to add all of the values on the diagonal lines above that time. It is quick and easy. In our particular example both the sending- and receiving-end voltages are approaching 150 volts which are the values shown in Figure 8-7, calculated by regular circuit theory. The line will end up with 150 volts all along the line in the steady state. This will happen very rapidly since T seconds in the normal case will be quite small in the range of pico or microseconds. If the reflection factors are negative at either end, the pulse returning will

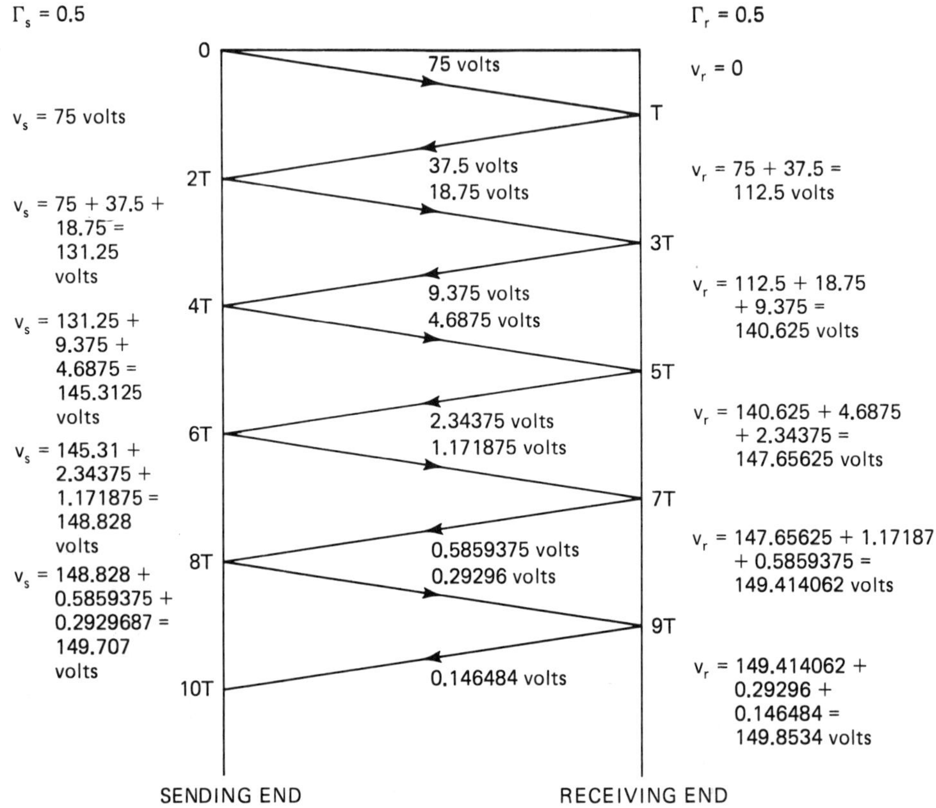

Figure 8-8. Voltage Bounce Diagram.

be negative and we would actually be subtracting that value from the values above the point. The results can then be plotted as a function of time as shown in Figure 8-9.

We can also produce a bounce diagram for the current by remembering that the reflection coefficients for the currents are the negatives of the reflection coefficients for the voltages and the initial pulse starting down the line is 1.5 amperes. The bounce diagram is shown in Figure 8-10. We can then plot i_s and i_r as functions of time.

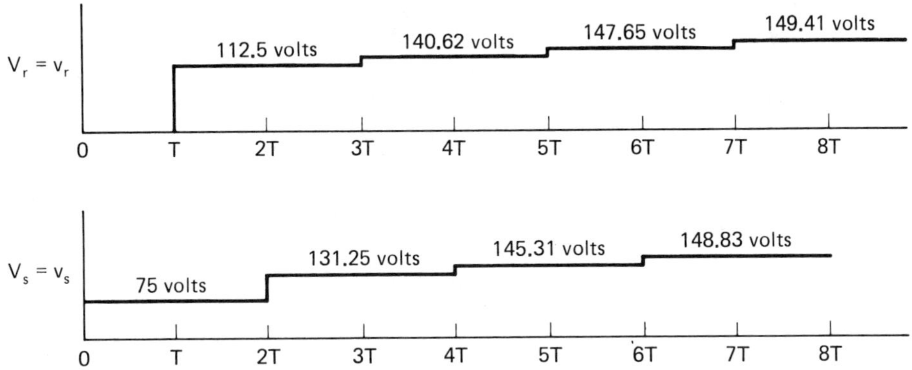

Figure 8-9. V_r and V_s as functions of time.

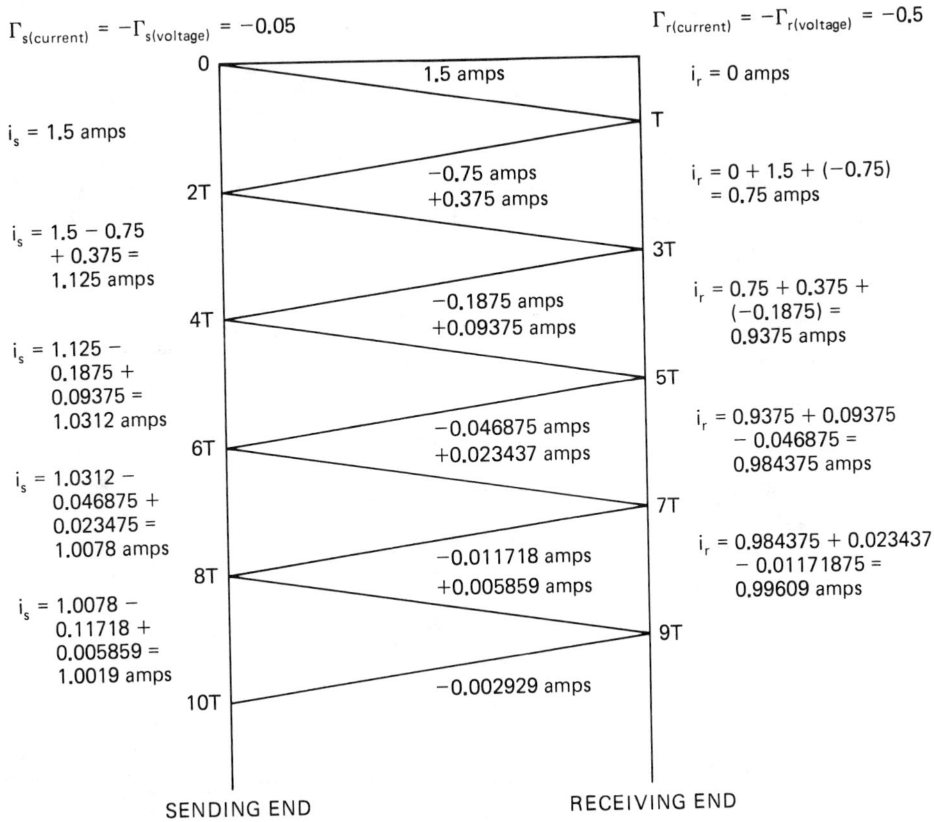

Figure 8-10. Current Bounce Diagram.

Both currents are approaching the steady-state value of 1 ampere. We do not actually have to produce a current bounce diagram to secure the values in Figure 8-11. At the receiving end $v_r = V_r = i_r R_r$, so once we have the receiving-end voltage we can find the receiving-end current. In our example, $i_r = V_r/150$.

Time Interval	v_r	i_r	
O-T	0.0	0.0	
T-3T	112.5	0.75	
3T-5T	140.625	0.9375	
5T-7T	147.656	0.9843	etc.

At the sending end it is a little more complicated but easy to do. Since $V_s = v_s = E_g - i_s R_g$, we can find for our example:

$$i_s = \frac{E_g - v_s}{R_g} = \frac{300 - v_s}{150}.$$

Knowing v_s from the voltage bounce diagram, we can determine i_s.

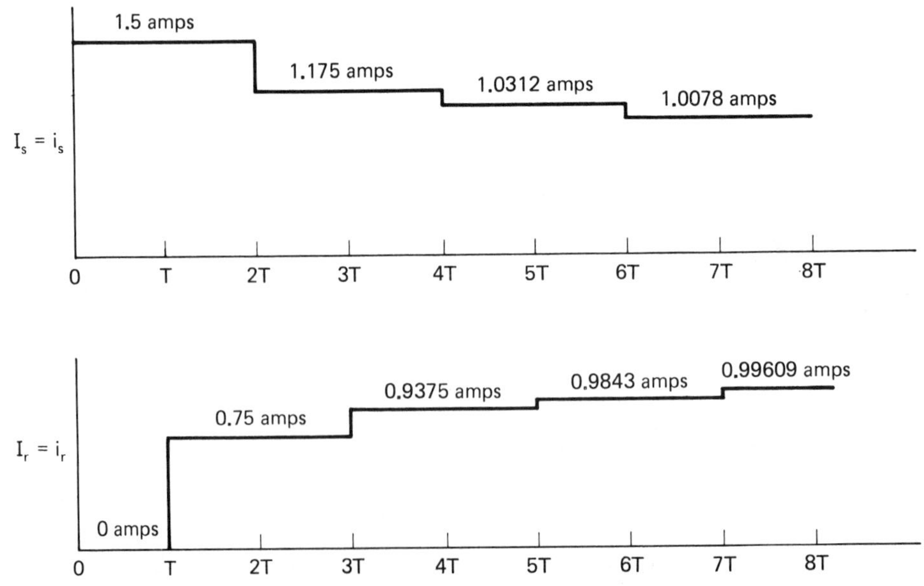

Figure 8-11. I_r and I_s as functions of time.

8.5 Another Example of the Bounce Diagram

Let us assume the following circuit:

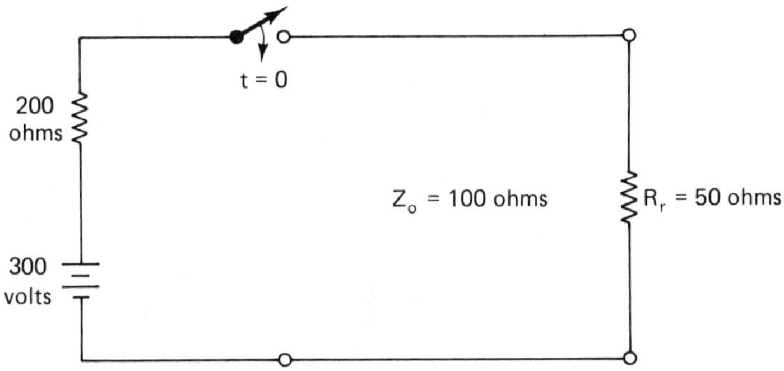

$$\Gamma_{r(\text{voltage})} = \frac{Z_r - Z_o}{Z_r + Z_o} = \frac{50 - 100}{50 + 100} = \frac{-50}{150} = -\frac{1}{3}. \qquad \Gamma_{r(\text{current})} = +\frac{1}{3}.$$

$$\Gamma_{s(\text{voltage})} = \frac{Z_g - Z_o}{Z_g + Z_o} = \frac{200 - 100}{200 + 100} = \frac{100}{300} = +\frac{1}{3}. \qquad \Gamma_{s(\text{current})} = -\frac{1}{3}.$$

At $t = 0$ our circuit becomes:

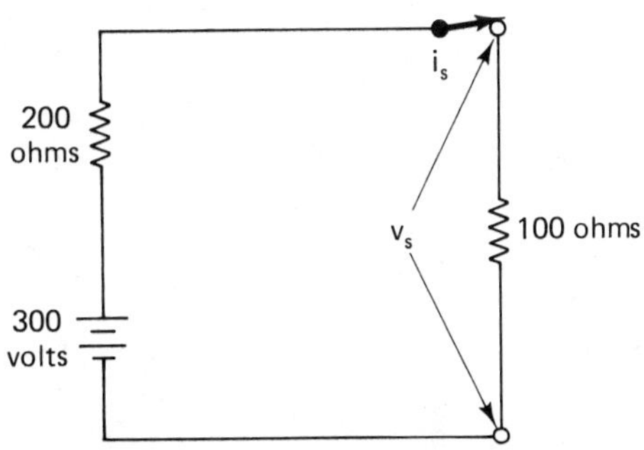

$$i_s = \frac{v_s}{Z_g + Z_o} = \frac{300}{200 + 100} = 1 \text{ ampere.}$$

$$v_s = i_s Z_o = (1)(100) = 100 \text{ volts.}$$

At $t = \infty$ our circuit becomes:

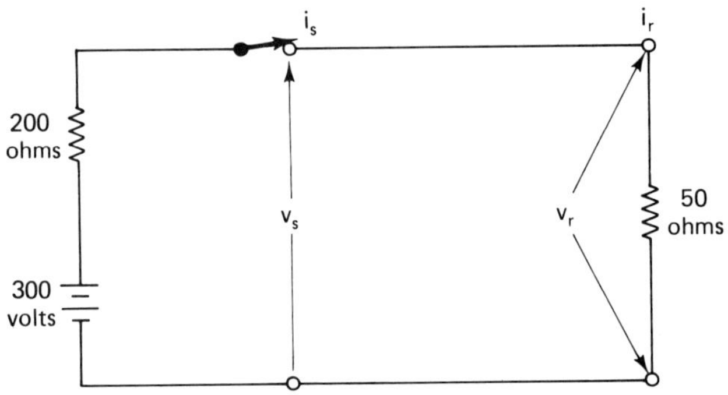

$$I_s = I_r = i_s = i_r = \frac{300}{250} = 1.2 \text{ amperes.}$$

$$v_r = V_r = v_s = V_s = (1.2)(50) = 60 \text{ volts} = 300 - (1.2)(200) = 300 - 240 = 60 \text{ volts.}$$

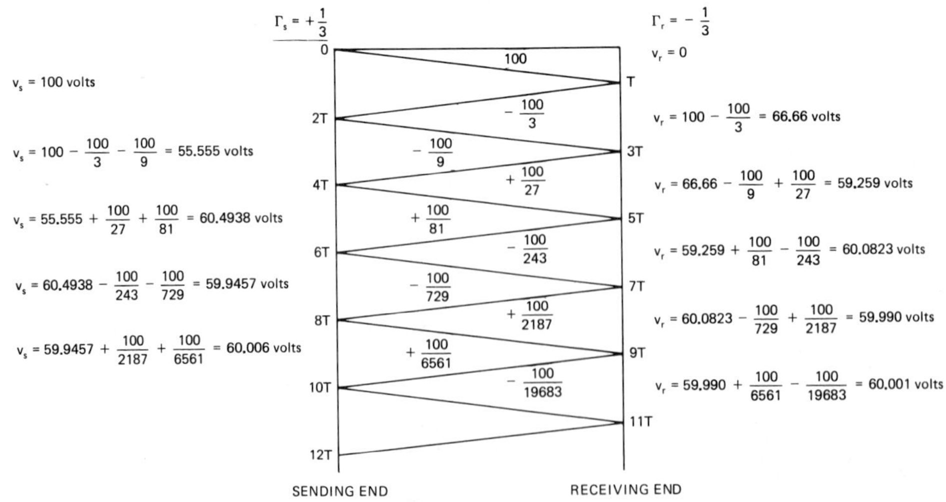

Voltage bounce diagram

As anticipated, both the sending- and receiving-end voltages are approaching the steady state value of 60 volts.

While we could find the sending- and receiving-end currents from the voltage bounce diagram let us construct the current bounce diagram as one more example.

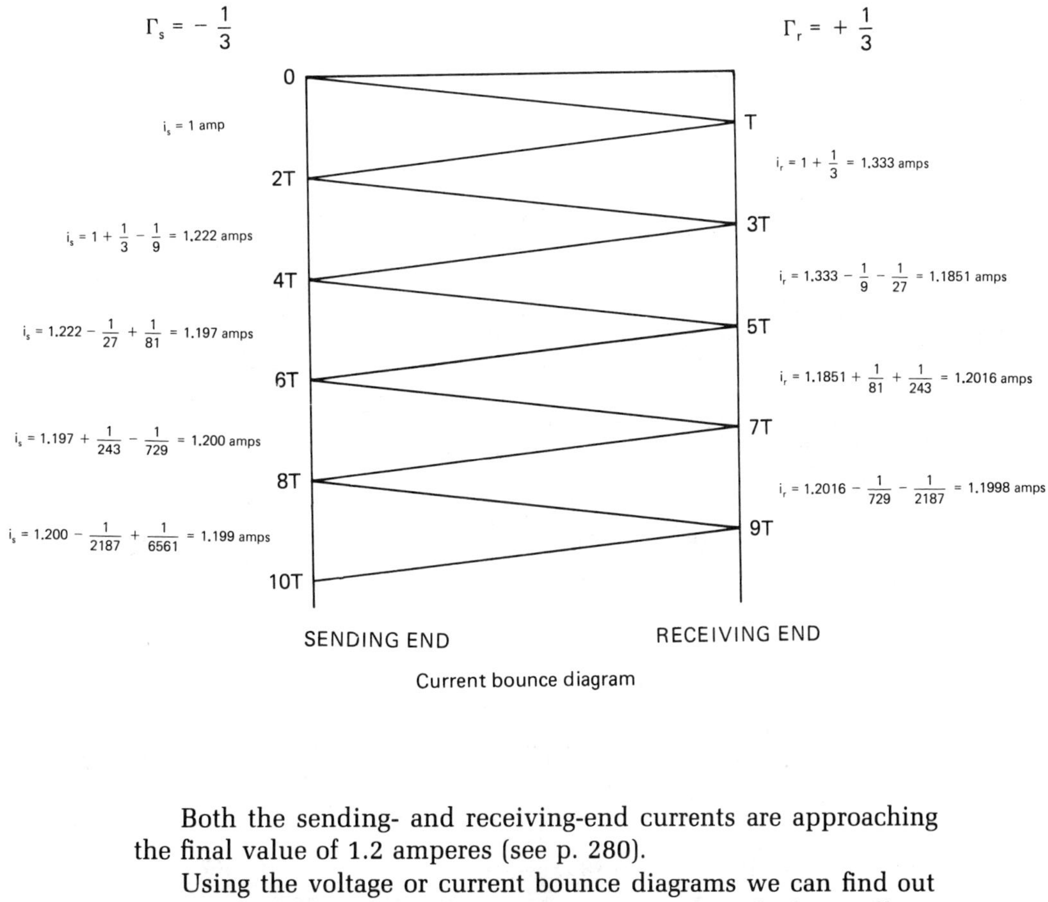

Current bounce diagram

Both the sending- and receiving-end currents are approaching the final value of 1.2 amperes (see p. 280).

Using the voltage or current bounce diagrams we can find out how the voltage or current varies at any point on the line. All we have to do is keep track of the time involved and add everything above the point as we proceed. For example, let us see how the voltage varies with time at the center of the line. Looking at the voltage bounce diagram shows us that the voltage is zero until $T/2$ seconds and then becomes 100 volts. It stays at 100 volts until $T/2 + T = 3T/2$ seconds when the pulse $-100/3$ appears which gives

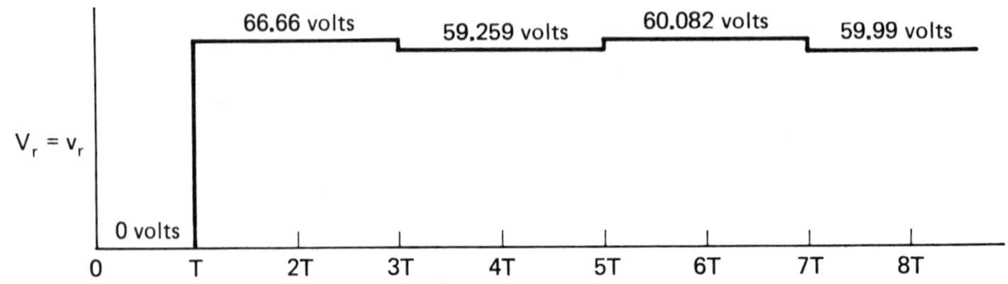
66.66 volts
59.259 volts
60.082 volts
59.99 volts
$V_r = v_r$
0 volts
0
T
2T
3T
4T
5T
6T
7T
8T

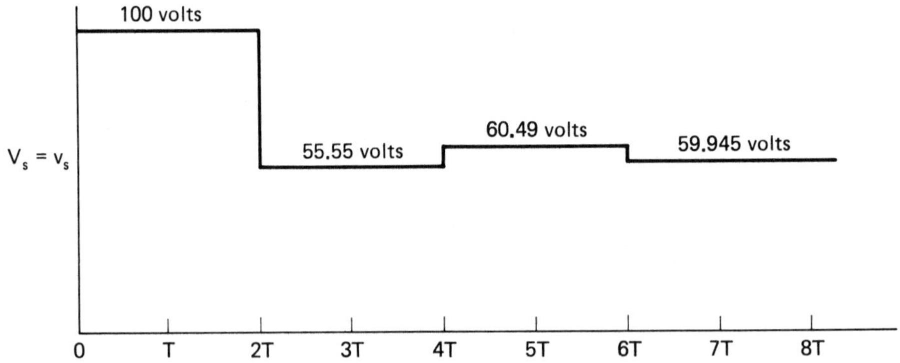
100 volts
55.55 volts
60.49 volts
59.945 volts
$V_s = v_s$
0
T
2T
3T
4T
5T
6T
7T
8T

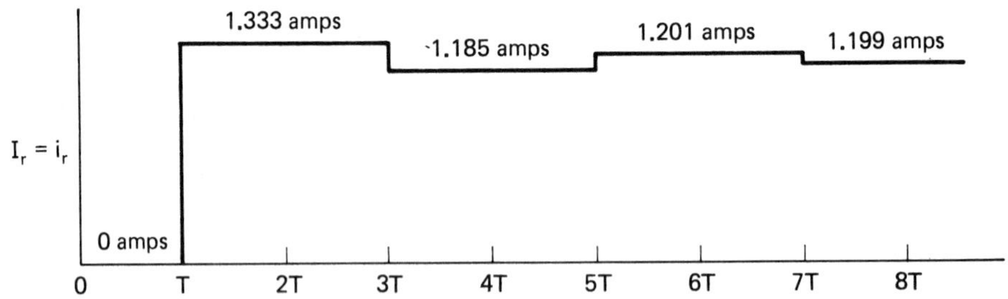
1.333 amps
1.185 amps
1.201 amps
1.199 amps
$I_r = i_r$
0 amps
0
T
2T
3T
4T
5T
6T
7T
8T

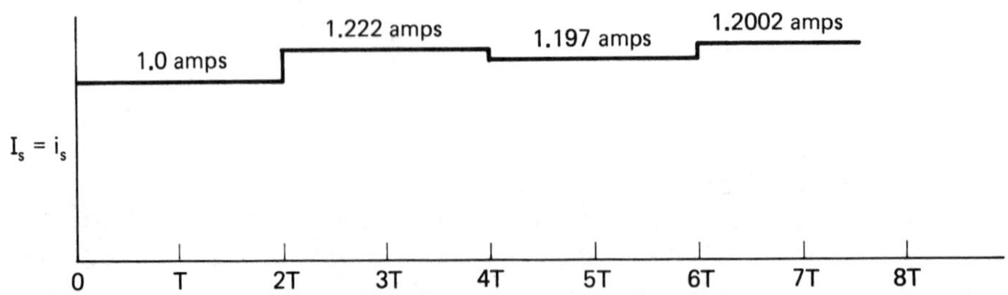
1.0 amps
1.222 amps
1.197 amps
1.2002 amps
$I_s = i_s$
0
T
2T
3T
4T
5T
6T
7T
8T

a value of 100 − 100/3 = 66.66 volts which will last for T seconds until the −100/9 pulse arrives. We can plot it as follows:

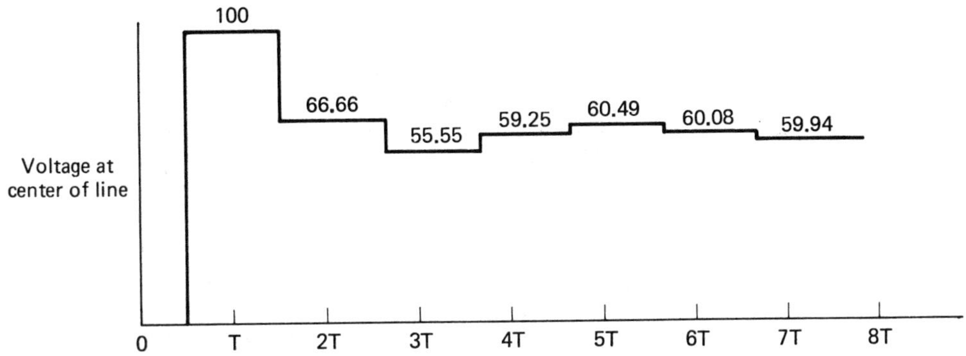

Here the voltage changes every T seconds instead of $2T$ seconds as was the case at the two ends of the line.

8.6 Chapter Summary

1. While sinusoidal waveshapes and those waveforms which can be resolved into sinusoidal waves of different frequencies represent a large number of important types of signals on a transmission line, not all waveforms are of a steady-state nature. Other signals may be non-periodic, random, or induced on the line by lightning, switching, faults, or rapid changes in loads. These transient-type signals are often put on the line as a signal; for instance, consider the signals in a digital computer.
2. By assuming a lossless case, one may approach the problem of a d. c.-type wave traveling down the line and achieve some understanding of what happens on the line even when it is lossy.
3. One of the basic concepts of a transmission line is that the instantaneous forward voltage at any point on the line is equal to the instantaneous forward current times the characteristic impedance. The instantaneous reverse voltage at any point on the line is the instantaneous reverse current times the negative of the characteristic impedance. In equation form this becomes:

$$\text{Forward wave.} \quad v_f = i_f Z_o.$$

$$\text{Reverse wave.} \quad v_r = i_r(-Z_o).$$

4. At the receiving end we have a reflection coefficient for the voltage.

$$\Gamma_{r(\text{voltage})} = \frac{v^-}{v^+} = \frac{Z_r - Z_o}{Z_r + Z_o} = \frac{R_r - R_o}{R_r + R_o}. \tag{8-17}$$

We also have a reflection coefficient for the current which is the negative of the reflection coefficient for the voltage.

$$\Gamma_{r(\text{current})} = \frac{i^-}{i^+} = -\Gamma_{r(\text{voltage})} = -\frac{R_r - R_o}{R_r + R_o}. \tag{8-18}$$

5. At the sending end we have two reflection coefficients, one for the voltage and one for the current which is the negative of the voltage coefficient. If R_g is the internal resistance of the source voltage, E_g, a constant d. c. voltage, we have:

$$\Gamma_{s(\text{voltage})} = \frac{R_g - R_o}{R_g + R_o}. \tag{8-19}$$

$$\Gamma_{s(\text{current})} = -\Gamma_{s(\text{voltage})} = -\frac{R_g - R_o}{R_g + R_o}. \tag{8-20}$$

6. When we consider the simple d. c.-type situation, with a d. c. voltage suddenly applied at the sending end, a pulse moves down the line at the phase velocity, v_p, and when it strikes the receiving end it is either completely absorbed if $R_r = R_o = Z_o$, or part of it is reflected back to the sending end. If any of it is reflected back towards the sending end, either it is completely absorbed at the sending end if $R_g = R_o = Z_o$ or part is reflected back towards the receiving end. If there are reflections at each end they continue back and forth until a steady-state type situation exists on the line.
7. At $t = 0$ the source sees the following circuit.

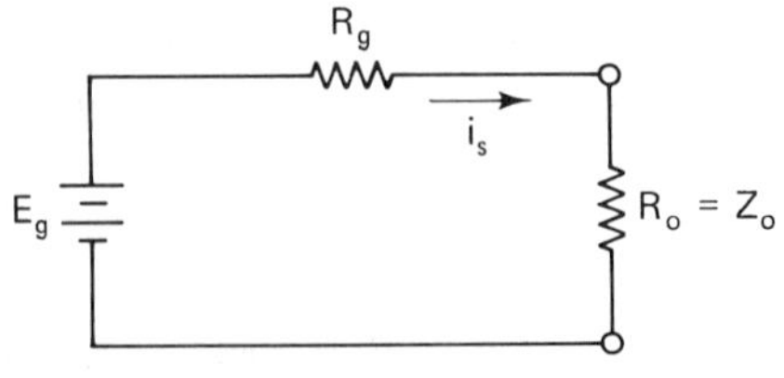

$$i_s = \frac{E_g}{R_g + R_o}.$$

$$v_s = \frac{E_g}{R_g + R_o}(R_o).$$

8. At $t = \infty$ the source sees the following circuit.

$$i = i_r = i_s = \frac{E_g}{R_g + R_r}.$$

$$v = v_r = v_s = \frac{E_g}{R_g + R_r}(R_r).$$

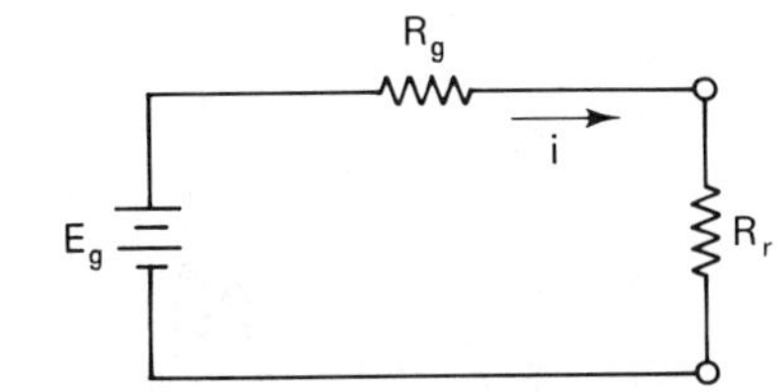

9. One can either trace the reflected signals back and forth on the line or one may use the Bounce diagram to make the solution more systematic and accurate.
10. On the Bounce diagram the voltage (and/or the current) may be found by adding those values that appear on the diagram above the time where the signal is desired.

8.7 Review and Study Questions

1. Under what special conditions was the Bounce diagram studied in this chapter? Consider the type of source, impedances, lossy or lossless, types of loads, and kind of pulse sent down the line.
2. Draw the equivalent circuit for the case where the time equals zero. Give the equations for the input voltage and current pulses.
3. Draw the equivalent circuit for the case when time is equal to infinity. What is the voltage all along the line from the sending end to the receiving end? What is the relationship between the sending-end and the receiving-end current?

4. Express the relationships between the forward voltage and current and the reverse voltage and current when instantaneous values are being considered. Which has the negative sign and why?
5. Give the expressions for the receiving-end reflection coefficients for both the voltage and the current.
6. Give the expressions for the sending-end reflection coefficients for both the voltage and the current.
7. Sketch the Bounce diagram for the voltage. What is placed on the left-hand side? on the right-hand side?
8. Indicate how you would find the receiving-end current as a function of time if you had the receiving-end voltage values.
9. Indicate how you would find the sending-end current as a function of time if you had the sending-end voltage values.
10. How do you guess the approach taken in this chapter would be modified if the signal sent down the line was caused by an a. c. source and the switch is closed at some point in the a. c. cycle?

Problems

8-1. Given the following lossless transmission line.

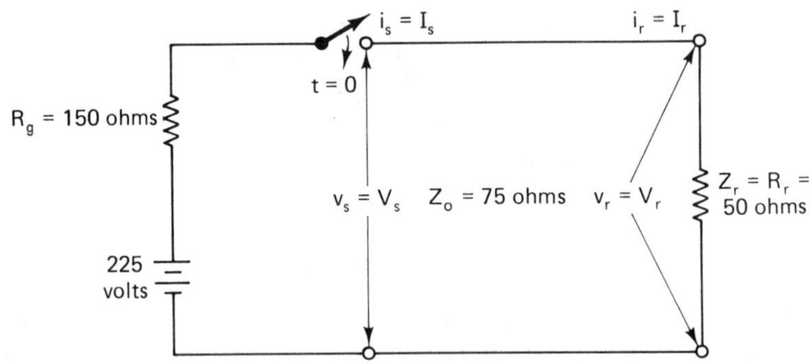

If it takes T seconds for a traveling wave to move from the sending end to the receiving end of the line, plot the following:

(a) Voltage at the receiving end as a function of time (up to $9T$ seconds). Write the value on the plot for each time interval.

(b) Plot the sending-end current for $9T$ seconds and write the values on the plot for each time interval.

(c) What are the final values of the load voltage and load current?

8-2. Use the circuit of problem 8-1 with the following values: $R_g = 200$ ohms; $Z_r = 50$ ohms; $Z_o = 100$ ohms; $E_g = 300$ volts.

(a) Draw the circuit for $t = 0$ seconds and for $t = \infty$ seconds.

(b) Plot the sending-end voltage, v_s, for $9T$ seconds. Write the value on the plot for each time interval.

(c) Plot the receiving-end current for $9T$ seconds. Write the value on the plot for each time interval.

(d) What are the final values of the load voltage and load current?

8-3. Use the circuit of problem 8-1 with the following values: $R_g = 50$ ohms; $Z_r = 300$ ohms; $Z_o = 100$ ohms; $E_g = 150$ volts.

(a) Plot v_r for $9T$ seconds.

(b) Plot i_r for $9T$ seconds.

(c) Plot i_s for $9T$ seconds.

(d) Plot v_s for $9T$ seconds.

(e) Plot the voltage halfway between the sending and receiving ends for $9T$ seconds.

(f) What are the final values of v_r, i_r, v_s, and i_s?

8-4. Repeat problem 8-3 if the load is 50 ohms instead of 300 ohms.

8-5. Use the circuit of problem 8-1 with the following values: $R_g = 25$ ohms; $R_r = 125$ ohms; $Z_o = 75$ ohms; $E_g = 200$ volts.

(a) Plot i_s for $9T$ seconds.

(b) Plot i_r for $9T$ seconds.

(c) Plot v_r for $9T$ seconds.

(d) Plot the current halfway between the sending and receiving ends for $9T$ seconds.

(e) What are the final values of v_r, v_s, i_r, and i_s?

8-6. Repeat problem 8-5 if the generator resistance is increased to 75 ohms.

8-7. Use the circuit of problem 8-1 with the following values: $R_g = 50$ ohms; $Z_r = 200$ ohms; $Z_o = 100$ ohms; $E_g = 225$ volts.

(a) Plot v_r and v_s for $9T$ seconds.

(b) Plot i_r i_s for $9T$ seconds.

(c) Plot the voltage halfway between the sending and receiving ends for $9T$ seconds.

(d) What are the final values of v_r, i_r, v_s, and i_s?

8-8. Repeat problem 8-7 if R_g is increased to 100 ohms.

8-9. Repeat problem 8-7 if Z_r is reduced to 100 ohms.

9

Line Parameters

9.1 Introduction

Up to this point in the text we have assumed that we know the parameters of the line, R, L, G, and C. The actual calculations of these parameters may be simple or complicated and, in some cases, impossible. Furthermore, their values are affected by factors which may or may not be controllable by the engineer. For instance, inductance and capacitance are always functions of the physical arrangement of the conductors so the engineer may increase L and decrease C by increasing the distance between the two wires of a two-wire line. The d. c. resistance may be changed by either changing the conductor material or by changing the size of the wire. The hardest parameter to calculate for most situations is G since it is really a function of the leakage current between conductors and the leakage current depends on factors which often are variable and impossible to calculate. Test measurements in the laboratory or in the field may be necessary when determining G.

The problem becomes more complex because currents tend to redistribute themselves in the conductors as the frequency increases. The current density across the cross-sectional area of the conductor becomes nonuniform, with more current tending to flow near the surface of the conductor and less in the middle. This not only changes the effective flux linkages in the conductor which in turn changes the inductance and inductive reactance but it changes the effective resistance of the conductor since the I^2R loss, in those areas where the current increases, increases more than the I^2R loss decreases in those areas where the current decreases. The net effect is an overall increase in the internal power loss and an effective increase in the resistance, so that the a. c. resistance is larger than the d. c. resistance.

This result of decreasing L and increasing R as the frequency increases due to the redistribution of the current in the conductor has become known as the *skin effect*. When conductors are close together, the current distribution in the conductors changes due to another effect known as the *proximity effect*. Again, since these processes are complicated and mathematically either difficult or impossible to solve, the actual determination of the transmission line parameters may be done by measurement.

The equations for finding the line parameters have, in some cases, been found in earlier courses under specialized conditions. For instance, in our fields course we often find the inductance of a coaxial cable, assuming that we have the d. c. or low-frequency case with uniform current density in the two conductors. If the radius of the inner conductor is a meters and the outer conductor has an inner radius of b meters and an outer radius of c meters, we can write for the inductance of the coaxial cable the following expression.

$$\frac{L}{\ell} = \frac{\mu}{8\pi} + \frac{\mu}{2\pi}\ln\frac{b}{a} + \frac{\mu}{2\pi(c^2 - b^2)^2}\left[c^4 \ln\frac{c}{b} - c^4 + \frac{c^4}{4} + c^2b^2 - \frac{b^4}{4}\right]. \quad (9\text{-}1)$$

The result is in henrys per meter.

$$\mu = \mu r \mu o. \quad (9\text{-}2)$$

In equation (9-1) the first term represents the internal inductance due to the internal flux linkages in the inner conductor; the second term represents the inductance due to the flux linkages between the two conductors and the last term represents the internal inductance due to the internal flux linkages in the outer conductor. But at high frequencies the flux linkages in the inner and outer conductors tend towards zero as the current in the inner conductor concentrates at its outer surface and the current in the outer conductor concentrates at its inner surface due to skin effect, and we are left with just the middle term of equation (9-1) at high frequencies.

In general, we can say that normally all of our parameters are related to the frequency as well as the physical configuration and the types of materials used. They may, also, under certain conditions, be dependent on other things such as temperature, humidity, air pressure, voltage levels, etc. In this chapter we will look at the line parameters for different types of conductor arrangements.

9.2 Conductor Arrangements

Some basic possible configurations are given in Figure 9-1.

In Figure 9-1, (a) is a two-conductor line, (b) is a coaxial cable,

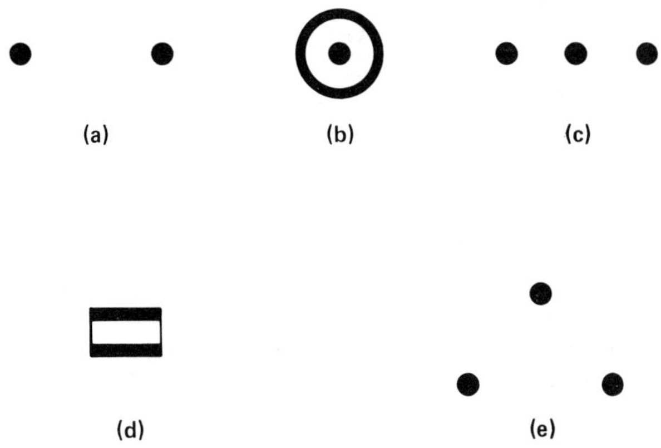

Figure 9-1. Basic line configurations.

(c) is a horizontal, three-wire, three-phase line, (d) is a strip line arrangement, and (e) is a three-wire, three-phase line with equal spacing between the wires. From these basic arrangements we can obtain other configurations as modifications. For example, the two conductors in (a) can be placed inside of a shield or the strip line in (d) could have different dimensions for the two elements. The two conductors in (a) could be placed next to each other and twisted. The physical arrangement will affect the parameters in ways which are easy to determine if we have the basic equations.

The two-conductor line in Figure 9-1(a) has an inductance per unit length given in many references as:

$$\frac{L}{\ell} \simeq \frac{\mu}{\pi} \ln \frac{D}{a} \text{ henrys per meter.} \tag{9-3}$$

D is the separation between the conductor centers and a is the radius of the conductors.

The capacitance may be found from the equation:

$$\frac{C}{\ell} \simeq \frac{\pi \varepsilon}{\ln \dfrac{D}{a}} \text{ farads per meter.} \tag{9-4}$$

In Chapter 1, ε_o was given as $(8.85415 \pm 0.00002) \times 10^{-12}$ farads per meter and we have been approximating this value by using $\dfrac{10^{-9}}{36\pi}$

because it is easy to remember and the error is less than two tenths of one percent. We should recall another item from Chapter 1, equation (1-26).

$$\varepsilon = \varepsilon_r \varepsilon_o. \tag{9-5}$$

When we look at equations (9-3) and (9-4), we notice that L and C may be adjusted by changing the separation between the two wires. If we increase the separation, we increase L and decrease C. If we place the conductors closer together, C increases and L becomes smaller. In equation (9-1) we see that the inductance of the coaxial cable due to the center term would remain constant if the ratio of the two radii a and b stays constant no matter what their actual magnitudes might be.

9.3 Resistance

The d. c. resistance for a conductor with a uniform cross-sectional area may be found from equation (9-6).

$$R = \frac{\rho L}{S} = \frac{L}{\sigma S} \text{ ohms.} \tag{9-6}$$

L is the length of the conductor in meters, S is the cross-sectional area in square meters, ρ is the resistivity in ohm-meters, and σ is the conductivity in mhos per meter or (ohm-meters)$^{-1}$. For standard annealed copper:

$$\rho = 1.72413 \times 10^{-8}. \qquad \sigma = 5.8 \times 10^{7}.$$

Tables are available for standard size, round conductors giving the resistance (normally in ohms per thousand feet in the U.S.A.), diameter, cross-sectional area, current-carrying capacity and other items which may be important for the engineer. In the U.S.A. the units in equation (9-6) are somewhat unique but they have been around for a long time and will probably be around for some time to come. R is still in ohms, but L is in feet and S is in circular mils. A circular mil is the area of the circle that is 0.001 inch in diameter. ρ has units of ohm-circular mil per foot. The circular mil area concept is convenient since the area in circular mils turns out to be the diameter in mils squared.

A typical wire table is given in Figure 9-2 (Knowlton, 1957, p. 263).

Wire Table.

A.W.G. No.	Diam. Mils	Lbs/ 1000 Ft.	Cross Section Cir Mils	Cross Section Sq. In.	D.C. Ohms/1000 Ft 25°C	D.C. Ohms/1000 Ft 65°C
0000	460	641.0	212,000	0.166	0.0500	0.0577
000	410	508.0	168,000	0.132	0.0630	0.0727
00	365	403.0	133,000	0.105	0.0795	0.0917
0	325	319.0	106,000	0.0829	0.1000	0.1160
1	289	253.0	83,700	0.0657	0.1260	0.1460
2	258	201.0	66,400	0.0521	0.1590	0.184
3	229	159.0	52,600	0.0413	0.201	0.232
4	204	126.0	41,700	0.0328	0.253	0.292
5	182	100.0	33,100	0.0260	0.319	0.369
6	162	79.5	26,300	0.0206	0.403	0.465
7	144	63.0	20,800	0.0164	0.508	0.586
8	128	50.0	16,500	0.0130	0.641	0.739
9	114	39.6	13,100	0.0103	0.808	0.932
10	102	31.4	10,400	0.00815	1.02	1.18
11	91	24.9	8,230	0.00647	1.28	1.48
12	81	19.8	6,530	0.00513	1.62	1.87
13	72	15.7	5,180	0.00407	2.04	2.36
14	64	12.4	4,110	0.00323	2.58	2.97
15	57	9.86	3,260	0.00256	3.25	3.75
16	51	7.82	2,580	0.00203	4.09	4.73
17	45	6.20	2,050	0.00161	5.16	5.96
18	40	4.92	1,620	0.00128	6.51	7.51
19	36	3.90	1,290	0.00101	8.21	9.48
20	32	3.09	1,020	0.000802	10.4	11.9
21	28.5	2.45	810	0.000636	13.1	15.1
22	25.3	1.94	642	0.000505	16.5	19.0
23	22.6	1.54	509	0.000400	20.8	24.0
24	20.1	1.22	404	0.000317	26.2	30.2
25	17.9	0.970	320	0.000252	33.0	38.1
26	15.9	0.769	254	0.000200	41.6	48.0
27	14.2	0.610	202	0.000158	52.5	60.6
28	12.6	0.484	160	0.000126	66.2	76.4
29	11.3	0.384	127	0.000099	83.4	96.3
30	10.0	0.304	101	0.000078	105.0	121.0

Figure 9-2. Wire table—standard annealed solid copper wire.

When a. c. is applied to a conductor there is a voltage drop due to the inductive reactance associated with the internal flux linkages. For d. c. and low-frequency a. c. this internal inductance is constant and independent of the size of the wire. It appears in equation (9-1) as the term associated with the center conductor of a coaxial cable.

$$\frac{L}{\ell} = \frac{\mu}{8\pi} \text{ henrys per meter.} \tag{9-7}$$

For a nonmagnetic material, the value would be 5×10^{-8} henrys per meter, and at low frequencies the reactance would also be a small value per meter. As the frequency increases the current density across the face of the conductor, which is constant for d. c., changes and becomes smaller at the center and increases towards the surface of the conductor. At high frequencies most of the current is near the surface and this causes a change in the effective resistance of the conductor. Furthermore, the internal flux linkages decrease so that the inductance associated with equation (9-7) tends toward zero. Since the power loss varies with the square of the current, the increase in the loss near the surface is greater than the decrease towards the center, causing an overall effect of increasing the total power loss in the conductor. The effective or a. c. resistance can be found from equation (9-8) if one is able to measure the power and the current.

$$R_{\text{a.c.}} = R_{\text{effective}} = \frac{P_{\text{a.c.}}}{I^2} \text{ ohms.} \tag{9-8}$$

$P_{\text{a.c.}}$ is the loss in watts at that frequency and I is the r.m.s. value of the current. The ratio of the a. c. to the d. c. resistance starts at 1.0 for d. c. and very low frequencies and increases as the frequency increases. While we are primarily interested in evaluating this increase in resistance because of skin effect it should be pointed out that if magnetic materials are in the vicinity of the conductors that there will be additional losses due to localized hysteresis and eddy current losses in the magnetic material. If the wires are close together there will be additional changes in the current densities due to the proximity effect. This will also affect the a. c. losses and increase the a. c. resistance.

The resistance changes with temperature and this should be taken into account before calculating the a. c. resistance. A convenient equation to use is (9-9).

$$\frac{R_{t2}}{R_{t1}} = \frac{A + t_2}{A + t_1}. \tag{9-9}$$

R_{t2} is the resistance at t_2 degrees centigrade, R_{t1} is the resistance at t_1 degrees centigrade and A is a constant, depending on the conductor material.

A = 234.5 for annealed 100% conductivity copper.
A = 241.5 for hard drawn 97.3% conductivity copper.
A = 228.1 for aluminum.

9.4 Proximity Effect

Skin effect and the proximity effect are both due to fields reacting on the currents flowing in the conductors. The proximity effect is mathematically complex so let us look at it from an intuitive viewpoint. Let us have two parallel conductors and isolate two small imaginary conductors A and A′ and two more small imaginary conductors B and B′ within the two conductors as shown in Figure 9-3.

Let us assume that each of the filamentary conductors A, A', B, and B' has a small radius of a meters. For a parallel wire line the inductance is given by equation (9-3) and indicates that the inductance increases as the separation, D, between the two conductors is increased. Since the separation between A and A' is larger than for B and B', the conductor pair A-A' will have more flux linkages and

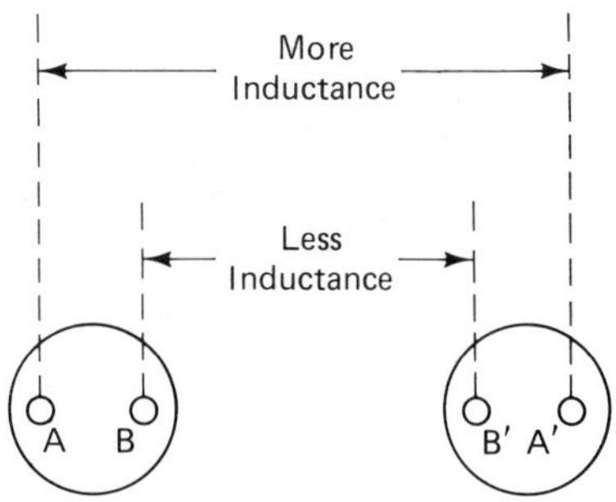

Figure 9-3. Two parallel conductors.

more inductance than the conductor pair, *B-B'*. The inductive reactance of pair *A-A'* will be larger and so more current will flow in the *B-B'* pair than in the *A-A'* pair. This causes the currents to be larger for the parts of the two conductors closest to each other. Fortunately, it only has an effect worth worrying about when the conductors are very close together, and normally it can be neglected.

9.5 Skin Effect

When d. c. current is flowing in a conductor, the current density, *J*, in amperes per square meter, is constant across the face of the conductor. Applying Ampere's Circuital Law we can find the *H* field within a conductor of radius *a* meters along the path with a radius of r meters as shown in Figure 9-4, where we are assuming that the conductor is centered on the z axis and the current is flowing into the page.

$$\oint \vec{H} \cdot d\vec{\ell} = I_{\text{linked}}. \tag{9-10}$$

$$H(2\pi r) = J(\text{area}) = \frac{I}{\pi a^2}(\pi r^2) =$$

$$H = \frac{rI}{2\pi a^2},\ r \leqslant a\ \frac{\text{amperes}}{\text{meter}}. \tag{9-11}$$

Our result is the equation of a straight line as shown in Figure 9-5.

The magnetic field intensity, *H*, is zero at the center and a maximum at the surface. The concentric magnetic flux lines within the

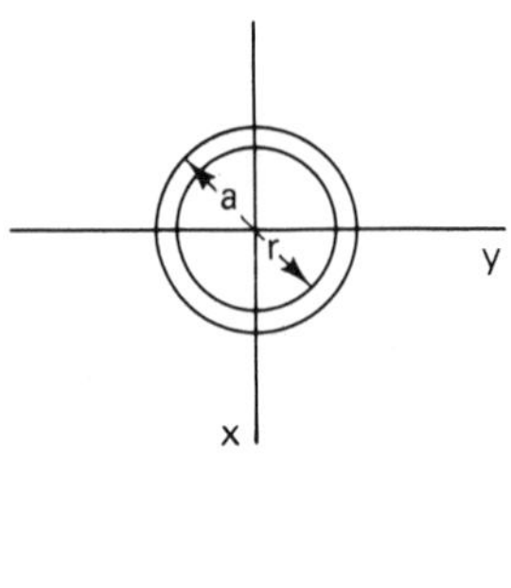

Figure 9-4. Conductor on the *z* axis.

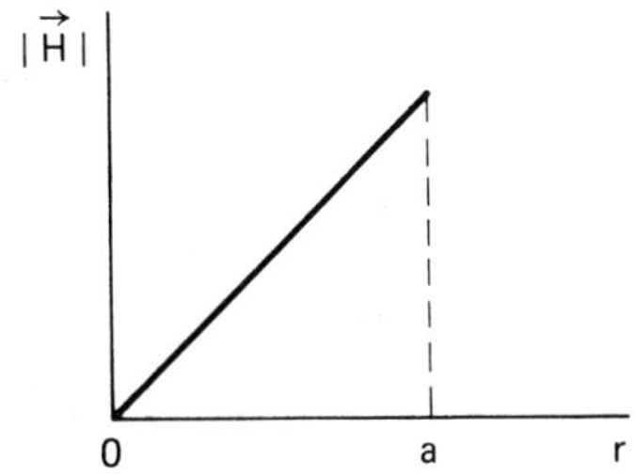

Figure 9-5. |H| versus *r* inside of conductor.

conductor link the current, with the current towards the center being linked by more flux lines than the current towards the surface, as shown in Figure 9-6. If conductor elements *A* and *B* have the same cross-sectional areas and *J* is constant, then the current in *A* is linked with more flux than the equal current in *B*. The inductance of the current element *A* is therefore larger than it is for *B* and its reactance is larger. The voltage drop per unit length of line must be equal, so with a larger inductance in the center and a smaller inductance towards the surface, the current redistributes itself so the smaller

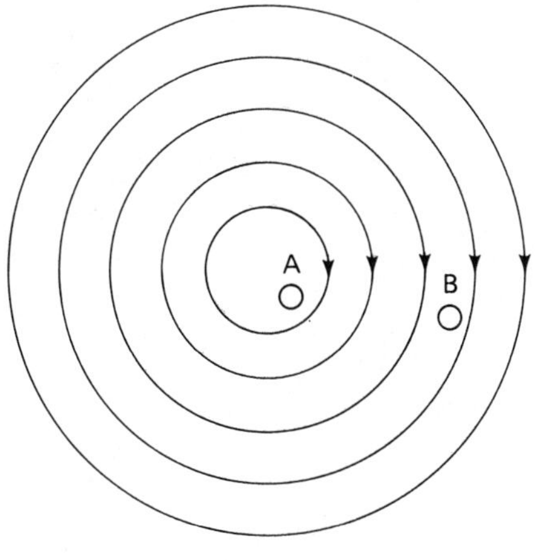

Figure 9-6. Flux linking current.

current towards the center is associated with the larger reactance whereas the larger current at the surface is associated with the smaller reactance.

A rough approximation can be made by dividing a conductor cross-sectional area into four equal areas. The areas in Figure 9-7 will be equal if the following is true.

$$r_2 = \sqrt{2}\, r_1.$$

$$r_3 = \sqrt{3}\, r_1.$$

$$r_4 = \sqrt{4}\, r_1.$$

$$A_1 = \pi r_1^{\,2}.$$

$$A_2 = \pi r_2^{\,2} - \pi r_1^{\,2} = \pi(\sqrt{2}\, r_1)^2 - \pi r_1^{\,2} = 2\pi r_1^{\,2} - \pi r_1^{\,2} = \pi r_1^{\,2} = A_1.$$

$$A_3 = \pi r_3^{\,2} - \pi r_2^{\,2} = \pi(\sqrt{3}\, r_1)^2 - \pi(\sqrt{2}\, r_1)^2 = 3\pi r_1^{\,2} - 2\pi r_1^{\,2} = \pi r_1^{\,2} = A_1 = A_2.$$

$$A_4 = \pi r_4^{\,2} - \pi r_3^{\,2} = \pi(\sqrt{4}\, r_1)^2 - \pi(\sqrt{3}\, r_1)^2 = 4\pi r_1^{\,2} - 3\pi r_1^{\,2} = \pi r_1^{\,2} = A_1 = A_2 = A_3.$$

Each section of the conductor would have the same d. c. resistance, since the cross-sectional areas are the same. Let us assume that each section has a d. c. resistance of 2 ohms. The four sections in parallel

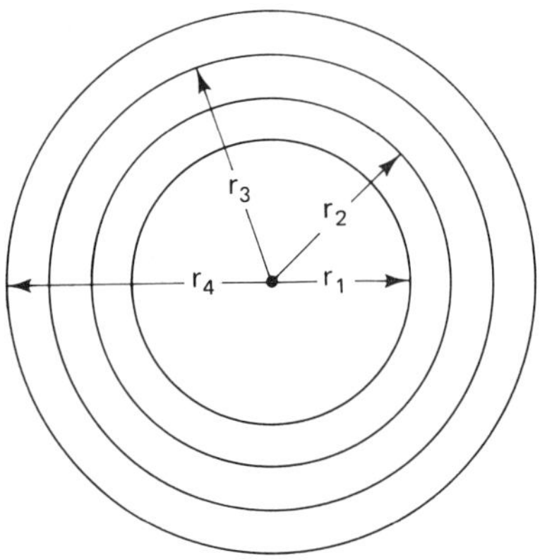

Figure 9-7. Equal areas.

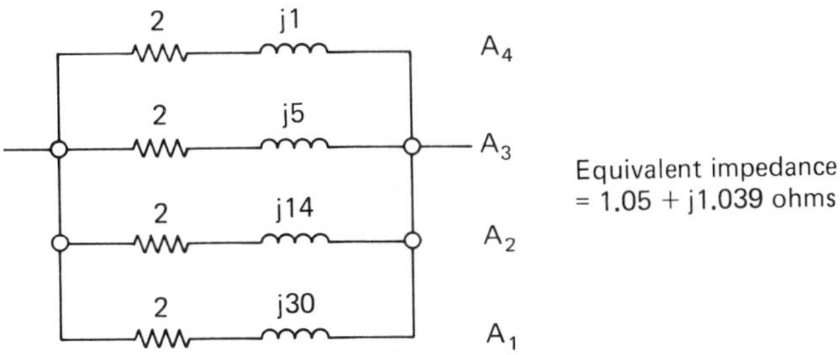

Figure 9-8. Conductor impedances in parallel.

would then have a total resistance of 0.5 ohms. If we apply a. c. to the conductor, the inner section will have the largest inductive reactance. For computational purposes let us assume the values as shown in Figure 9-8, with the lowest impedance given for the outside area, A_4.

When we calculate the equivalent impedance we find that it is $1.05 + j1.039$ ohms or $1.47 \angle 44.66^\circ$ ohms. The current in the center conductor (A_1), would be the smallest and in the outer conductor (A_4), the current would be the largest. The important result is that the d. c. resistance is 0.5 ohms but the a. c. resistance is 1.05 ohms or more than twice as large.

While the discussion above tries to give an intuitive picture of skin effect the equations are more interesting and much more fun. We will have to go back to electromagnetic field theory and start with some basic concepts. We will assume a round wire with a radius of a meters.

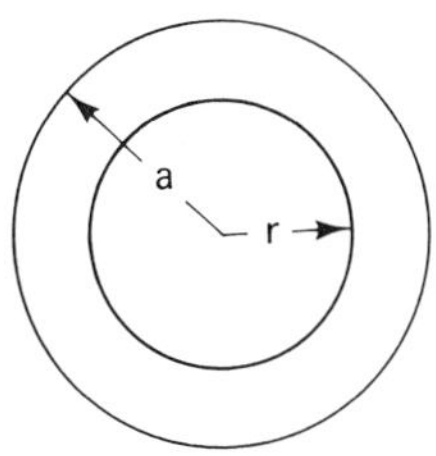

Figure 9-9. Conductor.

We will define the following quantities:

j_r = current density at a radius of r meters in amperes per square meter.

H_r = magnetic field intensity at a radius of r meters in amperes per meter.

B_r = magnetic flux density at a radius of r meters in webers per square meter.

ρ = resistivity of the wire in ohm-meters.

μ = the permeability of the wire which is assumed to be that of free space ($4\pi \times 10^{-7}$ henrys per meter).

The total current passing through the circle with a radius of r meters is:

$$I_r = \int j_r dS = \int_o^r j_r(2\pi r dr).$$

From Ampere's Circuital Law we have:

$$\oint \vec{H} \cdot d\vec{\ell} = I_{\text{linked}}, \text{ so at r: } H(2\pi r) = I_r = \int_o^r j_r(2\pi r dr).$$

$$H(\text{at r}) = \frac{1}{2\pi r}\int_o^r j_r(2\pi r dr) \text{ or } rH = \frac{1}{2\pi}\int_o^r j_r(2\pi r dr).$$

Differentiate.

$$r\frac{dH}{dr} + H\frac{dr}{dr} = \frac{1}{2\pi} j_r 2\pi r,$$

so

$$j_r = \frac{dH}{dr} + \frac{H}{r}. \tag{9-12}$$

In Figure 9-10 is shown a side view of the conductor. The flux going through the rectangle indicated in Figure 9-10 is:

$$\phi = BdS = \mu H(\ell dr) = \mu H(1dr) = \mu\ Hdr.$$

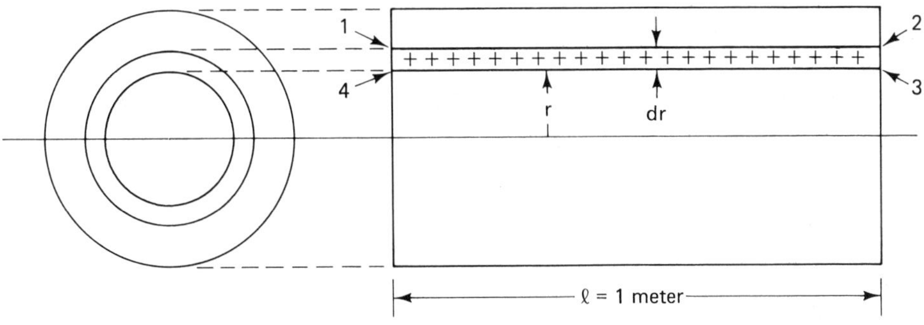

Figure 9-10. Side view of conductor.

If we consider the path 1-2-3-4-1 as one turn, we can calculate the voltage induced from Faraday's Law.

$$e = N\frac{d\phi}{dt} = (1)\frac{d\phi}{dt} = \frac{d}{dt}(\mu H dr) = \mu\, dr\frac{dH}{dt}. \tag{9-13}$$

The voltage drop along a path can be written as:

$$\text{Voltage drop} = RI = \frac{(\rho\ell)}{S}(j_r S) = j_r\rho\ell = j_r\rho(1) = j_r\rho.$$

If we let the current density be j_r along the path 4-3 in Figure 9-10 we can write for the current density along the path 1-2: $j_r + \frac{\partial j_r}{\partial r}dr$. Then, along path 4-3 we have a voltage drop of ρj_r and along path 1-2 we have a voltage drop of $\rho(j_r + \frac{\partial j_r}{\partial r}dr)$. The difference between these two values is the summation of the voltage drop around the rectangular path 1-2-3-4-1 or $\rho\frac{\partial j_r}{\partial r}dr$. This is equal to the voltage induced by the flux linking the rectangular path 1-2-3-4-1 changing with time (Faraday's Law).

$$\rho\frac{dj_r}{dr}dr = \mu\, dr\frac{dH}{dt},$$

or

$$\frac{dH}{dt} = \frac{\rho}{\mu}\frac{dj_r}{dr}. \tag{9-14}$$

From equation (9-12),

$$j_r = \frac{dH}{dr} + \frac{H}{r}, \quad \frac{dj_r}{dt} = \frac{d^2H}{dtdr} + \frac{1}{r}\frac{dh}{dt}. \tag{9-15}$$

From (9-14),

$$\frac{d^2H}{drdt} = \frac{\rho}{\mu}\frac{d^2j_r}{dr^2}.$$

Substitution in (9-15) gives:

$$\frac{dj_r}{dt} = \frac{\rho}{\mu}\frac{d^2j_r}{dr^2} + \frac{1}{r}\frac{\rho}{\mu}\frac{dj_r}{dr}$$

or

$$\frac{d^2j_r}{dr^2} + \frac{1}{r}\frac{dj_r}{dr} = \frac{\mu}{\rho}\frac{dj_r}{dt}. \tag{9-16}$$

Converting to the phasor form gives:

$$\frac{d^2J_r}{dr^2} + \frac{1}{r}\frac{dJ_r}{dr} = \frac{\mu}{\rho}\, j\omega J_r \tag{9-17}$$

or

$$\frac{d^2J_r}{dr^2} + \frac{1}{r}\frac{dJ_r}{dr} - \frac{j\omega\mu}{\rho}J_r = 0. \tag{9-18}$$

Equation (9-18) is a modified form of Bessel's equation of order zero. It can be put into the form:

$$\frac{d^2J_r}{dr^2} + \frac{1}{r}\frac{dJ_r}{dr} + n^2 J_r = 0, \tag{9-19}$$

where

$$n^2 = -\frac{j\omega\mu}{\rho} = -j\omega\sigma\mu = -j2\pi f\mu\sigma. \tag{9-20}$$

Let us define the quantity, δ, called the skin depth or the depth of penetration:

$$\delta = \frac{1}{\sqrt{\pi f \mu\sigma}}. \tag{9-21}$$

So,

$$\pi f\mu\sigma = \frac{1}{\delta^2}. \qquad n^2 = -j2\pi f\mu\sigma = -j\frac{2}{\delta^2}.$$

$$n = \sqrt{\frac{-2j}{\delta^2}} = \sqrt{-j}\,\frac{\sqrt{2}}{\delta}. \tag{9-22}$$

The solution for our differential equation can be written in the form:

$$J_r = AJ_o\,(\sqrt{-j\omega\mu\sigma r}) + BK_o\,(\sqrt{-j\omega\mu\sigma r}), \tag{9-23}$$

where J_o is a Bessel function of the first kind, order zero, and K_o is a Bessel function of the second kind, order zero.

If we look at the first term only,

$$A\,J_o\,(\sqrt{-j\omega\mu\sigma}\;r) = A\,J_o\,(\sqrt{-j2\pi f\mu\sigma}\;r) = A\,J_o\left(\sqrt{\frac{-j2}{\delta^2}}\;r\right) =$$

$$A\,J_o\left(\sqrt{-j}\,\frac{\sqrt{2}}{\delta}\;r\right) = A\,J_o(nr). \tag{9-24}$$

The solution can be obtained by assuming an infinite power series for J_r and substituting into equation (9-19). The result is:

$$J_r = A\,J_o(nr) = A\left(1 - \frac{n^2r^2}{2^2} + \frac{n^4r^4}{2^4(2!)^2} - \frac{n^6r^6}{2^6(3!)^2} + \frac{n^8r^8}{2^8(4!)^2} - \ldots\right) \tag{9-25}$$

Our n is complex so if we substitute in equation (9-25) we obtain:

$$J_r = A\left(1 + j\frac{\omega\mu\sigma r^2}{2^2} - \frac{\omega^2\mu^2\sigma^2 r^4}{2^4(2!)^2} - j\frac{\omega^3\mu^3\sigma^3 r^6}{2^6(3!)^2} + \ldots\right) \tag{9-26}$$

Equation (9-26) can be written as:

$$J_r = A(\text{ber}\sqrt{\omega\mu\sigma}\,r + j\,\text{bei}\sqrt{\omega\mu\sigma}\,r) \tag{9-27}$$

Since

$$\sqrt{\omega\mu\,\sigma} = \sqrt{2\pi f\mu\,\sigma} = \sqrt{\frac{2}{\delta^2}} = \frac{\sqrt{2}}{\delta},$$

we can write:

$$J_r = A\left(\text{ber}\frac{\sqrt{2}}{\delta}r + j\,\text{bei}\frac{\sqrt{2}}{\delta}r\right). \tag{9-28}$$

The ber function is the real part of equation (9-26) and the bei function is the imaginary part of equation (9-26). The ber and bei functions are tabulated for different values of the argument in various references. Dwight and McLachlan both have tables for arguments up to 10 and McLachlan gives series solutions for arguments both below and above 10 (Dwight, 1934, p. 238–239), (McLachlan, 1961, p. 138, p. 222–223).

Equation (9-19) has a second solution given as $K_o(\sqrt{-j\omega\mu\,\sigma}\,r)$, which can also be given as a series solution. With a complex argument the solution becomes:

$$B\,K_o\left(\sqrt{-j\omega\mu\,\sigma}\,r\right) = B\left(\text{ker}\frac{\sqrt{2}}{\delta}r + j\text{kei}\frac{\sqrt{2}}{\delta}r\right). \tag{9-29}$$

So our solution (9-23) can be written as:

$$J_r = A(\text{ber}\frac{\sqrt{2}}{\delta}r + j\text{bei}\frac{\sqrt{2}}{\delta}r) + B(\text{ker}\frac{\sqrt{2}}{\delta}r + j\text{kei}\frac{\sqrt{2}}{\delta}r). \tag{9-30}$$

Some values of ber, bei, ker, kei and their derivatives are given in Tables 9-1 and 9-2 (McLachlan, 1961, p. 222–223). Note that the derivatives are defined as is normally done.

Table 9–1
BER and BEI Bessel Functions

X	BER X	BEI X	BER′ X	BEI′ X
0	+1.0000	0.0000	0.0000	0.0000
0.1	+0.999998438	+0.002500000	–0.000062500	+0.049999974
0.2	+0.999975000	+0.009999972	–0.000499999	+0.099999167
0.3	+0.999873438	+0.022499684	–0.001687488	+0.149993672
0.4	+0.999600004	+0.039998222	–0.003999911	+0.199973334
0.5	+0.999023464	+0.062493218	–0.007812076	+0.249918621
0.6	+0.997975114	+0.089979750	–0.013498481	+0.299797507
0.7	+0.996248828	+0.122448939	–0.021433032	+0.349562345
0.8	+0.993601138	+0.159886230	–0.031988623	+0.399146758
0.9	+0.989751357	+0.202269363	–0.045536553	+0.448462528
1.0	+0.984381781	+0.249566040	–0.062445752	+0.497396511
2.0	+0.751734183	+0.972291627	–0.493067125	+0.917013613
2.5	+0.399968417	+1.457182044	–0.943583409	+0.998268847
3.0	–0.221380249	+1.937586785	–1.569846632	+0.880482324
3.5	–1.193598180	+2.283249967	–2.336059130	+0.435296178
4.0	–2.563416557	+2.292690323	–3.134653964	–0.491137441
4.5	–4.299086552	+1.686017204	–3.753681326	–2.052634662
5.0	–6.230082479	+0.116034382	–3.845339473	–4.354141
5.5	–7.973596451	–2.788980155	–2.907031958	–7.372913
6.0	–8.858315966	–7.334746541	–0.293079967	–10.846224
6.5	–7.866890928	–13.606512001	+4.717382012	–14.129423
7.0	–3.632930243	–21.239402580	+12.764522560	–16.041489
7.5	+5.454962184	–29.115711867	+24.130124710	–14.735602
8.0	+20.973955611	–35.016725165	+38.311325701	–7.660318
8.5	+43.935872751	–35.297700300	+53.441618430	+8.289519
9.0	+73.935729857	–24.712783168	+65.600770999	+36.299384
9.1	+80.576411145	–20.723569533	+67.144889467	+43.582976
9.2	+87.349952674	–15.976414197	+68.246178293	+51.459634
9.3	+94.208443358	–10.411661917	+68.831185381	+59.935547
9.4	+101.096359718	–3.969285324	+68.821113743	+69.011850
9.5	+107.950031881	+3.410573282	+68.131840035	+78.683888
9.6	+114.697114173	+11.786984189	+66.673989017	+88.940434
9.7	+121.256066255	+21.217531810	+64.353071286	+99.762855
9.8	+127.535651521	+31.757530896	+61.069692033	+111.124240
9.9	+133.434460262	+43.459152933	+56.719839030	+122.988479
10.0	+138.840465942	+56.370458554	+51.195258394	+135.309302

Table 9-2
KER and KEI Bessel Functions

X	KER X	KEI X	KER′ X	KEI′ X
0	+ ∞	−0.7853982	− ∞	0.0000
0.1	+2.4204740	−0.7768506	−9.9609593	+0.1459748
0.2	+1.7331427	−0.7581249	−4.9229485	+0.2229268
0.3	+1.3372186	−0.7331019	−3.2198652	+0.2742921
0.4	+1.0626239	−0.7038002	−2.3520699	+0.3095140
0.5	+0.8559059	−0.6715817	−1.8197998	+0.3332038
0.6	+0.6931207	−0.6374495	−1.4565386	+0.3481644
0.7	+0.5613783	−0.6021755	−1.1909433	+0.3563095
0.8	+0.4528821	−0.5663676	−0.9873351	+0.3590425
0.9	+0.3625148	−0.5305111	−0.8258687	+0.3574432
1.0	+0.2867062	−0.4949946	−0.6946039	+0.3523699
1.5	+0.0529349	−0.3313956	−0.2941816	+0.2956081
2.0	−0.0416645	−0.2024001	−0.1066010	+0.2198079
2.5	−0.0696880	−0.1106961	−0.0169298	+0.1488954
3.0	−0.0670292	−0.0511219	+0.0214762	+0.0920431
3.5	−0.0526393	−0.0160026	+0.0329886	+0.0509821
4.0	−0.03617885	+0.00219840	+0.03147849	+0.02391062
4.5	−0.02199988	+0.00972092	+0.02481454	+0.00771543
5.0	−0.01151173	+0.01118759	+0.01719340	−0.00081998
5.5	−0.00463216	+0.00971631	+0.01057633	−0.00444016
6.0	−0.00065304	+0.00721649	+0.00563171	−0.00522392
6.5	+0.001278080	+0.004723992	+0.002348995	−0.004600032
7.0	+0.001922022	+0.002700365	+0.000420510	−0.003459509
7.5	+0.001859888	+0.001266868	−0.000538787	−0.002295904
8.0	+0.001485834	+0.000369584	−0.000879724	−0.001336313
8.5	+0.001037349	−0.000114902	−0.000874656	−0.000646733
9.0	+0.000637164	−0.000319153	−0.000711231	−0.000208079
9.1	+0.000568055	−0.000336788	−0.000670745	−0.000145903
9.2	+0.000503046	−0.000348579	−0.000629326	−0.000091093
9.3	+0.000442203	−0.000355236	−0.000587517	−0.000043145
9.4	+0.000385540	−0.000357420	−0.000545789	−0.000001559
9.5	+0.000333029	−0.000355743	−0.000504544	+0.000034158
9.6	+0.000284604	−0.000350768	−0.000464122	+0.000064485
9.7	+0.000240168	−0.000343010	−0.000424806	+0.000089887
9.8	+0.000199598	−0.000332940	−0.000386830	+0.000110811
9.9	+0.000162751	−0.000320983	−0.000350379	+0.000127684
10.0	+0.000129466	−0.000307524	−0.000315597	+0.000140914

$$\text{ber}'\, x = \frac{d(\text{ber}\, x)}{dx}. \qquad \text{bei}'\, x = \frac{d(\text{bei}\, x)}{dx}, \text{ etc.}$$

If we consider a solid-round conductor and let r = 0 we have:

$$J_r = A(\text{ber}\, 0 + j\text{bei}\, 0) + B(\text{ker}\, 0 + j\text{kei}\, 0) = A\,(1 + j0) + B(\infty - j0.78539).$$

Since the current density must be finite the constant B must be zero for a solid-round conductor and we have:

$$J_r = A\left(\text{ber}\, \sqrt{2}\,\frac{r}{\delta} + j\text{bei}\, \sqrt{2}\,\frac{r}{\delta}\right). \tag{9-31}$$

Let us look at δ for copper.

$$\delta = \frac{1}{\sqrt{\pi f \mu\, \sigma}}. \qquad \sigma = 5.8 \times 10^7 \text{ mhos/meter,}$$

$$\mu = 4\pi \times 10^{-7} \text{ henrys/meter.}$$

$$\delta = \frac{1}{\sqrt{\pi f(4\pi \times 10^{-7})(5.8 \times 10^7)}} = \frac{0.06608}{\sqrt{f}} \text{ meters.} \tag{9-32}$$

If we set up a table we obtain:

Table 9-3
Skin Depth for Copper

Frequency (Hertz)	δ (Meters)	δ (Centimeters)	δ (Millimeters)
60	8.53×10^{-3}	0.85	8.53
1,000	2.089×10^{-3}	0.2089	2.089
1×10^6	66.085×10^{-6}	6.6×10^{-3}	0.066
100×10^6	6.608×10^{-6}	660.8×10^{-6}	6.6×10^{-3}
$1{,}000 \times 10^6$	2.089×10^{-6}	208.9×10^{-6}	2.08×10^{-3}

As the frequency goes up, δ, the skin depth, becomes quite small, so the term $\frac{r}{\delta}$ becomes quite large when r equals the radius of the conductor unless the wire is very small. When r = 0 we are at the center of the conductor, so:

$$J_r = A(\text{ber } 0 + j\text{bei } 0) = J_{r(\text{center})} = A(1 + j0) \text{ so } A = J_{r(\text{center})}$$

and

$$J_r = J_{r(\text{center})}\left(\text{ber } \sqrt{2}\,\frac{r}{\delta} + j\text{bei } \sqrt{2}\,\frac{r}{\delta}\right), \tag{9-33}$$

and the ratio of the current density at r = r to the current density at the center becomes:

$$\frac{J_r}{J_{r(\text{center})}} = \text{ber } \sqrt{2}\,\frac{r}{\delta} + j\text{bei } \sqrt{2}\,\frac{r}{\delta} \tag{9-34}$$

Let us look at a No. 10 copper wire which has a diameter of 101.9 mils or 2.588×10^{-3} meters. The radius, a, is 1.294×10^{-3} meters. At 60 hertz, $\delta = 8.53 \times 10^{-3}$, meters so:

$$\frac{a}{\delta} = \frac{1.294 \times 10^{-3}}{8.53 \times 10^{-3}} = 0.1517.$$

The ratio of the current density at the surface to the current density at the center becomes from equation (9-34):

$$\frac{J_{r(r = a)}}{J_{r(r = 0)}} = \text{ber } \sqrt{2}\,(0.1517) + j\text{bei } \sqrt{2}\,(0.1517) =$$

$$\text{ber } 0.2145 + j\text{bei } 0.2145 = 0.99996 + j\,0.011812 = 1.000030\,\underline{/0.676^\circ}.$$

The current density at the surface is practically the same as it is at the center of the conductor.

Let us change the frequency to 130,000 hertz.

$$\delta = \frac{0.06608}{\sqrt{f}} = \frac{0.06608}{\sqrt{130{,}000}} = 1.832 \times 10^{-4} \text{ meters.}$$

$$\frac{a}{\delta} = \frac{1.294 \times 10^{-3}}{1.832 \times 10^{-4}} = 7.06.$$

$$\frac{J_{r(r\,=\,a)}}{J_{r(r\,=\,0)}} = \text{ber}\,\sqrt{2}\,(7.06) + j\text{bei}\,\sqrt{2}\,(7.06) = \text{ber}\,9.985 + j\text{bei}\,9.985 =$$

$$138.0295 + j54.4337 = 148.37\,\underline{/21.52^\circ}.$$

At this frequency the current density at the surface is about 148 times as large as the current density at the center of the conductor and the two are 21.52 degrees out of phase with each other.

As the frequency goes up the ratio of the current density at the surface to the current density at the center of the conductor goes higher and higher. Most of the current in the conductor is flowing close to the surface. At low frequencies the current distribution is almost the same as it would be for the direct current case. Our No. 10 conductor, at a frequency of 60 hertz, had almost uniform current density across its cross-sectional area. The effect depends on the frequency, the conductor size, and the material.

We can express our result in a slightly different manner by finding the current density at the surface of the conductor in equation (9-28).

$$J_{r(r\,=\,a)} = A\left(\text{ber}\,\sqrt{2}\,\frac{a}{\delta} + j\text{bei}\,\sqrt{2}\,\frac{a}{\delta}\right)$$

or

$$A = \frac{J_{r(r\,=\,a)}}{\text{ber}\,\sqrt{2}\,\frac{a}{\delta} + j\text{bei}\,\sqrt{2}\,\frac{a}{\delta}}$$

so:

$$J_{r(r\,=\,r)} = J_{r(r\,=\,a)}\frac{\left(\text{ber}\,\sqrt{2}\,\frac{r}{\delta} + j\text{bei}\,\sqrt{2}\,\frac{r}{\delta}\right)}{\left(\text{ber}\,\sqrt{2}\,\frac{a}{\delta} + j\text{bei}\,\sqrt{2}\,\frac{a}{\delta}\right)}. \tag{9-35}$$

So,

$$\frac{J_{r(r\,=\,r)}}{J_{r(r\,=\,a)}} = \frac{\text{ber}\,\sqrt{2}\,\frac{r}{\delta} + j\text{bei}\,\sqrt{2}\,\frac{r}{\delta}}{\text{ber}\,\sqrt{2}\,\frac{a}{\delta} + j\text{bei}\,\sqrt{2}\,\frac{a}{\delta}}. \tag{9-36}$$

Let us look at our No. 10 conductor at a frequency of 500×10^3 hertz.

$$\delta = \frac{0.06608}{\sqrt{500 \times 10^3}} = 93.451 \times 10^{-6} \text{ meters.}$$

$$\frac{a}{\delta} = \frac{1.29 \times 10^{-3}}{93.451 \times 10^{-6}} = 13.84679.$$

$$\frac{J_{r(r\,=\,r)}}{J_{r(r\,=\,a)}} = \frac{\text{ber}\,\dfrac{\sqrt{2}\,r}{93.451 \times 10^{-6}} + \text{bei}\,\dfrac{\sqrt{2}\,r}{93.451 \times 10^{-6}}}{\text{ber}\,\sqrt{2}\,(13.84679) + j\text{bei}\,\sqrt{2}\,(13.84679)} = \frac{\text{ber}\,15.133 \times 10^3 r + j\text{bei}\,15.133 \times 10^3 r}{\text{ber}\,19.5823 + j\text{bei}\,19.5823}.$$

McLachlan gives equations for ber z and bei z when $z > 10$ (McLachlan, 1961, p. 222.)

$$\text{Ber}\,z \simeq \frac{0.3989 e^{z/\sqrt{2}}}{\sqrt{z}}\left[\sin(40.514z + 67.5)^\circ + \frac{1}{8z}\sin(40.514z + 22.5)^\circ\right] \quad (9\text{-}37)$$

$$\text{Bei}\,z \simeq \frac{0.3989 e^{z/\sqrt{2}}}{\sqrt{z}}\left[\sin(40.514z - 22.5)^\circ + \frac{1}{8z}\sin(40.514z - 67.5)^\circ\right] \quad (9\text{-}38)$$

Using these equations we find that:

$$\text{ber}\,19.5823 = 59.30 \times 10^3. \qquad \text{bei}\,19.5823 = 72.196 \times 10^3.$$

$$\frac{J_{r(r\,=\,r)}}{J_{r(r\,=\,a)}} = \frac{\text{ber}\,15.133 \times 10^3 r + j\text{bei}\,15.133 \times 10^3 r}{59.30 \times 10^3 + j72.196 \times 10^3} = \frac{\text{ber}\,15.133 \times 10^3 r + j\text{bei}\,15.133 \times 10^3 r}{93.428 \times 10^3 \,\underline{/50.60^\circ}}.$$

Let us find J_r at the center of the conductor in terms of the current density at the surface.

$$J_{\text{center }(r\,=\,0)} = J_{\text{surface }(r\,=\,a)} \frac{(\text{ber }0 + j\text{ bei }0)}{93.428 \times 10^3 \underline{/50.60^\circ}} =$$

$$J_{(\text{surface})}\,(10.703 \times 10^{-6} \underline{/-50.60^\circ}).$$

The current density at the center is very small when compared to the current density at the surface. Suppose we go to a place in the conductor where r is equal to the radius of the conductor minus the skin depth.

$$r = a - \delta = 1.29 \times 10^{-3} - 93.451 \times 10^{-6} = 1.1965 \times 10^{-3} \text{ meters.}$$

$$\frac{J_{(r\,=\,1.1965 \times 10^{-3})}}{J_{(\text{surface})}} =$$

$$\frac{\text{ber}(15.133 \times 10^3)(1.1965 \times 10^{-3}) + j\text{bei}(15.133 \times 10^3)(1.1965 \times 10^{-3})}{93.428 \times 10^3 \underline{/50.60^\circ}} =$$

$$\frac{\text{ber }18.107 + j\text{ bei }18.107}{93.428 \times 10^3 \underline{/50.60^\circ}} = \frac{33.814 \times 10^3 - j5.46 \times 10^3}{93.428 \times 10^3 \underline{/50.60^\circ}} =$$

$$366.65 \times 10^{-3} \underline{/-59.776^\circ}.$$

From this we see that the current density, δ units from the surface, is about four-tenths of the value at the surface or that most of the current is close to the surface and very little is flowing at the center of the conductor. At high frequencies it is a good approximation to

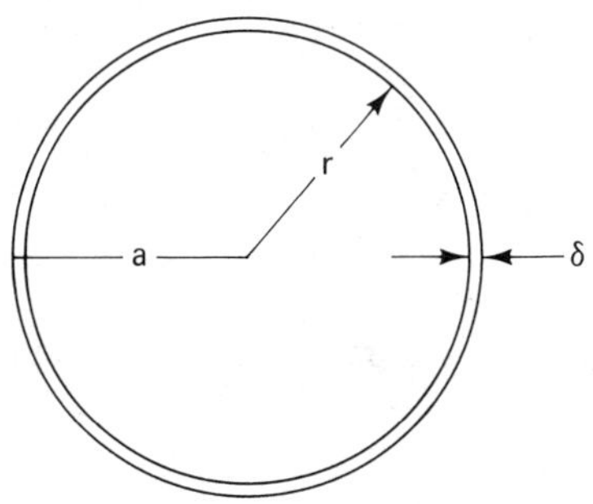

Figure 9-11. Skin depth.

assume that all of the current flow is in the circular ring with δ as the thickness as shown in Figure 9-11.

Area $\simeq 2\pi a\delta$, so

$$R_{ac} = \rho\frac{\ell}{A} \simeq \rho\frac{\ell}{2\pi a\delta}. \tag{9-39}$$

$$R_{dc} = \frac{\rho\ell}{\pi a^2} \text{ so } \frac{R_{ac}}{R_{dc}} \simeq \frac{\dfrac{\rho\ell}{2\pi a\delta}}{\dfrac{\rho\ell}{\pi a^2}} \simeq \frac{\rho\ell}{2\pi a\delta}\,\frac{\pi a^2}{\rho\ell} \simeq \frac{a}{2\delta}. \tag{9-40}$$

9.6 Inductance Changes Because of Skin Effect

As the frequency increases, the resistance and the inductance change as the flux linkages in the conductor reduce and as the current increases towards the surface and reduces in the center of the conductor. If all of the current were on the surface there would be no flux linkages internally and the conductor inductance and reactance would be zero.

Let us calculate the internal impedance of the conductor. Let $Z_{conductor} = R + j\omega L$, where R is the effective resistance and L is the effective inductance at a given frequency.

The total current in the wire is:

$$I = \int_0^a J_r 2\pi r dr = \int_0^a A\left(\text{ber}\,\sqrt{2}\,\frac{r}{\delta} + j\text{bei}\,\sqrt{2}\,\frac{r}{\delta}\right)2\pi r dr =$$

$$\int_0^a A\,\text{ber}\,\sqrt{2}\,\frac{r}{\delta}(2\pi r dr) + j\int_0^a A\,\text{bei}\,\sqrt{2}\,\frac{r}{\delta}(2\pi r dr).$$

By using series expansions it can be easily shown that:

$$\int x\,\text{ber}\,x\,dx = x\,\text{bei}'\,x. \qquad \int x\,\text{bei}\,x\,dx = -\,x\,\text{ber}'\,x.$$

$$I = \int_0^a A\,2\pi r\,\text{ber}\,\sqrt{2}\,\frac{r}{\delta}\,dr + j\int_0^a A\,2\pi r\,\text{bei}\,\sqrt{2}\,\frac{r}{\delta}\,dr.$$

$$\text{Let } x = \sqrt{2}\,\frac{r}{\delta}. \qquad dx = \frac{\sqrt{2}}{\delta}\,dr.$$

$$I = \int_o^a \delta^2\pi A\left(\sqrt{2}\,\frac{r}{\delta}\right) \text{ber}\,\sqrt{2}\,\frac{r}{\delta}\left(\frac{\sqrt{2}}{\delta}dr\right) + j\int_o^a \delta^2\pi A\left(\sqrt{2}\,\frac{r}{\delta}\right) \text{bei}\,\sqrt{2}\,\frac{r}{\delta}\left(\frac{\sqrt{2}}{\delta}\,dr\right) =$$

$$\left[\delta^2\pi A\,\sqrt{2}\,\frac{r}{\delta}\,\text{bei}'\,\sqrt{2}\,\frac{r}{\delta}\right]_o^a - j\left[\delta^2\pi A\,\sqrt{2}\,\frac{r}{\delta}\,\text{ber}'\,\sqrt{2}\,\frac{r}{\delta}\right]_o^a =$$

$$\pi A\delta\sqrt{2}\left[a\text{bei}'\,\sqrt{2}\,\frac{a}{\delta} - 0\right] - j\pi A\delta\,\sqrt{2}\left[a\text{ber}'\,\sqrt{2}\,\frac{a}{\delta} - 0\right] =$$

$$I = \pi A\,\sqrt{2}\,\delta a\left[\text{bei}\;2\,\frac{a}{\delta} - j\,\text{ber}'\;2\,\frac{a}{\delta}\right] = I_{\text{total}}.$$

At the surface,

$$V = I_s R_{\text{surface}} = J_s S\left(\rho\frac{\ell}{S}\right) = J_s\rho\ell = J_s\rho(1).$$

$$Z = \frac{V}{I} = \frac{\rho\left[A\left(\text{ber}\,\sqrt{2}\,\frac{a}{\delta} + j\,\text{bei}\,\sqrt{2}\,\frac{a}{\delta}\right)\right]}{\pi A\,\sqrt{2}\,\delta a\left[\text{bei}'\,\sqrt{2}\,\frac{a}{\delta} - j\,\text{ber}'\,\sqrt{2}\,\frac{a}{\delta}\right]} =$$

$$Z = \frac{\text{ber}\,\sqrt{2}\,\frac{a}{\delta} + j\,\text{bei}\,\sqrt{2}\,\frac{a}{\delta}}{\sqrt{2}\,\pi\sigma\delta a\left[\text{bei}'\,\sqrt{2}\,\frac{a}{\delta} - j\,\text{ber}'\,\sqrt{2}\,\frac{a}{\delta}\right]}. \tag{9-41}$$

Multiplying the numerator and denominator by the conjugate of the denominator gives:

$$Z = \frac{1}{\sqrt{2}\,\pi\sigma\delta a}\,\frac{\left(\text{ber}\,\sqrt{2}\,\frac{a}{\delta} + j\,\text{bei}\,\sqrt{2}\,\frac{a}{\delta}\right)\left(\text{bei}'\,\sqrt{2}\,\frac{a}{\delta} + j\,\text{ber}'\,\sqrt{2}\,\frac{a}{\delta}\right)}{\left(\text{bei}'\,\sqrt{2}\,\frac{a}{\delta}\right)^2 + \left(\text{ber}'\,\sqrt{2}\,\frac{a}{\delta}\right)^2} =$$

$$\frac{1}{\sqrt{2}\pi\sigma\delta a}\left[\frac{\left(\mathrm{ber}\,\sqrt{2}\,\frac{a}{\delta}\right)\left(\mathrm{bei}'\,\sqrt{2}\,\frac{a}{\delta}\right)-\left(\mathrm{bei}\,\sqrt{2}\,\frac{a}{\delta}\right)\left(\mathrm{ber}'\,\sqrt{2}\,\frac{a}{\delta}\right)}{\left(\mathrm{bei}'\,\sqrt{2}\,\frac{a}{\delta}\right)^{2}+\left(\mathrm{ber}'\,\sqrt{2}\,\frac{a}{\delta}\right)^{2}}\right]$$

$$+\,j\,\frac{1}{\sqrt{2}\pi\sigma\delta a}\left[\frac{\left(\mathrm{bei}\,\sqrt{2}\,\frac{a}{\delta}\right)\left(\mathrm{bei}'\,\sqrt{2}\,\frac{a}{\delta}\right)+\left(\mathrm{ber}\,\sqrt{2}\,\frac{a}{\delta}\right)\left(\mathrm{ber}'\,\sqrt{2}\,\frac{a}{\delta}\right)}{\left(\mathrm{bei}'\,\sqrt{2}\,\frac{a}{\delta}\right)^{2}+\left(\mathrm{ber}'\,\sqrt{2}\,\frac{a}{\delta}\right)^{2}}\right]=$$

$$R+j\omega L.$$

$$R=\frac{1}{\sqrt{2}\,\pi\sigma\delta a}\left[\frac{\left(\mathrm{ber}\,\sqrt{2}\,\frac{a}{\delta}\right)\left(\mathrm{bei}'\,\sqrt{2}\,\frac{a}{\delta}\right)-\left(\mathrm{bei}\,\sqrt{2}\,\frac{a}{\delta}\right)\left(\mathrm{ber}'\,\sqrt{2}\,\frac{a}{\delta}\right)}{\left(\mathrm{bei}\,\sqrt{2}\,\frac{a}{\delta}\right)^{2}+\left(\mathrm{ber}'\,\sqrt{2}\,\frac{a}{\delta}\right)^{2}}\right].\tag{9-42}$$

$$L=\frac{1}{(2\pi f)(\sqrt{2}\pi\sigma\delta a)}\left[\frac{\left(\mathrm{bei}\,\sqrt{2}\,\frac{a}{\delta}\right)\left(\mathrm{bei}'\,\sqrt{2}\,\frac{a}{\delta}\right)+\left(\mathrm{ber}\,\sqrt{2}\,\frac{a}{\delta}\right)\left(\mathrm{ber}'\,\sqrt{2}\,\frac{a}{\delta}\right)}{\left(\mathrm{bei}'\,\sqrt{2}\,\frac{a}{\delta}\right)^{2}+\left(\mathrm{ber}'\,\sqrt{2}\,\frac{a}{\delta}\right)^{2}}\right].\tag{9-43}$$

$$R_{dc}=\frac{\rho\ell}{\pi a^{2}}=\frac{\rho(1)}{\pi a^{2}}=\frac{1}{\sigma\pi a^{2}}.\qquad \sigma\pi a=\frac{1}{aR_{dc}}.$$

$$R=\frac{1}{\sqrt{2}\,\delta\,\frac{1}{aR_{dc}}}\left[\frac{\left(\mathrm{ber}\,\sqrt{2}\,\frac{a}{\delta}\right)\left(\mathrm{bei}'\,\sqrt{2}\,\frac{a}{\delta}\right)-\left(\mathrm{bei}\,\sqrt{2}\,\frac{a}{\delta}\right)\left(\mathrm{ber}'\,\sqrt{2}\,\frac{a}{\delta}\right)}{\left(\mathrm{bei}'\,\sqrt{2}\,\frac{a}{\delta}\right)^{2}+\left(\mathrm{ber}'\,\sqrt{2}\,\frac{a}{\delta}\right)^{2}}\right]=$$

$$R=\frac{aR_{dc}}{\delta\sqrt{2}}\left[\frac{\left(\mathrm{ber}\,\sqrt{2}\,\frac{a}{\delta}\right)\left(\mathrm{bei}'\,\sqrt{2}\,\frac{a}{\delta}\right)-\left(\mathrm{bei}\,\sqrt{2}\,\frac{a}{\delta}\right)\left(\mathrm{ber}'\,\sqrt{2}\,\frac{a}{\delta}\right)}{\left(\mathrm{bei}'\,\sqrt{2}\,\frac{a}{\delta}\right)^{2}+\left(\mathrm{ber}'\,\sqrt{2}\,\frac{a}{\delta}\right)^{2}}\right].\tag{9-44}$$

If $\frac{a}{\delta} = 100$,

$$R = \frac{100\,R_{dc}}{\sqrt{2}}\left[\frac{(\text{ber } 141.4)(\text{bei}'\ 141.4) - (\text{bei } 141.4)(\text{ber}'\ 141.4)}{(\text{bei } 141.4)^2 + (\text{ber}'\ 141.4)^2}\right].$$

McLachlan gives the following equations for $z > 10$ (McLachlan, 1961, p. 223).

$$\text{Ber}'\ z = \frac{0.3989e^{z/\sqrt{2}}}{\sqrt{z}}\left[\sin(40.514z + 112.5)^\circ - \frac{3}{8z}\sin(40.514z + 67.5)^\circ\right]. \tag{9-45}$$

$$\text{bei}'\ z = \frac{0.3989e^{z/\sqrt{2}}}{\sqrt{z}}\left[\sin(40.514z + 22.5)^\circ - \frac{3}{8z}\sin(40.514z - 22.5)^\circ\right]. \tag{9-46}$$

Using these two equations and equations (9-37) and (9-38), we obtain:

$\text{ber } 141.4 = 5.431 \times 10^{41}$. $\quad$ $\text{bei } 141.4 = -7.2048 \times 10^{41}$.

$\text{ber}'\ 141.4 = 8.915 \times 10^{41}$. $\quad$ $\text{bei}'\ 141.4 = -1.22883 \times 10^{41}$.

$R = R_{ac} = 49.96\,R_{dc}$.

If we use equation (9-40) we obtain:

$$R_{ac} = R_{dc}\frac{a}{2\delta} = R_{dc}\frac{100}{2} = 50\,R_{dc}.$$

So we obtain about the same answer from the approximate equation (9-40) as we do for the exact equation (9-44) when $a/\delta = 100$. This holds true down to about $a/\delta = 20$. Below this value we can use tables, graphs or the equation. We can go through the same type of procedure for the inductance.

$$L = \frac{1}{(2\pi f)(\sqrt{2}\,\pi\sigma\delta a)}\left[\frac{\left(\text{bei } \sqrt{2}\frac{a}{\delta}\right)\left(\text{bei}'\ \sqrt{2}\frac{a}{\delta}\right) + \left(\text{ber } \sqrt{2}\frac{a}{\delta}\right)\left(\text{ber}'\ \sqrt{2}\frac{a}{\delta}\right)}{\left(\text{bei}'\ \sqrt{2}\frac{a}{\delta}\right)^2 + \left(\text{ber}'\ \sqrt{2}\frac{a}{\delta}\right)^2}\right]. \tag{9-43}$$

The d. c. inductance $= L_{\text{low}} = \dfrac{\mu}{8\pi}$. $\quad \mu = 8\pi L_{\text{low}}$.

$$L = \frac{\mu}{2\pi\sqrt{2}\,(\pi f\mu\sigma)\delta a}\left[\frac{\left(\text{bei}\sqrt{2}\,\frac{a}{\delta}\right)\left(\text{bei}'\sqrt{2}\,\frac{a}{\delta}\right)+\left(\text{ber}\sqrt{2}\,\frac{a}{\delta}\right)\left(\text{ber}'\sqrt{2}\,\frac{a}{\delta}\right)}{\left(\text{bei}'\sqrt{2}\,\frac{a}{\delta}\right)^2+\left(\text{ber}'\sqrt{2}\,\frac{a}{\delta}\right)^2}\right].$$

$$\delta = \frac{1}{\sqrt{\pi f\mu\sigma}}. \qquad \delta^2 = \frac{1}{\pi f\mu\sigma},$$

so

$$L = \frac{\mu\,\delta^2}{2\pi\sqrt{2}\,\delta a}\left[\frac{\left(\text{bei}\sqrt{2}\,\frac{a}{\delta}\right)\left(\text{bei}'\sqrt{2}\,\frac{a}{\delta}\right)+\left(\text{ber}\sqrt{2}\,\frac{a}{\delta}\right)\left(\text{ber}'\sqrt{2}\,\frac{a}{\delta}\right)}{\left(\text{bei}'\sqrt{2}\,\frac{a}{\delta}\right)^2+\left(\text{ber}'\sqrt{2}\,\frac{a}{\delta}\right)^2}\right].$$

$$L = \frac{8\pi L_{\text{low}}\delta}{2\pi\sqrt{2}\,a}\left[\frac{\left(\text{bei}\sqrt{2}\,\frac{a}{\delta}\right)\left(\text{bei}'\sqrt{2}\,\frac{a}{\delta}\right)+\left(\text{ber}\sqrt{2}\,\frac{a}{\delta}\right)\left(\text{ber}'\sqrt{2}\,\frac{a}{\delta}\right)}{\left(\text{bei}'\sqrt{2}\,\frac{a}{\delta}\right)^2+\left(\text{ber}'\sqrt{2}\,\frac{a}{\delta}\right)^2}\right].$$

$$L = \frac{4\,L_{\text{low}}}{\sqrt{2}\,\frac{a}{\delta}}\left[\frac{\left(\text{bei}\sqrt{2}\,\frac{a}{\delta}\right)\left(\text{bei}'\sqrt{2}\,\frac{a}{\delta}\right)+\left(\text{ber}\sqrt{2}\,\frac{a}{\delta}\right)\left(\text{ber}'\sqrt{2}\,\frac{a}{\delta}\right)}{\left(\text{bei}'\sqrt{2}\,\frac{a}{\delta}\right)^2+\left(\text{ber}'\sqrt{2}\,\frac{a}{\delta}\right)^2}\right]. \tag{9-47}$$

$$\frac{L}{L_{\text{low}}} = \frac{L_{\text{ac}}}{L_{\text{low}}} = \frac{4}{\sqrt{2}\,\frac{a}{\delta}}\left[\frac{\left(\text{bei}\sqrt{2}\,\frac{a}{\delta}\right)\left(\text{bei}'\sqrt{2}\,\frac{a}{\delta}\right)+\left(\text{ber}\sqrt{2}\,\frac{a}{\delta}\right)\left(\text{ber}'\sqrt{2}\,\frac{a}{\delta}\right)}{\left(\text{bei}'\sqrt{2}\,\frac{a}{\delta}\right)^2+\left(\text{ber}'\sqrt{2}\,\frac{a}{\delta}\right)^2}\right]. \tag{9-48}$$

Again, if we use a/δ = 100 we would obtain:

$$\frac{L}{L_{low}} = \frac{4}{\sqrt{2}\,(100)}\left[\frac{(\text{bei }141.4)(\text{bei}'\ 141.4) + (\text{ber }141.4)(\text{ber}'\ 141.4)}{(\text{ber}'\ 141.4)^2 + (\text{bei}'\ 141.4)^2}\right] =$$

$$\frac{4}{\sqrt{2}\,(100)}(0.71) = 0.02.$$

The effective inductance is only 2/100 of the low frequency or d. c. inductance.

If we use $a/\delta = 0.707$,

$$\sqrt{2}\,\frac{a}{\delta} = \sqrt{2}\,(0.707) = 1.$$

$$\frac{L}{L_{low}} = \frac{4}{\sqrt{2}\,(0.707)}\left[\frac{(\text{bei }1.0)(\text{bei}'\ 1.0) + (\text{ber }1.0)\,(\text{ber}'\ 1.0)}{(\text{ber}'\ 1.0)^2 + (\text{bei}'\ 1.0)^2}\right] = 0.9974.$$

$$L \simeq L_{low} = \frac{\mu}{8\pi}.$$

The values of the effective R and L given by equations (9-42) and (9-43) are for one conductor. For a two-wire line we would multiply the R by 2 to obtain the R for the line. For L we would multiply by 2 and add the inductance due to the flux linkages between the two wires.

$$L_{total} = 2L_{conductor} + L_{spacing}. \tag{9-49}$$

9.7 Line Parameters

The inductance due to the flux linkages between the two wires shown in Figure 9-1(a) is derived in fields courses and is given by Hayt (Hayt, 1981, p. 438).

$$L_{spacing} = \frac{\ell\mu}{\pi}\cosh^{-1}\frac{D}{2a}, \tag{9-50}$$

where D = the center-to-center separation between the two wires and the conductors have a radius of a meters; ℓ = the length of the line. Since

$$\cosh^{-1} x = \ln(x + \sqrt{x^2 - 1}),$$

$$\frac{L_{\text{spacing}}}{\ell} = \frac{\mu}{\pi}\cosh^{-1}\frac{D}{2a} = \frac{\mu}{\pi}\ln\left[\frac{D}{2a} + \sqrt{\frac{(D)^2}{2a} - 1}\right].$$

if $D \gg a$ which is the normal like condition for power lines we have

$$\frac{L_{\text{spacing}}}{\ell} \simeq \frac{\mu}{\pi}\ln\left[\frac{D}{2a} + \frac{D}{2a}\right] \simeq \frac{\mu}{\pi}\ln\frac{D}{a}. \qquad (9\text{-}51)$$

At low frequencies we would then obtain the inductance of a two-wire line as equation (9-52) if we can neglect the low frequency skin effect.

$$\frac{L}{\ell} \simeq \frac{\mu}{\pi}\ln\frac{D}{a} + 2\frac{(\mu)}{8\pi} \simeq \left(\frac{\mu}{\pi}\ln\frac{D}{a} + \frac{\mu}{4\pi}\right) \text{henrys/meter.} \qquad (9\text{-}52)$$

If $\delta \ll a$, Hayt gives an equation for the approximate value (Hayt, 1981, p. 328).

$$\frac{L}{\ell} \simeq \frac{\mu}{\pi}\left(\frac{\delta}{2a} + \cosh^{-1}\frac{D}{2a}\right) \text{henrys/meter.} \qquad (9\text{-}53)$$

For the capacitance of a two-wire line, field theory gives us:

$$\frac{C}{\ell} = \frac{\pi\varepsilon}{\cosh^{-1}\dfrac{D}{2a}} \text{farads/meter,} \qquad (9\text{-}54)$$

or if $D \gg a$,

$$\frac{C}{\ell} \simeq \frac{\pi\varepsilon}{\ln\dfrac{D}{a}}. \qquad (9\text{-}55)$$

The conductance-per-unit length, G, is harder to calculate and/or determine. In fields courses, G can be found by assuming a constant value for the conductivity, σ, between the wires; G could be found as:

$$\frac{G}{\ell} = \frac{\pi\sigma}{\cosh^{-1}\dfrac{D}{2a}} \text{mhos/meter.} \qquad (9\text{-}56)$$

However, for an open-wire line the conductivity is a function of many different variables, such as temperature, pressure, humidity, etc. For power lines with wide separation it is often assumed to be zero.

For a coaxial cable where the conductivity of the dielectric material between the two conductors is not only usually constant but more apt to be known we have:

$$\frac{G}{\ell} = \frac{2\pi\sigma}{\ln \frac{b}{a}} \text{ mhos/meter.} \tag{9-57}$$

At high frequencies, when skin effect is causing zero flux linkages in the two conductors (current in the center conductor is approximately all at the outer surface and the current in the outer conductor is approximately all at the inner surface) we can write:

$$\frac{L}{\ell} = \frac{\mu}{2\pi} \ln \frac{b}{a} \text{ henrys/meter.} \tag{9-58}$$

$$\frac{C}{\ell} = \frac{2\pi\varepsilon}{\ln \frac{b}{a}} \text{ farads/meter.} \tag{9-59}$$

If the internal inductance is taken into account, Hayt gives the equations (Hayt, 1981, p. 437): If $\delta \ll a$ and $\delta \ll (c - b)$,

$$\frac{L}{\ell} = \frac{\mu}{2\pi} \left[\ln \frac{b}{a} + \frac{\delta}{2} \left(\frac{1}{a} + \frac{1}{b} \right) \right] \text{ henrys/meter.} \tag{9-60}$$

$$\frac{R}{\ell} = \frac{1}{2\pi\delta\sigma_c} \left(\frac{1}{a} + \frac{1}{b} \right) \text{ ohms/meter for both conductors.} \tag{9-61}$$

σ_c is the conductivity of the conductors.

9.8 Three-Phase Lines

Three-phase lines are handled in a special manner; the practicing engineer may find the parameters from tables given in handbooks or published by the manufacturers of power-transmission line conductors. Since power lines normally have long spans between towers, steel strands are often a part of the total conductor to add strength.

The presence of steel strands complicates the calculation of the overall line parameters since the current distribution in the conductors is affected, and additional losses arising because of hysteresis and eddy currents cause the effective resistance to increase. Another phenomenon comes into play since high-power transmission lines are always associated with high voltages. High voltages, in turn, are associated with high field strengths around the surfaces of the conductors and if the E field is large enough there will be a breakdown of the air in the vicinity of the wires where the field strength is the highest. A glow appears that is called "corona." Corona produces ozone and electromagnetic radiation, which is equivalent to electrical noise, and which interferes with the operation of such devices as radios and causes additional power losses on the line. The strength

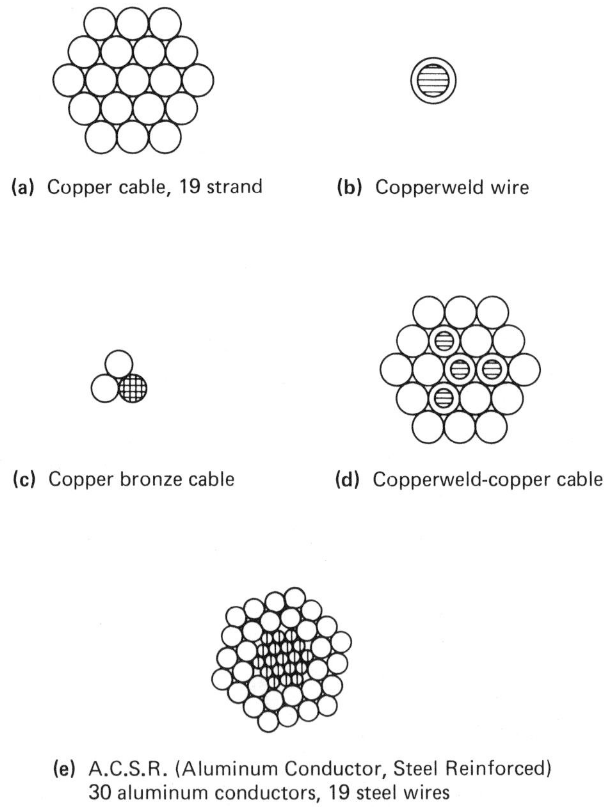

Figure 9-12. Types of conductors.

of the field at the conductor surface is reduced if larger-size wires are used; the wire size is often larger than necessary to carry the electrical power involved.

Various types of conductors are shown in Figure 9-12.

Power conductors are manufactured in various ways. In Figure 9-12, section (a) shows 19 strands of copper wire made up into a cable, (b) shows a copperweld wire made up of a steel core coated with copper. Section (c) shows two copper wires and one bronze wire as a cable. Power conductors may be made up of concentric copperweld wires for mechanical strength along with copper wires as shown in (d). Part (e) shows an ACSR conductor which is very popular for power transmission lines and consists of a steel core surrounded by aluminum wires. We will concentrate on the ACSR conductor which is often used for three-phase power transmission lines.

Let us start out by looking at a two-wire line. Equation (9-51) gives the inductance due to the external flux linkages for both wires.

$$\frac{L}{\ell} = \frac{\mu}{\pi} \ln \frac{D}{a},$$

so the inductance per wire would equal:

$$\frac{L}{\ell} = \frac{\mu}{2\pi} \ln \frac{D}{a} \text{ henrys/meter per conductor.} \qquad (9\text{-}62)$$

If we add the internal inductance for one wire and multiply by 1,609.34 to convert our equation to henrys per mile we obtain:

$$\frac{L}{\ell} = 1{,}609.34 \left[\frac{\mu}{2\pi} \ln \frac{D}{a} + \frac{\mu}{8\pi}\right] = 1{,}609.34 \left[\frac{4\pi \times 10^{-7}}{2\pi} \ln \frac{D}{a} + \frac{4\pi \times 10^{-7}}{8\pi}\right] =$$

$$321.868 \times 10^{-6} \ln \frac{D}{a} + 80.467 \times 10^{-6} =$$

$$321.868 \times 10^{-6} (2.302) \log \frac{D}{a}$$

$$+ 80.467 \times 10^{-6} = 740.296 \times 10^{-6} \log \frac{D}{a}$$

$$+ 80.467 \times 10^{-6},$$

so

so

$$\frac{L}{\ell} = 740.296 \times 10^{-6} \log \frac{D}{1} - 740.296 \times 10^{-6} \log a + 80.467 \times 10^{-6}.$$

When D is in feet the first term involving the log $D/1$ appears as the inductance due to the external flux linkages if the conductors had a spacing of D feet and a radius of one foot. We will assume a frequency of 60 hertz.

$$X_L = 2\pi f L = 2\pi(60)\,[740.296 \times 10^{-6} \log D] =$$
$$X_L = 0.27939 \log D \text{ ohms/mile.} \tag{9-63}$$

The second and third terms, $80.467 \times 10^{-6} - 740.296 \times 10^{-6} \log a$, must be associated with the internal inductance of a hypothetical conductor with a radius of one foot.

$$80.467 \times 10^{-6} + 740.296 \times 10^{-6} \log \frac{1}{a}.$$

The power engineer likes to express this in terms of the Geometric Mean Radius (G.M.R.) which came from the study of geometric mean distances. The G.M.R. is also known as the self geometric mean distance. For a round conductor,

$$\text{G.M.R.} = 0.7788\,a. \qquad a = \frac{\text{G.M.R.}}{0.7788}$$

While the study of geometric mean distances is interesting for the engineer who works in the power area, we will simply use the result.

$$\text{G.M.R.} = 0.7788\,a. \tag{9-64}$$

So we have:

$$80.467 \times 10^{-6} + 740.296 \times 10^{-6} \log \frac{0.7788}{\text{G.M.R.}} =$$

$$80.467 \times 10^{-6} + 740.296 \times 10^{-6} \log 0.7788 + 740.296 \times 10^{-6} \log \frac{1}{\text{G.M.R.}} =$$

$$80.467 \times 10^{-6} - 80.467 \times 10^{-6} + 740.296 \times 10^{-6} \log \frac{1}{\text{G.M.R.}} =$$

$$740.296 \times 10^{-6} \log \frac{1}{\text{G.M.R.}}.$$

We will define X_a, and assume a frequency of 60 hertz.

$$X_a = 2\pi 60\left(740.296 \times 10^{-6} \log \frac{1}{\text{G.M.R.}}\right)$$

$$= 0.2794 \log \frac{1}{\text{G.M.R.}}. \quad (9\text{-}65)$$

We will call this the conductor component. We will define X_L in equation (9-63) as the spacing component:

$$X_d = 0.2794 \log D. \quad (9\text{-}66)$$

We now have two components: a spacing component, X_d, and a conductor component, X_a. X_a is the inductive reactance for a conductor with a radius of one foot, and X_d is the inductive reactance involved with the spacing between conductors with a radius of one foot. The G.M.R. for each conductor configuration can be determined mathematically, but we do not have to go into any details since this information is given to us by the manufacturers and is also contained in various handbooks. Figure 9-13 comes from the *Aluminum Electrical Conductor Handbook* and gives X_a directly in ohms per mile for various ACSR conductors as well as the G.M.R. for each conductor. (*Aluminum Electrical Conductor Handbook*, 1982, p. 4-25.)

The spacing component, X_d, is given in Figure 9-14, which also comes from the *Aluminum Electrical Conductor Handbook* where it is repeated from a Westinghouse reference text (*Aluminum Electrical Conductor Handbook*, 1982, p. 3-14) (*Electrical Transmission and Distribution Reference Book*, 1964, p. 54).

Notice that the spacing component is zero for a separation of 1 foot and negative if the spacing is less than one foot.

To make the calculations easy, engineers did the same thing for the capacitive effect and broke it up into two components, a spacing component, X_d' and X_a', a conductor component. X_a' is given in Figure 9-13 and the spacing component, X_d' in Figure 9-15.

Figure 9-15 gives information contained in both of the references mentioned above. (*Aluminum Electrical Conductor Handbook*, 1982,

Bare Aluminum Conductors, Steel Reinforced (ACSR)
Electrical Characteristics of Multi-Layer Sizes

Code Word	Size Kcmil	Stranding Al./St.	Number of Aluminum Layers	Resistance				GMR	Phase-to-Neutral, 60 Hz Reactance at One Ft. Spacing	
				D.C.	AC–60 Hz					
				20°C Ohms/ Mile	25°C Ohms/ Mile	50°C Ohms/ Mile	75°C Ohms/ Mile	Feet	Inductive Ohms/Mile X_a	Capacitive Megohm-Miles X'_a
Waxwing	266.8	18/ 1	2	0.3398	0.347	0.382	0.416	0.0197	0.477	0.109
Partridge	266.8	26/ 7	2	0.3364	0.344	0.377	0.411	0.0217	0.465	0.107
Ostrich	300.	26/ 7	2	0.2993	0.306	0.336	0.366	0.0230	0.458	0.106
Merlin	336.4	18/ 1	2	0.2693	0.276	0.303	0.330	0.0221	0.463	0.106
Linnet	336.4	26/ 7	2	0.2671	0.273	0.300	0.327	0.0244	0.451	0.104
Oriole	336.4	30/ 7	2	0.2650	0.271	0.297	0.324	0.0255	0.445	0.103
Chickadee	397.5	18/ 1	2	0.2279	0.234	0.257	0.279	0.0240	0.452	0.103
Ibis	397.5	26/ 7	2	0.2260	0.231	0.254	0.277	0.0265	0.441	0.102
Lark	397.5	30/ 7	2	0.2243	0.229	0.252	0.274	0.0277	0.435	0.101
Pelican	477.	18/ 1	2	0.1899	0.195	0.214	0.233	0.0263	0.441	0.100
Flicker	477.	24/ 7	2	0.1889	0.194	0.213	0.232	0.0283	0.432	0.0992
Hawk	477.	26/ 7	2	0.1883	0.193	0.212	0.231	0.0290	0.430	0.0988
Hen	477.	30/ 7	2	0.1869	0.191	0.210	0.229	0.0304	0.424	0.0980
Osprey	556.5	18/ 1	2	0.1629	0.168	0.184	0.200	0.0284	0.432	0.0981
Parakeet	556.5	24/ 7	2	0.1620	0.166	0.183	0.199	0.0306	0.423	0.0969
Dove	556.5	26/ 7	2	0.1613	0.166	0.182	0.198	0.0313	0.420	0.0965
Eagle	556.5	30/ 7	2	0.1602	0.164	0.180	0.196	0.0328	0.415	0.0957
Peacock	605.	24/ 7	2	0.1490	0.153	0.168	0.183	0.0319	0.418	0.0957
Squab	605.	26/ 7	2	0.1485	0.153	0.167	0.182	0.0327	0.415	0.0953
Teal	605.	30/19	2	0.1475	0.151	0.166	0.181	0.0342	0.410	0.944
Rook	636.	24/ 7	2	0.1417	0.146	0.160	0.174	0.0327	0.415	0.0950
Grosbeak	636.	26/ 7	2	0.1411	0.145	0.159	0.173	0.0335	0.412	0.0946
Egret	636.	30/19	2	0.1403	0.144	0.158	0.172	0.0351	0.406	0.0937
Flamingo	666.6	24/ 7	2	0.1352	0.139	0.153	0.166	0.0335	0.412	0.0943
Crow	715.5	54/ 7	3	0.1248	0.128	0.141	0.153	0.0372	0.399	0.0920
Starling	715.5	26/ 7	2	0.1254	0.129	0.142	0.154	0.0355	0.405	0.0928
Redwing	715.5	30/19	2	0.1248	0.128	0.141	0.153	0.0372	0.399	0.0920
Drake	795.	26/ 7	2	0.1129	0.117	0.128	0.139	0.0375	0.399	0.0912
Mallard	795.	30/19	2	0.1122	0.116	0.127	0.138	0.0392	0.393	0.0904
Tern	795.	45/ 7	3	0.1143	0.119	0.130	0.141	0.0352	0.406	0.0925
Condor	795.	54/ 7	3	0.1135	0.117	0.129	0.140	0.0368	0.401	0.0917
Crane	874.5	54/ 7	3	0.1030	0.107	0.117	0.127	0.0387	0.395	0.0902
Canary	900.	54/ 7	3	0.1002	0.104	0.114	0.124	0.0392	0.393	0.0898
Rail	954.	45/ 7	3	0.09526	0.0994	0.109	0.118	0.0385	0.395	0.0897
Cardinal	954.	54/ 7	3	0.09452	0.0983	0.108	0.117	0.0404	0.389	0.0890
Ortolan	1033.5	45/ 7	3	0.08798	0.0922	0.101	0.110	0.0401	0.390	0.0886
Curlew	1033.5	54/ 7	3	0.08728	0.0910	0.0996	0.108	0.0420	0.385	0.0878
Bluejay	1113.	45/ 7	3	0.08161	0.0859	0.0939	0.102	0.0416	0.386	0.0874
Finch	1113.	54/19	3	0.08138	0.0851	0.0931	0.101	0.0436	0.380	0.0867
Bunting	1192.5	45/ 7	3	0.07619	0.0805	0.0880	0.0954	0.0431	0.382	0.0864

Figure 9-13. ACSR table.

Separation Component (X_d) of Inductive Reactance at 60 Hz[1] ohms Per Conductor Per Mile[1]

Separation of Conductors

		Inches										
feet	0	1	2	3	4	5	6	7	8	9	10	11
0	—	−0.3015	−0.2174	−0.1682	−0.1333	−0.1062	−0.0841	−0.0654	−0.0492	−0.0349	−0.0221	−0.0106
1	0	0.0097	0.0187	0.0271	0.0349	0.0423	0.0492	0.0558	0.0620	0.0679	0.0735	0.0789
2	0.0841	0.0891	0.0938	0.0984	0.1028	0.1071	0.1112	0.1152	0.1190	0.1227	0.1264	0.1299
3	0.1333	0.1366	0.1399	0.1430	0.1461	0.1491	0.1520	0.1549	0.1577	0.1604	0.1631	0.1657
4	0.1682	0.1707	0.1732	0.1756	0.1779	0.1802	0.1825	0.1847	0.1869	0.1891	0.1912	0.1933
5	0.1953	0.1973	0.1993	0.2012	0.2031	0.2050	0.2069	0.2087	0.2105	0.2123	0.2140	0.2157
6	0.2174	0.2191	0.2207	0.2224	0.2240	0.2256	0.2271	0.2287	0.2302	0.2317	0.2332	0.2347
7	0.2361	0.2376	0.2390	0.2404	0.2418	0.2431	0.2445	0.2458	0.2472	0.2485	0.2498	0.2511
8	0.2523											
9	0.2666											
10	0.2794											
11	0.2910											
12	0.3015											
13	0.3112											
14	0.3202											
15	0.3286											
16	0.3364											
17	0.3438											
18	0.3507											
19	0.3573											
20	0.3635											
21	0.3694											
22	0.3751											
23	0.3805											
24	0.3856											
25	0.3906											
26	0.3953											
27	0.3999											
28	0.4043											
29	0.4086											
30	0.4127											
31	0.4167											
32	0.4205											
33	0.4243											
34	0.4279											
35	0.4314											
36	0.4348											
37	0.4382											
38	0.4414											
39	0.4445											
40	0.4476											
41	0.4506											
42	0.4535											
43	0.4564											
44	0.4592											
45	0.4619											
46	0.4646											
47	0.4672											
48	0.4697											
49	0.4722											

(1) From formula: at 60 Hz

$Xd = 0.2794 \log_{10} d$

d = separation in feet

From: *Electrical Transmission and Distribution Reference Book,* Westinghouse Electric Corporation, 1964.

Figure 9-14. Spacing component.

Separation Component (X'_d) of Capacitive Reactance at 60 Hz (1) Megohm-Miles Per Conductor

Separation of Conductors

	inches											
feet	0	1	2	3	4	5	6	7	8	9	10	11
0	–	−0.0737	−0.0532	−0.0411	−0.0326	−0.0260	−0.0206	−0.0160	−0.0120	−0.0085	−0.0054	−0.0026
1	0	0.0024	0.0046	0.0066	0.0085	0.0103	0.0120	0.0136	0.0152	0.0166	0.0180	0.0193
2	0.0206	0.0218	0.0229	0.0241	0.0251	0.0262	0.0272	0.0282	0.0291	0.0300	0.0309	0.0318
3	0.0326	0.0334	0.0342	0.0350	0.0357	0.0365	0.0372	0.0379	0.0385	0.0392	0.0399	0.0405
4	0.0411	0.0417	0.0423	0.0429	0.0435	0.0441	0.0446	0.0452	0.0457	0.0462	0.0467	0.0473
5	0.0478	0.0482	0.0487	0.0492	0.0497	0.0501	0.0506	0.0510	0.0515	0.0519	0.0523	0.0527
6	0.0532	0.0536	0.0540	0.0544	0.0548	0.0552	0.0555	0.0559	0.0563	0.0567	0.0570	0.0574
7	0.0577	0.0581	0.0584	0.0588	0.0591	0.0594	0.0598	0.0601	0.0604	0.0608	0.0611	0.0614
8	0.0617											
9	0.0652											
10	0.0683											
11	0.0711											
12	0.0737											
13	0.0761											
14	0.0783											
15	0.0803											
16	0.0823											
17	0.0841											
18	0.0858											
19	0.0874											
20	0.0889											
21	0.0903											
22	0.0917											
23	0.0930											
24	0.0943											
25	0.0955											
26	0.0967											
27	0.0978											
28	0.0989											
29	0.0999											
30	0.1009											
31	0.1019											
32	0.1028											
33	0.1037											
34	0.1046											
35	0.1055											
36	0.1063											
37	0.1071											
38	0.1079											
39	0.1087											
40	0.1094											
41	0.1102											
42	0.1109											
43	0.1116											
44	0.1123											
45	0.1129											
46	0.1136											
47	0.1142											
48	0.1149											
49	0.1155											

(1) From formula: for 60 Hz
$X'_d = 0.06831 \log_{10} d$
d = Separation in feet

From: *Electrical Transmission and Distribution Reference Book,* Westinghouse Electric Corporation, 1964.

Figure 9-15. Spacing component.

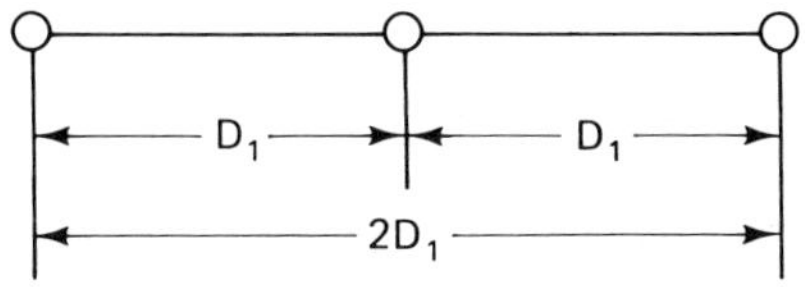

Figure 9-16. Three-wire line—horizontal.

p. 3-18) (*Electrical Transmission and Distribution Reference Book,* 1964, p. 55.)

For three-phase, three-wire transmission lines, we still use these same tables, but D is calculated as the geometric mean distance which is defined as the mth root of the product of m distances. For instance, in Figure 9-16 we would obtain:

$$D = \sqrt[3]{(D_1)(D_1)(2D_1)} = \sqrt[3]{2D_1^3} = D_1\sqrt[3]{2} = D = 1.26\ D_1.$$

If our line had an equilateral configuration as in Figure 9-17, we would obtain:

$$D = \sqrt[3]{(D_1)(D_1)(D_1)} = \sqrt[3]{D_1{}^3} = D_1.$$

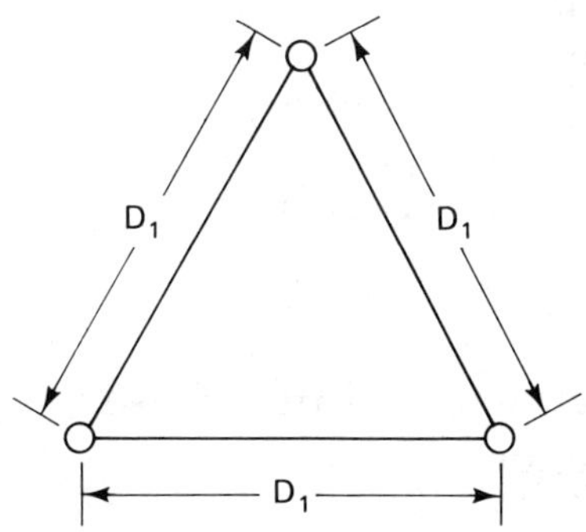

Figure 9-17. Three-wire line—triangular.

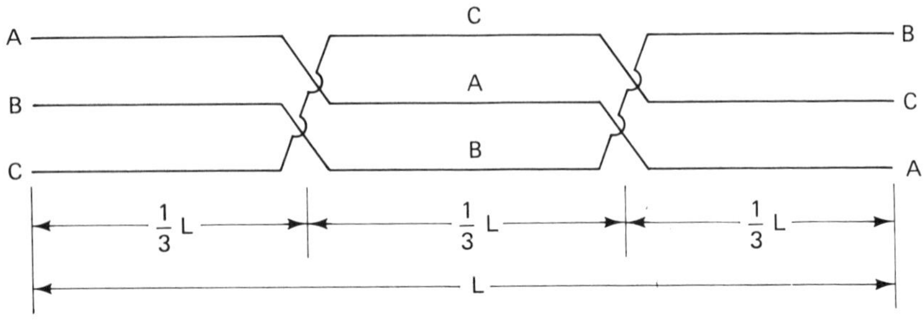

Figure 9-18. Transposed line.

The geometric mean distance, D, is called the equivalent spacing and applies when the line is transposed. A transposed line is one where the conductors are strung along the line so that their relative positions are changed every so often as shown in Figure 9-18.

Transposing the line is a common practice since it balances out the inductive and capacitive effects if we have a horizontal line such as illustrated in Figure 9-16. This would balance the voltage drops in each of the phases. If we had a spacing such as shown in Figure 9-17 the inductive and capacitive effects would be balanced with respect to the other two wires but we would have an unbalanced relationship to ground if we did not transpose. Another reason for transposing power lines is to reduce the induced voltages in other parallel lines such as telegraph and telephone lines.

Once we have the type of conductor and the spacing chosen we can calculate the parameters of our three-phase line. The values calculated will be the value per conductor or per-phase (line-to-neutral) since we always work on a wye basis.

9.9 Example Solution of a Three-Phase Line with ACSR Conductors

Let us assume the following conditions for our line: Spacing between conductors is 10 feet for the configuration shown in Figure 9-16. Conductor: 266.8 MCM with 26 aluminum strands and 7 steel strands (Partridge conductor). Temperature is 25°C. Line is 100 miles long.

Find:

(a) Geometric mean distance; (b) X_d; (c) X_d'; (d)X_a; (e) X_a'; (f) Z; (g) Y; (h) ZY.

Solution

$$D = \sqrt[3]{(10)(10)(20)} = \sqrt[3]{2{,}000} = 12.60 \text{ feet.}$$

$$0.60 \text{ feet} = 7.2 \text{ inches, so } D = 12 \text{ feet, } 7.2 \text{ inches.}$$

From Figure 9-14:

$$X_d = 0.3073 \text{ ohms/mile by extrapolation.}$$

or

$$X_d = 0.2794 \log 12.6 = 0.3074 \text{ ohms/mile.}$$

From Figure 9-13 we obtain:

$X_a = 0.465$ ohms/mile. $X_a' = 0.1074$ megohm-miles.

From Figure 9-15 we obtain:

$$X_d' = 0.075 \text{ megohm-miles.}$$

So,

$$X = X_d + X_a = 0.307 + 0.465 = 0.772 \text{ ohms/mile.}$$

$$X = X_d' + X_a' = 0.075 + 0.1074 = 0.1824 \text{ megohm-miles.}$$

Notice that the tables are in megohm-miles for C and ohms/mile for L. Let us find our z. The effective a.c. resistance from Figure 9-13 is 0.3452 ohms/mile so we would have:

$z = 0.3452 + j\,0.772 = 0.8456 \underline{/65.908°}$.

$Z = z\ell = (0.8456 \underline{/65.908°})\,(100) = 84.56 \underline{/65.908°}$.

$$\frac{0.1824 \times 10^6 \text{ ohm-miles}}{100 \text{ miles}} = 1.824 \times 10^3 \text{ ohms} = Z_{(\text{shunt capacitance})}$$

$$Y = \frac{1}{Z} = \frac{1}{1.824 \times 10^3} = 5.482 \times 10^{-4} \text{ mhos} = 5.482 \times 10^{-4} \underline{/90^\circ} \text{ mhos.}$$

$$ZY = (84.56 \underline{/65.908^\circ})(5.482 \times 10^{-4} \underline{/90^\circ}) = 0.046 \underline{/155.908^\circ}.$$

9.10 Chapter Summary

1. Transmission line parameters may be easy to find for simple configurations and low frequency cases but may be quite difficult as the frequency goes up and the environment involving the line becomes more complex.
2. *Skin* and *proximity* effects are the result of a field phenomenon which causes the current in the conductor to redistribute itself as the frequency increases. Both effects cause an increase in the effective a. c. resistance and a decrease in the effective internal inductance.
3. The resistance of a wire is not only affected by skin effect but is a function of the temperature.
4. The effective resistance can be found by taking the total power loss and dividing it by the r.m.s. value of the current squared.

$$R_{\text{a.c.}} = R_{\text{effective}} = \frac{P_{\text{a.c.}}}{I^2}. \qquad (9\text{-}8)$$

5. The mathematical process of finding the skin effect involves various Bessel functions but the calculations are made simple if tables of various Bessel functions are available.
6. A convenient concept for working with skin effect is the quantity, δ, called the skin depth or the depth of penetration.

$$\delta = \frac{1}{\sqrt{\pi f \mu \sigma}}. \qquad (9\text{-}21)$$

7. At high frequencies where the skin effect is quite pronounced so that a/δ is approximately 20 or greater it becomes convenient to assume that all of the current is flowing in a cross-sectional area of a conductor determined by δ. Under these conditions the effective resistance of a round conductor with a radius of a meters becomes:

$$R_{\text{a.c.}} = \frac{a}{2\delta} R_{\text{d.c.}}. \qquad (9\text{-}40)$$

Under these conditions it is safe to assume that the internal flux linkages have gone to zero and the effective internal inductance is zero.

8. When finding the parameters of a line the engineer should use all the help he has available to him through information supplied not only by handbooks but by the manufacturers of various types of conductors. For example, the Aluminum Association puts out excellent information for power-line engineers.
9. The parameters of three-phase lines and single-phase, two-wire power lines may normally be found quite easily from tables. They involve such ideas as the geometric mean radius, the geometric mean distance and the equivalent spacing.

9.11 Review and Study Questions

1. Why are there three terms for calculating the inductance of a coaxial cable? Under what conditions would equation (9-1) be used as it is given? Under what conditions is equation (9-1) modified and why?
2. Explain what factors might affect the following transmission line parameters: (a) R, (b) G, (c) L, (d) C.
3. Describe the physical reason for *skin effect*.
4. Describe the physical reason for the *proximity effect*.
5. What are the results on the line parameters of *skin effect* and the *proximity effect*?
6. Why is the a.c. resistance larger than the d.c. resistance?
7. Give sketches of various types of transmission line conductor configurations. Include a twisted-pair telephone line; a coaxial cable; a strip line; a three-phase, three-wire line; and a three-phase, three-wire line with protecting ground wires.
8. What do you obtain if you multiply equation (9-4) by equation (9-3)? What do you obtain if you multiply equation (9-59) by equation (9-58)? What is your conclusion?
9. Define skin depth.
10. If the *skin effect* is quite pronounced what is the effective cross-sectional area of a round conductor with a radius of a meters at very high frequencies?
11. How is a conductor with a radius of one foot used in this chapter? Is it a real conductor? Explain.

12. What is *Ampere's Circuital Law*? How is it used in discussing the development of *skin effect* in conductors?
13. Have you ever run across the ber and bei functions before? In what type of situation?
14. What is a transposed line? Why is a line transposed?
15. What is a spacing component? What is a conductor component? How are they used in this chapter?
16. How do you change megohm-miles to mhos per mile?
17. If a power line engineer is talking about Chickadee what might he be referring to?

Problems

9-1. A two-wire line has the following distributed parameters at a frequency of 1,500 hertz: R = 2.00 ohms/km, L = 2.0 millihenrys/km, C = 0.050 microfarads/km, G = 0.05 micromhos/km.
(a) Find the characteristic impedance, Z_o, the attenuation constant, α, and the phase constant, β, at 1,500 hertz.
(b) Determine the phase velocity, v_p.

9-2. A two-wire telephone line has the following parameters at a voice frequency of 1,000 hertz: $R = 7.0$ ohms/km; $L = 2.0 \times 10^{-3}$ henrys/km; $G = 5 \times 10^{-8}$ mhos/km; $C = 5 \times 10^{-8}$ farads/km.
(a) Find the characteristic impedance, Z_o, the attenuation constant, α, and the phase constant, β, at 1,000 hertz.
(b) Determine the phase velocity, v_p.
(c) Determine the wavelength, λ.

9-3. A telephone toll cable has the following parameters at 1,000 hertz: $R = 50.0$ ohms/km; $L = 500 \times 10^{-6}$ henrys/km; $G = 500 \times 10^{-9}$ mhos/km; $C = 30 \times 10^{-9}$ farads/km. Determine:
(a) α, β, Z_o.
(b) v_p.
(c) λ.

9-4. Two No. 2 wires are separated by a distance of 1 foot between centers in air. Neglecting skin effect calculate the following:
(a) Inductance per unit length.
(b) Capacitance per unit length.
(c) Using the d.c. resistance as found in Figure 9-2 at 65° C and assuming that $G = 0$, calculate at 60 hertz the following: (1) α (2) β (3) Z_o (4) v_p.

9-5. Repeat problem 9-4 assuming the separation is one inch instead of one foot.

9-6. Repeat problem 9-4 for two No. 20 wires.

9-7. What is the self-inductance of one wire at 30 hertz when we neglect skin effect for the following conductors?
(a) No. 000;
(b) No. 4;
(c) No. 10.

9-8. A standard annealed copper wire has a resistance of 22.64 ohms at 48° F. What is the resistance at 100° F?

9-9. A 97.3% conductivity copper wire has a resistance of 1.5 ohms at 60° C. What is its resistance at 120°C?

9-10. A standard annealed copper wire has a resistance of 40 ohms at 50° C. What is the temperature of the wire when the resistance is 43.2 ohms?

9-11. One equation that is often given for the ber function (McLachlan, 1961, p. 222) is

$$\text{ber } z = 1 - \frac{z^4}{2^2 4^2} + \frac{z^8}{2^2 4^2 6^2 8^2} - \frac{z^{12}}{2^2 4^2 6^2 8^2 10^2 12^2} + \ldots$$

(a) Calculate the value of ber 0.8 and compare with the value in Table 9-1.
(b) Calculate the value of ber 5.5 and compare with the value in Table 9-1.
(c) How many terms did you use in (a) and (b)?

9-12. One equation that is often given for the bei function (McLachlan, 1961, p. 222) is

$$\text{bei } z = \frac{z^2}{2^2} - \frac{z^6}{2^2 4^2 6^2} + \frac{z^{10}}{2^2 4^2 6^2 8^2 10^2} - \ldots$$

(a) Calculate the value of bei 0.8 and compare with the value given in Table 9-1.
(b) Calculate the value of bei 5.5 and compare with the value given in Table 9-1.
(c) How many terms did you use in (a) and (b)?

9-13. Using equation (9-37) calculate ber 20.

9-14. Using equation (9-38) calculate bei 20.

9-15. Calculate a/δ for a copper wire under the following circumstances:
(a) No. 10 wire, frequency 6 megahertz.
(b) No. 10 wire, frequency 100 megahertz.

9-16. Repeat problem 9-15 for a No. 14 wire for the following circumstances:
(a) 60 hertz.
(b) 100 megahertz.
(c) 1 gigahertz.

9-17 A three-phase power line is 103 miles long, uses 556.5 MCM, 24/7 ACSR conductor and the spacing between the conductors is as follows:

Assuming the temperature is 50°C:
(a) Find the series impedance, z, per mile.
(b) Assuming $G = 0$, find the shunt admittance per mile, y.
(c) Determine Z and Y.
(d) What is α, β, and Z_o?

9-18. A three-phase power line is 117 miles long, uses 266.8 MCM, 26/7 ACSR conductors; the spacing between the conductors is as follows:

Assuming the temperature is 50°C, determine:
(a) The series impedance, z, per mile.
(b) The shunt admittance, y, per mile (Assume $G = 0$).
(d) Z and Y.
(d) α, β, and Z_o.

9-19. Repeat problem 9-18 if the conductor is 795 MCM, 26/7 ACSR and the spacing is 17.35 feet.

9-20. Repeat problem 9-18 if the conductor is 605 MCM, 26/7 ACSR and the spacing is 23 feet.

10

Special Topics

10.1 Introduction

In the last chapter we looked at some of the methods for determining the line parameters. Since the propagation constant, γ, and the characteristic impedance, Z_o, are functions of the line constants and the frequency, it is possible to adjust the values of γ and Z_o within certain constraints by adjusting the values of the line parameters. We have already discussed the possibility of moving the wires of a parallel wire line closer together to increase C and decrease L. If we change the ratio of b to a for a coaxial cable, L and C can be changed. Frequency is always a factor affecting the effective a.c. resistance and L because of skin and proximity effects. Of course, there are always the situations where almost nothing can be done about the line parameters. The telephone engineer trying to squeeze many twisted-pair wire circuits into a cable for underground service wants the wires as close together as possible so he is tied to a situation where he has high capacitance and low inductance and usually a small conductor with its high value of resistance.

Unless we have certain idealized conditions, γ and Z_o are affected by frequency. This, in turn, affects the operation of the line with respect to the distortion of signals being propagated from the sending to the receiving end. If faithful reproduction of the signal is mandatory then the engineer must take steps to insure this result. Ideally we would obtain zero distortion in our signal if the fundamental and all the higher frequency components traveled at the same speed and were attenuated at the same rate. This can be accomplished if a certain relationship exists between R, L, C, and G. Under normal conditions this relationship does not exist unless the engineer takes the necessary steps in the design of the line. Since this involves increased costs it is fortunate that this type of procedure is not necessary except in very rare cases.

It is possible to make some approximations about γ and Z_o, depending on the relative values of R, L, C, and G. These approximations are often useful to the engineer working on a line where the approximations make good sense for the particular situation. Often they can give the engineer a quick picture of what is going to happen on the line for a given assumed condition for the particular problem.

We have been talking about a voltage reflection coefficient at the receiving end but we can also talk about a reflection coefficient at any point on the line. All we have to do is replace the line at that point by its impedance, looking towards the load, at that point. If we replace the line towards the load end by its impedance at that point, nothing changes from that point to the sending end. The sending-end source would not be aware of any change on the line and up to the point where we put the impedance all the voltages, currents and powers would remain the same. The new reflection coefficient will, however, be related to the original one at the load end using the load impedance. If the line is lossless, the magnitude will be the same as the magnitude at the receiving end. We have already used this result when we worked with Smith charts.

It is interesting to look at the maximum value of the voltage reflection coefficient. Since it is the ratio of a reflected component to an incident component, one would naturally assume that the maximum value would be unity; however, it turns out, as we shall see in this chapter, that under theoretical conditions we may have a reflection coefficient greater than one. It would require a perfect reactance and a very special Z_o in order to do it.

Since measurements are an important part of engineering work, we have mentioned in the text that certain things about our line be ascertained from measurement. In this chapter we will be discussing the finding of an unknown load impedance in terms of some measurements which may be made in the laboratory on high frequency lines with rather simple techniques. It turns out that all we have to measure is the VSWR, the distance to the load from a voltage minimum (or voltage maximum) and the distance between two adjacent voltage minimums.

We will also look in this chapter at the possibility of impedance-matching through the use of "so-called transformers," which are really transmission lines that are either 1/8 or 1/4 wavelength long.

10.2 Approximations for α, β, Z_o—Low-Loss Case, $G = 0$

Let us look at γ and Z_o in more detail. Repeating equations (1-38) and (1-39) gives:

$$\gamma = \alpha + j\beta = \sqrt{(R + j\omega L)(G + j\omega L)}. \quad (1\text{-}38)$$

$$Z_o = \sqrt{\frac{(R + j\omega L)}{(G + j\omega C)}}. \quad (1\text{-}39)$$

Let us define a low-loss line as one where $G = 0$ and $R \ll \omega L$.

$$\gamma = \sqrt{(R + j\omega L)(0 + j\omega C)} = \sqrt{-\omega^2 LC + j\omega RC} = \alpha + j\beta$$

or,

$$-\omega^2 LC + j\omega RC = \alpha^2 + 2j\alpha\beta - \beta^2 = \alpha^2 - \beta^2 + 2j\alpha\beta.$$

Equating the real and imaginary terms gives:

$$\alpha^2 - \beta^2 = -\omega^2 LC. \quad (10\text{-}1)$$

$$2\alpha\beta = \omega RC. \quad (10\text{-}2)$$

$$\alpha = \frac{\omega RC}{2\beta} \quad (10\text{-}3)$$

If α is small, so that $\alpha \ll \beta$, $-\beta^2 \simeq -\omega^2 LC$ and

$$\beta \simeq \omega\sqrt{LC}. \quad (10\text{-}4)$$

This is the same result as we had for the lossless line. So,

$$\alpha = \frac{\omega RC}{2\beta} \simeq \frac{\omega RC}{2\omega\sqrt{LC}} \simeq \frac{RC}{2\sqrt{LC}} \simeq \frac{R}{2}\sqrt{\frac{C^2}{LC}} \simeq \frac{R}{2}\sqrt{\frac{C}{L}} \simeq \frac{R}{2}\frac{1}{\sqrt{\frac{L}{C}}} \simeq \alpha. \quad (10\text{-}5)$$

$$Z_o = \sqrt{\frac{(R + j\omega L)}{(G + j\omega C)}} = \sqrt{\frac{(R + j\omega L)}{(0 + j\omega C)}}. \quad \text{If } R \ll \omega L,\ Z_o \simeq \sqrt{\frac{L}{C}}. \quad (10\text{-}6)$$

Then,

$$\alpha \simeq \frac{R}{2Z_o}. \tag{10-7}$$

10.3 Approximations for α, β, Z_o—Low-Loss Case, *G* and *R* not Zero.

If we assume that G and R are not equal to zero we can write:

$$\begin{aligned}
\gamma &= \sqrt{(R + j\omega L)(G + j\omega C)} = \\
&\sqrt{(j\omega L)(j\omega C)\left(\frac{R}{J\omega L} + 1\right)\left(\frac{G}{j\omega C} + 1\right)} = \\
&j\omega\sqrt{LC}\left[\left(1 + \frac{R}{j\omega L}\right)\left(1 + \frac{G}{j\omega C}\right)\right]^{1/2} = \\
&j\omega\sqrt{LC}\left[\left(1 - \frac{jR}{\omega L}\right)\left(1 - \frac{jG}{\omega C}\right)\right]^{1/2} = \\
&\alpha + j\beta = j\omega\sqrt{LC}\left[1 - j\left(\frac{R}{\omega L} + \frac{G}{\omega C}\right) - \frac{RG}{\omega L\omega C}\right]^{1/2}.
\end{aligned}$$

If $R \ll \omega L$ and $G \ll \omega C$,

$$\alpha + j\beta \simeq j\omega\sqrt{LC}\left[1 - j\left(\frac{R}{\omega L} + \frac{G}{\omega C}\right)\right]^{1/2}.$$

Since

$$(1 + x)^{1/2} = 1 + \frac{x}{2} + \frac{{}^1/_2({}^1/_2 + 1)x^2}{1 \cdot 2} + \ldots$$

$$\gamma = \alpha + j\beta \simeq j\omega\sqrt{LC}\left[1 - j\frac{\left(\dfrac{R}{\omega L} + \dfrac{G}{\omega C}\right)}{2} + \ldots\right.$$

If we neglect the higher-order terms we have:

$$\alpha + j\beta \simeq j\omega\sqrt{LC} + \frac{\omega\sqrt{LC}}{2}\left[\frac{R}{\omega L} + \frac{G}{\omega C}\right].$$

So

$$\alpha + j\beta \simeq \frac{\sqrt{LC}}{2}\left[\frac{R}{L} + \frac{G}{C}\right] + j\omega\sqrt{LC},$$

and

$$\alpha \simeq \frac{\sqrt{LC}}{2}\left[\frac{R}{L} + \frac{G}{C}\right] \simeq \frac{1}{2}\left[R\sqrt{\frac{LC}{L^2}} + G\sqrt{\frac{LC}{C^2}}\right]$$
$$\simeq \frac{1}{2}\left[R\sqrt{\frac{C}{L}} + G\sqrt{\frac{L}{C}}\right] \simeq \frac{1}{2}\left[\frac{R}{\sqrt{\frac{L}{C}}} + G\sqrt{\frac{L}{C}}\right]. \quad (10\text{-}8)$$

Again,

$$\beta \simeq \omega\sqrt{LC}. \quad (10\text{-}9)$$

$$Z_o = \sqrt{\frac{(R + j\omega L)}{(G + j\omega C)}} \simeq \sqrt{\frac{L}{C}}. \quad (10\text{-}10)$$

and α becomes:

$$\alpha \simeq \frac{1}{2}\left[\frac{R}{Z_o} + GZ_o\right]. \quad (10\text{-}11)$$

These results are called the high-frequency approximations.

10.4 Distortionless Line, *R/L* = *G/C*

A very special case results when $R/L = G/C$. We have what has been called in the past, a distortionless line. This technique was used by telephone companies for many years to reduce the distortion on their lines where the conductors were close together with the resultant high C and low L values.

$$\gamma = \alpha + j\beta = \sqrt{(R + j\omega L)(G + j\omega C)}$$
$$= \sqrt{(L)\left(\frac{R}{L} + j\omega\right)(C)\left(\frac{G}{C} + j\omega\right)}.$$

If

$$\frac{R}{L} = \frac{G}{C}, \quad \alpha + j\beta = \sqrt{LC\left(\frac{R}{L} + j\omega\right)\left(\frac{R}{L} + j\omega\right)} = \sqrt{LC\left(\frac{R}{L} + j\omega\right)^2}$$
$$= \sqrt{LC}\left(\frac{R}{L} + j\omega\right) = \frac{R}{L}\sqrt{LC} + j\omega\sqrt{LC} = \alpha + j\beta.$$

Thus,

$$\alpha = \frac{R}{L}\sqrt{LC} = \frac{G}{C}\sqrt{LC}. \tag{10-12}$$

$$\beta = \omega\sqrt{LC}. \tag{10-13}$$

Notice that this is the one case where we have the same value for β as we had for the lossless case even though we have losses in the line:

$$Z_o = \sqrt{\frac{(R + j\omega L)}{(G + j\omega C)}} = \sqrt{\frac{L\left(\frac{R}{L} + j\omega\right)}{C\left(\frac{G}{C} + j\omega\right)}}.$$

If $\frac{R}{L} = \frac{G}{C}$ we have:

$$Z_o = \sqrt{\frac{L}{C}\frac{\left(\frac{R}{L} + j\omega\right)}{\left(\frac{R}{L} + j\omega\right)}} = \sqrt{\frac{L}{C}}. \tag{10-14}$$

Under these special conditions, Z_o is real as it was in the lossless case. We can now rewrite the expression for α.

$$\alpha = \frac{R}{L}\sqrt{LC} = R\sqrt{\frac{LC}{L^2}} = R\sqrt{\frac{C}{L}} = \frac{R}{\sqrt{\frac{L}{C}}} = \frac{R}{Z_o}. \tag{10-15}$$

$$\alpha = \frac{G}{C}\sqrt{LC} = G\sqrt{\frac{LC}{C^2}} = G\sqrt{\frac{L}{C}} = GZ_o. \tag{10-16}$$

Now let us investigate how these results affect the operation of a transmission line. Going back to equation (1-24) we have:

$$v_p = \frac{\omega}{\beta},$$

or, in our case:

$$v_p = \frac{\omega}{\omega\sqrt{LC}} = \frac{1}{\sqrt{LC}}. \tag{10-17}$$

The phase velocity is independent of the frequency if $R/L = G/C$. But notice that α and Z_o are also independent of frequency. Let us assume that we are sending a signal down the line that has a fundamental and third harmonic component as shown in Figure 10-1.

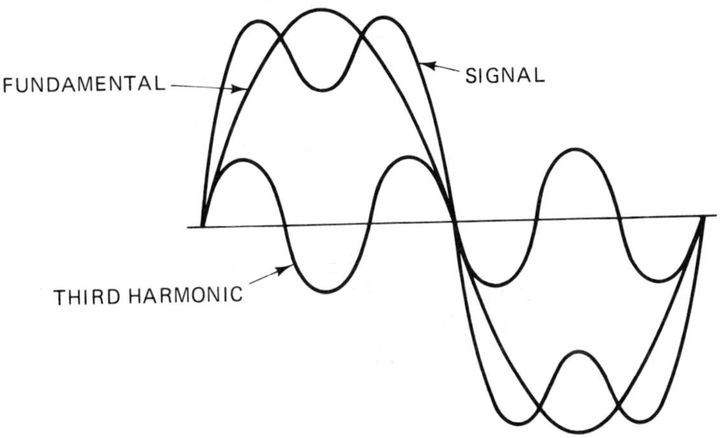

If v_p and α are independent of frequency, the two waves will travel down the line at the same velocity and attenuate at the same rate so that at the receiving end we would have a signal with the same wave shape and relative positions of the fundamental and third harmonic components. But if the two components travel at different velocities and attenuate differently we might end up with a wave shape as indicated in Figure 10-2.

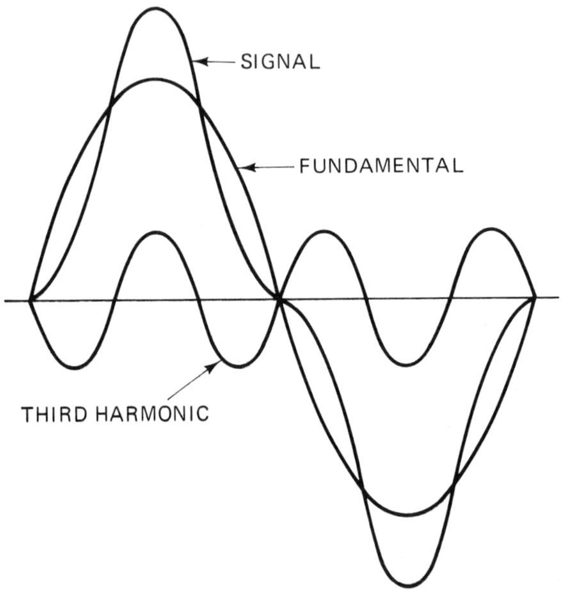

Now the signal at the receiving end looks considerably different from the signal put in at the sending end. We have distortion unless α and v_p are independent of frequency. To achieve a distortionless line engineers added fixed inductance coils called *loading* coils to telephone lines. It turns out that the distortion is less even if the exact value of L was not achieved in order to satisfy the relationship that $R/L = G/C$.

Suppose we calculate some values for a line where:

$$\frac{R}{L} = \frac{G}{C}. \tag{10-18}$$

Let us assume that $R = 1$ ohm/meter. $G = 11.11 \times 10^{-6}$ mhos/meter. $L = 1 \times 10^{-6}$ henrys/meter. $C = 11.11 \times 10^{-12}$ farads/meter.

$$\frac{R}{L} = \frac{G}{C}. \qquad \frac{1}{1 \times 10^{-6}} = 1 \times 10^{6} = \frac{11.11 \times 10^{-6}}{11.11 \times 10^{-12}}.$$

We have satisfied equation (10-18) and after calculating the values we are interested in we have:

ω	α	β	v_p	Z_o
100	3.33×10^{-3}	333.33×10^{-9}	3×10^{8}	$300 \underline{/0^\circ}$
1,000	3.33×10^{-3}	3.33×10^{-6}	3×10^{8}	$300 \underline{/0^\circ}$
100,000	3.33×10^{-3}	333.33×10^{-6}	3×10^{8}	$300 \underline{/0^\circ}$
100×10^{6}	3.33×10^{-3}	333.33×10^{-3}	3×10^{8}	$300 \underline{/0^\circ}$

α stays constant, v_p stays constant, and Z_o stays real and constant. Notice that β does change but is equal to a constant times the frequency.

10.5 Approximations for α, β, Z_o, $R \gg \omega L$, $G \ll \omega C$

Let us look at a case that might occur at low frequencies with high values of R and C and a low value of L. (Two small wires next to each other.)

$$\gamma = \alpha + j\beta = \sqrt{(R + j\omega L)(G + j\omega C)} \simeq \sqrt{R(j\omega C)} \simeq \sqrt{R\omega C}\,\underline{/90^\circ}$$
$$\simeq \sqrt{R\omega C}\,\underline{/45^\circ} \simeq \frac{\sqrt{R\omega C}}{\sqrt{2}} + j\frac{\sqrt{R\omega C}}{\sqrt{2}} = \alpha + j\beta.$$

$$\alpha \simeq \beta \simeq \frac{\sqrt{R\omega C}}{\sqrt{2}}. \qquad (10\text{-}19)$$

Recalling equation (1-52):

$$\lambda = \frac{2\pi}{\beta} \text{ or } \beta = \frac{2\pi}{\lambda} = \alpha,$$

so

$$\alpha = 2\pi \frac{\text{nepers}}{\text{wavelength}}.$$

This would be a very high attenuation, as we can see by looking at the term $V^{-}e^{-\alpha d}$. Let $d = \lambda$ and $\alpha = 2\pi/\lambda$.

$$V^{-}\, e^{-2\pi\lambda/\lambda} = V^{-}e^{-2\pi} = 0.00186\, V^{-},$$

which shows that the magnitude would drop from 1.0 to 0.00186 times V^{-} in one wavelength. This would attenuate the signal to practically nothing in only a few wavelengths.

$$Z_o = \sqrt{\frac{(R + j\omega L)}{(G + j\omega C)}} \simeq \sqrt{\frac{R}{j\omega C}} \simeq \sqrt{\frac{R}{\omega C}\underline{/90^\circ}} \simeq \sqrt{\frac{R}{\omega C}}\,\underline{/-45^\circ}.$$

$$Z_o \simeq \sqrt{\frac{R}{\omega C}}\,\underline{/-45^\circ} \simeq \frac{1}{\sqrt{2}}\sqrt{\frac{R}{\omega C}} - j\frac{1}{\sqrt{2}}\sqrt{\frac{R}{\omega C}}. \tag{10-20}$$

We end up with a Z_o that has a capacitive reactance equal in magnitude to the resistive component.

10.6 Approximations for α, β, Z_o, $R \ll \omega L$, $G \gg \omega C$

Let us look at one more case.

$$Z_o = \sqrt{\frac{(R + j\omega L)}{(G + j\omega C)}} \simeq \sqrt{\frac{j\omega L}{G}} \simeq \sqrt{\frac{\omega L}{G}\underline{/90^\circ}} \simeq \sqrt{\frac{\omega L}{G}}\,\underline{/45^\circ}$$

$$\simeq \frac{1}{\sqrt{2}}\sqrt{\frac{\omega L}{G}} + j\frac{1}{\sqrt{2}}\sqrt{\frac{\omega L}{G}} \simeq Z_o.$$

$$\gamma = \sqrt{(R + j\omega L)(G + j\omega C)} \simeq \sqrt{(j\omega L)(G)} \simeq \sqrt{\omega LG\,\underline{/90^\circ}}$$

$$\simeq \sqrt{\omega LG}\,\underline{/45^\circ} \simeq \frac{\sqrt{\omega LG}}{\sqrt{2}} + j\frac{\sqrt{\omega LG}}{\sqrt{2}} \simeq \alpha + \beta. \qquad \alpha \simeq \beta \simeq \frac{\sqrt{\omega LG}}{\sqrt{2}}. \tag{10-21}$$

$\beta = \dfrac{2\pi}{\lambda} = \alpha$, so $\alpha = 2\pi\dfrac{\text{nepers}}{\text{wavelength}}$ and once again we would have very high attenuation as we did in Section 10.5, where $R \gg \omega L$, $G \ll \omega C$.

10.7 Reflection Coefficient Anywhere on the Line

We have had the equations (1-32) and (1-37).

$$V = Ae^{-\gamma x} + Be^{\gamma x} \text{ and } I = \frac{A}{Z_o}e^{-\gamma x} - \frac{B}{Z_o}e^{\gamma x}.$$

If we let $A = V_1$ and $B = V_2$, we obtain:

$$V = V_1 e^{-\gamma x} + V_2 e^{\gamma x} = V_1 e^{-\gamma(\ell - d)} + V_2 e^{\gamma(\ell - d)} = V_1 e^{-\gamma\ell} e^{\gamma d} + V_2 e^{\gamma\ell} e^{-\gamma d}.$$

$$I = \frac{V_1}{Z_o} e^{-\gamma x} - \frac{V_2}{Z_o} e^{\gamma x} = \frac{V_1}{Z_o} e^{-\gamma(\ell - d)} - \frac{V_2}{Z_o} e^{\gamma(\ell - d)} = \frac{V_1}{Z_o} e^{-\gamma\ell} e^{\gamma d} - \frac{V_2}{Z_o} e^{\gamma\ell} e^{-\gamma d}.$$

So we have:

$$V = V_1 e^{-\gamma\ell} e^{\gamma d} + V_2 e^{\gamma\ell} e^{-\gamma d}. \tag{10-22}$$

$$I = \frac{V_1}{Z_o} e^{-\gamma\ell} e^{\gamma d} - \frac{V_2}{Z_o} e^{\gamma\ell} e^{-\gamma d}. \tag{10-23}$$

We have also had equations (3-5) and (3-6).

$$V = V^+ e^{\gamma d} + V^- e^{-\gamma d} \text{ and } I = \frac{V^+}{Z_o} e^{\gamma d} - \frac{V^-}{Z_o} e^{-\gamma d}.$$

Comparing these equations with equations (10-22) and (10-23) gives:

$$V^+ = V_1 e^{-\gamma\ell}. \tag{10-24}$$

$$V^- = V_2 e^{\gamma\ell}. \tag{10-25}$$

Using equation (3-9), we have:

$$\Gamma_r = \frac{V^-}{V^+} = \frac{V_2 e^{\gamma\ell}}{V_1 e^{-\gamma\ell}} = \frac{V_2}{V_1} e^{2\gamma\ell} = \frac{Z_r - Z_o}{Z_r + Z_o}. \tag{10-26}$$

At $x = 0$, $V_s = V_1 + V_2$, and at $d = 0$, $V_r = V^+ + V^-$.

In Figure 10-3 we can indicate the following components:

V_1 = forward component at the sending end.
V_2 = reverse component at the sending end.
$V^+ = V_1 e^{-\gamma\ell}$ = forward component at the receiving end.
$V^- = V_2 e^{\gamma\ell}$ = reverse component at the receiving end.

Let us now define a voltage reflection coefficient anywhere on the line at d meters from the receiving end or x meters from the sending end. At this point we would have an impedance, Z, looking towards the load. If we chopped off the line at this point and replaced everything to the right of this point with the impedance Z, we could find the voltage reflection coefficient for the new load impedance,

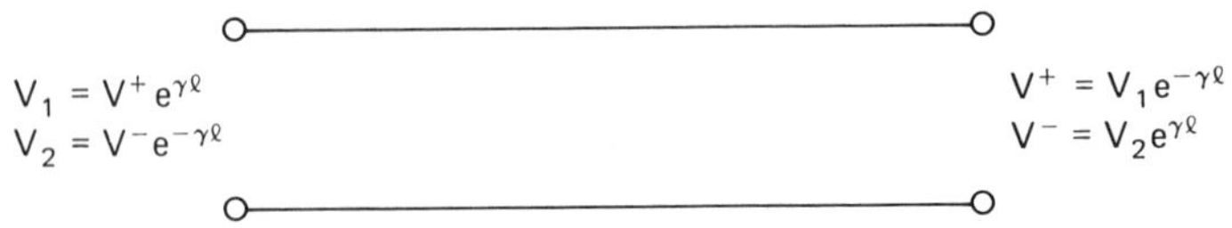

Z. But the sending end would not be aware of any change on the line.

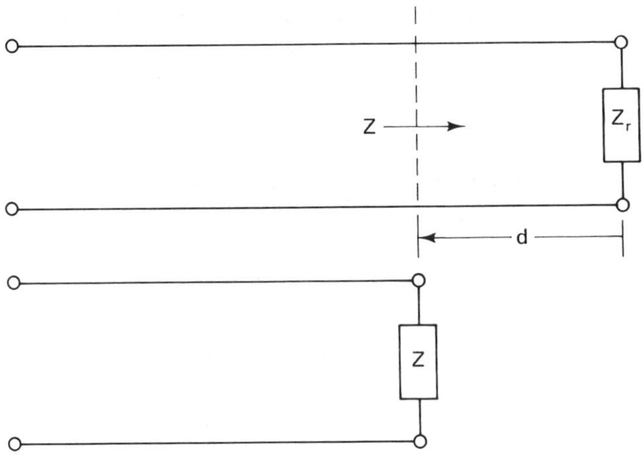

Figure 10-4. Replacing part of line by the impedance, *Z*.

Let us call the voltage reflection coefficient at d, Γ_d. We could just as well call it Γ_x.

$$\Gamma_d = \frac{Z - Z_o}{Z + Z_o} \tag{10-27}$$

We now use equation (6-7) for the impedance anywhere on the line.

$$Z = Z_o \left[\frac{e^{\gamma d} + \Gamma_r e^{-\gamma d}}{e^{\gamma d} - \Gamma_r e^{-\gamma d}} \right].$$

$$\Gamma_d = \frac{Z_o\left[\dfrac{e^{\gamma d} + \Gamma_r e^{-\gamma d}}{e^{\gamma d} - \Gamma_r e^{-\gamma d}}\right] - Z_o}{Z_o\left[\dfrac{e^{\gamma d} + \Gamma_r e^{-\gamma d}}{e^{\gamma d} - \Gamma_r e^{-\gamma d}}\right] + Z_o} = \frac{e^{\gamma d} + \Gamma_r e^{-\gamma d} - e^{\gamma d} + \Gamma_r e^{-\gamma d}}{e^{\gamma d} + \Gamma_r e^{-\gamma d} + e^{\gamma d} - \Gamma_r e^{-\gamma d}} =$$

$$\frac{2\Gamma_r e^{-\gamma d}}{2e^{\gamma d}} = \Gamma_r e^{-2\gamma d} = \frac{V^-}{V^+}e^{-2\gamma d} = \Gamma_d. \qquad (10\text{-}28)$$

Or

$$\Gamma_d = \frac{V_2 e^{\gamma\ell}}{V_1 e^{-\gamma\ell}}e^{-2\gamma d} = \frac{V_2}{V_1}e^{2\gamma\ell}e^{-2\gamma d} = \frac{V_2}{V_1}e^{2\gamma(\ell-d)} =$$

$$\frac{V_2}{V_1}e^{2\gamma x} = \Gamma_d = \Gamma_x. \qquad (10\text{-}29)$$

Let us look at equation (10-28).

$$\Gamma_d = \frac{V^-}{V^+}e^{-2\gamma d} = \Gamma_r e^{-2(\alpha+j\beta)d} = \Gamma_r e^{-2\alpha d}e^{-2j\beta d} = |\Gamma_r|\,\underline{/\phi}\; e^{-2\alpha d}e^{-2j\beta d} =$$

$$|\Gamma_r|e^{j\phi}e^{-2\alpha d}e^{-2j\beta d} = |\Gamma_r|\,e^{-2\alpha d}e^{j(\phi\,-\,2\beta d)} = |\Gamma_r|\,e^{-2\alpha d}\,\underline{/\phi\text{-}2\beta d}. \qquad (10\text{-}30)$$

If the line is lossless,

$$\Gamma_d = |\Gamma_r|\,\underline{/\phi\text{-}2\beta d} = \left|\frac{V^-}{V^+}\right|\,\underline{/\phi\text{-}2\beta d}.$$

The magnitude of the voltage reflection coefficient is constant all along the line if the line is lossless. We used this result when working with the Smith chart in Chapter 4 when all of the impedances around the circle we drew had the same magnitude for Γ_r. If we assume that $\phi = 0°$, we would have:

$$\Gamma_d = |\Gamma_r|\,\underline{/\text{-}2\beta d}.$$

When $2\beta d = 2\pi$ radians our phasor, Γ_d, goes around the circle one time, or

$$d = \frac{2\pi}{2\beta} = \frac{\pi}{\beta}.$$

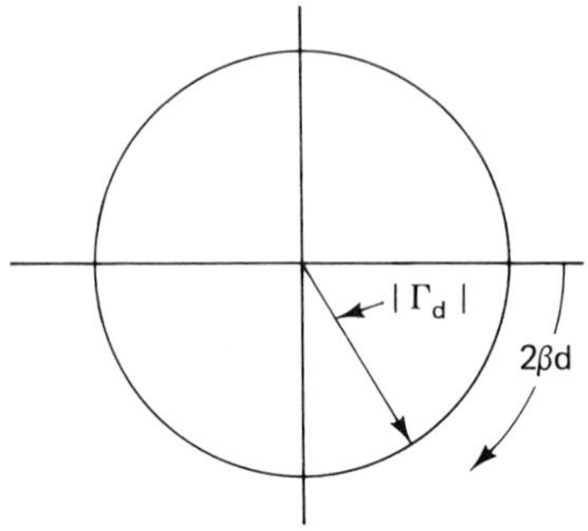

Figure 10-5. Reflection coefficient on lossless line.

Since

$$\lambda = \frac{2\pi}{\beta}, \quad \frac{\lambda}{2} = \frac{\pi}{\beta} = d.$$

We go around once for each half wavelength or twice for each wavelength. If the line is not lossless, the effective magnitude of the voltage reflection coefficient is reduced by a factor of $e^{-2\alpha d}$. Since, from equation (3-15),

$$\text{VSWR} = \frac{1 + |\Gamma_r|}{1 - |\Gamma_r|},$$

we can write, if the line is lossless:

$$\text{VSWR} = \frac{1 + |\Gamma_d|}{1 - |\Gamma_d|}, \tag{10-31}$$

and the VSWR is constant all along the line. We also used this result when we worked with the Smith chart.

10.8 Maximum Value of the Voltage Reflection Coefficient

Since the voltage reflection coefficient was originally defined in terms of the reflected component of the phasor voltage and the incident component of the phasor voltage at the receiving end, it would seem that the maximum value would be unity. However, let us look at the voltage reflection coefficient in a little more detail, starting again with equation (3-9).

$$\Gamma_r = \frac{Z_r - Z_o}{Z_r + Z_o} = \frac{\dfrac{Z_r}{Z_o} - 1}{\dfrac{Z_r}{Z_o} + 1}.$$

Let

$$\frac{Z_r}{Z_o} = \left|\frac{Z_r}{Z_o}\right| \underline{/\alpha},$$

where we are using α as the symbol for an angle and not for the attenuation constant.

$$\Gamma_r = \frac{\left|\dfrac{Z_r}{Z_o}\right| \underline{/\alpha} - 1}{\left|\dfrac{Z_r}{Z_o}\right| \underline{/\alpha} + 1} = \frac{\left|\dfrac{Z_r}{Z_o}\right| (\cos\alpha + j \sin\alpha) - 1}{\left|\dfrac{Z_r}{Z_o}\right| (\cos\alpha + j \sin\alpha) + 1} =$$

$$\Gamma_r = \frac{\left|\dfrac{Z_r}{Z_o}\right| \cos\alpha - 1 + j\left|\dfrac{Z_r}{Z_o}\right| \sin\alpha}{\left|\dfrac{Z_r}{Z_o}\right| \cos\alpha + 1 + j\left|\dfrac{Z_r}{Z_o}\right| \sin\alpha},$$

so

$$|\Gamma_r|^2 = \frac{\left[\left|\dfrac{Z_r}{Z_o}\right| \cos\alpha - 1\right]^2 + \left[\left|\dfrac{Z_r}{Z_o}\right| \sin\alpha\right]^2}{\left[\left|\dfrac{Z_r}{Z_o}\right| \cos\alpha + 1\right]^2 + \left[\left|\dfrac{Z_r}{Z_o}\right| \sin\alpha\right]^2} = \tag{10-32}$$

$$\frac{\left|\dfrac{Z_r}{Z_o}\right|^2 - 2\left|\dfrac{Z_r}{Z_o}\right| \cos\alpha + 1}{\left|\dfrac{Z_r}{Z_o}\right|^2 + 2\left|\dfrac{Z_r}{Z_o}\right| \cos\alpha + 1} = |\Gamma_r|^2.$$

If we look at the equation for Z_o,

$$Z_o = \sqrt{\frac{(R + j\omega L)}{(G + j\omega C)}} = |Z_o| \underline{/\theta};$$

the extremes for the angle of Z_o will occur when:

$$\text{Case 1.} \quad Z_o = \sqrt{\frac{R}{j\omega C}} = \sqrt{\frac{R}{\omega C} \underline{/-90^\circ}} = \sqrt{\frac{R}{\omega C}} \underline{/-45^\circ}.$$

$$\text{Case 2.} \quad Z_o = \sqrt{\frac{j\omega L}{G}} = \sqrt{\frac{\omega L}{G} \underline{/90^\circ}} = \sqrt{\frac{\omega L}{G}} \underline{/+45^\circ}.$$

So the angle of Z_o is theoretically between -45° and $+45^\circ$. Z_r, in general, can have an angle between -90° and $+90^\circ$ so the angle, α, of Z_r/Z_o can have an angle between -135° and $+135^\circ$ if we look at the theoretical extremes.

$$\frac{|Z_r| \underline{/-90^\circ}}{|Z_o| \underline{/+45^\circ}} = \frac{|Z_r|}{|Z_o|} \underline{/-135^\circ},$$

or

$$\frac{|Z_r| \underline{/+90^\circ}}{|Z_o| \underline{/-45^\circ}} = \frac{|Z_r|}{|Z_o|} \underline{/+135^\circ}.$$

When we look at equation (10-32) we see that the magnitude of $|\Gamma_r|^2$ and thus the magnitude of $|\Gamma_r|$ would be greatest when $\cos\alpha$ has its largest negative value, which would occur when $\alpha = +$ or -135°.

$$\cos 135^\circ = \cos(-135^\circ) = -0.707 = -\frac{1}{\sqrt{2}}.$$

$$|\Gamma_r|^2_{max} = \frac{\left|\frac{Z_r}{Z_o}\right|^2 - 2\left|\frac{Z_r}{Z_o}\right| \frac{(-1)}{\sqrt{2}} + 1}{\left|\frac{Z_r}{Z_o}\right|^2 + 2\left|\frac{Z_r}{Z_o}\right| \frac{(-1)}{\sqrt{2}} + 1} = \frac{\left|\frac{Z_r}{Z_o}\right|^2 + \sqrt{2}\left|\frac{Z_r}{Z_o}\right| + 1}{\left|\frac{Z_r}{Z_o}\right|^2 - \sqrt{2}\left|\frac{Z_r}{Z_o}\right| + 1}. \qquad (10\text{-}33)$$

Now let us take the derivative of Γ_r^2 with respect to $\left|\frac{Z_r}{Z_o}\right|$ to determine the value of $\left|\frac{Z_r}{Z_o}\right|$ that gives the maximum value of $|\Gamma_r|$.

$$\frac{d|\Gamma_r|^2}{d\left|\frac{Z_r}{Z_o}\right|} =$$

$$\frac{\left[\left|\frac{Z_r}{Z_o}\right|^2 - \sqrt{2}\left|\frac{Z_r}{Z_o}\right| + 1\right]\left[2\left|\frac{Z_r}{Z_o}\right| + \sqrt{2}\right] - \left[\left|\frac{Z_r}{Z_o}\right|^2 + \sqrt{2}\left|\frac{Z_r}{Z_o}\right| + 1\right]\left[2\left|\frac{Z_r}{Z_o}\right| - \sqrt{2}\right]}{\left[\left|\frac{Z_r}{Z_o}\right|^2 - \sqrt{2}\left|\frac{Z_r}{Z_o}\right| + 1\right]^2}.$$

When we multiply out the terms in the numerator and set it equal to zero we obtain:

$$2\sqrt{2} = 2\sqrt{2}\left|\frac{Z_r}{Z_o}\right|^2. \qquad 1 = \left|\frac{Z_r}{Z_o}\right|^2, \text{ so } \left|\frac{Z_r}{Z_o}\right| = \pm 1 = +1.$$

To obtain a maximum value of Γ_r, we find that:

$$\left|\frac{Z_r}{Z_o}\right| = 1 \text{ and } \alpha \text{ must be } \pm 135°.$$

Substitution in equation (10-32) gives:

$$|\Gamma_r|^2_{max} = \frac{1 - 2(1)\left(\frac{-1}{\sqrt{2}}\right) + 1}{1 + 2(1)\left(\frac{-1}{\sqrt{2}}\right) + 1} = \frac{2 + \sqrt{2}}{2 - \sqrt{2}} = \frac{3.414}{0.58578} = 5.8284.$$

$$|\Gamma_r|_{max} = \sqrt{5.8284} = 2.4142 = \frac{|V^-|}{|V^+|}, \text{ or } |V^-| = 2.4142\,|V^+|.$$

We see that the maximum value of the voltage reflection coefficient is greater than one. This appears strange since it indicates that the phasor value of the reflected component of voltage is larger than the phasor value of the incident component of voltage. But notice that it takes some unique values of the load impedance and the characteristic impedance to achieve this result.

10.9 Example of a Case Where the Magnitude of the Voltage Reflection Coefficient Is Greater Than Unity

Let us assume the following values for our example:

$$Z_r = 10 \angle 90^\circ = 0 + j10. \qquad Z_o = 10 \angle -45^\circ = 7.071 - j7.071.$$

$$\Gamma_r = \frac{Z_r - Z_o}{Z_r + Z_o} = \frac{j10 - (7.071 - j7.071)}{j10 + (7.071 - j7.071)} = \frac{-7.071 + j17.071}{7.071 + j2.93} =$$
$$\frac{18.50 \angle 112.5^\circ}{7.66 \angle 22.5^\circ} = 2.414 \angle 90^\circ = \frac{|V^-|}{|V^+|} \angle 90^\circ.$$

To achieve this result requires a situation that is not physically realizable. We would have to have a characteristic impedance that came from a line with pure R and C, no G or L, and a perfectly pure inductive load.

Let us look at a case that we considered in Section 10.5, where $R \gg \omega L$ and $G \ll \omega C$. We will assume the following conditions: $R = 25$ ohms/meter; $L = G = 0$. $C = 1 \times 10^{-6}$ farads/meter; $\omega = 10^6$ radians/second.

$$Z_o = \sqrt{\frac{(R + j\omega L)}{(G + j\omega C)}} = \sqrt{\frac{25}{j(10^6)(1 \times 10^{-6})}} = \sqrt{-j25} =$$
$$\sqrt{25 \angle -90^\circ} = 5 \angle -45^\circ = 3.535 - j3.535.$$

$$\gamma = \sqrt{(R + j\omega L)(G + j\omega C)} = \sqrt{(25)(j)(10^6)(10^{-6})} =$$
$$\sqrt{25 \angle 90^\circ} = 5 \angle 45^\circ = 3.535 + j3.535.$$

So $\alpha = 3.535$ and $\beta = 3.535$.

Let the length of the line be 0.2λ and the load impedance, Z_r, equal to $j5$ ohms. $0.2(360^\circ) = 72^\circ$.

$$\Gamma_r = \frac{Z_r - Z_o}{Z_r + Z_o} = \frac{j5 - (3.535 - j3.535)}{j5 + (3.535 - j3.535)} = 2.414 \angle 90^\circ = \frac{V^-}{V^+}.$$

$$V^- = 2.414\, V^+ \angle 90^\circ = j2.424\, V^+.$$

Let us look at this line by using the phasor solution:

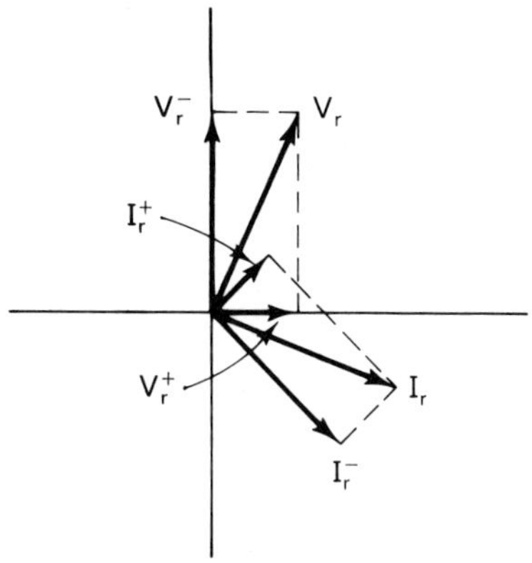

$$V_r = V^+ \underline{/0^\circ} + j2.414\ V^+ = V^+\ (1 + j2.414) = 2.613\ V^+ \underline{/67.498^\circ}.$$

To make the solution easier, let us assume that V^+ at the receiving end is 10 volts, r.m.s.

$$V_r = (2.613)(10) \underline{/67.498^\circ} = 26.13 \underline{/67.498^\circ} \text{ volts}.$$

$$I^+ = \frac{V^+}{Z_o} = \frac{10 \underline{/0^\circ}}{5 \underline{/-45^\circ}} = 2 \underline{/45^\circ} = 1.414 + j1.414.$$

$$I^- = -\frac{V^-}{Z_o} = \frac{-2.414(10)\underline{/90^\circ}}{5 \underline{/-45^\circ}} = -4.828 \underline{/135^\circ} = 4.828 \underline{/-45^\circ} = 3.414 - j3.414.$$

$$I_r = I^+ + I^- = 1.414 + j1.414 + 3.414 - j3.414 = 4.828 - j2 = 5.225 \underline{/-22.5^\circ} \text{ amperes}.$$

$$P_r = V_r I_r \cos\theta_r = (26.13)(5.225)\cos 90^\circ = 0 \text{ watts}.$$

$$Q_r = V_r I_r \sin\theta_r = (26.13)(5.225)\sin 90^\circ = 136.53 \text{ vars}.$$

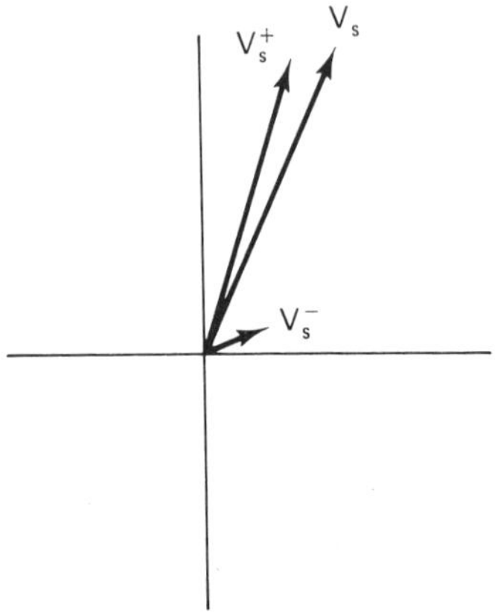

$$\lambda = \frac{2\pi}{\beta} = \frac{2\pi}{3.535} = 1.77 \text{ meters.}$$

$$0.2\,\lambda = 0.2(1.77) = 0.355 \text{ meters.}$$

$$e^{\alpha\ell} = e^{(3.535)(0.355)} = e^{1.26} = 3.51.$$

$$e^{-\alpha\ell} = e^{-1.26} = 0.285.$$

$$|V^+|e^{\alpha\ell} = 10(3.51) = 35.1.$$

$$|V^-|e^{-\alpha\ell} = (10)(2.414)(0.285) = 6.87.$$

$$I^+_{(\text{sending end})} = \frac{V^+_{(\text{sending end})}}{Z_o} = \frac{35.1\,\underline{/72^\circ}}{5\,\underline{/-45^\circ}} =$$

$$7.027\,\underline{/117^\circ} = -3.190 + j6.261 \text{ amperes.}$$

$$I^-_{(\text{sending end})} = -\frac{V^-_{(\text{sending end})}}{Z_o} = \frac{-6.87\,\underline{/18^\circ}}{5\,\underline{/-45^\circ}} =$$

$$-1.373\,\underline{/63^\circ} = -0.6236 - j1.224 = 1.373\,\underline{/-117^\circ} \text{ amperes.}$$

$$V_s = V^+ + V^- = 35.1\,\underline{/72^\circ} + 6.87\,\underline{/18^\circ} = 39.56\,\underline{/63.92^\circ} \text{ volts.}$$

$$I_s = I^+ + I^- = 7.027\,\underline{/117^\circ} - 1.373\,\underline{/63^\circ} = 6.317\,\underline{/127.13^\circ} \text{ amperes.}$$

$$Z_s = \frac{39.56\,\underline{/63.92^\circ}}{6.317\,\underline{/127.13^\circ}} = \frac{V_s}{I_s} = 6.263\,\underline{/-63.20^\circ} = 2.823 - j5.59 \text{ ohms.}$$

$$P_s = V_sI_s\cos\theta_s = (39.56)(6.317)\cos 63.21^\circ = 112.6 \text{ watts.}$$

$$Q_s = V_sI_s\sin\theta_s = (39.56)(6.317)\sin 63.21^\circ = 223.08 \text{ vars.}$$

10.10 Load Impedance in Terms of the Voltage Reflection Coefficient

Let us start out with equation (3-8).

$$\frac{Z_r}{Z_o} = \frac{1 + \Gamma_r}{1 - \Gamma_r}.$$

Let $Z_r = R_r + jX_r$ and let Z_o be real. We can then write:

$$\frac{Z_r}{Z_o} = \frac{R_r}{Z_o} + j\frac{X_r}{Z_o} = \frac{1 + \Gamma_r}{1 - \Gamma_r}. \tag{10-34}$$

It should be pointed out that, when we specify that the characteristic impedance is real, we could have either the lossy or the lossless case since we have found out that even in the lossy case it is possible for Z_o to be real. Let

$$\Gamma_r = |\Gamma_r| \underline{/\phi} = |\Gamma_r| (\cos\phi + j \sin\phi),$$

so

$$\frac{Z_r}{Z_o} = \frac{1 + |\Gamma_r| (\cos\phi + j \sin\phi)}{1 - |\Gamma_r| (\cos\phi + j \sin\phi)} =$$

$$\left(\frac{1 + |\Gamma_r| \cos\phi + j\,|\Gamma_r| \sin\phi}{1 - |\Gamma_r| \cos\phi - j\,|\Gamma_r| \sin\phi}\right)\left(\frac{1 - |\Gamma_r| \cos\phi + j\,|\Gamma_r| \sin\phi}{1 - |\Gamma_r| \cos\phi + j\,|\Gamma_r| \sin\phi}\right) =$$

$$\frac{1 - |\Gamma_r|^2 + j(2|\Gamma_r|\sin\phi)}{1 - 2|\Gamma_r| \cos\phi + |\Gamma_r|^2} =$$

$$\frac{R_r}{Z_o} + j\,\frac{X_r}{Z_o}.$$

So

$$\frac{R_r}{Z_o} = \frac{1 - |\Gamma_r|^2}{1 + |\Gamma_r|^2 - 2|\Gamma_r| \cos\phi}. \tag{10-35}$$

$$\frac{X_r}{Z_o} = \frac{2|\Gamma_r| \sin\phi}{1 + |\Gamma_r|^2 - 2|\Gamma_r| \cos\phi}. \tag{10-36}$$

10.11 Example of Determining an Unknown Load Impedance from Laboratory Measurements

Let us use the results found in Section 10.10 to find an unknown load impedance from laboratory measurements. In the laboratory we measure the following items: VSWR = 3.50. Distance between two successive voltage minimums = 45.5 centimeters. The first voltage minimum is 36.8 centimeters from the load. The characteristic impedance of the line is known to be 50 ohms, real. We may assume that the line is practically lossless.

Solution

Distance between voltage minimums $= \frac{\lambda}{2} = 45.5$ cm.

$$\lambda = 2(45.5) = 91 \text{ cm} = 0.91 \text{ meters.}$$

$$|\Gamma_r| = \frac{\text{VSWR} - 1}{\text{VSWR} + 1}.$$

Equation (3-24)

$$|\Gamma_r| = \frac{3.5 - 1}{3.5 + 1} = \frac{2.5}{4.5} = 0.556.$$

$$\frac{d_{(\text{voltage minimum})}}{\lambda} = \frac{36.8}{91} = 0.404.$$

Using equation (3-29) we find:

$$\phi = 4\pi \frac{d_{\text{min}}}{\lambda} - \pi \pm 2n\pi.$$

$$\phi = 4\pi(0.404) - \pi \pm 2n\pi = 5.077 - \pi \pm 2n\pi = 1.94 \pm 2n\pi.$$

With $n = 0$, $\phi = 1.94$ radians or 111.15°.

$$\frac{R_r}{Z_o} = \frac{1 - |\Gamma_r|^2}{1 + |\Gamma_r|^2 - 2|\Gamma_r| \cos\phi} = \frac{1 - (0.556)^2}{1 + (0.556)^2 - 2(0.556)\cos 111.15^\circ} = 0.405.$$

$$\frac{X_r}{Z_o} = \frac{2|\Gamma_r|\sin\phi}{1 + |\Gamma_r|^2 - 2|\Gamma_r|\cos\phi} = \frac{2(0.556)\sin 111.15°}{1 + (0.556)^2 - 2(0.556)\cos 111.15°} = +0.608.$$

$$R_r = (0.405)Z_o = 0.405(50) = 20.25 \text{ ohms}.$$

$$X_r = (0.608)Z_o = 0.608(50) = 30.40 \text{ ohms}.$$

While the use of equations is all right, it should be pointed out here that, as we have seen before, the use of the Smith chart for this example would have been even easier.

10.12 Quarter-Wave Transformer

Let us start with equation (5-29).

$$Z = Z_o\frac{(Z_r + jZ_o\tan\beta d)}{(Z_o + jZ_r\tan\beta d)},$$

and let $\beta d = \frac{\pi}{2}$ radians which is equivalent to 90°, or 1/4 wavelength.

$$Z = Z_o\left[\frac{\frac{Z_r}{\tan\beta d} + jZ_o}{\frac{Z_o}{\tan\beta d} + jZ_r}\right] = Z_o\left[\frac{\frac{Z_r}{\tan 90°} + jZ_o}{\frac{Z_o}{\tan 90°} + jZ_r}\right] = Z_o\left[\frac{0 + jZ_o}{0 + jZ_r}\right] = \frac{Z_o^2}{Z_r} = Z.$$

This relationship allows us to match a load to a transmission line much as we did in circuits with a transformer: thus it is called a quarter-wave transformer. Suppose we had a 300-ohm load and we wished to match it to a 75-ohm transmission line.

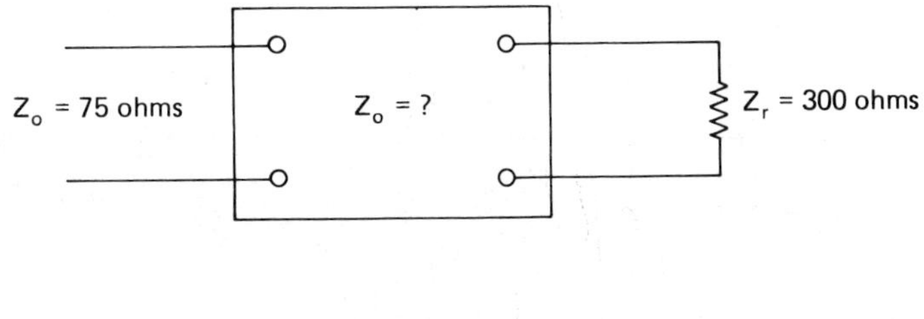

$$Z = \frac{Z_o^2}{Z_r}. \qquad 75 = \frac{Z_o^2}{300}.$$

$$Z_o^2 = 75(300) = 22{,}500. \qquad Z_o = \sqrt{22{,}500} = 150 \text{ ohms}.$$

We could have a match by inserting a quarter wavelength of a line with a characteristic impedance of 150 ohms.

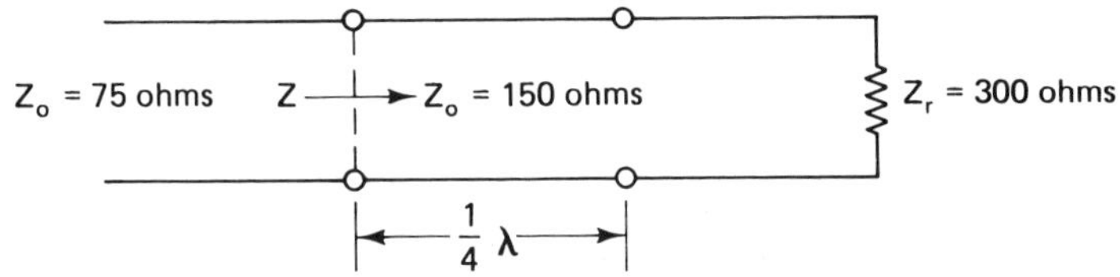

Using equation (5-6) the input impedance, Z, can be found as:

$$Z = Z_o\frac{(Z_r\cos\beta d + jZ_o\sin\beta d)}{(Z_o\cos\beta d + jZ_r\sin\beta d)} = \frac{150(300\cos 90^\circ + j150\sin 90^\circ)}{(150\cos 90^\circ + j300\sin 90^\circ)} = \frac{150(j150)}{j300} = 75 \text{ ohms}.$$

So the line with the characteristic impedance of 75 ohms has a load of 75 ohms and it is matched.

10.13 Lossless Transformer, One-Eighth Wavelength Long

If we have a line one-eighth wavelength long we can terminate it in a pure resistance of any value and the input impedance will have a magnitude equal to the characteristic impedance.

$\frac{1}{8}\lambda$ is equivalent to 45 degrees or $\pi/4$ radians.

$$Z = Z_o\frac{(Z_r\cos\beta d + jZ_o\sin\beta d)}{(Z_o\cos\beta d + jZ_r\sin\beta d)} = Z_o\frac{Z_r\cos 45^\circ + jZ_o\sin 45^\circ)}{(Z_o\cos 45^\circ + jZ_r\sin 45^\circ)}$$

$$= Z_o\frac{(0.707Z_r + j0.707Z_o)}{(0.707Z_o + j0.707Z_r)},$$

so the magnitude of the impedance becomes:

$$|Z| = \left| Z_o \frac{\sqrt{0.5Z_r^2 + 0.5Z_o^2}}{\sqrt{0.5Z_o^2 + 0.5Z_r^2}} \right| = |Z_o|.$$

10.14 Chapter Summary

1. Approximations may be made for γ and Z_o depending on the engineering situation that exists.
   (a) Low loss—$G = 0$; $R \ll \omega L$; $\alpha \ll \beta$.

$$\alpha \simeq \frac{R}{2}\frac{1}{\sqrt{LC}} \simeq \frac{R}{2Z_o}.$$

$$\beta \simeq \omega\sqrt{LC}. \qquad Z_o \simeq \sqrt{\frac{L}{C}}.$$

   (b) Low loss—$G \neq 0$; $R \ll \omega L$; $G \ll \omega C$.

$$\alpha \simeq \frac{1}{2}\left[\frac{R}{\sqrt{\frac{L}{C}}} + G\sqrt{\frac{L}{C}}\right] \simeq \frac{1}{2}\left[\frac{R}{Z_o} + GZ_o\right].$$

$$\beta \simeq \omega\sqrt{LC}. \qquad Z_o \simeq \sqrt{\frac{L}{C}}.$$

   (c) Distortionless line.

$$\frac{R}{L} = \frac{G}{C}.$$

$$\alpha = \frac{R}{L}\sqrt{LC} = \frac{G}{C}\sqrt{LC} = \frac{R}{Z_o} = GZ_o. \qquad \beta = \omega\sqrt{LC}. \qquad Z_o = \sqrt{\frac{L}{C}}.$$

$$v_p = \frac{1}{\sqrt{LC}}.$$

All quantities are independent of frequency except as R and L are changed by skin and proximity effects.

(d) High attenuation—$R \gg \omega L$; $G \ll \omega C$; $L \simeq 0$.

$$\alpha \simeq \beta \simeq \frac{\sqrt{R\omega C}}{\sqrt{2}}.$$

$$Z_o \simeq \sqrt{\frac{R}{\omega C}} \underline{/-45^\circ} \simeq \frac{1}{\sqrt{2}}\sqrt{\frac{R}{\omega C}} - j\frac{1}{\sqrt{2}}\sqrt{\frac{R}{\omega C}}.$$

(e) High attenuation—$R \ll \omega L$; $G \gg \omega C$.

$$\alpha \simeq \beta \simeq \frac{\sqrt{\omega LG}}{\sqrt{2}}.$$

$$Z_o \simeq \sqrt{\frac{\omega L}{G}} \underline{/45^\circ} \simeq \frac{1}{\sqrt{2}}\sqrt{\frac{\omega L}{G}} + j\frac{1}{\sqrt{2}}\sqrt{\frac{\omega L}{G}}.$$

2. The distortionless line is one where the attenuation constant, α, phase constant, β, the characteristic impedance, Z_o, and the phase velocity, v_p, are all independent of frequency. This will be so if $R/L = G/C$ and the values of the parameters are not affected by skin or proximity effects. Under these conditions, all the frequency components of a signal travel at the same speed and are attenuated at the same rate so that the relative positions and magnitudes of the frequency components remain the same from one end of the line to the other. Thus the input signal is faithfully reproduced at the output of the line.
3. At any point, d, on the line where the impedance looking towards the load is Z we can define a voltage reflection coefficient.

$$\Gamma_d = \frac{Z - Z_o}{Z + Z_o}.$$

Γ_d is related to the voltage reflection coefficient at the receiving end, Γ_r, by the relation:

$$\Gamma_d = |\Gamma_r| e^{-2\alpha d} \underline{/\phi - 2\beta d},$$

where

$$\Gamma_r = |\Gamma_r| \underline{/\phi}.$$

If the line is lossless,

$$\Gamma_d = |\Gamma_r| \underline{/\phi\text{-}2\beta d}.$$

4. The theoretical maximum value of the voltage reflection coefficient occurs when:

$$\frac{Z_r}{Z_o} = \left|\frac{Z_r}{Z_o}\right| \underline{/\alpha} = 1 \underline{/\pm 135°}.$$

If this is true, the maximum value becomes 2.414 which means that the phasor value of the reflected component of the voltage is 2.414 times as large as the phasor value of the incident component. However, the practical limit for the voltage reflection coefficient is unity.

5. If the load impedance is $Z_r = R_r + jX_r$ and $\Gamma_r = |\Gamma_r| \underline{/\phi}$, with Z_o real,

$$\frac{R_r}{Z_o} = \frac{1 - |\Gamma_r|^2}{1 + |\Gamma_r|^2 - 2|\Gamma_r| \cos\phi}.$$

$$\frac{X_r}{Z_o} = \frac{2\,|\Gamma_r| \sin\phi}{1 + |\Gamma_r|^2 - 2|\Gamma_r| \cos\phi}.$$

These equations may be used to determine an unknown load impedance by making certain measurement on the line. These measurements are the VSWR, the distance between two voltage minimums and the distance from a voltage minimum to the load.

6. A quarter-wave transformer may be used to match the value of load resistance to a line with a characteristic impedance that is different from the load resistance by using a quarter-wave line with a special calculated value of characteristic impedance.
7. An eighth-wavelength transformer can be terminated in any value of load resistance and still have an input impedance with a magnitude equal to the characteristic impedance of the line providing the line is approximately lossless.

10.15 Review and Study Questions

1. How do we define a low-loss line? How does it differ from a lossless line?
2. What happens when $R/L = G/C$? Can you derive the equations for α, β, Z_o, and v_p under these conditions? Is α equal to zero? What is a line operating under these conditions called? Why might this condition be important and under what conditions?
3. Under what conditions discussed in this chapter is the attenuation very high? Can you think of engineering situations where these conditions might exist?
4. What is the relationship between the forward component of the phasor voltage at the sending and the receiving ends of the line? How are the phasor components of the reverse component of voltage related at each end of the line?
5. Define the voltage reflection coefficient at any point, x or d, on the line. How is it related to the voltage reflection coefficient at the receiving end?
6. On a lossless line how does the magnitude of the voltage reflection coefficient at one point on the line compare to its magnitude at any other point on the line? How is this affected if the line is lossy?
7. Under what conditions will the voltage reflection coefficient be a maximum value? Can this happen in the "real" world?
8. Describe how one might determine an unknown load impedance by making measurements in the laboratory on a high-frequency line. Give the equations necessary. How would the same type of determination be made on the Smith chart? Which do you think would be easier?
9. What is a quarter-wave transformer? How might it be used?
10. What is an eighth-wave transformer? How might it be used? Does it help in matching a load resistance to a transmission line?
11. Can you derive equation (10-35) and (10-36) from scratch?
12. What are the possible angles for the ratio of any load impedance to the characteristic impedance? Are the extreme values practical?
13. What do you obtain when you multiply any complex number by its conjugate?
14. What happens to a signal on a line if the phase velocity is independent of frequency?

Problems

10-1. A line has a resistance of 0.5 ohms per meter, an L of 10^{-2} henrys per meter and a C of 1×10^{-6} farads per meter. G may be assumed to be zero. $\omega = 10^3$ radians per second.
   (a) Determine α, β, and Z_o by using the exact equations, (1-38) and (1-39).
   (b) Repeat part (a) by using the approximate equations (10-5), (10-4) and (10-6).
   (c) What is the percent error in using the approximate equations?

10-2. A telephone line has the following parameters per mile: $R = 52$ ohms; $L = 0.05$ henrys; $C = 0.05 \times 10^{-6}$ farads; $G = 1.0 \times 10^{-6}$ mhos.
   (a) Determine the amount of loading required to make this line distortionless. (Assume the coil added has zero resistance.)
   (b) Determine v_p before the loading was added for the following frequencies:
      (1) 100 hertz.
      (2) 1 kilohertz.
      (3) 1 megahertz.
   (c) Determine v_p after the loading is added for the same frequencies as in (b).

10-3. A lossless line has a Z_o of 50 ohms and is terminated in a Z_r of $40 + j30$ ohms. Determine the magnitude and angle of the reflection coefficient at the following points on the line:
   (a) At the load.
   (b) 0.1 wavelengths from the load.
   (c) 0.25 wavelengths from the load.
   (d) 0.5 wavelengths from the load.
   (e) 1 wavelength from the load.

10-4. A line has the following parameters per mile: $R = 50$ ohms; $L = 0.04$ henrys; $C = 0.02 \times 10^{-6}$ farads; $G = 2.0 \times 10^{-6}$ mhos. If the line is terminated in a load of $200 + j200$ ohms and the frequency is 100 kilohertz determine the following:
   (a) The voltage reflection coefficient at the load.
   (b) The voltage reflection coefficient 0.25 wavelengths from the load.
   (c) The voltage reflection coefficient 1.0 wavelengths from the load.

10-5. A transmission line has the following parameters per meter: $R = 100$ ohms; $L = G = 0$; $C = 1 \times 10^{-6}$ farads; $f = 0.1 \times 10^6$ hertz.

(a) Determine the value of Z_r for a maximum value of the voltage reflection coefficient.

(b) If the line is 2.6 wavelengths long and the input voltage is 100 volts, r.m.s., determine the load voltage with the value of Z_r determined in (a).

(c) Find I_r under these conditions.

10-6. A lossless line is tested in the laboratory with the following results: $Z_o = 100$ ohms; wavelength = 88 cm; the first voltage minimum is 22 cm from the load; VSWR = 3.0.

(a) Determine the load impedance.

(b) Determine the voltage reflection coefficient.

10-7. A lossless line has a characteristic impedance of 50 ohms, real, and is 14.58 meters long. The wavelength is 3 meters. With an unknown load on the line the VSWR was measured to be 2.2 and the first voltage minimum was 0.8865 meter from the load. Determine the load impedance.

10-8. A 50-ohm, lossless transmission line is found to have a VSWR of 1.39. The first voltage minimum is 5 cm from the load and the next voltage minimum is 30 cm from the load. Determine the load impedance.

10-9. A lossless transmission line has a Z_o of 100 ohms. It is operated at a frequency so that the wavelength is 2 meters with a VSWR of 4.0. The first voltage maximum is 145 volts, r.m.s., and occurs at 0.75 meter from the load. Determine:

(a) The load voltage.

(b) The load impedance.

10-10. Measurements were made on a lossless 50-ohm slotted line in the laboratory. The first voltage minimum is located 22.5 cm from the load and the second is located 67.5 cm from the load. Its value is 20 volts, r.m.s. A voltage maximum of 80 volts, r.m.s., is located 45 cm from the load. Assuming that v_p is the speed of light, determine:

(a) The VSWR.

(b) The frequency.

(c) The load impedance.

(d) The voltage reflection coefficient.

10-11. A 50-ohm lossless, coaxial, slotted line is used to determine a load impedance in the laboratory. It is determined that the VSWR is 2.5 and the first voltage maximum is 0.179 wavelength from the load. Determine the load impedance.

10-12. A lossless transmission line has a Z_o of 100 ohms. It has a wavelength of 3 meters and a VSWR of 3.0. The first voltage minimum is 0.73 meter from the load and has a magnitude of 10 volts, r.m.s. Determine:

(a) The load impedance.

(b) The load voltage.

10-13. A slotted-line measurement on a lossless 75-ohm line gives a VSWR of 2.1 with the first voltage minimum located 0.175 wavelength from the load. What is the load impedance?

11

Power Relationships

11.1 Introduction

The efficiency of a transmission line may or may not be important, depending on the engineering considerations involved. If the engineer is transmitting large amounts of energy over long distances, efficiency might be the most important criterion in designing the line. In general, increased efficiency on a power line demands higher voltages and consequently increased costs. The power line engineer has no real control over the effective load impedance since it is determined by consumers whose demands are variable from minute to minute and from year to year. Within a limited range the vars on the line may be controlled by the use of static capacitors or rotating synchronous machines which again involve increased costs. Since the I^2R losses on a power line represent the primary concern as far as efficiencies are concerned, the power engineer will take those steps which will make economic sense to reduce the current in the line. Increasing the voltage and reducing the vars will have this effect.

The engineer who is primarily interested in signal transmission will be interested mainly in accurate reproduction of the signal at the receiving end, whether it represents a medical signal or the information contained in a television picture. Audio signals are not so critical since the ear has a tendency to forgive distortion resulting from phase shift and attenuation. Often efficiency is of minor importance since the energy involved is usually very small and inexpensive to generate.

Many engineers who start with low levels of signal strength are concerned about maximum power transfer from one point to another point and so they are not really concerned with efficiency as a number, but rather efficiency in terms of minimum losses on the line so that the maximum power is transferred from the sending end to the receiving end. In this chapter we will find out that the power transmitted down the line consists of three components in the general case. One component involves the forward voltage and forward current and is the component which furnishes power to the load. A second component of power involves the reverse voltage and reverse current and reduces the power going towards the load. The third component of power results from an interaction between the forward voltage and the reverse current and the reverse voltage and the forward current. This component either adds or subtracts from the power moving down the line depending on the spot on the line being considered. As before, if the line is matched, the only power going toward the load will be the forward power due to the forward voltage and forward current. This is associated with a VSWR of unity, no standing waves, and maximum power transfer. If the VSWR is greater than unity there will be standing waves, power reflected from the load back down the line toward the sending end, and increased losses on the line.

When we have been considering the voltage and current relationships in the text up to this point we have often been talking about the lossless case which is approached in practice in certain situations. The lossless line is simply the approximation for the low-loss lossy line. Standing waves on the low-loss line look like the lossless case over a limited region of the line. If the line contains many wavelengths in its physical length we would notice a difference as we moved along the line. But even as we approximate the actual situation we must always remember that all lines are lossy and consume energy in the series R and the shunt G of the line. In this chapter we will look at the higher frequency case where we are concerned with the power on the line more for the effect on the operation of the line than its effect on the efficiency of the line.

11.2 Decibels

Suppose we look at the power, P_1, at a point on a line where x equals x_1, and the power, P_2, at a point on the line where x equals x_2. At x_1 we will have an impedance of $R_1 + jX_1$ and at x_2 we will have an impedance of $R_2 + jX_2$ when looking towards the load. Let the voltages at the two points be V_1 and V_2. Our situation is illustrated in Figure 11-1.

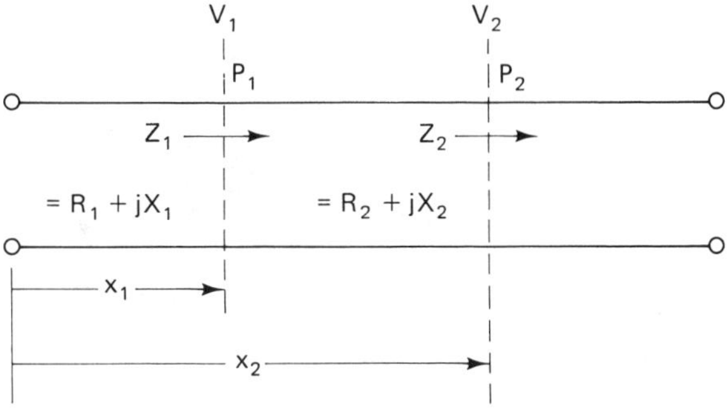

Figure 11-1. Study of power relationships.

If the line is chopped off at x_2 and replaced with Z_2 the power fed to the load and that portion of the line to the right of x_2 would be represented by the power loss in R_2. The same thing would be true if we chopped the line off at x_1 and replaced everything to the right by Z_1. Now the power going to the right of x_1 could be represented by the power loss in R_1. Let us define a decibel in the following way.

$$\text{decibel} = db = 10 \log \frac{P_1}{P_2} \tag{11-1}$$

The decibel is really defined as 1/10 of a bel, which is the log of P_1 divided by P_2. The difference between the powers at the two points would represent the power loss in the line between x_1 and x_2. If P_1 were 50 watts and P_2 were 25 watts, the power loss would be 25 watts. If we substituted in equation (11-1) we would obtain:

$$db = 10 \log \frac{50}{25} = 10 \log 2 = 10(0.301) = 3.01\ db.$$

Suppose we define the decibel in a different fashion as given in equation (11-2).

$$db = 10 \log \frac{P_2}{P_1}. \tag{11-2}$$

Now we would obtain for the given condition assumed above:

$$db = 10 \log \frac{25}{50} = 10 \log (0.5) = -3.01\ db.$$

The minus sign could be used as an indicator that P_2 is smaller than P_1 and that the power on the line is being attenuated as it proceeds down the line. But it is common to talk about a positive attenuation on the line so we will define the attenuation in *db* as in equation (11-1).

$$\text{Positive attenuation in } db = 10 \log \frac{P_1}{P_2}. \tag{11-3}$$

This equation will give a positive answer if P_2 is less than P_1. Electrical engineers often define the decibel as:

$$db = 10 \log \frac{P_1}{P_2} = 10 \log \frac{V_1{}^2/R_1}{V_2{}^2/R_2}. \tag{11-4}$$

This definition would be true in our case if X_1 and X_2 were both zero, so that V_1 was across R_1 and V_2 across R_2. Under this special condition we would have:

$$db = 10 \log \frac{V_1{}^2 R_2}{V_2{}^2 R_1}$$

and if $R_1 = R_2$ we would obtain:

$$db = 20 \log \frac{V_1}{V_2}. \tag{11-5}$$

Electrical engineers often use this equation to find "so-called" decibels even if R_1 does not equal R_2. Here we are expressing the decibel in terms of the ratio of two voltages and it is no longer related to the fundamental equation concerning power unless we have the special conditions that $Z_1 = R_1$, $Z_2 = R_2$ and $R_1 = R_2$, which would be very rare on a transmission line.

If we have only a forward wave which is expressed by equation (1-37) we can write it as:

$$V = Ae^{-\gamma x} = Ae^{-(\alpha + j\beta)x} = Ae^{-\alpha x}\, e^{-j\beta x}$$

so we have:

$$v(t) = \mathcal{R}\,[Ae^{-\alpha x}e^{-j\beta x}e^{j\omega t}] = Ae^{-\alpha x}\cos(\omega t - \beta x),$$

and the forward wave is attenuated by a factor of $e^{-\alpha x}$ as it moves down the line. If we consider the phasor value:

$$\frac{|V_1|}{|V_2|} = \frac{|Ae^{-\alpha x_1}|}{|Ae^{-\alpha x_2}|} = |e^{-\alpha x_1}\,e^{+\alpha x_2}| = |e^{\alpha(x_2 - x_1)}| = |e^{\alpha\Delta x}|,$$

where $\Delta x = x_2 - x_1$.

α is in nepers per meter so $\alpha\Delta x$ is in nepers.

Let $\alpha\Delta x = N$(nepers).

$$db = 20\log\frac{V_1}{V_2} = 20\log e^{\alpha\Delta x} = 20\log e^N = 20\,N\log e = 20N\log 2.718 =$$

$$20N(0.43429) = 8.6858N,$$

so we obtain the relationship that:

$$1\text{ neper} = 8.6858\text{ decibels.} \tag{11-6}$$

So,

$$20\,db \times 1\,\frac{\text{neper}}{8.6858db} = 2.302\text{ nepers,}$$

and

$$5\text{ nepers} \times \frac{8.6858\,db}{1\text{ neper}} = 43.42\,db.$$

A neper is obviously larger than a decibel. This is only good, remember, if $R_1 = R_2, X_1 = X_2 = 0$; we are discussing the case where $Z_1 = Z_2$ and $Z_r = Z_o$. If there are reflections on the line our relationships are much more complicated.

11.3. General Power Equations

Let us look at the complex power on a transmission line at a point x meters from the sending end or *d* meters from the receiving end. In Chapter 10 we derived the equations for the voltage reflection coefficient at this point by replacing all of the line from this point to the load by the equivalent impedance at that point. This equation was (10-29).

$$\Gamma_d = \Gamma_x = \frac{V_2 e^{2\gamma x}}{V_1}.$$

Also we have:

$$V = V_1 e^{-\gamma x} + V_2 e^{\gamma x}. \tag{11-7}$$

Which is obtained by replacing *A* by V_1 and *B* by V_2 in equation (1-32). Using a little algebra:

$$V = V_1 e^{-\gamma x}\left(1 + \frac{V_2 e^{\gamma x}}{V_1 e^{-\gamma x}}\right) = V_1 e^{-\gamma x}\left(1 + \frac{V_2 e^{2\gamma x}}{V_1}\right).$$

$$V = V_1 e^{-\gamma x}(1 + \Gamma_d). \tag{11-8}$$

When we replace *A* by V_1 and *B* by V_2 in equation (1-37) we obtain:

$$I = \frac{V_1 e^{-\gamma x}}{Z_o} - \frac{V_2 e^{\gamma x}}{Z_o}. \tag{11-9}$$

So,

$$I = \frac{V_1 e^{-\gamma x}}{Z_o}\left(1 - \frac{V_2 e^{2\gamma x}}{V_1}\right).$$

$$I = \frac{V_1 e^{-\gamma x}}{Z_o}(1 - \Gamma_d). \tag{11-10}$$

The complex power at x can be written:

$$\text{Complex power} = \text{C. P.} = VI^* = \text{watts} + j\text{vars}. \tag{11-11}$$

$$\text{C. P.} = \left[V_1 e^{-\gamma x}(1 + \Gamma_d)\right]\left[\frac{V_1}{Z_o} e^{-\gamma x}(1 - \Gamma_d)\right]^* =$$

$$\left[Z_o\frac{V_1}{Z_o}\,e^{-(\alpha\,+\,j\beta)x}(1+\Gamma_d)\right]\left[\frac{V_1}{Z_o}\,e^{-(\alpha\,+\,j\beta)x}(1-\Gamma_d)\right]^*=$$

$$Z_o\left[\frac{V_1}{Z_o}\,e^{-\alpha x}e^{-j\beta x}(1+\Gamma_d)\right]\left[\frac{V_1}{Z_o}\,e^{-\alpha x}e^{-j\beta x}(1-\Gamma_d)\right]^*=$$

$$Z_o\left[\frac{V_1}{Z_o}\,e^{-\alpha x}e^{-j\beta x}(1+\Gamma_d)\right]\left[\frac{V_1{}^*}{Z_o{}^*}\,e^{-\alpha x}e^{+j\beta x}(1-\Gamma_d)^*\right].$$

Since

$$[1-(a+jb)]^*=[1-a-jb]^*=[1-a+jb]=$$
$$[1-(a-jb)]=[1-(a+jb)^*],\ (1-\Gamma_d)^*=(1-\Gamma_d{}^*)$$

and we have:

$$\text{C. P.}=Z_o\left[\frac{V_1}{Z_o}\,e^{-\alpha x}e^{-j\beta x}(1+\Gamma_d)\right]\left[\frac{V_1{}^*}{Z_o{}^*}\,e^{-\alpha x}e^{j\beta x}(1-\Gamma_d{}^*)\right]=$$

$$Z_o\,\frac{|V_1|^2}{|Z_o|^2}\,e^{-2\alpha x}(1+\Gamma_d)(1-\Gamma_d{}^*)=$$

$$Z_o\,\frac{|V_1|^2}{|Z_o|^2}\,e^{-2\alpha x}\,[1-|\Gamma_d|^2+\Gamma_d-\Gamma_d{}^*].$$

Since:

$$(a+jb)-(a+jb)^*=a+jb-(a-jb)=a+jb-a+jb=2jb,$$
$$\Gamma_d-\Gamma_d{}^*=2j(\text{imaginary part of }\Gamma_d).$$

$$\text{C. P.}=(R_o+jX_o)\,\frac{|V_1|^2}{|Z_o|^2}\,e^{-2\alpha x}\,[1-|\Gamma_d|^2+2jIm(\Gamma_d)]=$$

$$\text{C. P.}=\frac{|V_1|^2}{|Z_o|^2}\,e^{-2\alpha x}\,R_o\left[1-|\Gamma_d|^2-\frac{2\,X_o}{R_o}\,Im(\Gamma_d)\right]$$
$$+\,j\,\frac{|V_1|^2}{|Z_o|^2}\,e^{-2\alpha x}X_o\left[1-|\Gamma_d|^2+2\,\frac{R_o}{X_o}\,Im(\Gamma_d)\right] \qquad (11\text{-}12)$$

The real power, in watts, $P_{real}=P_r$, is equal to:

$$P_r=\frac{|V_1|^2}{|Z_o|^2}\,e^{-2\alpha x}\,R_o\left[1-|\Gamma_d|^2-2\,\frac{X_o}{R_o}\,Im(\Gamma_d)\right]. \qquad (11\text{-}13)$$

And the reactive power, in vars, Q, is equal to:

$$Q = \frac{|V_1|^2}{|Z_o|^2} e^{-2\alpha x} X_o \left[1 - |\Gamma_d|^2 + 2 \frac{R_o}{X_o} \mathrm{Im}(\Gamma_d) \right]. \quad (11\text{-}14)$$

We will concentrate on equation (11-13). For the real power to be positive, or zero, we have the condition that:

$$\left[|\Gamma_d|^2 + 2 \frac{X_o}{R_o} \mathrm{Im}(\Gamma_d) \right] \leqslant 1.$$

If the quantity is equal to one the real power equals zero; if it is less than one the real power is a positive quantity. Let us look at:

$$|\Gamma_d|^2 + 2 \frac{X_o}{R_o} \mathrm{Im}(\Gamma_d). \quad (11\text{-}15)$$

Let Z (the impedance looking toward the load) at x or $d = R + jX$ and $Z_o = R_o + jX_o$.

$$\Gamma_d = \frac{Z - Z_o}{Z + Z_o} = \frac{R + jX - (R_o + jX_o)}{R + jX + (R_o + jX_o)} = \frac{(R - R_o) + j(X - X_o)}{(R + R_o) + j(X + X_o)}.$$

Multiplying by the conjugate of the denominator gives:

$$\Gamma_d = \frac{(R - R_o) + j(X - X_o)}{(R + R_o) + j(X + X_o)} \; \frac{(R + R_o) - j(X + X_o)}{(R + R_o) - j(X + X_o)}.$$

So

$$\Gamma_d = \frac{R^2 - R_o^2 + X^2 - X_o^2 + j(2XR_o - 2X_oR)}{(R + R_o)^2 + (X + X_o)^2}. \quad (11\text{-}16)$$

And so the imaginary part of

$$\Gamma_d = \frac{2XR_o - 2X_oR}{(R + R_o)^2 + (X + X_o)^2}. \quad (11\text{-}17)$$

$$\Gamma_d = \frac{(R - R_o) + j(X - X_o)}{(R + R_o) + j(X + X_o)}$$

so

$$|\Gamma_d| = \frac{\sqrt{(R - R_o)^2 + (X - X_o)^2}}{\sqrt{(R + R_o)^2 + (X + X_o)^2}}$$

and

$$|\Gamma_d|^2 = \frac{(R - R_o)^2 + (X - X_o)^2}{(R + R_o)^2 + (X + X_o)^2}. \tag{11-18}$$

Substitution in equation (11-15) gives:

$$|\Gamma_d|^2 + 2\frac{X_o}{R_o} Im(\Gamma_d) =$$

$$\frac{(R + R_o)^2 + (X - X_o)^2}{(R + R_o)^2 + (X + X_o)^2} + 2\frac{X_o}{R_o}\left[\frac{2XR_o - 2X_oR}{(R + R_o)^2 + (X + X_o)^2}\right].$$

After some algebraic manipulation we obtain:

$$|\Gamma_d|^2 + 2\frac{X_o}{R_o} Im(\Gamma_d) = \frac{(R + R_o)^2 + (X + X_o)^2}{(R + R_o)^2 + (X + X_o)^2} - \frac{\left(4RR_o + 4\frac{X_o{}^2R}{R_o}\right)}{(R + R_o)^2 + (X + X_o)^2} =$$

$$1 - \frac{\left(4RR_o + 4\frac{X_o{}^2R}{R_o}\right)}{(R + R_o)^2 + (X + X_o)^2}. \tag{11-19}$$

If we substitute this result into equation (11-13) for the real power we obtain:

$$P_r = \frac{|V_1|^2}{|Z_o|^2} e^{-2\alpha x} R_o \left[1 - \left(|\Gamma_d|^2 + 2\frac{X_o}{R_o} Im(\Gamma_d)\right)\right] =$$

$$\frac{|V_1|^2}{|Z_o|^2} e^{-2\alpha x} R_o \left[1 - \left\{1 - \frac{4RR_o + \frac{4X_o{}^2R}{R_o}}{(R + R_o)^2 + (X + X_o)^2}\right\}\right] =$$

$$\frac{|V_1|^2}{|Z_o|^2} e^{-2\alpha x} R_o \left[\frac{4\,RR_o + \frac{4X_o{}^2R}{R_o}}{(R + R_o)^2 + (X + X_o)^2}\right] = P_r. \tag{11-20}$$

Since R and R_o are always positive and X_o^2 is always positive, the real power, P_r, is either positive or zero. It will be zero when $R = 0$. Equation (11-20) can also be rewritten as:

$$P_r = \frac{|V_1|^2}{|Z_o|^2} e^{-2\alpha x} \frac{4R(R_o^2 + X_o^2)}{(R + R_o)^2 + (X + X_o)^2} = \frac{|V_1|^2 e^{-2\alpha x} 4R}{(R + R_o)^2 + (X + X_o)^2}. \tag{11-21}$$

Again we see that the real power is zero when R equals zero or it is positive; it can never be negative.

In equation (11-13) we see that the real power is due to three components and it is interesting to determine where they come from.

$$\frac{|V_1|^2}{|Z_o|^2} e^{-2\alpha x} R_o(1). \tag{11-22}$$

$$\frac{|V_1|^2}{|Z_o|^2} e^{-2\alpha x} R_o(-|\Gamma_d|^2). \tag{11-23}$$

$$\frac{|V_1|^2}{|Z_o|^2} e^{-2\alpha x} R_o\left[-2\frac{X_o}{R_o} Im(\Gamma_d)\right]. \tag{11-24}$$

The three components are involved with the terms,

$$1, \ -|\Gamma_d|^2 \text{ and } -\frac{2X_o}{R_o} Im(\Gamma_d).$$

Let us rewrite equations (11-7) and (11-9).

$$V = V_1 e^{-\gamma x} + V_2 e^{\gamma x}. \qquad I = \frac{V_1}{Z_o} e^{-\gamma x} - \frac{V_2}{Z_o} e^{\gamma x}.$$

Let us write

$$I_1 = \frac{V_1}{Z_o} \text{ and } I_2 = -\frac{V_2}{Z_o},$$

and rewrite our current equation.

$$I = I_1 e^{-\gamma x} + I_2 e^{\gamma x}.$$

So

$$V = V_1e^{-\alpha x}e^{-j\beta x} + V_2e^{\alpha x}e^{j\beta x},$$

and

$$I = I_1e^{-\alpha x}e^{-j\beta x} + I_2e^{\alpha x}e^{j\beta x}.$$

$$\text{C. P.} = VI^* = [V_1e^{-\alpha x}e^{-j\beta x} + V_2e^{\alpha x}e^{j\beta x}]\,[I_1e^{-\alpha x}e^{-j\beta x} + I_2e^{\alpha x}e^{j\beta x}]^* =$$
$$[V_1e^{-\alpha x}e^{-j\beta x} + V_2e^{\alpha x}e^{j\beta x}]\,[I_1{}^*e^{-\alpha x}e^{j\beta x} + I_2{}^*e^{\alpha x}e^{-j\beta x}] =$$
$$V_1I_1{}^*e^{-2\alpha x} + V_1I_2{}^*e^{-2j\beta x} + V_2I_1{}^*e^{2j\beta x} + V_2I_2{}^*e^{2\alpha x}. \tag{11-25}$$

The first term in equation (11-25) is involved with the forward component of the voltage and the forward component of the current.

$$V_1I_1{}^*e^{-2\alpha x} = V_1\frac{(V_1)^*}{(Z_o)^*}\,e^{-2\alpha x} = V_1\,\frac{V_1{}^*}{Z_o{}^*}\,e^{-2\alpha x}\,\frac{Z_o}{Z_o} = (R_o + jX_o)\,\frac{V_1V_1{}^*}{Z_oZ_o{}^*}\,e^{-2\alpha x} =$$
$$(R_o + jX_o)\,\frac{|V_1|^2}{|Z_o|^2}\,e^{-2\alpha x}.$$

The real power in this expression is:

$$R_o\,\frac{|V_1|^2}{|Z_o|^2}\,e^{-2\alpha x},$$

which is the same as equation (11-22) or the first term in equation (11-13).

If we look at the last term in equation (11-25) we have:

$$V_2I_2{}^*e^{2\alpha x} = V_2\frac{(-V_2)^*}{(Z_o)^*}\,e^{2\alpha x} = -\frac{V_2V_2{}^*}{Z_o{}^*}\,e^{2\alpha x}\left[\frac{Z_o}{Z_o}\right] =$$
$$-(R_o + jX_o)\,\frac{|V_2|^2}{|Z_o|^2}\,e^{2\alpha x}.$$

and the real power is:

$$-\,R_o\,\frac{|V_2|^2}{|Z_o|^2}\,e^{2\alpha x},$$

but

$$\Gamma_d = \frac{V_2}{V_1} e^{2\gamma x} = \frac{V_2}{V_1} e^{2\alpha x} e^{2j\beta x}$$

so

$$V_2 = \frac{\Gamma_d V_1}{e^{2\alpha x} e^{2j\beta x}} =$$

$$V_2 = \Gamma_d V_1 e^{-2\alpha x} e^{-2j\beta x}$$

and

$$|V_2| = |\Gamma_d| \, |V_1| e^{-2\alpha x}$$

so the real power term now becomes:

$$- R_o \frac{|\Gamma_d|^2 |V_1|^2 \, e^{-4\alpha x} e^{2\alpha x}}{|Z_o|^2} = - R_o \frac{|V_1|^2}{|Z_o|^2} e^{-2\alpha x} |\Gamma_d|^2 =$$

$$\frac{|V_1|^2}{|Z_o|^2} e^{-2\alpha x} R_o \, [-|\Gamma_d|^2],$$

so we see that the power associated with the reverse voltage, V_2, and the reverse current, I_2, gives the component of power indicated in equation (11-23) or the second term in equation (11-13).

Finally we have the two terms in the middle of equation (11-25).

$$V_1 I_2^* e^{-2j\beta x} + V_2 I_1^* e^{j2\beta x}.$$

Here we have one component associated with the forward component of the voltage and the reverse component of the current and a second term associated with the reverse component of voltage with the forward component of the current. Let us look at these two terms in detail.

$$V_1 \frac{(-V_2)^*}{(Z_o)^*} e^{-2j\beta x} + V_2 \frac{(V_1)^*}{(Z_o)^*} e^{2j\beta x} =$$

$$\left[\frac{V_2 V_1^*}{Z_o^*} e^{2j\beta x} - \frac{V_1 V_2^*}{Z_o^*} e^{-2j\beta x} \right] \frac{Z_o}{Z_o} =$$

$$(R_o + jX_o) \left[\frac{V_2 V_1^*}{Z_o Z_o^*} e^{2j\beta x} - \frac{V_1 V_2^*}{Z_o Z_o^*} e^{-2j\beta x} \right]. \qquad (11\text{-}25)$$

But

$$V_2 = \Gamma_d V_1 e^{-2\alpha x} e^{-2j\beta x}$$

and

$$V_2{}^* = \Gamma_d{}^* V_1{}^* e^{-2\alpha x} e^{2j\beta x}.$$

So equation (11-25) becomes:

$$(R_o + jX_o)\left[\frac{\Gamma_d V_1 e^{-2\alpha x}\, e^{-2j\beta x} V_1{}^* e^{2j\beta x}}{|Z_o|^2} - \frac{V_1 \Gamma_d{}^* V_1{}^* e^{-2\alpha x}\, e^{2j\beta x} e^{-2j\beta x}}{|Z_o|^2}\right] =$$

$$(R_o + jX_o)\left[\Gamma_d \frac{|V_1|^2}{|Z_o|^2}\, e^{-2\alpha x} - \Gamma_d{}^* \frac{|V_1|^2}{|Z_o|^2}\, e^{-2\alpha x}\right] =$$

$$(R_o + jX_o)\left[\frac{|V_1|^2}{|Z_o|^2}\, e^{-2\alpha x}\right][\Gamma_d - \Gamma_d{}^*]. \qquad (11\text{-}26)$$

But we have already seen that:

$$\Gamma_d - \Gamma_d{}^* = 2j\,[Im(\Gamma_d)]$$

so equation (11-26) becomes:

$$(R_o + jX_o)\left[\frac{|V_1|^2}{|Z_o|^2}\, e^{-2\alpha x}\, 2jIm(\Gamma_d)\right]$$

and now the real term becomes:

$$-X_o \frac{|V_1|^2}{|Z_o|^2}\, e^{-2\alpha x}\, 2Im(\Gamma_d) = \frac{|V_1|^2}{|Z_o|^2}\, e^{-2\alpha x} R_o \left[-\frac{2X_o}{R_o}\, Im(\Gamma_d)\right],$$

which is the same as equation (11-24) and the last term in equation (11-13). Thus in the power equation we have one term associated with the forward voltage and forward current, one term with the reverse voltage and reverse current, and one term that is associated with the interaction between the forward voltage with the reverse current and the reverse voltage with the forward current. If Z_o is real, so that X_o is zero, we see that the power equation has only the two terms: one associated with the forward power and the other associated with the reverse power; there is no power associated with the interaction between the forward and reverse components. If the line is matched we would have only the forward component of voltage and the forward component of current and only the power associated with the forward waves.

So we have the two equations:

$$P_r = \frac{|V_1|^2}{|Z_o|^2} e^{-2\alpha x} R_o \left[1 - |\Gamma_d|^2 - 2\frac{X_o}{R_o} \operatorname{Im}(\Gamma_d)\right]. \tag{11-13}$$

$$P_r = \frac{|V_1|^2}{|Z_o|^2} e^{-2\alpha x} R_o \left[\frac{4RR_o + \dfrac{4X_o^2R}{R_o}}{(R + R_o)^2 + (X + X_o)^2}\right]. \tag{11-20}$$

And the expressions in the brackets are:

$$\left[1 - |\Gamma_d|^2 - 2\frac{X_o}{R_o} \operatorname{Im}(\Gamma_d)\right]. \tag{11-27}$$

$$\left[\frac{4\,RR_o + \dfrac{4X_o^2R}{R_o}}{(R + R_o)^2 + (X + X_o)^2}\right]. \tag{11-28}$$

We also recall that equation (11-20) can be written as:

$$\frac{|V_1|^2 e^{-2\alpha x}\, 4R}{(R + R_o)^2 + (X + X_o)^2}. \tag{11-21}$$

11.4 Some Examples of Power on a Transmission Line

Let us start out by looking at the values given by equations (11-27) and (11-28). The answers are, of course, not in watts, but for the moment we are not interested in the actual power.

Example A

Let us assume that: $Z_o = 35.353 - j35.353$ ohms. $Z = j50$ ohms.

$$\Gamma_d = \frac{Z - Z_o}{Z + Z_o} = \frac{j50 - (35.353 - j35.353)}{j50 + 35.353 - j35.353} = \frac{-35.353 + 35.353}{35.353 + j14.647} = 2.414 \underline{/90^\circ}.$$

$$1 - |\Gamma_d|^2 - \frac{2X_o}{R_o} \mathrm{Im}(\Gamma_d) = 1 - (2.414)^2 - 2\frac{(-35.353)(2.414)}{35.353} =$$

$$1 - 5.828 + 4.828 = 0.$$

This result was expected since the impedance Z has no real term and could not absorb any real power.

Example B

Let us assume that: $Z_o = 40 - j30 = 50 \underline{/-36.868^\circ}$ ohms. $Z = 50 + j0$ ohms.

$$\Gamma_d = \frac{Z - Z_o}{Z + Z_o} = \frac{50 - (40 - j30)}{50 + 40 - j30} = \frac{10 + j30}{90 - j30} = 0.333 \underline{/90^\circ} = 0 + j0.333.$$

$$1 - |\Gamma_d|^2 - 2\frac{X_o}{R_o} \mathrm{Im}(\Gamma_d) = 1 - (0.333)^2 - \frac{2(-30)(0.333)}{40} =$$

$$1 - 0.111 + 0.5 = 1.388$$

Or we could use:

$$\frac{4RR_o + \dfrac{4X_o{}^2R}{R_o}}{(R + R_o)^2 + (X + X_o)^2} = \frac{(4)(50)(40) + \dfrac{4(-30)^2 50}{40}}{(50 + 40)^2 + (0 - 30)^2} = \frac{8{,}000 + 4{,}500}{8{,}100 + 900} =$$

$$\frac{12{,}500}{9{,}000} = 1.388.$$

Here we notice that we now have all three components of power present on our line.

Example C

Let us assume that: $Z_o = 50$ ohms. $Z = 50$ ohms.

$$\Gamma_d = \frac{Z - Z_o}{Z + Z_o} = \frac{50 - 50}{50 + 50} = \frac{0}{100} = 0.$$

$$1 - |\Gamma_d|^2 - \frac{2X_o}{R_o} Im(\Gamma_d) = 1 - 0 - \frac{2(0)(0)}{50} = 1 - 0 - 0 = 1.$$

Here we have only the power associated with the forward voltage and the forward current, with the other two terms coming out with a value of zero because the line is matched at the point where $Z = Z_o$.

11.5 Power Relationships on the Transmission Line for the Lossless Case

Let us repeat our power equation (11-13).

$$P_r = \frac{|V_1|^2}{|Z_o|^2} e^{-2\alpha x} R_o \left[1 - |\Gamma_d|^2 - \frac{2X_o}{R_o} Im(\Gamma_d)\right].$$

If we assume the line is lossless, then α equals zero and X_o equals zero and our power equation becomes equation (11-29).

$$P_r = \frac{|V_1|^2}{|Z_o|^2} R_o \, [1 - |\Gamma_d|^2]. \tag{11-29}$$

From our discussion in Section 11.3, we know that the forward power (power associated with the forward voltage and forward current) is:

$$P_{\text{forward}} = \frac{|V_1|^2}{|Z_o|^2} R_o \, (1). \tag{11-30}$$

If the line is lossless,

$$P_{\text{forward}} = \frac{|V_1|^2}{|R_o|^2} R_o = \frac{V_1{}^2}{R_o}. \tag{11-31}$$

The reverse power (power associated with the reverse voltage and the reverse current) is:

$$P_{\text{reverse}} = \frac{|V_1|^2}{|Z_o|^2} R_o \, [- \; |\Gamma_d|^2] = \frac{|V_1|^2}{|R_o|^2} R_o \, [- \; |\Gamma_d|^2] =$$

$$\frac{|V_1|^2}{R_o} - |\Gamma_d|^2 = P_{\text{reflected}}. \tag{11-32}$$

If we take the ratio of the reflected (reverse) power to the forward (incident) power and concern ourselves only with the magnitudes we would have:

$$\frac{P_{\text{reflected}}}{P_{\text{incident}}} = \frac{|V_1|^2 \, |\Gamma_d|^2}{\frac{|V_1|^2}{R_o} \; R_o} = |\Gamma_d|^2. \tag{11-33}$$

We can now set up a table for the ratio of the reflected power to the incident power in terms of the reflection coefficient. Remember that on a lossless line the voltage reflection coefficient magnitude is constant all along the line.

The values are rather simple to calculate as just the square of a number but it so happens we can use the Smith chart to find the answer very quickly. Figure 11-2 shows the table given on the lower right-hand corner of the Smith chart. The top scale is the voltage reflection coefficient magnitude and the next scale down gives the square of these values or the ratio of the reflected power to the incident power. On the Smith chart it is indicated as *PWR*.

Table 11-1
Table of Ratio of Reflected to Incident Power versus Γ_d for Lossless Line

Voltage Reflection Coefficient	*Power-Reflected Power-Incident*
0	0
0.1	0.01
0.2	0.04
0.3	0.09
0.4	0.16
0.5	0.25
0.6	0.36
0.7	0.49
0.8	0.64
0.9	0.81
1.0	1.00

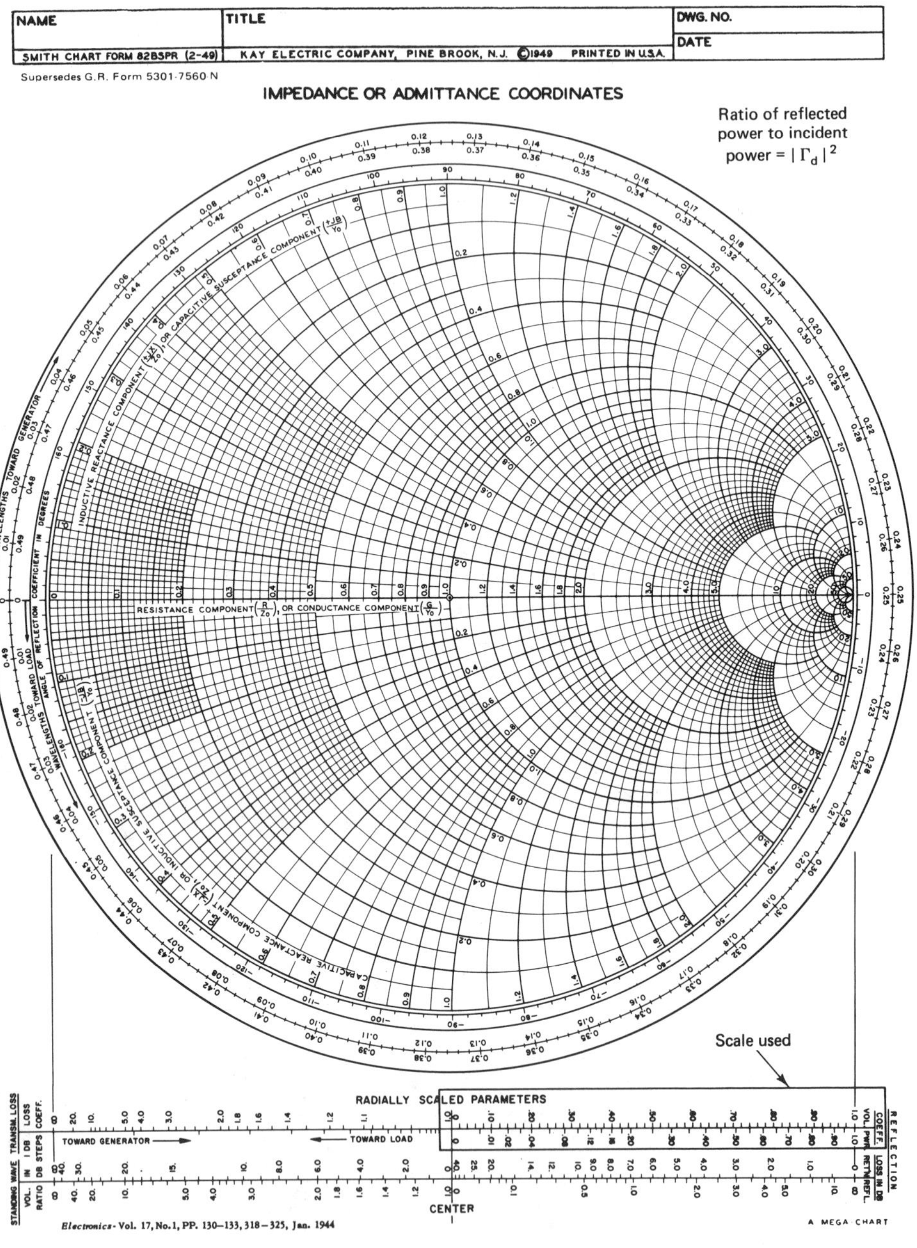

Figure 11-2. Ratio of reflected power to incident power = $|\Gamma_d|^2$.

11.6 A Different Approach to the Power on the Transmission Line for the Lossless Case

If we have the lossless case we can compute the power at the point of a voltage maximum or voltage minimum in a very simple way, since at either place the impedance looking toward the load is a pure resistance.

$$Z_{max} = R_{max} = \frac{V_{max}}{I_{min}} = \frac{|V^+| + |V^-|}{\frac{|V^+|}{Z_o} - \frac{|V^-|}{Z_o}} = R_o\left[\frac{|V^+| + |V^-|}{|V^+| - |V^-|}\right] =$$

$$R_o\frac{V_{max}}{V_{min}} = R_o(\text{VSWR}) = R_{max}. \tag{11-34}$$

$$V_{max} = I_{min}R_{max} = I_{min}R_o(\text{VSWR}) \text{ so } I_{min} = \frac{V_{max}}{R_o(\text{VSWR})}.$$

So the power at V_{max} becomes:

$$P \text{ at } V_{max} = |V_{max}|\,|I_{min}| = |V_{max}|\frac{|V_{max}|}{R_o(\text{VSWR})} =$$

$$\frac{|V_{max}|^2}{R_o(\text{VSWR})} = |I_{min}|^2 R_o\text{ VSWR} = P \text{ at } V_{max}. \tag{11-35}$$

$$Z_{min} = R_{min} = \frac{V_{min}}{I_{max}} = \frac{|V^+| - |V^-|}{\frac{|V^+|}{Z_o} + \frac{|V^-|}{Z_o}} = R_o\left[\frac{|V^+| - |V^-|}{|V^+| + |V^-|}\right] = R_o\frac{V_{min}}{V_{max}} =$$

$$R_o\frac{1}{\frac{V_{max}}{V_{min}}} = \frac{R_o}{(\text{VSWR})} = R_{min}. \tag{11-36}$$

$$V_{min} = I_{max}R_{min} = \frac{I_{max}R_o}{(\text{VSWR})} \text{ so } I_{max} = \frac{V_{min}(\text{VSWR})}{R_o}.$$

$$P \text{ at } V_{min} = |V_{min}|\,|I_{max}| = \frac{|V_{min}|\,|V_{min}|\,(\text{VSWR})}{R_o} =$$

$$\frac{|V_{min}|^2\,(\text{VSWR})}{R_o} = \frac{|I_{max}|^2 R_o}{(\text{VSWR})} = P \text{ at } V_{min}. \tag{11-37}$$

By this time it should be unnecessary to point out that these equations are really derived with the phasor diagrams in mind. Again with the phasor diagram in mind we can write:

$$V_{max} = |V^+| + |V^-|. \qquad I^+ = \frac{V^+}{R_o}. \qquad I^- = -\frac{V^-}{R_o}.$$

$$|I^+| = \frac{|V^+|}{R_o}. \qquad |I^-| = \frac{|V^-|}{R_o} \text{ so } V_{max} = |I^+|R_o + |I^-|R_o = R_o\,[|I^+| + |I^-|] =$$

$$R_o\,|I_{max}| = V_{max}. \tag{11-38}$$

$$V_{min} = |V^+| - |V^-| = |I^+|\,R_o - |I^-|R_o = R_o\,[|I^-| - |I^-|] = R_o\,|I_{min}|. \tag{11-39}$$

$$P \text{ at } V_{max} = |V_{max}|\,|I_{min}| = R_o\,|I_{max}|\,|I_{min}| = R_o(|I^+| + |I^-|)(|I^+| - |I^-|) =$$

$$R_o\,|I^+|^2 - R_o\,|I^-|^2 = P \text{ at } V_{max}. \tag{11-40}$$

The forward power and reflected power can be defined in terms of:

$$\text{Forward power} = |I^+|^2 R_o. \tag{11-41}$$

$$\text{Reflected power} = |I^-|^2 R_o. \tag{11-42}$$

Also:

$$P \text{ at } V_{min} = |V_{min}|\,|I_{max}| = |V_{min}|\,\frac{|V_{max}|}{R_o} = \frac{(|V^+| - |V^-|)(|V^+| + |V^-|)}{R_o} =$$

$$\frac{|V^+|^2}{R_o} - \frac{|V^-|^2}{R_o} = P. \tag{11-43}$$

Again we can say:

$$\text{Forward power} = \frac{|V^+|^2}{R_o}. \tag{11-44}$$

$$\text{Reflected power} = \frac{|V^-|^2}{R_o}. \tag{11-45}$$

Since:

$$\frac{|V^-|}{|V^+|} = |\Gamma_r| = |\Gamma_d|. \qquad |V^-| = |\Gamma_d|\,|V^+|.$$

This gives us another way to derive equation (11-33).

$$\frac{\text{Power (reflected)}}{\text{Power (incident)}} = \frac{\dfrac{|V^-|^2}{R_o}}{\dfrac{|V^+|^2}{R_o}} = \frac{|V^-|^2}{|V^+|^2} = \frac{|V^+|^2\,|\Gamma_d|^2}{|V^+|^2} = |\Gamma_d|^2. \quad (11\text{-}33)$$

11.7 Examples of Using the Power Equations

Let us take a couple of examples to illustrate the use of the power equations.

Example A

Given: Lossless line with the following specified: $Z_r = 100 + j100 = 141.421 \underline{/-45^\circ}$ ohms; $Z_o = 50$ ohms; $\ell = 2.2\lambda$; $V_s = 100 \underline{/0^\circ}$ volts, r.m.s. Find: The components of the power.

Solution

$$2.2\lambda = 2.2(360^\circ) = 792^\circ.$$

$$Z_s = Z_o\left[\frac{Z_r\cos\beta\ell + jZ_o\sin\beta\ell}{Z_o\cos\beta\ell + jZ_r\sin\beta\ell}\right] =$$

$$50\,\frac{(141.42\,\underline{/45^\circ})\cos 792^\circ + j(50)\sin 792^\circ}{(50)\cos 792^\circ + (1\,\underline{/90^\circ})(141.42\,\underline{/45^\circ})\sin 792^\circ} =$$

$$33.985\,\underline{/-61.45^\circ} = 16.244 - j29.85 \text{ ohms.}$$

$$I_s = \frac{V_s}{Z_s} = \frac{100\,\underline{/0^\circ}}{33.985\underline{/-61.45^\circ}} = 2.942\,\underline{/61.45^\circ} \text{ amperes.}$$

$$P_s = V_sI_s\cos\theta_s = (100)(2.942)\cos 61.45^\circ = 140.646 \text{ watts.}$$

$$V_r = V_s\cos\beta\ell - jI_sZ_o\sin\beta\ell =$$

$$(100\,\underline{/0^\circ})\cos 792^\circ - (1\,\underline{/90^\circ})(2.942\,\underline{/61.45^\circ})(50)\sin 792^\circ =$$

$$153.805 - j66.881 = 167.717\,\underline{/-23.50^\circ} \text{ volts.}$$

$$I_r = \frac{V_r}{Z_r} = \frac{167.717\,\underline{/-23.50^\circ}}{141.421\,\underline{/45^\circ}} = 1.186\,\underline{/-68.50^\circ} \text{ amperes.}$$

$$P_r = V_r I_r \cos\theta_r = (167.717)(1.186)\cos 45^\circ = 140.64 \text{ watts.}$$

$$\Gamma_s = \frac{Z_s - Z_o}{Z_s + Z_o} = \frac{16.244 - j29.85 - 50}{16.244 - j29.85 + 50} = 0.620 \angle{-114.255^\circ} = \frac{V_2}{V_1}.$$

Here we are finding V_1 since it is the phasor voltage specified in our power equations.

$$V_s = 100 = V_1 + V_2 = V_1 + 0.620V_1 \angle{-114.255^\circ}.$$

$$100 = V_1(1 + 0.620 \angle{-114.255^\circ}) = V_1\, 0.935 \angle{-37.88^\circ}.$$

$$|V_1| = \frac{100}{0.935} = 106.899 \text{ volts.}$$

Now we can use our equation (11-13).

$$P_r = \frac{|V_1|^2}{|Z_o|^2} e^{-2\alpha x} R_o \left[1 - |\Gamma_d|^2 - 2\frac{X_o}{R_o} Im(\Gamma_d)\right].$$

Since the line is lossless we do not have to worry about the attenuation term.

$$P_r = \frac{(106.899)^2}{50^2}(1)(50)\,[1 - (0.620)^2 - 2(0)] =$$

$$228.55\,[1 - 0.3846] = 228.55\,(0.6153) = 140.646 \text{ watts.}$$

$$\text{Power}_{(\text{incident})} = \frac{|V_1|^2}{|Z_o|^2} R_o\,(1) = \frac{(106.899)^2}{50^2}(50) = 228.55 \text{ watts.}$$

$$\text{Power}_{(\text{reflected})} = \frac{|V_1|^2}{|Z_o|^2} R_o\,[-|\Gamma_d|^2] = \frac{(106.899)^2}{50^2}(50)\,[-(0.620)^2] =$$

$$-87.904 \text{ watts.}$$

$$P_{real} = 228.55 - 87.904 = 140.646 \text{ watts.}$$

Since this is a lossless line we only have the two components of power to deal with; the forward power is 228.550 watts and the reflected power is 87.904 watts and the difference is the net power going to the load.

Example B

Given: A lossy line with the following specified: $\ell = 10$ meters; $z = 34 + j399$ ohms per meter; $y = 6 \times 10^{-6} + j40 \times 10^{-3}$ mhos per meter; $V_r = 100$ volts, r.m.s.; $Z_r = 200$ ohms. Find: The components of the power at the sending end, at the load end and halfway down the line.

Solution

$$\gamma = \sqrt{zy} = \sqrt{(400.446\,\underline{/85.129^\circ})(40 \times 10^{-3}\,\underline{/89.991^\circ})} =$$

$$4.002\,\underline{/87.56^\circ} = 0.170 + j3.998 = \alpha + j\beta.$$

$\alpha = 0.170$ nepers/meter.

$\beta = 3.998$ radians/meter.

$$Z_o = \sqrt{\frac{z}{y}} = \sqrt{\frac{400.446\,\underline{/85.129^\circ}}{40 \times 10^{-3}\,\underline{/89.991^\circ}}} = 100.05\,\underline{/-2.430^\circ} =$$

$$99.965 - j4.243 \text{ ohms.}$$

$$\Gamma_r = \frac{Z_r - Z_o}{Z_r + Z_o} = \frac{200 - (99.955 - j4.243)}{200 + 99.965 - j4.243} = 0.333\,\underline{/3.239^\circ}.$$

$\beta\ell = (3.9980)(10) = 39.98$ radians which is equivalent to $2{,}291.03^\circ$.

$$\frac{2{,}291.03^\circ}{360^\circ} = 6.3639 \text{ wavelengths,}$$

so we have 6 revolutions plus $0.3639(360^\circ) = 131.03^\circ$

$2\beta\ell = 79.972$ radians, which is equivalent to 4,582.06 degrees or 262.06°.

$$e^{\alpha\ell} = e^{(0.170)(10)} = e^{1.70} = 5.493.$$

$$e^{-\alpha\ell} = \frac{1}{e^{\alpha\ell}} = 0.182. \qquad e^{-2\alpha\ell} = e^{-2(0.170)(10)} = 0.0331.$$

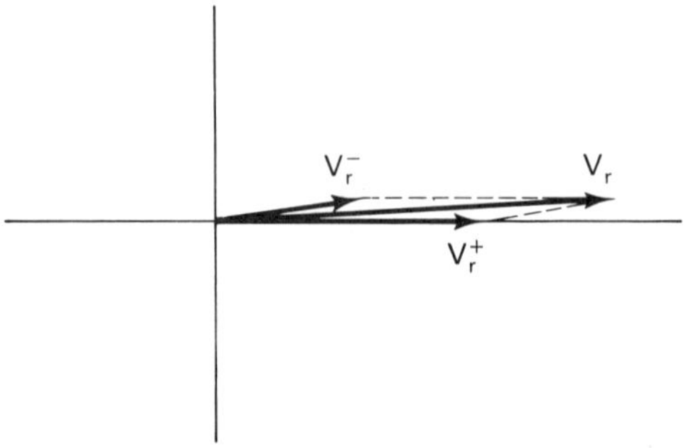

$$V_r = V^+ \angle 0^\circ + 0.333V^+ \angle 3.239^\circ = 1.333V^+ \angle 0.810^\circ = 100\angle 0.810^\circ.$$

$$|V^+| = \frac{100}{1.333} = 74.998 \text{ volts.}$$

$$|V^-| = (0.333)(74.998) = 25.031 \text{ volts.}$$

$$|V_s{}^+| = 74.998e^{\alpha\ell} = 74.998(5.493) = 412.016 = |V_1|.$$

$$|V_s{}^-| = 25.031e^{-\alpha\ell} = 25.031(0.182) = 4.556 = |V_2|.$$

$$V_s = 412.016 \angle 131.03^\circ + 4.556 \angle 3.239^\circ - 131.03^\circ = 411.156 \angle 131.653^\circ \text{ volts.}$$

$$I_s \frac{V_s{}^+}{Z_o} - \frac{V_s{}^-}{Z_o} = \frac{412.016 \angle 131.03^\circ}{100.055 \angle -2.430^\circ} - \frac{4.556 \angle -127.79^\circ}{100.055 \angle -2.430^\circ} = 4.126 \angle 132.84^\circ \text{ amperes.}$$

$$P_s = V_sI_s\cos\theta_s = (411.156)(4.126)\cos 1.187^\circ = 1{,}696.454 \text{ watts.}$$

Let us first find the components of the power at the sending end.

$$\Gamma_s = \frac{V_2}{V_1} = \frac{V_s{}^-}{V_s{}^+} = \frac{4.556 \angle -127.790^\circ}{412.016 \angle 131.030^\circ} = 0.0110 \angle -258.82^\circ =$$

$$0.011 \angle 101.179^\circ = -0.002 + j0.0108.$$

With $x = 0$.

$$P_s = \frac{|V_1|^2}{|Z_o|^2} R_o \left[1 - |\Gamma_s|^2 - 2\frac{X_o}{R_o}\text{Im}(\Gamma_s)\right] =$$

$$\frac{(412.016)^2}{(100.055)^2}(99.965)\left[1 - (0.0110)^2 - \frac{2(-4.243)}{99.965}(0.0108)\right] =$$

$$1{,}695.10\ (1 - 0.000122 + 0.00092) = 1{,}695.10(1.000798) = 1{,}696.45 \text{ watts.}$$

$$P_s = (1{,}695.1 - 0.2073 + 1.56) \text{ watts.}$$

We see that at the sending end the power in the forward wave is 1,695.1 watts and in the reverse wave it is −0.2073 watts and in the component due to the interaction between the forward voltage and reverse current and the reverse voltage and forward current it is a positive 1.56 watts.

Let us repeat our work half way down the line or when $\ell = 5$ meters. $\beta\ell = (3.998)(5) = 19.993$ radians, so we have 1,445.51 degrees or a net rotation of 65.515°.

$$e^{\alpha\ell} = e^{(0.170)(5)} = e^{0.851} = 2.343.$$

$$e^{-\alpha\ell} = 0.426.$$

At x = 5 meters, $V^+ = e^{\alpha\ell}(V_r^{\ +}) = 175.78$ volts, and $V^- = 10.679$ volts.

$$V_{x\ =\ 5\ \text{meters}} = 175.78\ \underline{/65.515^\circ} + 10.679\ \underline{/3.239^\circ - 65.515^\circ} = 169.452\ \underline{/62.660^\circ} \text{ volts.}$$

$$\Gamma_{d(\text{at } x\ =\ 5\ \text{meters})} = \frac{V^-}{V^+} = \frac{10.679\ \underline{/-62.275^\circ}}{175.786\ \underline{/65.515^\circ}} = 0.0607\ \underline{/-127.79^\circ} = -0.0372 - j0.0480.$$

$$e^{-2\alpha x} = e^{-2(0.170)(5)} = e^{-1.703} = 0.1820.$$

$$I_{\text{at } x\ =\ 5\ \text{meters}} = \frac{175.786\ \underline{/65.515^\circ}}{100.055\ \underline{/-2.430^\circ}} - \frac{10.679\ \underline{/-62.275^\circ}}{100.055\ \underline{/-2.430^\circ}} = 1.824\ \underline{/70.596^\circ} \text{ amperes.}$$

$$P_{\text{at } x\ =\ 5\ \text{meters}} = VI\cos\theta = (169.452)(1.824)\cos 7.93^\circ = 306.16 \text{ watts.}$$

$$P = \frac{|V_1|^2}{|Z_o|^2}e^{-2\alpha x}R_o\left[1 - |\Gamma_d|^2 - \frac{2X_o}{R_o}\text{Im}(\Gamma_d)\right].$$

$$P = \frac{(412.016)^2}{(100.055)^2}(0.182)(99.965)\left[1 - (0.0607)^2 - \frac{2(-4.243)}{99.965}(-0.048)\right] = 308.557(1 - 0.00369 - 0.00407) = (308.557 - 1.14 - 1.26) \text{ watts.}$$

$$P = 308.557\ (0.992) = 306.16 \text{ watts.}$$

Here the power in the forward wave is 308.557 watts, in the reverse wave it is −1.14 watts, and for the last term we now have a value of −1.26 watts. The last term in the power equation will change sign as the *j* term for the voltage reflection coefficient at that point becomes plus or minus as we move down the line.

Let us go to the load and recalculate:

$$P = \frac{V^2}{R} = \frac{100^2}{200} = 50 \text{ watts.}$$

$$\Gamma_r = 0.333 \angle 3.239° = 0.333 + j0.01886.$$

$$P = \frac{|V_1|^2}{|Z_o|^2} e^{-2\alpha x} R_o \left[1 - |\Gamma_r|^2 - \frac{2X_o}{R_o} Im(\Gamma_r)\right] =$$

$$\frac{(412.016)^2}{(100.055)^2}(0.03313)(99.965)\left[1 - (0.333)^2 - 2\frac{(4.243)}{99.965}(0.01886)\right] =$$

$$56.1669\ (1 - 0.1113 + 0.0016) = (56.1669 - 6.25 + 0.0898) = 50.00 \text{ watts.}$$

Now the power in the forward wave is 56.166 watts; in the reverse wave we have −6.25 watts; the last term is positive again but only 0.089 watts.

Let us set up a table showing how the three components of the power vary as we move down the line.

As we study the figures we see that the forward or incident power starts off at 1,695.10 watts at the sending end and is attenuated due to the losses in the line until only 56.166 watts arrives at the load end. At the load end 6.25 watts is reflected; as it travels towards the sending end it is attenuated and has a value of only 0.207 watts as it reaches the sending end. The power associated with the interaction between the forward and reflected components changes sign and either adds or subtracts as one moves down the line.

Table 11-2
Power on Lossy 10-Meter Line as a Function of the Distance (x) from the Sending End (Power Given in Watts)

x	Forward Power	Reflected Power	Interactive Power	Total Power
0	1,695.10	−0.2073	+1.56	1,696.45
5	308.557	−1.14	−1.26	306.16
10	56.166	−6.25	+0.089	50.00

11.8 Chapter Summary

1. The decibel is defined primarily in terms of two powers, P_1 and P_2.

$$\text{decibel} = \text{db} = 10 \log \frac{P_1}{P_2}. \tag{11-1}$$

If P_1 and P_2 represent the real power at two points on a transmission line, then the decibel equation may be used to specify the difference or the attenuation in power between the two points. Since power is V^2/R we can simplify things by assuming that the impedence on the line is purely resistive and that the voltage on the line at that point is across that R. Only under the assumption that $R_1 = R_2$ can we say that db $= 20 \log V_1/V_2$.

2. Since α is in nepers per unit distance, α times distance equals nepers. If we equate decibels and nepers we find that 1 neper = 8.6858 decibels.
3. The real power on a transmission line may be written as:

$$P_r = \frac{|V_1|^2}{|Z_o|^2} e^{-2\alpha x} R_o \left[1 - |\Gamma_d|^2 - \frac{2X_o}{R_o} Im(\Gamma_d)\right] \text{ watts.}$$

The three components are the result of the interactions between the forward and reverse components of voltage with the forward and reverse components of current.

$$\frac{|V_1|^2}{|Z_o|^2} e^{-2\alpha x} R_o$$

is due to the forward component of voltage and the forward component of current and is either positive or zero.

$$-|\Gamma_d|^2 \frac{|V_1|^2}{|Z_o|^2} e^{-2\alpha x} R_o$$

is due to the reverse component of voltage and the reverse component of current and is either zero or negative.

$$-2\frac{X_o}{R_o} Im(\Gamma_d) \frac{|V_1|^2}{|Z_o|^2} e^{-2\alpha x} R_o$$

is due to the forward component of voltage interacting with the reverse component of current and the reverse component of voltage interacting with the forward component of current. It may be zero or positive or negative depending on the location on the line where the power is being calculated since $Im(\Gamma_d)$ stands for the imaginary part of the voltage reflection coefficient at the point on the line being considered and the imaginary part changes sign as one moves down the line. It is zero if the line is lossless ($X_o = 0$) or if the imaginary part of the voltage reflection coefficient happens to be zero at that point on the line.

4. The reactive power on the line may be written as:

$$Q = \frac{|V_1|^2}{|Z_o|^2} e^{-2\alpha x} X_o \left[1 - |\Gamma_d|^2 + 2\frac{R_o}{X_o} Im(\Gamma_d) \right] \text{ vars.}$$

It also consists of three parts which are due to the interaction between the various components of voltage and current. Notice that if the line is lossless ($X_o = 0$) we have only one component:

$$Q = \frac{|V_1|^2}{R_o} 2Im(\Gamma_d),$$

which can be positive, zero, or negative depending on the location on the line which determines the imaginary part of Γ_d.

5. For the lossless case we have a simple expression.

$$\frac{P_{\text{reflected}}}{P_{\text{incident}}} = |\Gamma_d|^2.$$

This may be found very quickly from the Smith chart.

6. In the general lossy case the real power starts from the sending end and is attenuated as it moves towards the receiving end. If the line is not matched the power that is reflected is attenuated as it moves towards the sending end.

11.9 Review and Study Questions

1. Define a decibel in terms of power. Define a decibel in terms of voltage. Define a decibel in terms of current.
2. What is the relationship between decibel and a neper? Which is larger? Can nepers per meter be converted to decibels per meter? If so, how?

3. What is attenuation? What is the difference between positive and negative attenuation?
4. Prove that $(1 - \Gamma_d)^* = (1 - \Gamma_d^*)$.
5. Prove that $\Gamma_d - \Gamma_d^* = 2j$(imaginary part of Γ_d).
6. Define real power, reactive power, and complex power and give the units of each.
7. What kind of real power does one obtain by the interaction of a fundamental component of voltage and a third harmonic component of current? How does this compare with the interaction between a forward component of voltage and a reverse component of current?
8. Describe the three components of real power on a transmission line. Where does each component come from?
9. How is the real power on the line affected if the line is matched? How is it affected if the line is lossless and terminated in an open circuit?
10. Describe the three components of reactive power on a transmission line. Where does each component come from?
11. How is the reactive power on the line affected if the line is matched? How is it affected if the line is lossless and terminated in a short circuit?
12. How does the imaginary part of Γ_d vary with distance?
13. Express the efficiency of a line in terms of the received power and the losses.
14. When is efficiency of little or no importance on a line? When might it be very important?
15. How is the efficiency affected by the presence of standing waves on the line?

Problems

11-1. A transmission line has an attenuation factor of 0.15 nepers per meter. Express this in the following units:
   (a) Nepers per foot.
   (b) Nepers per kilometer.
   (c) *Db* per meter.
   (d) *Db* per mile.

11-2. A matched line has a voltage of 200 volts, r.m.s. at the load and 230 volts, r.m.s. at the input of the line. It is 1,000 meters long.

(a) Determine the total voltage attenuation in decibels.
(b) Determine the attenuation constant, α, in nepers per meter.

11-3. A line has an attenuation factor of 3.0×10^{-2} nepers per meter.
(a) Find the attenuation in db for 100 meters of line.
(b) Find the attenuation in db for 1,000 meters of line.
(c) Find the attenuation in db for 10,000 meters of line.

11-4. A transmission line has an attenuation factor of 0.010 nepers per mile and is a matched line with a resistive load.
(a) If the efficiency of power transmission is 50% what is the length of the line?
(b) For the line in (a) what is the attenuation in decibels?

11-5. A lossy line has the following parameters per kilometer: $R = 6.8$ ohms; $G = 6.8 \times 10^{-8}$ mhos; $C = 0.00565 \times 10^{-6}$ farads; $L = 2.11 \times 10^{-3}$ henrys; frequency = 1,000 hertz; $Z_r = 400 + j100$ ohms; $\ell = 1{,}000$ kilometers; $V_s = 100$ volts, r.m.s.
(a) Find α, β, and Z_o.
(b) Determine V_r, I_r, and I_s.
(c) Find the input power.
(d) Find the load power.
(e) Find the real power at the midpoint of the line.
(f) Separate the power found in (e) into its three components.

11-6. The problem in 11-5 is terminated in its characteristic impedance. Repeat problem 11-5 under the new conditions.

11-7. A lossless line has a Z_o of 100 ohms and is 3.4 wavelengths long. The input voltage is 100 volts, r.m.s. and the load impedance is $75 - j75$ ohms. Determine:
(a) The load voltage.
(b) The load current.
(c) The input current.
(d) The input power.
(e) The power to the load.
(f) The three components of the power at 1.7 wavelengths from the sending end.

11-8. A lossless line can be described as follows: $Z_o = 50$ ohms. $Z_r = 60 - j80$ ohms. $\ell = 7.85$ wavelengths. $V_s = 20$ volts, r.m.s. Determine:
(a) The input power.
(b) The load power.
(c) The voltage reflection coefficient at the load.

(d) The ratio of the reflected power to the incident power.
(e) The reflected power when $x = 6$ wavelengths from the input end.

11-9. A lossless line has a Z_o of 50 ohms and is terminated in a load of $30 + j30$ ohms. The input voltage is 10 volts, r.m.s., and the line is 2.85 wavelengths long. Determine:
(a) The load voltage.
(b) The load current.
(c) The input current.
(d) The input power.
(e) The reflected power at the midpoint of the line.
(f) The incident power at the midpoint of the line.

12

Wave Guides

12.1 Introduction

When frequencies reach up towards the gigahertz range, coaxial cables have high attenuation and this makes their use restricted to very short physical distances. The student should be very careful when using coaxial cables at these high frequencies since some of the older cables were very lossy and output might be only a fraction of the input even if the cable is only a few feet long. Even the newer low-loss coaxial cables have relatively high attenuation. It has become common practice to use other methods of transmitting electrical signals at very high frequencies that do not produce these high losses. One technique that has been very useful to the engineer for many years is to employ a hollow tube, which is either circular or rectangular in cross section and which is known as a "Wave Guide."

In Chapter 1 we briefly discussed the fact that a two-wire line is essentially a "wave guide" in that the energy moving along the line can be considered as being contained in the electromagnetic field which the line is guiding to the receiving end through the region around the line. All forms of transmission lines can be thought of as "wave guides" but in this chapter we are reserving the term for those hollow tubes which are coated on the inside with a thin coating of a good conductor. Like coaxial cables they, by their very construction, are shielded from external fields and do not radiate energy through their walls. Their attenuation is less and their power capacity is greater than a coaxial cable but their use is restricted to very high frequencies since their dimensions are related to the wavelength being used. As we will discover, the physical dimensions become very large and impractical as the frequency is decreased.

The basis for considering the wave guide is Maxwell's equations, so we must return to field theory for an understanding of their operation. As suggested previously, the wave guide acts only as a container for the electromagnetic fields that are moving down the guide. The student will find, however, that this is another case where field theory, when properly applied, gives practical results that are useful in the engineering world. When the theory says that the wave guide will not propagate below a certain frequency the student will find, in the laboratory, that this is exactly true. This is strong justification for the study of field theory as a background for much of what happens in most areas of the electrical engineering profession.

Employing Maxwell's equations gives us some equations which are solved by a common technique of assuming a product solution. The solutions obtained are complex in complete form but we are able to find out useful information without looking at the long, involved solutions. The solutions, with boundary values substituted in, tell us that rectangular wave guides have two basic modes of operation, namely the *TM* and *TE* modes. These, in turn, are broken down into many possible modes of operation but fortunately only a few principal modes are normally used. So we have TM_{mn} and TE_{mn} modes of operation with m and n being positive integers, 0, 1, 2, etc. We will also find out that the possible modes of operation are determined by the physical dimensions of the wave guide and how they are excited. For each wave guide size there is a critical (cutoff) frequency below which the wave guide will not propagate the energy down the guide. The smaller the guide the higher the frequency becomes but unfortunately, the lower the frequency the larger the guide becomes. This restricts the use of wave guides to very high frequencies—usually 300 megahertz and higher. It also becomes evident that the electromagnetic wave in the guide has a different wavelength than the same frequency wave would have in free space and becomes longer than the corresponding wave in free space.

All through this text we have been talking about the phase velocity, v_p, but when dealing with wave guides and the transmission of intelligence as a signal in the guide it becomes convenient to define a new velocity called the group velocity, v_g. The interesting aspect is that the phase velocity in a wave guide now becomes greater than the velocity of light; the group velocity, which is in reality the speed of the useful signal, becomes less than the velocity of light. A detailed study of the field picture within the guide would illustrate the various aspects of the phase and group velocity but the pictures are so complicated that we will look at the two velocities only in a mathematical sense.

12.2. Basic Solutions of Maxwell's Equations for a Rectangular Wave Guide

We will concern ourselves with one particular wave guide—namely a rectangular guide parallel to the z-axis with the widest dimension along the x-axis as shown in Figure 12-1.

We will assume that the interior of the wave guide is filled with a material that is linear, homogeneous and isotropic and the material we will select is air. In order to deal with wave guides we will have to return to our field theory course and Maxwell's equations:

$$\nabla \times \vec{H} = \vec{J} + \frac{\partial \vec{D}}{\partial t} \tag{12-1}$$

$$\nabla \times \vec{E} = \frac{\partial \vec{B}}{\partial t} \tag{12-2}$$

$$\nabla \cdot \vec{D} = \rho. \tag{12-3}$$

$$\nabla \cdot \vec{B} = 0. \tag{12-4}$$

We also have the auxiliary equations:

$$\vec{D} = \varepsilon \vec{E}. \tag{12-5}$$

$$\vec{B} = \mu \vec{H}. \tag{12-6}$$

$$\vec{J} = \sigma \vec{E}. \tag{12-7}$$

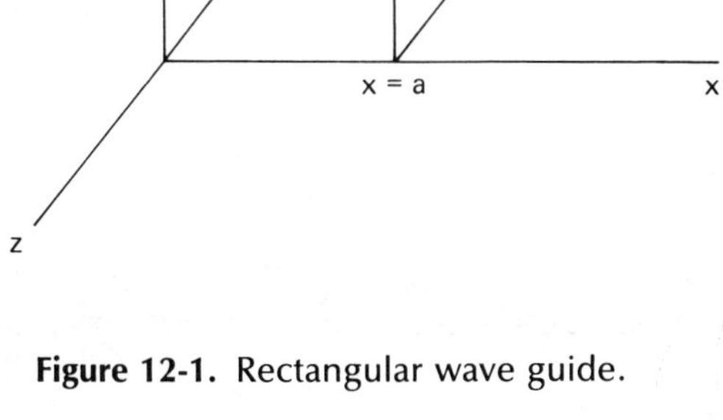

Figure 12-1. Rectangular wave guide.

$\vec{H}$ is the magnetic field intensity vector in amperes per meter.
$\vec{J}$ is the current density vector in amperes per square meter.
$\vec{D}$ is the electric flux density vector in coulombs per square meter.
$\vec{E}$ is the electric field intensity vector in volts per meter.
$\vec{B}$ is the magnetic flux density vector in webers per square meter.
ρ is the volume charge density in coulombs per cubic meter.
ε is the permittivity in farads per meter.
μ is the permeability in henrys per meter.
σ is the conductivity in mhos per meter.

For the material inside our wave guide, air, we will take $\varepsilon = \varepsilon_r\varepsilon_o = (1)\varepsilon_o$, since ε_r for air is 1, so ε will equal 8.8419×10^{-12} farads per meter inside of the guide. Also for the air inside of the guide we will have $\mu = \mu_r\mu_{o'}, = (1)\mu_o = 4\pi \times 10^{-7}$ henrys per meter since $\mu_r = 1$ for air.

By substitution we obtain:

$$\nabla \times \vec{H} = \sigma\vec{E} + \varepsilon\frac{\partial \vec{E}}{\partial t}. \tag{12-8}$$

$$\nabla \times \vec{E} = -\mu\frac{\partial \vec{H}}{\partial t}. \tag{12-9}$$

$$\nabla \cdot \vec{E} = \frac{\rho}{\varepsilon}. \tag{12-10}$$

$$\nabla \cdot \vec{H} = 0. \tag{12-11}$$

In addition to assuming that the material inside the guide is air, we will also assume that the volume charge density, ρ, and the conductivity, σ, are equal to zero. This gives us:

$$\nabla \times \vec{H} = \varepsilon_o\frac{\partial \vec{E}}{\partial t}. \tag{12-12}$$

$$\nabla \times \vec{E} = -\mu_o\frac{\partial \vec{H}}{\partial t}. \tag{12-13}$$

$$\nabla \cdot \vec{E} = 0. \tag{12-14}$$

$$\nabla \cdot \vec{H} = 0. \tag{12-15}$$

Now we will assume that all our quantities are sinusoidal (or cosinusoidal) in form and convert our equations to the phasor form.

$$\nabla \times \vec{H}_p = j\omega\varepsilon_o\vec{E}_p. \qquad (12\text{-}16)$$

$$\nabla \times \vec{E}_p = -j\omega\mu_o\vec{H}_p. \qquad (12\text{-}17)$$

$$\nabla \cdot \vec{E}_p = 0. \qquad (12\text{-}18)$$

$$\nabla \cdot \vec{H}_p = 0. \qquad (12\text{-}19)$$

To save energy and your instructor's chalk, let us drop the p with the understanding that we are dealing with phasors from here on in our work.

Taking the curl of equation (12-17) gives:

$$\nabla \times \nabla \times \vec{E} = \nabla \times (-j\omega\mu_o\vec{H}) = -j\omega\mu_o(\nabla \times \vec{H}) = -j\omega\mu_o(j\omega\varepsilon_o\vec{E}) =$$
$$\omega^2\mu_o\varepsilon_o\vec{E} = \nabla \times \nabla \times \vec{E} = \nabla(\nabla \cdot \vec{E}) - \nabla^2\vec{E},$$

but

$$\nabla \cdot \vec{E} = 0,$$

so

$$\omega^2\mu_o\varepsilon_o\vec{E} = -\nabla^2\vec{E},$$

or

$$\nabla^2\vec{E} + \omega^2\mu_o\varepsilon_o\vec{E} = 0. \qquad (12\text{-}20)$$

We can repeat the process for equation (12-16).

$$\nabla \times \nabla \times \vec{H} = \nabla \times j\omega\varepsilon_o\vec{E} = j\omega\varepsilon_o\nabla \times \vec{E} = j\omega\varepsilon_o(-j\omega\mu_o\vec{H}) =$$
$$\omega^2\mu_o\varepsilon_o\vec{H} = \nabla \times \nabla \times \vec{H} = \nabla(\nabla \cdot \vec{H}) - \nabla^2\vec{H},$$

but

$$\nabla \cdot \vec{H} = 0,$$

so

$$\omega^2\mu_o\varepsilon_o\vec{H} = -\nabla^2\vec{H}$$

or

$$\nabla^2\vec{H} + \omega^2\mu_o\varepsilon_o\vec{H} = 0. \qquad (12\text{-}21)$$

Thus we have the two equations (12-20) and (12-21) which are similar in form. Let us start out with equation (12-21).

$$\nabla^2 H_x\,\vec{a}_x + \nabla^2 H_y\,\vec{a}_y + \nabla^2 H_z\,\vec{a}_z + \omega^2\mu_o\varepsilon_o(H_x\,\vec{a}_x + H_y\,\vec{a}_y + H_z\,\vec{a}_z) = 0.$$

$\vec{a}_x$, $\vec{a}_y$, and $\vec{a}_z$ are unit vectors in the cartesian coordinate system and they might be written as $\vec{i}$, $\vec{j}$, and $\vec{k}$ by some authors. Expanding gives:

$$\left[\frac{\partial^2 H_x}{\partial x^2} + \frac{\partial^2 H_x}{\partial y^2} + \frac{\partial^2 H_x}{\partial z^2}\right]\vec{a}_x + \left[\frac{\partial^2 H_y}{\partial x^2} + \frac{\partial^2 H_y}{\partial y^2} + \frac{\partial^2 H_y}{\partial z^2}\right]\vec{a}_y$$
$$+ \left[\frac{\partial^2 H_z}{\partial x^2} + \frac{\partial^2 H_z}{\partial y^2} + \frac{\partial^2 H_z}{\partial z^2}\right]\vec{a}_z + \omega^2\mu_o\varepsilon_o H_x\,\vec{a}_x + \omega^2\mu_o\varepsilon_o H_y\,\vec{a}_y$$
$$+ \omega^2\mu_o\varepsilon_o H_z\,\vec{a}_z = 0. \tag{12-22}$$

We would obtain exactly the same form for $\vec{E}$.

$$\left[\frac{\partial^2 E_x}{\partial x^2} + \frac{\partial^2 E_x}{\partial y^2} + \frac{\partial^2 E_x}{\partial z^2}\right]\vec{a}_x + \left[\frac{\partial^2 E_y}{\partial x^2} + \frac{\partial^2 E_y}{\partial y^2} + \frac{\partial^2 E_y}{\partial z^2}\right]\vec{a}_y$$
$$+ \left[\frac{\partial^2 E_z}{\partial x^2} + \frac{\partial^2 E_z}{\partial y^2} + \frac{\partial^2 E_z}{\partial z^2}\right]\vec{a}_z +$$
$$+ \omega^2\mu_o\varepsilon_o E_z\,\vec{a}_z = 0. \tag{12-23}$$

If we equate the x, y, and z components in equations (12-22) and (12-23) we obtain six equations.

$$\frac{\partial^2 H_x}{\partial x^2} + \frac{\partial^2 H_x}{\partial y^2} + \frac{\partial^2 H_x}{\partial z^2} + \omega^2\mu_o\varepsilon_o H_x = 0. \tag{12-24}$$

$$\frac{\partial^2 H_y}{\partial x^2} + \frac{\partial^2 H_y}{\partial y^2} + \frac{\partial^2 H_y}{\partial z^2} + \omega^2\mu_o\varepsilon_o H_y = 0. \tag{12-25}$$

$$\frac{\partial^2 H_z}{\partial x^2} + \frac{\partial^2 H_z}{\partial y^2} + \frac{\partial^2 H_z}{\partial z^2} + \omega^2\mu_o\varepsilon_o H_z = 0. \tag{12-26}$$

$$\frac{\partial^2 E_x}{\partial x^2} + \frac{\partial^2 E_x}{\partial y^2} + \frac{\partial^2 E_x}{\partial z^2} + \omega^2\mu_o\varepsilon_o E_x = 0. \tag{12-27}$$

$$\frac{\partial^2 E_y}{\partial x^2} + \frac{\partial^2 E_y}{\partial y^2} + \frac{\partial^2 E_y}{\partial z^2} + \omega^2\mu_o\varepsilon_o E_y = 0. \tag{12-28}$$

$$\frac{\partial^2 E_z}{\partial x^2} + \frac{\partial^2 E_z}{\partial y^2} + \frac{\partial^2 E_z}{\partial z^2} + \omega^2\mu_o\varepsilon_o E_z = 0. \tag{12-29}$$

Let us start out with equation (12-26) and assume a product solution. Notice that a solution for (12-26) will have the same type of form as the solution for all the other equations, since they are all similar in form.

$$\frac{\partial^2 H_z}{\partial x^2} + \frac{\partial^2 H_z}{\partial y^2} + \frac{\partial^2 H_z}{\partial z^2} = -\omega^2 \mu_o \varepsilon_o H_z \tag{12-26}$$

We will assume the solution, $H_z = \mathscr{X}\mathscr{Y}\mathscr{Z}$ where $\mathscr{X}$ is a function of x only, $\mathscr{Y}$ is a function of y only, and $\mathscr{Z}$ is a function of z only: $\mathscr{X} = f(x)$ only; $\mathscr{Y} = f(y)$ only; $\mathscr{Z} = f(z)$ only.

So

$$\frac{\partial H_z}{\partial x} = \frac{\partial \mathscr{X}}{\partial x}\mathscr{Y}\mathscr{Z}, \text{ and } \frac{\partial^2 H_z}{\partial x^2} = \frac{\partial^2 \mathscr{X}}{\partial x^2}\mathscr{Y}\mathscr{Z}. \tag{12-30}$$

$$\frac{\partial H_z}{\partial y} = \mathscr{X}\frac{\partial \mathscr{Y}}{\partial y}\mathscr{Z}, \text{ and } \frac{\partial^2 H_z}{\partial y^2} = \mathscr{X}\frac{\partial^2 \mathscr{Y}}{\partial y^2}\mathscr{Z}. \tag{12-31}$$

$$\frac{\partial H_z}{\partial z} = \mathscr{X}\mathscr{Y}\frac{\partial \mathscr{Z}}{\partial z}, \text{ and } \frac{\partial^2 H_z}{\partial z^2} = \mathscr{X}\mathscr{Y}\frac{\partial^2 \mathscr{Z}}{\partial z^2}. \tag{12-32}$$

Substituting back into equation (12-26) gives:

$$\frac{\partial^2 H_z}{\partial x^2} + \frac{\partial^2 H_z}{\partial y^2} + \frac{\partial^2 H_z}{\partial z^2} = \frac{\partial^2 \mathscr{X}}{\partial x^2}\mathscr{Y}\mathscr{Z} + \mathscr{X}\frac{\partial^2 \mathscr{Y}}{\partial y^2}\mathscr{Z} + \mathscr{X}\mathscr{Y}\frac{\partial^2 \mathscr{Z}}{\partial z^2} = -\omega^2 \mu_o \varepsilon_o \mathscr{X}\,\mathscr{Y}\,\mathscr{Z}. \tag{12-33}$$

Dividing through by $\mathscr{X}\,\mathscr{Y}\,\mathscr{Z}$ gives:

$$\frac{1}{\mathscr{X}}\frac{\partial^2 \mathscr{X}}{\partial x^2} + \frac{1}{\mathscr{Y}}\frac{\partial^2 \mathscr{Y}}{\partial y^2} + \frac{1}{\mathscr{Z}}\frac{\partial^2 \mathscr{Z}}{\partial z^2} = -\omega^2 \mu_o \varepsilon_o. \tag{12-34}$$

The first term on the left is a function of x only, the second term is a function of y only, and the third term is a function of z only. The only way we can add up a function of x and a function of y and a function of z and then have the sum equal to a constant would be for the functions to be constants themselves. Since our goal is equations relating to the operation of a rectangular wave guide, we will report only the successful attempts to solve the equation and pick the right constants to start with. We will define these constants as follows:

$$\frac{1}{\mathscr{X}}\frac{\partial^2\mathscr{X}}{\partial x^2} = -A^2. \qquad (12\text{-}35)$$

$$\frac{1}{\mathscr{Y}}\frac{\partial^2\mathscr{Y}}{\partial y^2} = -B^2. \qquad (12\text{-}36)$$

$$\frac{1}{\mathscr{Z}}\frac{\partial^2\mathscr{Z}}{\partial z^2} = \gamma^2. \qquad (12\text{-}37)$$

So substituting in equation (12-34) we have:

$$-A^2 - B^2 + \gamma^2 = -\omega^2\mu_o\varepsilon_o. \qquad (12\text{-}38)$$

We can now solve equations (12-35), (12-36), and (12-37) by normal mathematical techniques.

$$\frac{\partial^2 X}{\partial x^2} = -A^2\mathscr{X}.$$

$$\frac{d^2\mathscr{X}}{dx^2} + A^2\mathscr{X} = 0.$$

$$(D^2 + A^2)\mathscr{X} = 0.$$

$D^2 + A^2 = 0$. $D^2 = -A^2$, so $D = \pm jA$. We can write a solution for $\mathscr{X}$ in either of two forms.

$$\mathscr{X} = C_7e^{jAx} + C_8e^{-jAx} = C_1\sin Ax + C_2\cos Ax. \qquad (12\text{-}39)$$

C_7, C_8, C_1, and C_2 are the usual constants used with solutions of this type. Also,

$$\frac{\partial^2\mathscr{Y}}{\partial y^2} = -B^2\mathscr{Y}.$$

$$\frac{d^2\mathscr{Y}}{dy^2} + B^2\mathscr{Y} = 0.$$

$$(D^2 + B^2)\mathscr{Y} = 0.$$

$$D^2 + B^2 = 0. \qquad D^2 = -B^2, \text{ so } D = \pm jB.$$

So we can write a solution for Y in either of two forms.

$$\mathscr{Y} = C_9e^{jBy} + C_{10}e^{-jBy} = C_3\sin By + C_4\cos By. \qquad (12\text{-}40)$$

And finally,

$$\frac{\partial^2 \mathcal{Z}}{\partial z^2} = \gamma^2 \mathcal{Z}.$$

$$\frac{d^2 \mathcal{Z}}{dz^2} - \gamma^2 \mathcal{Z} = 0.$$

$$(D^2 - \gamma^2)\mathcal{Z} = 0.$$

$D^2 - \gamma^2 = 0$; $D^2 = \gamma^2$, so $D = \pm\gamma$. And we can write a solution for Z as:

$$\mathcal{Z} = C_5 e^{\gamma z} + C_6 e^{-\gamma z}. \tag{12-41}$$

Thus, one possible form of our solution could be:

$$H_z = \mathcal{X}\mathcal{Y}\mathcal{Z} = (C_1 \sin Ax + C_2 \cos Ax)(C_3 \sin By + C_4 \cos By)(C_5 e^{\gamma z} + C_6 e^{-\gamma z}). \tag{12-42}$$

We could repeat the process for the other five equations, (12-24), (12-25), (12-27), (12-28), and (12-29), and we would obtain solutions with identical forms which would differ only in the variables and constants. Let us look at the last part of equation (12-42) and see what happens when we convert H_z to a function of time.

$$H_z(t) = \mathcal{R}[(C_1 \sin Ax + C_2 \cos Ax)(C_3 \sin By + C_4 \cos By)(C_5 e^{\gamma z} e^{j\omega t} + C_6 e^{-\gamma z} e^{j\omega t})].$$

Since the last term has some familiar terms, let us let $\gamma = \alpha + j\beta$ just for kicks and see what happens.

$$H_z(t) = \mathcal{R}[(C_1 \sin Ax + C_2 \cos Ax)(C_3 \sin By + C_4 \cos By)\ (C_5 e^{(\alpha + j\beta)z} e^{j\omega t} + C_6 e^{-(\alpha + j\beta)z} e^{j\omega t})]$$

Rearranging the last part of the equation gives:

$$H_z(t) = \mathcal{R}[(C_1 \ldots\ldots)(C_3 \ldots\ldots)(C_5 e^{\alpha z} e^{j(\omega t + \beta z)} + C_6 e^{-\alpha z} e^{j(\omega t - \beta z)})] = C_{20} e^{\alpha z} \cos(\omega t + \beta z) + C_{21} e^{-\alpha z} \cos(\omega t - \beta z). \tag{12-43}$$

Equation (12-43) will be correct, providing that C_1, C_2, C_3, C_4, C_5, and C_6 are real if:

$$\gamma = \alpha + j\beta. \tag{12-44}$$

Thus we end up with two traveling waves; one going in the positive z direction and one going in the negative z direction if equation (12-44) holds true. This appears to be a normal assumption since we have energy moving through the space of our wave guide and our experience with conductor-type transmission lines would tend to lead us to believe the situation should be similar here.

Let us consider only the wave traveling in the positive z direction all by itself, since the two waves are operating independently of each other in any situation which is linear.

$$H_z = (C_1 \sin Ax + C_2 \cos Ax)(C_3 \sin By + C_4 \cos By)(C_6 e^{-\gamma z}). \tag{12-45}$$

$$\frac{\partial H_z}{\partial z} = -\gamma(C_1 \sin Ax + C_2 \cos Ax)(C_3 \sin By + C_4 \cos By)(C_6 e^{-\gamma z}) = -\gamma H_z. \tag{12-46}$$

In a similar fashion we would have:

$$\frac{\partial H_y}{\partial z} = -\gamma H_y. \tag{12-47}$$

$$\frac{\partial H_x}{\partial z} = -\gamma H_x. \tag{12-48}$$

Now we will go back to equations (12-16) and (12-17).

$$\nabla \times \vec{H} = j\omega\varepsilon_o \vec{E}. \qquad \nabla \times \vec{E} = -j\omega\mu_o \vec{H}.$$

From these two equations we obtain:

$$\frac{\partial H_z}{\partial y} - \frac{\partial H_y}{\partial z} = j\omega\varepsilon_o E_x. \tag{12-49}$$

$$\frac{\partial H_x}{\partial z} - \frac{\partial H_z}{\partial x} = j\omega\varepsilon_o E_y. \tag{12-50}$$

$$\frac{\partial H_y}{\partial x} - \frac{\partial H_x}{\partial y} = j\omega\varepsilon_o E_z. \tag{12-51}$$

$$\frac{\partial E_z}{\partial y} - \frac{\partial E_y}{\partial_z} = -j\omega\mu_o H_x. \qquad (12\text{-}52)$$

$$\frac{\partial E_x}{\partial z} - \frac{\partial E_z}{\partial x} = -j\omega\mu_o H_y. \qquad (12\text{-}53)$$

$$\frac{\partial E_y}{\partial x} - \frac{\partial E_x}{\partial y} = -j\omega\mu_o H_z. \qquad (12\text{-}54)$$

Substituting the values from equations (12-46), (12-47), and (12-48) gives:

$$\frac{\partial H_z}{\partial y} - (-\gamma H_y) = j\omega\varepsilon_o E_x. \qquad (12\text{-}55)$$

$$-\gamma H_x - \frac{\partial H_z}{\partial x} = j\omega\varepsilon_o E_y. \qquad (12\text{-}56)$$

(12-51) $$\frac{\partial H_y}{\partial z} - \frac{\partial H_x}{\partial y} = j\omega\varepsilon_o E_z.$$

Because the equations are similar in form we would also have:

$$\frac{\partial E_z}{\partial y} - (-\gamma E_y) = j\omega\mu_o H_x. \qquad (12\text{-}57)$$

$$-\gamma E_x - \frac{\partial E_z}{\partial x} = -j\omega\mu_o H_y. \qquad (12\text{-}58)$$

(12-54) $$\frac{\partial E_y}{\partial x} - \frac{\partial E_x}{\partial z} = -j\omega\mu_o H_z.$$

Let us take equation (12-55) and equation (12-58) and apply a little algebra.

$$E_x = \frac{1}{j\omega\varepsilon_o}\left[\frac{\partial H_z}{\partial y} + \gamma H_y\right]. \qquad (12\text{-}59)$$

$$H_y = \frac{1}{j\omega\mu_o}\left[\gamma E_x + \frac{\partial E_z}{\partial x}\right]. \qquad (12\text{-}60)$$

So,

$$E_x = \frac{1}{j\omega\varepsilon_o}\left[\frac{\partial H_z}{\partial y} + \gamma\frac{1}{j\omega\mu_o}\left(\gamma E_x + \frac{\partial E_z)}{\partial x}\right)\right] =$$

$$\frac{1}{j\omega\varepsilon_o}\frac{\partial H_z}{\partial y} + \frac{\gamma^2 E_x}{-\omega^2\mu_o\varepsilon_o} + \frac{\gamma}{-\omega^2\mu_o\varepsilon_o}\frac{\partial E_z}{\partial x} = E_x.$$

$$E_x\left[1 + \frac{\gamma^2}{\omega^2\mu_o\varepsilon_o}\right] = \frac{-j}{\omega\varepsilon_o}\frac{\partial H_z}{\partial y} - \frac{\gamma}{\omega^2\mu_o\varepsilon_o}\frac{\partial E_x}{\partial x}.$$

$$E_x\left[\frac{\omega^2\mu_o\varepsilon_o + \gamma^2}{\omega^2\mu_o\varepsilon_o}\right] = \frac{-j}{\omega\varepsilon_o}\frac{\partial H_z}{\partial y} - \frac{\gamma}{\omega^2\mu_o\varepsilon_o}\frac{\partial E_z}{\partial x}. \quad (12\text{-}61)$$

Again reporting the successful attempts we will define a new parameter, h, as follows:

$$\omega^2\mu_o\varepsilon_o + \gamma^2 = h^2. \quad (12\text{-}62)$$

We had equation (12-38),

$$-A^2 - B^2 + \gamma^2 = -\omega^2\mu_o\varepsilon_o$$

so

$$A^2 + B^2 - \gamma^2 = \omega^2\mu_o\varepsilon_o = h^2 - \gamma^2,$$

so

$$A^2 + B^2 = h^2 = \omega^2\mu_o\varepsilon_o + \gamma^2. \quad (12\text{-}63)$$

Substituting back into equation (12-61) gives:

$$E_x\frac{h^2}{\omega^2\mu_o\varepsilon_o} = \frac{-j}{\omega\varepsilon_o}\frac{\partial H_z}{\partial y} - \frac{\gamma}{\omega^2\mu_o\varepsilon_o}\frac{\partial E_z}{\partial x},$$

or

$$E_x = \frac{-\gamma}{h^2}\frac{\partial E_z}{\partial x} - \frac{j\omega\mu_o}{h^2}\frac{\partial H_z}{\partial y}. \quad (12\text{-}64)$$

If we take equations (12-59) and (12-60) and go the other way we would have:

$$H_y = \frac{-\gamma}{h^2}\frac{\partial H_z}{\partial y} - \frac{j\omega\varepsilon_o}{h^2}\frac{\partial E_z}{\partial x}. \quad (12\text{-}65)$$

If we had started with equations (12-56) and (12-58) we would have ended up with:

$$E_y = \frac{-\gamma}{h^2}\frac{\partial E_z}{\partial y} + \frac{j\omega\mu_o}{h^2}\frac{\partial H_z}{\partial x}. \tag{12-66}$$

$$H_x = \frac{-\gamma}{h^2}\frac{\partial H_z}{\partial x} + \frac{j\omega\varepsilon_o}{h^2}\frac{\partial E_z}{\partial y}. \tag{12-67}$$

(12-64) $$E_x = \frac{-\gamma}{h^2}\frac{\partial E_z}{\partial x} - \frac{j\omega\mu_o}{h^2}\frac{\partial H_z}{\partial y}.$$

(12-65) $$H_y = \frac{-\gamma}{h^2}\frac{\partial H_z}{\partial y} - \frac{j\omega\varepsilon_o}{h^2}\frac{\partial E_z}{\partial x}.$$

(12-62) $$h^2 = \omega^2\mu_o\varepsilon_o + \gamma^2.$$

It turns out that the above equations allow for two special classes of solutions.

I. $E_z = 0,\ H_z \neq 0.$

This means that the two remaining $\vec{E}$ components, E_x and E_y are transverse to the z direction of propagation. This is called the transverse electric mode or the TE mode.

II. $E_z \neq 0,\ H_z = 0.$

This means that the two remaining $\vec{H}$ components, H_x and H_y, are transverse to the z direction of travel. This is called the transverse magnetic mode or the TM mode.

12.3 The *TM* Mode of Operation.

Let us look first at the TM mode of operation when $H_z = 0$. Our equations under these conditions become:

$$E_x = \frac{-\gamma}{h^2}\frac{\partial E_z}{\partial x}. \tag{12-68}$$

$$E_y = \frac{-\gamma}{h^2}\frac{\partial E_z}{\partial y}. \tag{12-69}$$

$$H_x = \frac{j\omega\varepsilon_o}{h^2}\frac{\partial E_z}{\partial y}. \tag{12-70}$$

$$H_y = \frac{-j\omega\varepsilon_o}{h^2}\frac{\partial E_z}{\partial x}. \tag{12-71}$$

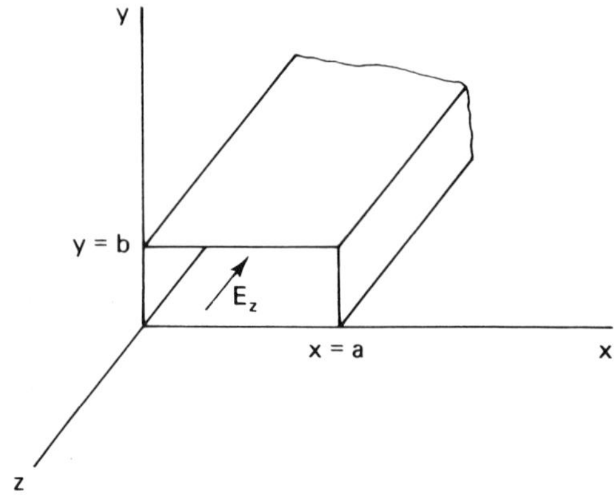

Figure 12-2.

We will look at the forward component only so we can write:

$$E_z = (C_1 \sin Ax + C_2 \cos Ax)(C_3 \sin By + C_4 \cos By)(C_6 e^{-\gamma z}). \quad (12\text{-}72)$$

At the surface of a good conductor the tangential component of $\vec{E}$ is equal to zero.

$$E_{tan} = 0 \text{ at } x = 0,\ y = 0, \text{ at } y = b \text{ and } x = a.$$

Using the boundary conditions we can now evaluate the constants.

$E_z = 0$ at $x = 0$.

$E_z = (0 + C_2)(C_3 \sin By + C_4 \cos By)C_6 e^{-\gamma z} = 0.$ Therefore, $C_2 = 0$.

$E_z = 0$ at $y = 0$.

$E_z = (C_1 \sin Ax)(0 + C_4)C_6 e^{-\gamma z} = 0.$ Therefore, $C_4 = 0$.

$E_z = 0$ at $x = a$.

$E_z = (C_1 \sin Aa)(C_3 \sin By)C_6 e^{-\gamma z} = 0.$

$$A = \frac{m\pi}{a}, \text{ with } m = 0, 1, 2, \text{ etc.} \tag{12-73}$$

The last boundary condition gives us:

$$E_z = 0 \text{ at } y = b.$$

$$E_z = C_1 \sin \frac{m\pi x}{a} C_3 \sin Bb\, C_6 e^{-\gamma z} = 0.$$

The sin $Aa = 0$ when $Aa = m\pi$ with $m = 0, 1, 2, 3, \ldots$ Therefore,

$$B = \frac{n\pi}{b}, \text{ with } n = 0, 1, 2, \text{ etc.} \tag{12-74}$$

So,

$$E_z = K \sin \frac{m\pi}{a} x \sin \frac{n\pi}{b} y\, e^{-\gamma z}, \text{ with } m \text{ and } n = 0, 1, 2, 3, \text{ etc.} \tag{12-75}$$

However, when we look at equation (12-75) we notice that neither m nor n can be zero since this would wipe out our solution; E_z would be zero always and everywhere and so would the E_x, E_y, H_x, and H_y components which depend on E_z. So for the *TM* mode of operation we have to modify our statement; m and n are integer values of 1, 2, 3, 4, etc., and neither may ever be equal to zero.

Going back to equation (12-63) gives:

$$A^2 + B^2 = h^2 = \omega^2\mu_o\varepsilon_o + \gamma^2 = \left[\frac{m\pi}{a}\right]^2 + \left[\frac{n\pi}{b}\right]^2.$$

Or

$$\gamma^2 = \left[\frac{m\pi}{a}\right]^2 + \left[\frac{n\pi}{b}\right]^2 - \omega^2\mu_o\varepsilon_o.$$

Solving for γ we can come up with an old friend:

$$\gamma = \sqrt{\left[\frac{m\pi}{a}\right]^2 + \left[\frac{n\pi}{b}\right]^2 - \omega^2\mu_o\varepsilon_o} = \alpha + j\beta. \tag{12-76}$$

By studying equation (12-76), we arrive at some important results since the terms under the square root sign can either be positive or negative. γ is a real number if $\omega^2\mu_o\varepsilon_o < \left[\frac{m\pi}{a}\right]^2 + \left[\frac{n\pi}{b}\right]^2$. Under such conditions we would have only attenuation and no propagation, as we can see by looking at the situation in more detail. Under such a condition β would be zero and we would have only the α term.

$C_6e^{-\gamma z} = C_6e^{-\alpha z}$ and our function of time would have a form as follows:

$$E_z(t) = \mathcal{R}\left[K \sin \frac{m\pi}{a} x \sin \frac{n\pi}{b} y\, e^{-\alpha z}e^{j\omega t}\right] =$$

$$K \sin \frac{m\pi}{a} x \sin \frac{n\pi}{b} y\, e^{-\alpha z}\cos \omega t. \tag{12-77}$$

From our previous work we see that we do not have a traveling wave equation for $E_z(t)$. However, if we assume the opposite condition:

$$\omega^2\mu_o\varepsilon_o > \left[\frac{m\pi}{a}\right]^2 + \left[\frac{n\pi}{b}\right]^2, \tag{12-78}$$

the quantity under the square root sign is negative and we end up with γ being an imaginary number.

$$\gamma = j\sqrt{\omega^2\mu_o\varepsilon_o - \left[\frac{m\pi}{a}\right]^2 - \left[\frac{n\pi}{b}\right]^2} = 0 + j\beta, \tag{12-79}$$

where

$$\beta = \sqrt{\omega^2\mu_o\varepsilon_o - \left[\frac{m\pi}{a}\right]^2 - \left[\frac{n\pi}{b}\right]^2}. \tag{12-80}$$

Now we obtain propagation since we would have the following results: $C_6e^{-\gamma z} = C_6e^{-j\beta z}$ and our function of time would have a form as follows:

$$E_z(t) = \mathcal{R}\left[K \sin \frac{m\pi}{a} \times \sin \frac{n\pi}{b} y\, e^{-j\beta z}e^{j\omega t}\right] =$$

$$K \sin \frac{m\pi}{a} \times \sin \frac{n\pi}{b} y \cos (\omega t - \beta z). \tag{12-81}$$

Now we have the expression for a traveling wave moving in the positive z direction.

The separation between propagation and attenuation occurs when:

$$\omega^2\mu_o\varepsilon_o = \left[\frac{m\pi}{a}\right]^2 + \left[\frac{n\pi}{b}\right]^2. \qquad (12\text{-}82)$$

The frequency at which this occurs is called the cutoff frequency, ω_c.

$$\omega_c{}^2\mu_o\varepsilon_o = \left[\frac{m\pi}{a}\right]^2 + \left[\frac{n\pi}{b}\right]^2,$$

$$\omega_c = \frac{1}{\sqrt{\mu_o\varepsilon_o}}\sqrt{\left[\frac{m\pi}{a}\right]^2 + \left[\frac{n\pi}{b}\right]^2} = 2\pi f_c. \qquad (12\text{-}83)$$

$$f_c = \frac{1}{2\pi\sqrt{\mu_o\varepsilon_o}}\sqrt{\left[\frac{m\pi}{a}\right]^2 + \left[\frac{n\pi}{b}\right]^2}. \qquad (12\text{-}84)$$

Notice that we have been assuming that $\mu = \mu_o$ and $\varepsilon = \varepsilon_o$; if this were not true we would have equation (12-85).

$$f_c = \frac{1}{2\pi\sqrt{\mu\varepsilon}}\sqrt{\left[\frac{m\pi}{a}\right]^2 + \left[\frac{n\pi}{b}\right]^2} = \frac{1}{2\sqrt{\mu\varepsilon}}\sqrt{\left[\frac{m}{a}\right]^2 + \left[\frac{n}{b}\right]^2}. \qquad (12\text{-}85)$$

However, under the assumed conditions of an air dielectric, we have:

$$\frac{1}{\sqrt{\mu_o\varepsilon_o}} = \frac{1}{\sqrt{(4\pi \times 10^{-7})\dfrac{10^{-9}}{36\pi}}} = 3 \times 10^8 \text{ meters/second} = c \text{ (velocity of light)}.$$

So equation (12-84) becomes:

$$f_c = \frac{c}{2}\sqrt{\left[\frac{m}{a}\right]^2 + \left[\frac{n}{b}\right]^2} = \frac{c}{\lambda_c}. \qquad (12\text{-}86)$$

λ_c is defined as the cut-off wavelength.

$$\lambda_c = \frac{2}{\sqrt{\left[\frac{m}{a}\right]^2 + \left[\frac{n}{b}\right]^2}}. \qquad (12\text{-}87)$$

$$f_c^{\,2} = \frac{1}{4\pi^2\mu\varepsilon}\left[\left[\frac{m\pi}{a}\right]^2 + \left[\frac{n\pi}{b}\right]^2\right], \text{ so } \left[\frac{m\pi}{a}\right]^2 + \left[\frac{n\pi}{b}\right]^2 = f_c^{\,2}4\pi^2\mu\varepsilon,$$

and

$$\beta = \sqrt{\omega^2\mu\varepsilon - \left[\frac{m\pi}{a}\right]^2 - \left[\frac{n\pi}{b}\right]^2} = \sqrt{\omega^2\mu\varepsilon - 4\pi^2 f_c^{\,2}\mu\varepsilon} =$$

$$\sqrt{\omega^2\mu\varepsilon - \frac{(2\pi f)(2\pi f)\ \mu\varepsilon f_c^{\,2}}{f^2}} = \sqrt{\omega^2\mu\varepsilon - \omega^2\mu\varepsilon\frac{f_c^{\,2}}{f^2}} =$$

$$\omega\sqrt{\mu\varepsilon}\sqrt{1 - \left(\frac{f_c}{f}\right)^2} = \beta. \qquad (12\text{-}88)$$

Again, if $\mu = \mu_o$ and $\varepsilon = \varepsilon_o$,

$$\beta = \omega\sqrt{\mu_o\varepsilon_o}\sqrt{1 - \left[\frac{f_c}{f}\right]^2} = \frac{2\pi f}{c}\sqrt{1 - \left[\frac{f_c}{f}\right]^2}. \qquad (12\text{-}89)$$

In free space, $c = f\lambda$, so $\frac{f}{c} = \frac{1}{\lambda}$ and

$$\beta = \frac{2\pi}{\lambda}\sqrt{1 - \left[\frac{f_c}{f}\right]^2}. \qquad (12\text{-}90)$$

In equation (12-90), λ is the free space wavelength.

Using the same definition as we did earlier in the text:

$$v_p = \frac{\omega}{\beta} = \frac{\omega}{\sqrt{\omega^2\mu\varepsilon - \left[\frac{m\pi}{a}\right]^2 - \left[\frac{n\pi}{b}\right]^2}} = \qquad (12\text{-}91)$$

$$\frac{\omega}{\omega\sqrt{\mu\varepsilon}\sqrt{1 - \left[\frac{f_c}{f}\right]^2}} = \frac{1}{\sqrt{\mu\varepsilon}\sqrt{1 - \left[\frac{f_c}{f}\right]^2}} = v_p. \qquad (12\text{-}92)$$

Again, for free space conditions:

$$v_p = \frac{c}{\sqrt{1 - \left[\frac{f_c}{f}\right]^2}}. \tag{12-93}$$

If we look at this equation, we see that when $f = f_c$, or the frequency becomes equal to the critical frequency, that the phase velocity, v_p, becomes infinite. If f is greater than f_c, v_p is greater than c. This looks impossible; having something moving at a velocity greater than the speed of light doesn't make sense. As it turns out, information or energy cannot go faster than the speed of light, so for the moment we will define a new quantity which we will not try to interpret physically and which we will call the group velocity, v_g.

$$v_g = \frac{1}{\frac{d\beta}{d\omega}} = \frac{1}{\frac{d}{d\omega}\left[\omega^2\mu\varepsilon - \left[\frac{m\pi}{a}\right]^2 - \left[\frac{n\pi}{b}\right]^2\right]^{1/2}} =$$

$$\frac{1}{\frac{1}{2}\left[\omega^2\mu\varepsilon - \left[\frac{m\pi}{a}\right]^2 - \left[\frac{n\pi}{b}\right]^2\right]^{-1/2} 2\omega\mu\varepsilon} = \frac{\sqrt{\omega^2\mu\varepsilon - \left[\frac{m\pi}{a}\right]^2 - \left[\frac{n\pi}{b}\right]^2}}{\omega\mu\varepsilon} =$$

$$\frac{1}{v_p\mu\varepsilon} = v_g. \tag{12-94}$$

And again, when $\mu = \mu_o$ and $\varepsilon = \varepsilon_o$, we have:

$$v_g = \frac{1}{v_p\,\mu_o\,\varepsilon_o} = \frac{c^2}{v_p} \text{ or } v_g v_p = c^2. \tag{12-95}$$

The v_g will be less than the speed of light when v_p is greater than the speed of light or $v_p > c$ for all $f > f_c$.

Let us define the wavelength in the guide as:

$$\lambda_g = \frac{v_p}{f}. \tag{12-96}$$

So,

$$\lambda_g = \frac{v_p}{f} = \frac{\omega}{\beta f} = \frac{2\pi f}{\beta f} = \frac{2\pi}{\beta}. \tag{12-97}$$

This we recognize as the same expression as we had before for our transmission lines.

$$\lambda_g = \frac{2\pi}{\sqrt{\omega^2\mu\varepsilon - \left[\frac{m\pi}{a}\right]^2 - \left[\frac{n\pi}{b}\right]^2}} = \frac{2\pi}{\omega\sqrt{\mu\varepsilon}\sqrt{1 - \left[\frac{f_c}{f}\right]^2}}. \tag{12-98}$$

If $\mu = \mu_o$ and $\varepsilon = \varepsilon_o$,

$$\lambda_g = \frac{2\pi c}{2\pi f\sqrt{1 - \left[\frac{f_c}{f}\right]^2}} = \frac{c}{f\sqrt{1 - \left[\frac{f_c}{f}\right]^2}} = \frac{\lambda}{\sqrt{1 - \left[\frac{f_c}{f}\right]^2}}. \tag{12-99}$$

In equation (12-99), λ is the free space value of the wavelength. We notice that λ_g is greater than λ for all frequencies greater than the critical frequency, or $\lambda_g > \lambda$ for all $f > f_c$.

The wave guide presents some new physical concepts for us to think about. Above the critical frequency, the wavelength in the guide is larger than the wavelength in free space and the phase velocity, v_p, is also greater. At the critical frequency the wavelength in the guide becomes infinite as does the phase velocity.

$$\lambda_g\sqrt{1 - \left[\frac{f_c}{f}\right]^2} = \lambda. \qquad \lambda_g^2\left[1 - \left[\frac{f_c}{f}\right]^2\right] = \lambda^2.$$

$$\frac{1}{\lambda^2}\left[1 - \left[\frac{f_c}{f}\right]^2\right] = \frac{1}{\lambda_g^2}.$$

But

$$\lambda_c = \frac{c}{f_c}. \tag{12-100}$$

Also, we have $f_c = \dfrac{c}{\lambda_c}$, and $f = \dfrac{c}{\lambda}$, so

$$\frac{1}{\lambda^2}\left[1 - \frac{c^2}{\lambda_c^{\,2}}\frac{\lambda^2}{c^2}\right] = \frac{1}{\lambda_g^{\,2}}, \text{ or } \frac{1}{\lambda^2} - \frac{1}{\lambda_c^{\,2}} = \frac{1}{\lambda_g^{\,2}}.$$

$$\frac{1}{\lambda^2} = \frac{1}{\lambda_g^{\,2}} + \frac{1}{\lambda_c^{\,2}}. \qquad (12\text{-}101)$$

This gives us a relationship between the wavelength in free space, λ, in the guide, λ_g, and the cut-off wavelength, λ_c.

Under the assumed conditions the phasor value of E_z becomes from equation (12-75):

$$E_z = K \sin\frac{m\pi x}{a} \sin\frac{n\pi y}{b}\, e^{-j\beta z}, \qquad (12\text{-}102)$$

with $m = 1, 2, 3, \ldots$ and $n = 1, 2, 3, \ldots$

Now it is apparent that E_z will be different for different values of m and n. It is common practice to designate the type of *TM* wave in terms of m and n, by indicating their values and calling it the *mn* mode of operation.

$$TM_{mn} = \text{mode of operation.}$$

Since neither m nor n can be zero, we have no TM_{00} mode and the lowest mode of operation is the TM_{11} mode.

As functions of time we could have for the forward wave the following values:

$$E_z(t) = \mathcal{R}\,[(K \sin Ax \sin By\, e^{-j\beta z} e^{j\omega t})] =$$

$$K \sin Ax \sin By \cos(\omega t - \beta z). \qquad (12\text{-}103)$$

$$E_x = \frac{-\gamma}{h^2}\frac{\partial E_z}{\partial x}. \qquad \gamma = j\beta.$$

$$E_x = \frac{-j\beta}{h^2} KA \cos Ax \sin By\, e^{-j\beta z}.$$

$$E_x(t) = \mathcal{R}\left[\frac{-j\beta\, KA}{h^2} \cos Ax \sin By\, e^{-j\beta z} e^{j\omega t}\right] =$$

$$\mathcal{R}\left[\frac{-j\beta\, KA}{h^2} \cos Ax \sin By\left(\cos(\omega t - \beta z) + j \sin(\omega t - \beta z)\right)\right] =$$

$$\frac{\beta\, KA}{h^2} \cos Ax \sin By \sin(\omega t - \beta z) = E_x(t). \tag{12-104}$$

Repeating the process we obtain:

$$E_y(t) = \frac{\beta KB}{h^2} \sin Ax \cos By \sin(\omega t - \beta z). \tag{12-105}$$

$$H_x(t) = -\frac{\omega\varepsilon\, KB}{h^2} \sin Ax \cos By \sin(\omega t - \beta z). \tag{12-106}$$

$$H_y(t) = \frac{\omega\varepsilon\, KA}{h^2} \cos Ax \sin By \sin(\omega t - \beta z). \tag{12-107}$$

Since $A = \frac{m\pi}{a}$, $m = 1, 2, 3, \ldots$ and $B = \frac{n\pi}{b}$, $n = 1, 2, 3, \ldots$ and

$$h^2 = A^2 + B^2 = \left[\frac{m\pi}{a}\right]^2 + \left[\frac{n\pi}{b}\right]^2,$$

we can write:

$$E_z(t) = K \sin \frac{m\pi x}{a} \sin \frac{n\pi y}{b} \cos(\omega t - \beta z). \tag{12-108}$$

$$E_x(t) = \frac{\beta\, K}{\left[\left[\frac{m\pi}{a}\right]^2 + \left[\frac{n\pi}{b}\right]^2\right]} \frac{m\pi}{a} \cos \frac{m\pi x}{a} \sin \frac{n\pi y}{b} \sin(\omega t - \beta z). \tag{12-109}$$

$$E_y(t) = \frac{\beta\, K}{\left[\left[\frac{m\pi}{a}\right]^2 + \left[\frac{n\pi}{b}\right]^2\right]} \frac{n\pi}{b} \sin \frac{m\pi x}{a} \cos \frac{n\pi y}{b} \sin(\omega t - \beta z). \tag{12-110}$$

$$H_x(t) = \frac{-\omega\varepsilon K}{\left[\left[\frac{m\pi}{a}\right]^2 + \left[\frac{n\pi}{b}\right]^2\right]} \frac{n\pi}{b} \sin\frac{m\pi x}{a} \cos\frac{n\pi y}{b} \sin(\omega t - \beta z). \tag{12-111}$$

$$H_y(t) = \frac{\omega\varepsilon K}{\left[\frac{m\pi}{a}\right]^2 + \left[\frac{n\pi}{b}\right]^2} \frac{m\pi}{a} \cos\frac{m\pi x}{a} \sin\frac{n\pi y}{b} \sin(\omega t - \beta z). \tag{12-112}$$

$$H_z(t) = 0. \tag{12-113}$$

As you might suspect, by now, visualizing the fields would be very difficult. Sarbacher and Edson published a text in 1947 that has an excellent treatment of wave guides and should be consulted by anyone who wishes a more detailed study. For the different modes of operation all we have to do is insert the proper values of m and n in the above equations.

12.4. The *TE* Mode of Operation.

The *TE* mode of operation is one that is often used in practice, particularly the TE_{10} mode. We recall that for the *TE* mode, E_z is zero and H_z is not zero. If we go back to our equations we have:

(12-64) $$E_x = \frac{-\gamma}{h^2}\frac{\partial E_z}{\partial x} - \frac{j\omega\mu_o}{h^2}\frac{\partial H_z}{\partial y}.$$

(12-66) $$E_y = \frac{-\gamma}{h^2}\frac{\partial E_z}{\partial y} + \frac{j\omega\mu_o}{h^2}\frac{\partial H_z}{\partial x}.$$

(12-67) $$H_x = \frac{-\gamma}{h^2}\frac{\partial H_z}{\partial x} + \frac{j\omega\varepsilon_o}{h^2}\frac{\partial E_z}{\partial y}.$$

(12-65) $$H_y = \frac{-\gamma}{h^2}\frac{\partial H_z}{\partial y} - \frac{j\omega\varepsilon_o}{h^2}\frac{\partial E_z}{\partial x}.$$

With $E_z = 0$, these equations become:

$$E_x = \frac{-j\omega\mu_o}{h^2}\frac{\partial H_z}{\partial y}. \tag{12-114}$$

$$E_y = \frac{j\omega\mu_o}{h^2}\frac{\partial H_z}{\partial x}. \qquad (12\text{-}115)$$

$$H_x = \frac{-\gamma}{h^2}\frac{\partial H_z}{\partial x}. \qquad (12\text{-}116)$$

$$H_y = \frac{-\gamma}{h^2}\frac{\partial H_z}{\partial y}. \qquad (12\text{-}117)$$

We have already found the form of the solution for H_z when we wrote the equation for the forward wave in (12-45).

$H_z = (C_1\sin Ax + C_2\cos Ax)(C_3\sin By + C_4\cos By)C_6e^{-\gamma z}$, so using (12-114) gives:

$$E_x = \frac{-j\omega\mu_o}{h^2}(C_1\sin Ax + C_2\cos Ax)(BC_3\cos By - BC_4\sin By)C_6e^{-\gamma z}.$$

But again we have the boundary conditions that $E_{tan} = 0$ at $y = 0 = E_x = 0$.

Therefore, C_3 must equal 0. Also $E_x = 0$ at $y = b$.

$$E_x = 0 = \frac{-j\omega\mu_o}{h^2}(C_1\sin Ax + C_2\cos A_x)(-BC_4\sin Bb)C_6e^{-\gamma z}.$$

We find the same conditions that we had for the TM mode; namely, that $\sin Bb = 0$, when $Bb = n\pi$, where $n = 0, 1, 2, \ldots$, or $B = \frac{n\pi}{b}$, which is the same equation (12-74) that we obtained for the TM mode. Now we have:

$$E_y = \frac{j\omega\mu_o}{h^2}(AC_1\cos Ax - AC_2\sin A_x)(C_4\cos By)C_6e^{-\gamma z}.$$

But $E_{tan} = 0 = E_y$ at $x = 0$. Therefore, C_1 has to be 0. Also, $E_y = 0$ at $x = a$. So,

$$E_y = \frac{j\omega\mu_o}{h^2}(-AC_2\sin Aa)(C_4\cos By)C_6e^{-\gamma z}.$$

As before, sin $Aa = 0$ when $Aa = m\pi$, $m = 0, 1, 2, \ldots,$ or $A = \frac{m\pi}{a}$, which is the same as equation (12-73) for the TM mode. So we now have the following results.

$$H_z = C_2 \cos Ax\, C_4 \cos By\, C_6 e^{-\gamma z},$$

or

$$H_z = K' \cos Ax \cos By\, e^{-\gamma z}. \quad (12\text{-}118)$$

$$H_z = K' \cos \frac{m\pi x}{a} \cos \frac{n\pi y}{b} e^{-\gamma z}. \quad (12\text{-}119)$$

$$E_x = \frac{-j\omega\mu_o}{h^2}\frac{\partial H_z}{\partial y} = \frac{j\omega\mu_o}{h^2} K' B\cos Ax \sin By\, e^{-\gamma z}. \quad (12\text{-}120)$$

$$E_y = \frac{j\omega\mu_o}{h^2}\frac{\partial H_z}{\partial x} = -j\omega\mu_o K' A\sin Ax \cos By\, e^{-\gamma z}. \quad (12\text{-}121)$$

$$H_x = \frac{-\gamma}{h^2}\frac{\partial H_z}{\partial x} = \frac{\gamma}{h^2} K' A \sin Ax \cos By\, e^{-\gamma z}. \quad (12\text{-}122)$$

$$H_y = \frac{-\gamma}{h^2}\frac{\partial H_z}{\partial y} = \frac{\gamma}{h^2} K' B \cos Ax \sin By\, e^{-\gamma z}. \quad (12\text{-}123)$$

We have propagation when $\gamma = j\beta$, and from the previous work we have $A = \frac{m\pi}{a}$, with $m = 0, 1, 2, 3$, etc., and $B = \frac{n\pi}{b}$, with $n = 0, 1, 2, 3$, etc.

As we did for the TM mode we can now substitute and obtain:

$$H_z = K' \cos \frac{m\pi x}{a} \cos \frac{n\pi y}{b} e^{-j\beta z}. \quad (12\text{-}124)$$

$$H_x = \frac{j\beta K'}{h^2}\frac{m\pi}{a} \sin \frac{m\pi x}{a} \cos \frac{n\pi y}{b} e^{-j\beta z}. \quad (12\text{-}125)$$

$$H_y = \frac{j\beta K'}{h^2}\frac{n\pi}{b} \cos \frac{m\pi x}{a} \sin \frac{n\pi y}{b} e^{-j\beta z}. \quad (12\text{-}126)$$

$$E_x = \frac{j\omega\mu_o K'}{h^2}\frac{n\pi}{b}\cos\frac{m\pi x}{a}\sin\frac{n\pi y}{b}e^{-j\beta z}. \qquad (12\text{-}127)$$

$$E_y = \frac{-j\omega\mu_o K'}{h^2}\frac{m\pi}{a}\sin\frac{m\pi x}{a}\cos\frac{n\pi y}{b}e^{-j\beta z}. \qquad (12\text{-}128)$$

$$E_z = 0. \qquad (12\text{-}129)$$

By looking at these equations, we can see that both m and n cannot be zero at the same time, since we would end up with only an H_z term and all the rest would be zero. In other words, we cannot have a TE_{00} mode. But we can have a TE_{01} or a TE_{10} mode of operation.

Let us look at the TE_{10} mode of operation, since it is an easy mode to produce. With $n = 0$ and $m = 1$, we have:

$$H_z = K'\cos\frac{\pi x}{a}e^{-j\beta z}. \qquad (12\text{-}130)$$

$$H_x = \frac{j\beta}{h^2}\frac{\pi}{a}\sin\frac{x}{a}e^{-j\beta z}. \qquad (12\text{-}131)$$

$$H_y = 0. \qquad (12\text{-}132)$$

$$E_x = 0. \qquad (12\text{-}133)$$

$$E_y = \frac{-j\omega\mu_o}{h^2}K'\frac{\pi}{a}\sin\frac{x}{a}e^{-j\beta z}. \qquad (12\text{-}134)$$

$$E_z = 0. \qquad (12\text{-}135)$$

$$h^2 = \left[\frac{\pi}{a}\right]^2. \qquad (12\text{-}136)$$

We can still use our previous equations that had nothing to do with whether the wave was in a TM or a TE mode of operation. So for the TE_{10} mode we have:

$$\lambda_c = \frac{2}{\sqrt{\left[\frac{m}{a}\right]^2 + \left[\frac{n}{b}\right]^2}} = \frac{2}{\sqrt{\left[\frac{1}{a}\right]^2}} = \frac{2}{\frac{1}{a}} = 2a. \qquad (12\text{-}137)$$

The dimension a for our rectangular wave guide turns out to be one half of the cutoff wavelength.

$$\beta = \sqrt{\omega^2\mu_o\varepsilon_o - \left[\frac{m\pi}{a}\right]^2 - \left[\frac{n\pi}{b}\right]^2} = \sqrt{\omega^2\mu_o\varepsilon_o - \frac{\pi^2}{a^2}}. \quad (12\text{-}138)$$

$$(12\text{-}90) \qquad \beta = \frac{2\pi}{\lambda}\sqrt{1 - \left[\frac{f_c}{f}\right]^2}.$$

$$f_c = \frac{c}{2}\sqrt{\left[\frac{m}{a}\right]^2 + \left[\frac{n}{b}\right]^2} = \frac{c}{2}\sqrt{\left(\frac{1}{a}\right)^2} = \frac{c}{2a}. \quad (12\text{-}139)$$

$$v_p = \frac{\omega}{\beta} = \frac{\omega}{\sqrt{\omega^2\mu_o\varepsilon_o - \frac{\pi^2}{a^2}}}. \quad (12\text{-}140)$$

$$(12\text{-}93) \qquad v_p = \frac{c}{\sqrt{1 - \left(\frac{f_c}{f}\right)^2}}.$$

$$(12\text{-}99) \qquad \lambda_g = \frac{\lambda}{\sqrt{1 - \left[\frac{f_c}{f}\right]^2}}.$$

Table 12-1 lists some important characteristics and dimensions of different wave guides. Let us look at the X-band since this type of wave guide is often available from government surplus and therefore available to university laboratories. We will assume that it is operated in the TE_{10} mode.

Table 12-1. Wave Guide Specifications.

Common Band Letter	FREQUENCY RANGE (KMC)	OUTER DIMENSIONS INCHES	INNER DIMENSIONS INCHES	CUT OFF FREQ.* TE_{10}	PRACTICAL CW POWER RATING **	WAVEGUIDE WAVELENGTH INCHES ***
P	.32 - .49	23.37x11.87	23.0x11.5	.256	152.0-218.4	61.84-28.29
N	.39 - .54	21.37x10.87	21.0x10.5	.281	126.3-180.6	56.70-26.28
J	.41 - .64	18.25x9.25	18.0x9.0	.328	93.4-131.9	48.13-22.67
H	.49 - .75	15.25x7.75	15.0x7.5	.393	67.6-93.3	40.46-18.50
F	.64 - .96	11.75x6.00	11.5x5.75	.513	35.0-53.8	30.89-14.55
E	.75 - 1.15	10.00x5.125	9.75x4.875	.605	27.0-38.5	26.77-12.51
D	.95 - 1.5	7.95x4.10	7.70x3.85	.766	17.2-24.1	20.39-7.617
L	1.12 - 1.7	6.66x3.41	6.50x3.25	.908	11.9-17.2	17.93-8.593
M	1.7 - 2.6	4.46x2.31	4.30x2.15	1.375	5.2-7.5	11.81-5.361
S	2.6 - 3.95	3.00x1.50	2.84x1.34	2.080	2.2-3.2	7.54-3.51
C	3.95-5.85	2.00x1.00	1.87x0.87	3.155	0.94-1.32	4.956-2.395
A	5.85-8.2	1.50x0.75	1.37x0.62	4.285	0.56-0.71	2.975-1.691
B	7.05-10.0	1.25x0.625	1.12x0.49	5.260	0.35-0.46	2.514-1.388
X	8.2 -12.4	1.00x0.50	.90x0.40	6.560	0.20-0.29	2.397-1.121
G	12.4 -18.0	.702x.391	.622x.311	9.490	0.12-0.16	1.478-0.771
K	18.0 -26.5	.500x.250	.420x.170	14.080	0.043-0.058	1.055-0.525
T	26.5-40.0	.360x.220	.280x.140	21.100	0.022-0.031	0.734-0.347
V	33.0 -50.0	.304x.192	.224x.112	26.350	0.014-0.020	0.595-0.278
W	50.0 -75.0	.228x.154	.148x0.074	39.900	0.0063-0.0090	0.392-0.186
Z	60.0- 90.0	.202x.141	.122x0.061	48.400	0.0042-0.0060	0.334-0.156

* In KMC ** Lowest to Highest Frequency Megawatts *** Lowest to Highest Freq.

$a = 0.9$ inches $= 2.286$ centimeters $= 0.02286$ meters.

$\lambda_c = 2a = 2(0.02286) = 0.0457$ meters $= 4.57$ cm.

$f_c = \dfrac{c}{2a} = \dfrac{3 \times 10^8}{0.0457} = 6.5619 \times 10^9$ hertz $= 6{,}561.0$ megahertz $=$ 6.561 gigahertz.

In order to propagate a wave down the guide, the frequency must be higher than 6.561 gigahertz. Suppose we assume that we are operating at a frequency of 8 gigahertz (8×10^9 hertz). The wavelength in free space would be

$$\lambda = \frac{c}{f} = \frac{3 \times 10^8}{8 \times 10^9} = 37.5 \times 10^{-3} \text{ meters} = 3.75 \text{ cm.}$$

$$\beta = \frac{2\pi}{\lambda}\sqrt{1 - \left[\frac{f_c}{f}\right]^2} = \frac{2\pi}{37.5 \times 10^{-3}}\sqrt{1 - \left[\frac{6.561}{8}\right]^2} = 95.85 \text{ radians/meter.}$$

$$v_p = \frac{\omega}{\beta} = \frac{2(8 \times 10^9)}{95.85} = 5.24 \times 10^8 \text{ meters/second,}$$

which is larger than c.

$$v_g = \frac{c^2}{v_p} = \frac{(3 \times 10^8)^2}{5.24 \times 10^8} = 1.71 \times 10^8 \text{ meters/second.}$$

The wavelength in the guide is:

$$\lambda_g = \frac{\lambda}{\sqrt{1 - \left[\frac{f_c}{f}\right]^2}} = \frac{3.75 \text{ cm}}{\sqrt{1 - \left[\frac{6.56}{8}\right]^2}} = 6.55 \text{ cm.}$$

We can also use equation (12-101).

$$\frac{1}{\lambda^2} = \frac{1}{\lambda_g^2} + \frac{1}{\lambda_c^2}. \qquad \frac{1}{(3.75)^2} = \frac{1}{(\lambda_g)^2} + \frac{1}{(4.571)^2}.$$

Solving gives $\lambda_g = 6.55$ m.

If we repeat our calculations for a lower frequency, such as 7.2 gigahertz, and compare our new results we would have the following:

X-Band, TE_{10} Mode

Freq. gHz	Cutoff Freq. gHz	Cutoff Wavelength	λ	β	v_p	v_g	λ_g
8.0	6.56	4.57 cm	3.75 cm	95.85	1.75c	0.57c	6.55 cm
7.2	6.56	4.57 cm	4.16 cm	62.07	2.43c	0.41c	10.12 cm

As the frequency is reduced towards the cutoff frequency the wavelength in the guide increases, v_p increases, v_g decreases, and the ratio of the wavelength in the guide to the wavelength in free space increases.

If we calculate the cutoff frequency for several different *TE* modes we would have the following results. TE_{10}, 6,561.6 mHz; TE_{20}, 13,123.3 mHz; TE_{01}, 14,763.7 mHz; TE_{11}, 16,156.2 mHz. The mode with the lowest cutoff frequency is called the principal mode of operation.

The method of exciting the wave guide into operating in a particular mode of operation is to place a small antenna in the guide. The locations and configurations of the antennas will determine the mode of operation. Figure 12-3 shows a couple of different schemes.

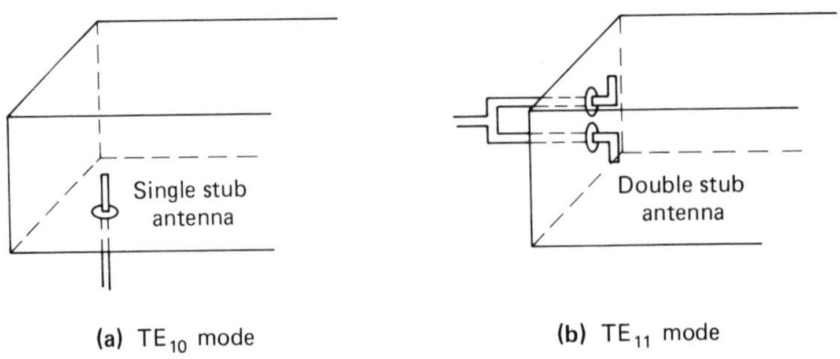

Figure 12-3. Method of exciting wave guides into different modes of operation.

12.5 Group and Phase Velocity

We have defined phase velocity in terms of the velocity at which a constant value of a sine wave moves down the line. For a transmission line we were dealing with voltages and currents but for a wave guide we have been discussing the *E* and *H* components of an electromagnetic wave traveling down the guide. We defined v_g as the group velocity and indicated what it was strictly by equation (12-94).

(12-94) $$v_g = \frac{1}{\frac{d\beta}{d\omega}}.$$

We have previously discussed some special cases on transmission lines where the phase velocity, v_p, is independent of frequency. For example, in the lossless case we had:

$$\beta = \omega\sqrt{LC}, \text{ so } v_p = \frac{\omega}{\beta} = \frac{\omega}{\omega\sqrt{LC}} = \frac{1}{\sqrt{LC}}.$$

The distortionless line also had a phase velocity that was independent of frequency. Under the condition that this is the case, we describe the line by saying it is nondispersive. When we describe waves traveling through some medium we talk about the medium as being dispersive or nondispersive. In a nondispersive medium the phase velocity has a maximum value of the speed of light but in a dispersive medium it is possible that v_p is greater than the speed of light. It turns out that the answer is that energy or, in the case of electrical engineering situations, signals or intelligence, cannot be propagated at a velocity exceeding the velocity of light so that the phase velocity cannot be representing the propagation of energy or intelligence in a dispersive medium.

The term *group velocity* arises when we are discussing wave propagation in a medium where waves of different frequencies travel at different phase velocities, that is, a dispersive medium. Assuming that our signal has different frequency components and they travel at different phase velocities; this gives the result that, as the wave travels, the signal shape changes and after some travel in space the signal will not be the same as the one that was transmitted. Dispersion takes place. Since most students are familiar with modulation let us take for our example an amplitude-modulated wave as indicated by equation (12-141).

$$v = V_m(1 + m\cos 2\pi f_\ell t)\cos 2\pi f_h t. \tag{12-141}$$

The m = modulation index or factor; f_ℓ = the modulating frequency which is the low frequency; f_h = the carrier or high frequency; m varies from 1 to 0 and $f_h \gg f_\ell$; V_m stands for the maximum value of the voltage.

$$\text{Let } \delta = \frac{f_\ell}{f_h} = \frac{2\pi f_\ell}{2\pi f_h} = \frac{\omega_\ell}{\omega_h}. \tag{12-142}$$

It is assumed that δ is a very small number, an incremental value.

$$v = V_m(1 + m\cos\omega_\ell t)\cos\omega_h t = V_m(1 + m\cos\delta\omega_h t)\cos\omega_h t.$$

Now let us convert this into a traveling wave going in the positive z direction only. We have the terms, $(\omega_h t - \beta z)$ and $\delta(\omega_h t - \beta z)$.

$$v = V_m[1 + m\cos\delta(\omega_h t - \beta z)]\cos(\omega_h t - \beta z) =$$
$$V_m[1 + m\cos(\delta\omega_h t - \delta\beta z)]\cos(\omega_h t - \beta z). \tag{12-143}$$

Let $\omega = \omega_h$.

$$v = V_m[1 + m\cos(\delta\omega t - \delta\beta z)\cos(\omega t - \beta z)] =$$
$$V_m[\cos(\omega t - \beta z) + m\cos(\delta\omega t - \delta\beta z)\cos(\omega t - \beta z)].$$

Since $\cos x\cos y = \frac{1}{2}[\cos(x + y) + \cos(x - y)]$ we can write:

$$\cos(\omega t - \beta z)\cos(\delta\omega t - \delta\beta z) =$$
$$\frac{1}{2}[\cos(\omega t - \beta z + \delta\omega t - \delta\beta z) + \cos(\omega t - \beta z - \delta\omega t + \delta\beta z)].$$

Using this we can write:

$$v = V_m\left\{\cos(\omega t - \beta z) + \frac{m}{2}[\cos(\omega t - \beta z) + (\delta\omega t - \delta\beta z)] + \cos[(\omega t - \beta z) - (\delta\omega t - \delta\beta z)]\right\}.$$

$$v = \frac{V_m}{2}\{\cos(\omega t - \beta z) + m\cos[(\omega t - \beta z) + (\delta\omega t - \delta\beta z)]\} +$$

$$\frac{V_m}{2}\{\cos(\omega t - \beta z) + m\cos[(\omega t - \beta z) - (\delta\omega t - \delta\beta z)]\} =$$

$$\frac{V_m}{2}\{\cos(\omega t - \beta z) + m\cos[(\omega + \delta\omega)t - (\beta + \delta\beta)z]\} +$$

$$\frac{V_m}{2}\{\cos(\omega t - \beta z) + m\cos[(\omega - \delta\omega)t - (\beta - \delta\beta)z]\} = v =$$

$$\frac{V_m}{2}\{\cos(\omega t - \beta z) + m\cos[\omega(1 + \delta)t - \beta(1 + \delta)z]\} +$$

$$\frac{V_m}{2}\{\cos(\omega t - \beta z) + m\cos[\omega(1 - \delta)t - \beta(1 - \delta)z]\}. \qquad (12\text{-}144)$$

Let us assume that $m = 1$ (100% modulation).

$$v = \frac{V_m}{2}\{\cos(\omega t - \beta z) + \cos[\omega(1 + \delta)t - \beta(1 + \delta)z]\} +$$

$$\frac{V_m}{2}\{\cos(\omega t - \beta z) + \cos\,[\omega(1 - \delta)t - \beta(1 - \delta)z]\}.$$

We can use another trigonometric identity.

$$\cos A + \cos B = 2\cos\frac{(A - B)}{2}\cos\frac{(A + B)}{2}.$$

So we can write:

$$\cos(\omega t - \beta z) + \cos[\omega(1 + \delta)t - \beta(1 + \delta)z] =$$

$$2\cos\left[\frac{\omega t - \beta z - \omega t - \delta\omega t + \beta z + \delta\beta z}{2}\right]\cos\left[\frac{\omega t - \beta z + \omega t + \delta\omega t - \beta z - \delta\beta z}{2}\right] =$$

$$2\cos\left[\frac{-\delta\omega t + \delta\beta z}{2}\right]\cos\left[\frac{2\omega t + \delta\omega t - 2\beta z - \delta\beta z}{2}\right] =$$

$$2\cos\left[\frac{\delta\beta z - \omega\delta t}{2}\right]\cos\left[\left(\omega + \frac{\delta\omega}{2}\right)t - \left(\beta + \frac{\delta\beta}{2}\right)z\right].$$

In a similar manner we can obtain:

$$\cos(\omega t - \beta z) + \cos[\omega(1 - \delta)t + \beta(1 - \delta)z] =$$
$$2 \cos \frac{(\omega\delta t - \delta\beta z)}{2} \cos\left[\left(\omega - \frac{\delta\omega}{2}\right)t - \left(\beta - \frac{\delta\beta}{2}\right)z\right].$$

So finally we have:

$$v = \frac{V_m}{2} 2 \cos \frac{1}{2}(\delta\beta z - \omega\delta t)\cos\left[\left(\omega + \frac{\delta\omega}{2}\right)t - \left(\beta - \frac{\delta\beta}{2}\right)z\right] +$$
$$\frac{V_m}{2} 2 \cos \frac{1}{2}(\omega\delta z - \delta\beta z)\cos\left[\left(\omega - \frac{\omega\delta}{2}\right)t - \left(\beta - \frac{\delta\beta}{2}\right)z\right]. \quad (12\text{-}145)$$

Inspecting the equation carefully we notice that we have a traveling wave with the amplitude varying slowly according to the relation: $\cos \frac{1}{2} (\omega\delta t - \delta\beta z)$. We will have a constant point on the envelope when $\delta\omega t - \delta\beta z$ = a constant, or $\delta\omega dt - \delta\beta dz = 0$. Solving, we have: $\frac{\delta\omega}{\delta\beta} = \frac{dz}{dt} = \frac{1}{\frac{\delta\beta}{\delta\omega}}$ = velocity in the z direction of a constant point on the envelope. This is what we previously called the group velocity, v_g.

$$v_g = \frac{dz}{dt} = \frac{1}{\frac{\delta\beta}{\delta\omega}} = \frac{1}{\frac{\partial\beta}{\partial\omega}}. \quad (12\text{-}146)$$

Since

$$v_p = \frac{\omega}{\beta}, \; \omega = \beta v_p.$$

If v_p is independent of ω, $\partial\omega = v_p \, \partial\beta$, and we find that:

$$v_p = \frac{\partial\omega}{\partial\beta} = \frac{1}{\frac{\partial\beta}{\partial\omega}} = v_g.$$

In a nondispersive medium $v_p = v_g$.

12.6 Additional Comments About Wave Guides

Up to now the student will have noticed that no mention has been made of the return wave which was referred to in equation (12-44). Also we have not mentioned the term, characteristic impedance, Z_o, in connection with our treatment of the wave guide. Since the wave guide operation is really due to a field phenomenon with no direct reference to voltages and currents, the term, "characteristic impedance" has no meaning in the sense we have been using it. Since the guide will normally be energized with a system such as those indicated in Figure 12-3 by some sort of small-antenna type stub, the stub becomes a matching device connecting the source or load with the wave guide. In practice the device used to connect the wave guide to some other sort of device or transmission line is called a "coupler" and it may couple the wave guide to a 50-ohm coaxial cable connector so that if the load is 50 ohms the wave guide is matched to the load through the coupler. The coupler may couple a particular guide to a 100-ohm connector, or any other value depending on the design of the coupler. If, however, the coupler is connected to a load which does not match it, this will be reflected back into the guide and there will be the return wave and the resultant standing waves in the guide. However, since the guide itself does not possess the normal characteristic impedance, it will be the match at the output coupler which will determine if the guide has a standing wave in it or not.

Wave guides are prepared for measurement work in the laboratory by having slots cut into them in the same manner that coaxial cables have slots in them. By inserting the proper probe into the slots in either case it is possible to sample the fields and determine if there are standing waves or not. These devices are called "slotted" lines for both cases; that is, we can have a slotted wave guide or a slotted coaxial cable for measurement purposes. Even though we do not think of a characteristic impedance or even a reflection coefficient for the wave guide we still can measure and take steps to eliminate the standing waves in the guide. In the laboratory the high-frequency signal generators are usually modulated with a low-frequency square wave so that measurements might be made on the slotted wave guide.

12.7 Chapter Summary

1. The guide considered in this chapter has a rectangular cross-sectional area with a long dimension of *a* meters and a short dimension of *b* meters.

2. Maxwell's equations are used with the assumptions that within the wave guide there is zero volume charge density and zero conductivity for the air dielectric. This results in a number of partial differential equations which can be solved by assuming a product solution. The product solution indicates two traveling waves moving in opposite directions in the guide. When the forward component is considered and the boundary conditions are taken into account, equations are obtained for different modes of operation.
3. The two modes of operation are the TM_{mn} and the TE_{mn} where m and n are positive integers, 0, 1, 2, etc.
4. The principal mode of operation is the lowest mode possible. This turns out to be TM_{11} for the TM mode and the TE_{10} for the TE mode. The mode of operation for a particular size wave guide is determined by the method of excitation.
5. The basic equations for both the TM_{mn} and the TE_{mn} modes of operation are:

$$f_{c(\text{critical frequency})} = \frac{1}{2\sqrt{\mu\varepsilon}}\sqrt{\left[\frac{m}{a}\right]^2 + \left[\frac{n}{b}\right]^2}.$$

$$f_{c(\text{in air})} = \frac{c}{2}\sqrt{\left[\frac{m}{a}\right]^2 + \left[\frac{n}{b}\right]^2} = \frac{c}{\lambda_c}.$$

$$\lambda_{c(\text{cut-off wavelength})} = \frac{2}{\sqrt{\left[\frac{m}{a}\right]^2 + \left[\frac{n}{b}\right]^2}}.$$

$$\beta_{(\text{during propagation})} = \sqrt{\omega^2\mu\varepsilon - \left[\frac{m\pi}{a}\right]^2 - \left[\frac{n\pi}{b}\right]^2}.$$

$$\beta_{(\text{in air during propagation})} = \frac{2\pi f}{c}\sqrt{1 - \left[\frac{f_c}{f}\right]^2}.$$

In free space, $\lambda = \dfrac{c}{f}$.

$$\beta = \frac{2\pi}{\lambda}\sqrt{1 - \left[\frac{f_c}{f}\right]^2}.$$

$$\lambda_g = \frac{\lambda}{\sqrt{1 - \left[\frac{f_c}{f}\right]^2}},$$

or

$$\frac{1}{\lambda^2} = \frac{1}{\lambda_g^{\,2}} + \frac{1}{\lambda_c^{\,2}}.$$

$$v_p = \frac{1}{\sqrt{\mu\varepsilon}\sqrt{1 - \left[\frac{f_c}{f}\right]^2}}.$$

$$v_{p(\text{in air})} = \frac{c}{\sqrt{1 - \left[\frac{f_c}{f}\right]^2}}.$$

In air:

$$v_g = \frac{c^2}{v_p} = \frac{1}{\frac{\partial\beta}{\partial\omega}} = \frac{\lambda}{\sqrt{1 - \left[\frac{f_c}{f}\right]^2}}.$$

6. Wave guide operation is normally specified in terms of bands of operation such as *H*, *M*, *S*, *X*, etc.
7. In a wave guide the intelligence being transmitted in the wave travels at the group velocity, v_g. The phase velocity, v_p, which is associated with the wavelength in the guide, is larger than the velocity of light and $v_p v_g = c^2$ with an air dielectric in the guide.
8. As one approaches the critical frequency from a higher frequency, the phase velocity and the wavelength in the guide become larger and larger.

9. A dispersive medium is one where waves of different frequencies travel at different phase velocities.
10. When dealing with wave guides there is no use for the term *characteristic impedance*, as such. The matching of the wave guide with a load is determined by the type of load. If it is matched there will be no return wave and no standing wave. However if the load is not matched there will be a standing wave in the guide.

12.8 Review and Study Questions

1. Give Maxwell's equations in vector form. Give them in partial differential equation form. Change the differential equation form into the phasor form.
2. What conditions are assumed in this chapter for the dielectric in the guide? How would the results be changed by assuming a more general case?
3. What type of solution was used in this chapter for the differential equations? When we go to the phasor form of solutions what assumptions are we making?
4. What boundary conditions were assumed at the surfaces of the wave guide?
5. What is the critical frequency? How is it determined?
6. What is the difference between the TM_{11} mode of operation and the TE_{11} mode of operation?
7. Why is there no TM_{00} mode of operation?
8. How does the group velocity, the phase velocity and the wavelength in the guide vary as the frequency moves upward from the critical frequency?
9. How does a TE_{01} mode of operation compare with a TE_{10} mode?
10. What is a dispersive medium?
11. What is the group velocity? How is it related to the phase velocity?
12. How are wave guides terminated at each end?
13. What is the normal frequency range for an "L" band wave guide? How does this differ for an "X" band wave guide?
14. What are the relative advantages and disadvantages in comparing wave guides with coaxial cables?
15. What is the characteristic impedance of a wave guide?
16. A student needs an attenuator for his work at 300 megahertz and the only thing handy is an attenuator for an "X" band wave guide; he connects the wave guide attenuator up to coaxial-

wave guide couplers and tries to use it in his work. What is wrong with this approach?

17. Why would it be difficult to measure the distance between two voltage minimums in a slotted line wave guide just above the critical frequency?

Problems

12-1. Determine the cutoff wavelength and the cutoff frequency for the following wave guides operating in the TM_{11} mode.
(a) S Band.
(b) A Band.
(c) G Band.
(d) K Band.

12-2. An X-Band wave guide with inside dimensions of 0.9″ × 0.4″ is to be operated with an air dielectric in the TM_{11} mode at 1 gigahertz above its critical frequency. Determine:
(a) The cutoff frequency.
(b) The cutoff wavelength.
(c) The wavelength in the guide.
(d) The wavelength in free space.
(e) The phase velocity.
(f) The group velocity.

12-3. An X-Band wave guide has internal dimensions of 0.9″ × 0.4″. It is operating in the TE_{10} mode with an air dielectric at 10 gigahertz. Determine:
(a) The phase constant.
(b) The cutoff frequency.
(c) The cutoff wavelength.
(d) The wavelength in the guide.
(e) The phase velocity.
(f) The group velocity.
(g) The wavelength outside of the guide.

12-4. An S-Band wave guide is operating in the TE_{10} mode at 1 gigahertz above its critical frequency with an air dielectric. Determine:
(a) The phase constant.
(b) The cutoff wavelength.
(c) The cutoff frequency.
(d) The wavelength in the guide.
(e) The phase velocity.

(f) The group velocity.
(g) The wavelength in free space.

12-5. An X-Band wave guide, operating in the TE_{10} mode, is used for measurement work in the laboratory. The following measurements were obtained:

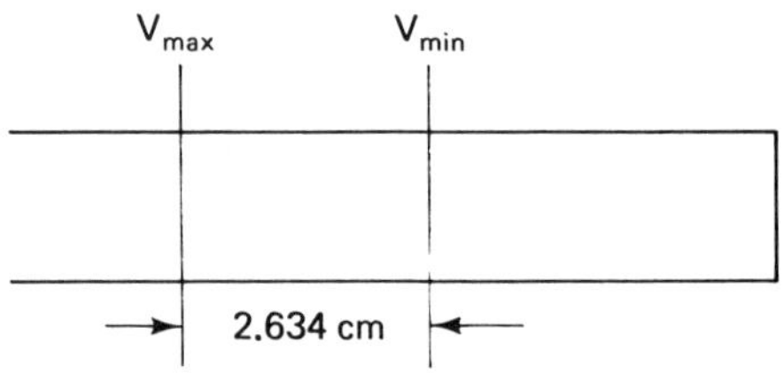

(a) Determine the frequency.
(b) Determine the wavelength in free space.

12-6. An X-Band wave guide has internal dimensions of 0.9″ × 0.4″. It is operating in the TE_{10} mode. The dielectric is air. Find the cutoff frequency of the principal mode and the maximum frequency that can be used to the exclusion of all other modes.

12-7. For each of the following wave guides operating in the TE_{10} mode, at a frequency of 1 gigahertz above the critical frequency, determine:
(a) The critical frequency.
(b) The critical wavelength.
(c) The phase velocity with an air dielectric: X-Band; S-Band; K-Band.

12-8. Determine the critical frequency for the P-Band wave guide operating in the TE_{10} mode.

12-9. An X-Band wave guide is operating with an air dielectric in the TE_{10} mode at a frequency of 100 kilohertz above the critical frequency. Determine:
(a) The wavelength in the guide.
(b) The wavelength in free space.
(c) The phase velocity.
(d) The group velocity.

12-10. Repeat problem 12-9 for a K-Band wave guide.

13

Measurements

13.1 Introduction

When working with transmission lines there are a number of items important to the engineer, both before and after construction. During the design process the engineer would like to know what the line parameters such as R, G, L, and C are going to be when the line is finished. Related items such as α, β, and Z_o are also of interest for many types of lines. For many situations the parameters may be determined from a knowledge of the materials used and the physical configuration of the line. In other cases the determination ahead of time may be difficult or impossible since factors such as temperature, humidity, air pressure, voltage level, wind, manufacturing defects and the effects of frequency as related to skin and proximity effects are often variable and unknown. It may be that the line parameters must be determined from measurement after the line has been constructed or that a sample section will have to be used prior to construction to determine the necessary information. After the line is in operation the items which may be of interest are the voltage, current, power, line losses, vars and the VSWR.

At 60 hertz, the measurements on the line after construction are made with the usual 60 hertz type instrumentation. No matter what level of voltages or currents are involved the engineer can, with potential and current transformers, use the regular a.c. type voltmeters, ammeters and wattmeters. In addition the power engineer will usually be interested in the reactive power so a kilovar meter often is adjacent to the kilowatt meter. With the power systems in this country so interconnected with each other, the accurate measurement of the energy going back and forth between power companies is an economic necessity. An important consideration for a power company will be the losses in its own system developed by the transmitting (wheeling) of power across its system to another power company. Since power frequency meters are well known and highly developed, the power engineer normally does not have a need for very many special types of instrumentation. Obviously there is no need for devices to measure the VSWR on the system but there may be a need for measuring the effects of transients on the system developed by faults, lightning strokes or other types of abnormal operation. Measurements may be dictated by other considerations also, such as the future use of the power line for communication between points on the line, for higher frequency uses to monitor the voltages, currents and power on the line at various points, and for load control of the customers. The superimposing of higher frequencies on the power frequencies makes the line serve dual functions and the engineer can consider each case individually since the frequencies are isolated from each other by circuitry that keeps the power frequencies out of the higher frequency circuits and vice versa.

As frequencies climb one has to abandon the common electromagnetic-type meters and go to electronic instrumentation to measure voltage, current and power, as well as other items of interest. The development of modern oscilloscopes that will operate up into the gigahertz range of frequencies has made them extremely valuable in measurement work. The measurement work in the range of frequencies from 60 hertz up towards the gigahertz range has been of considerable interest to engineers for many decades. The development of quicker, more accurate instrumentation continues at a rapid pace. As one goes up the frequency scale one arrives at the point where it becomes impossible to make measurements with the usual instruments with which the student is normally familiar from basic circuits courses. Obviously a frequency is finally reached where one can physically measure the distance between voltage minimums on a line. At these higher frequencies the student will find new types of instruments such as standing wave meters, power meters, slotted lines, and various types of bridges and impedance and admittance meters.

In this chapter we are going to concentrate on measurements which are unique by using instruments that are normally not seen by students unless they take a laboratory course in microwave measurements. All through the text we have referred to measurements of a special type as the topics arose so in this chapter we will look at some special techniques not previously discussed. In our previous work we have seen that a great deal of information is known about a transmission line and its load by noting the VSWR on the line and the locations of either the voltage minimums or the voltage maximums with respect to the physical location of the load. These measurements do not involve the more common instruments usually used in a basic circuits or electronics laboratory when one is working at lower frequencies. Students must adjust their thinking and their procedures and spend more time becoming familiar with the instrumentation. For instance, the output of the usual high-frequency signal generator is often very small and can be lost by attenuation in a coaxial cable connection without the student realizing what is happening. Furthermore he or she often will make numerous measurements without ever seeing a voltmeter, ammeter or wattmeter. In measuring voltages, for instance, the actual magnitude of the voltage often is of no importance but instead the ratio of two voltages may be extremely important if one is looking at the VSWR.

Important measurements at very high frequencies involve such things as the determination of the VSWR, unknown load impedances, power, input impedance of a line, frequency characteristics with respect to attenuation as a function of frequency, proper placement of stub tuners and many other things which may be of special interest in a unique engineering application. The measurement of R, L, C, and G along with the propagation constant and the characteristic impedance presents addition problems to the engineer who may need this information before, during or after construction. Since R and L are especially affected by frequency variations, their measurement may be of particular importance in those cases where they are difficult to calculate. We have, from time to time, mentioned the fact that G is often impossible to calculate and in some engineering applications the only way to determine it would be by measurement.

As students approach measurement techniques on transmission lines they should relate their previous experiences in measurement work to what needs to be done on the line and adjust, as need be, to new and different approaches to electrical measurements. Read the manufacturer's instructions very carefully since many of the instruments are not only delicate and easily damaged by improper use but have accuracies which are affected by the way they are used.

13.2 Determination of Line Parameters

Line parameters can be determined by performing two tests—the Short-Circuit and the Open-Circuit tests. This is under the assumption that we have the instrumentation to measure the impedance parameter. We will go back to equation (6-20).

$$\text{(6-20)} \qquad Z = Z_o\left[\frac{(Z_r \cosh \gamma d + Z_o \sinh \gamma d)}{(Z_o \cosh \gamma d + Z_r \sinh \gamma d)}\right].$$

When $d = \ell$, $Z = Z_s = Z_{\text{input}}$.

$$Z_{\text{input}} = Z_o\frac{(Z_r \cosh \gamma\ell + Z_o \sinh \gamma\ell)}{(Z_o \cosh \gamma\ell + Z_r \sinh \gamma\ell)}.$$

$$\frac{Z_{\text{input}}}{Z_o} = \frac{\dfrac{Z_r \cosh \gamma\ell}{Z_o \cosh \gamma\ell} + \dfrac{Z_o \sinh \gamma\ell}{Z_o \cosh \gamma\ell}}{\dfrac{Z_o \cosh \gamma\ell}{Z_o \cosh \gamma\ell} + \dfrac{Z_r \sinh \gamma\ell}{Z_o \cosh \gamma\ell}} = \frac{\dfrac{Z_r}{Z_o} + \tanh \gamma\ell}{1 + \dfrac{Z_r}{Z_o}\tanh \gamma\ell}. \qquad (13\text{-}1)$$

$$\frac{Z_{\text{input}}}{Z_o} = \frac{\dfrac{Z_r}{Z_o} + \tanh(\alpha + j\beta)\ell}{1 + \dfrac{Z_r}{Z_o}\tanh(\alpha + j\beta)\ell}. \qquad (13\text{-}2)$$

If we have a given length of line, ℓ, and measure the input impedance with the receiving end shorted, we have with $Z_r = 0$ the short-circuited input impedance, Z_{sc}.

$$\frac{Z_{sc}}{Z_o} = \frac{\tanh(\alpha + j\beta)\ell}{1}, \text{ or } Z_{sc} = Z_o\tanh(\alpha + j\beta)\ell. \qquad (13\text{-}3)$$

If we take the same line, remove the short and measure the input impedance with the receiving end open-circuited we have with $Z_r = \infty$,

$$\frac{Z_{oc}}{Z_o} = \frac{\dfrac{1}{Z_o} + \dfrac{\tanh(\alpha + j\beta)\ell}{Z_r}}{\dfrac{1}{Z_r} + \dfrac{\tanh(\alpha + j\beta)\ell}{Z_o}} = \frac{\dfrac{1}{Z_o}}{\dfrac{1}{Z_o}\tanh(\alpha + j\beta)\ell}$$

or

$$Z_{oc} = \frac{Z_o}{\tanh(\alpha + j\beta)\ell}, \qquad (13\text{-}4)$$

or

$$Z_{oc} = Z_o\coth(\alpha + j\beta)\ell. \qquad (13\text{-}5)$$

If we multiply our two equations together we have:

$$Z_{sc}Z_{oc} = \frac{Z_o^2 \tanh(\alpha + j\beta)\ell}{\tanh(\alpha + j\beta)\ell} = Z_o^2.$$

So

$$Z_{sc}Z_{oc} = Z_o^2. \tag{13-6}$$

$$Z_o = \sqrt{Z_{sc}Z_{oc}}. \tag{13-7}$$

If we can measure the open- and short-circuit input impedance we can determine the characteristic impedance of the line using equation (13-7).

If we divide equation (13-3) by equation (13-5) we have:

$$\frac{Z_{sc}}{Z_{oc}} = \frac{Z_o \tanh(\alpha + j\beta)\ell}{Z_o \coth(\alpha + j\beta)\ell} = \tanh^2(\alpha + j\beta)\ell$$

or

$$\tanh(\alpha + j\beta)\ell = \sqrt{\frac{Z_{sc}}{Z_{oc}}} = M. \tag{13-8}$$

$$\tanh \gamma\ell = \frac{\dfrac{e^{\gamma\ell} - e^{-\gamma\ell}}{2}}{\dfrac{e^{\gamma\ell} + e^{-\gamma\ell}}{2}} = M = \frac{e^{\gamma\ell} - e^{-\gamma\ell}}{e^{\gamma\ell} + e^{-\gamma\ell}} = \frac{e^{\gamma\ell}(1 - e^{-2\gamma\ell})}{e^{\gamma\ell}(1 + e^{-2\gamma\ell})}.$$

$$M = \frac{1 - e^{-2\gamma\ell}}{1 + e^{-2\gamma\ell}}, \text{ or } M + Me^{-2\gamma\ell} = 1 - e^{-2\gamma\ell}.$$

$$e^{-2\gamma\ell}(M + 1) = 1 - M.$$

$$e^{-2\gamma\ell} = \frac{1 - M}{1 + M} = \frac{1 - \sqrt{\dfrac{Z_{sc}}{Z_{oc}}}}{1 + \sqrt{\dfrac{Z_{sc}}{Z_{oc}}}},$$

or

$$e^{2\gamma\ell} = \frac{1 + \sqrt{\dfrac{Z_{sc}}{Z_{oc}}}}{1 - \sqrt{\dfrac{Z_{sc}}{Z_{oc}}}}$$

or

$$2\gamma\ell = \ln\left[\frac{1 + \sqrt{\dfrac{Z_{sc}}{Z_{oc}}}}{1 - \sqrt{\dfrac{Z_{sc}}{Z_{oc}}}}\right].$$

$$\gamma\ell = \frac{1}{2}\ln\left[\frac{1 + \sqrt{\dfrac{Z_{sc}}{Z_{oc}}}}{1 - \sqrt{\dfrac{Z_{sc}}{Z_{oc}}}}\right] = (\alpha + j\beta)\ell. \tag{13-9}$$

$$\text{Let } \ln\left[\frac{1 + \sqrt{\dfrac{Z_{sc}}{Z_{oc}}}}{1 - \sqrt{\dfrac{Z_{sc}}{Z_{oc}}}}\right] = \ln Ae^{j\phi} = \ln A$$

$$\ln A + j(\phi + 2n\pi). \tag{13-10}$$

Here n equals an integer, 1, 2, 3, etc.

So

$$(\alpha + j\beta)\ell = \alpha\ell + j\beta\ell = \frac{1}{2}\,[\ln A + j(\phi + 2n\pi)],$$

and

$$\alpha\ell = \frac{1}{2}\ln A \text{ and } \alpha = \frac{1}{2\ell}\ln A = \frac{1}{2\ell}\ln\left[\frac{1 + \sqrt{\dfrac{Z_{sc}}{Z_{oc}}}}{1 - \sqrt{\dfrac{Z_{sc}}{Z_{oc}}}}\right]. \tag{13-11}$$

$$\beta\ell = \frac{1}{2}(\phi + 2n\pi), \text{ so } \beta = \frac{1}{2\ell}(\phi + 2n\pi), \text{ where } n \text{ is an integer.} \tag{13-12}$$

For α we can determine a unique value by using the natural log of the magnitude indicated in equation (13-11). But for β we obtain a number of values depending on n and it may be difficult at times to determine which n to use.

We can also go back to equation (13-8),

$$\tanh(\alpha + j\beta)\ell = \sqrt{\frac{Z_{sc}}{Z_{oc}}}.$$

$$\alpha\ell + j\beta\ell = \tanh^{-1}\sqrt{\frac{Z_{sc}}{Z_{oc}}}. \tag{13-13}$$

We can then use the identity:

$$\tanh^{-1}(A + jB) = \frac{1}{2}\tanh^{-1}\frac{2A}{1 + A^2 + B^2} + j\frac{1}{2}\tan^{-1}\frac{2B}{1 - A^2 - B^2}. \tag{13-14}$$

13.3 Example of Determining Line Parameters from Measurement

Let us assume that we have measured the open- and short-circuit impedances for a line that is 10 meters long.

$$Z_{sc} = 100.97\,\underline{/-6.19^\circ}. \qquad Z_{oc} = 99.145\,\underline{/1.33^\circ}.$$

Find: α, β, and Z_o.

Solution

$Z_o = \sqrt{Z_{sc}Z_{oc}} = \sqrt{(100.97\,\underline{/-6.19^\circ})(99.145)\,\underline{/1.33^\circ})} =$ $100.05\,\underline{/-2.43^\circ} = 99.96 - j4.24$ ohms.

$$\sqrt{\frac{Z_{sc}}{Z_{oc}}} = \sqrt{\frac{100.97\,\underline{/-6.19^\circ}}{99.145\,\underline{/1.33^\circ}}} = 1.009\,\underline{/-3.75^\circ} = 1.007 - j0.066.$$

$$\left[\frac{1 + \sqrt{\dfrac{Z_{sc}}{Z_{oc}}}}{1 - \sqrt{\dfrac{Z_{sc}}{Z_{oc}}}}\right] = \left[\frac{1 + 1.007 - j0.066}{1 - 1.007 + j0.066}\right] = 30.18 \underline{/-97.94^\circ}.$$

Therefore, A = 30.18 and $\phi = -97.94^\circ$.

$$\alpha = \frac{1}{2\ell} \ln A = \frac{1}{2(10)} \ln 30.18 = 0.17 \text{ nepers/meter.}$$

$$\beta = \frac{1}{2\ell}(\phi + 2n\pi). \qquad \phi = -97.94^\circ \text{ or } -1.709 \text{ radians.}$$

$$\beta = \frac{1}{2(10)}(-1.709 + 2n\pi).$$

For different values of n we obtain different values of β.

n	β	n	β
0	−0.085	8	2.43
1	0.228	9	2.74
2	0.543	10	3.06
3	0.86	11	3.28
4	1.11	12	3.68
5	1.48	13	3.99
6	1.80	14	4.31
7	2.11	15	4.63

To determine the right value of β we might use other measurements such as the wavelength on the line. Suppose we measured the distance between voltage minimums and found the wavelength to be about 1.6 meters.

$$\beta = \frac{2\pi}{\lambda} = \frac{2\pi}{1.6} = 3.93.$$

So it would appear that the proper value of n is 13 and that β is equal to 3.99 radians per meter.

If we use equation (13-13).

$$\alpha\ell + j\beta\ell = \tanh^{-1}\sqrt{\frac{Z_{sc}}{Z_{oc}}} = \tanh^{-1}(1.007 - j0.066) =$$

$$\frac{1}{2}\tanh^{-1}\frac{2A}{1 + A^2 + B^2} + j\frac{1}{2}\tan^{-1}\frac{2B}{1 - A^2 - B^2} =$$

$$\frac{1}{2}\tanh^{-1}\frac{(2)(1.007)}{1 + (1.007)^2 + (0.066)^2} + j\frac{1}{2}\tan^{-1}\frac{(2)(0.066)}{1 - (1.007)^2 - (0.066)^2} =$$

$$1.625 + j\,0.716 = \alpha\ell + j\beta\ell.$$

So we have:

$$\alpha\ell = \alpha(10) = 1.625. \qquad \alpha = 0.1625.$$

$$\beta\ell = \beta(10) = 0.716. \qquad \beta = 0.0716.$$

So we have obtained approximately the same value for α, but again the value of β is in doubt and we see that we should be careful about just substituting into equations.

13.4 Determination of the Load Impedance from Measurement-Shift of the Voltage Minimum Point

We have already seen earlier in the text how we can determine the load impedance for the low loss or lossless cases by measuring the distance from the load to a voltage maximum or minimum and using either equations or the Smith chart. We would need to measure the VSWR so let us discuss this for a moment. The VSWR is the ratio of the maximum value of the voltage to the minimum value of the voltage so any technique that measures these two values may be used. At high frequencies when the wavelength is within reasonable values for measurement in a laboratory a favorite way to make measurements is to use a slotted line. This usually takes the form of a coaxial cable with a slot cut in the outer conductor. A probe is placed in the slot to sample the field between the two conductors and this is converted to an equivalent voltage on the meter. Provisions for moving the probe back and forth are made and a metric scale is placed along the slot so that distances can be determined. By taking the ratio of the maximum meter reading to the minimum meter reading we can determine the VSWR. Notice that we do not have to actually determine the absolute values of the voltages—just their ratio. We can do the same thing when using wave guides by cutting a slot in the wave guide.

One technique which can be used to determine an unknown load impedance involves the following procedure. We determine a voltage minimum point with the load connected and then we replace the load with a short circuit and see how the voltage minimum shifts. To start with let us assume that the load is capacitive. (A resistance in series with a capacitor.) We are assuming a lossless or very low loss situation. For a capacitive load we know that the angle of the voltage reflection coefficient is negative. We also see from the phasor diagram that we have a voltage minimum first as we move down the line from the load. We will let θ be the angle from the load to the voltage minimum. This is illustrated in Figure 13-1.

If we plotted the magnitude of the voltage as a function of the distance from the load we would have a situation as illustrated in Figure 13-2.

If we replace the load by a short circuit our pattern would look like Figure 13-3.

Let us combine the curve from Figure 13-2 with the curve from Figure 13-3 giving us Figure 13-4.

From Figure 13-4 we see that the voltage minimum shifts to the right (towards the load) by a distance of Δd. Let us go back to an equation in Chapter 5.

$$Z = Z_o \frac{(Z_r + j\,Z_o \tan \beta d)}{(Z_o + j\,Z_r \tan \beta d)}. \qquad (5\text{-}29)$$

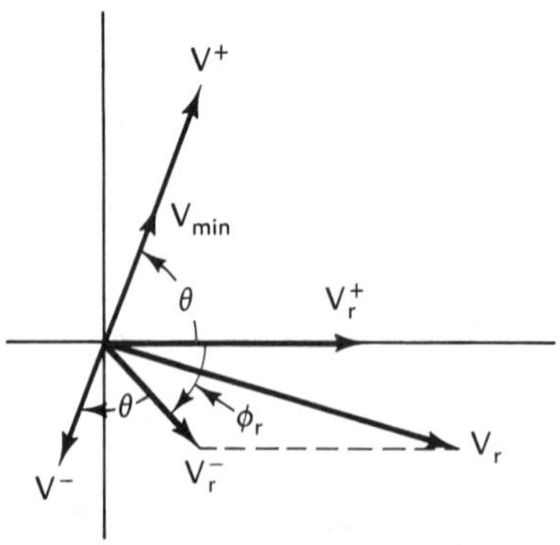

ϕ_r = angle of Γ_r

$\theta + \theta + \phi_r = 180°$

$2\theta = 180° - \phi_r$

$$\theta = \frac{180° - \phi_r}{2}$$

Figure 13-1. Voltage minimum at Θ degrees from load with Z_r capacitive.

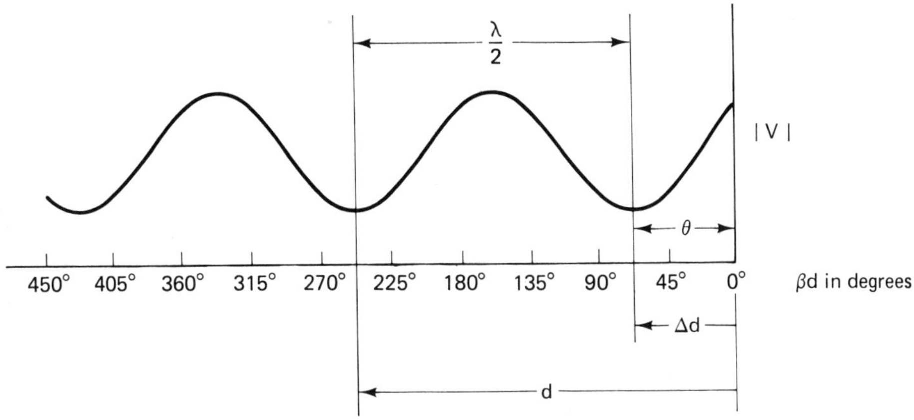

Figure 13-2. Standing wave pattern for voltage.

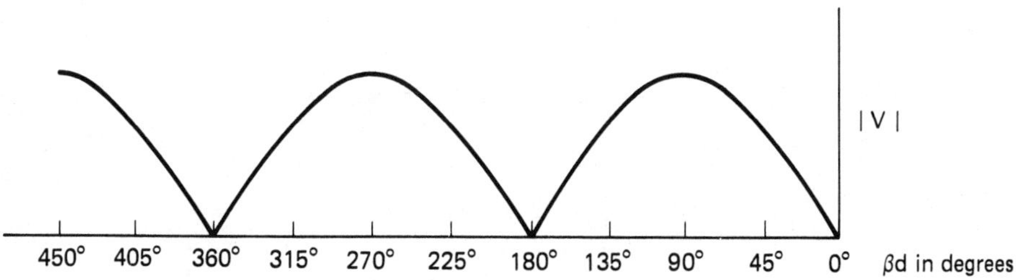

Figure 13-3. Standing wave pattern for voltage with short circuit as load.

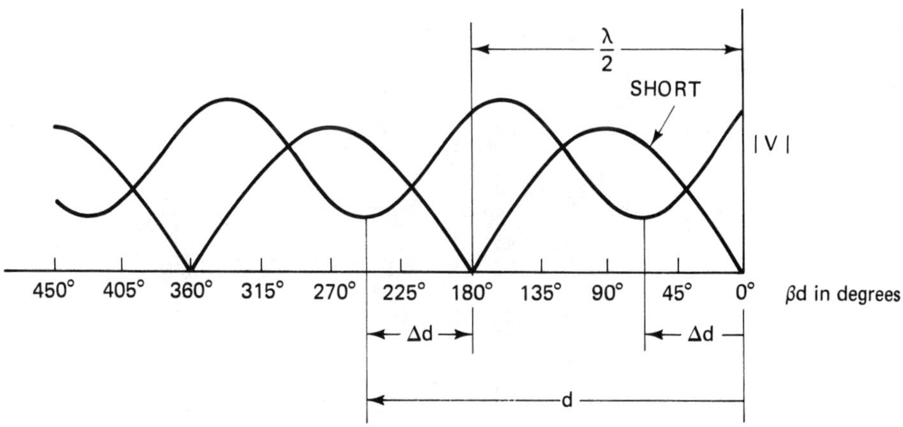

Figure 13-4. Combined pattern.

With a little algebra we obtain:

$$Z_r = Z_o\frac{(Z - Z_o\tan \beta d)}{(Z_o - Z \tan \beta d)}. \tag{13-15}$$

If we let Z equal the impedance at d, where we have a voltage minimum we have:

$$Z = Z_{min} = R_{min} = \frac{Z_o}{\text{VSWR}}. \tag{3-23}$$

$$Z_r = Z_o\frac{\left[\frac{Z_o}{\text{VSWR}} - jZ_o \tan\beta d\right]}{\left[Z_o - j\frac{Z_o}{\text{VSWR}}\tan\beta d\right]} = Z_o\frac{\left[\frac{1}{\text{VSWR}} - j\tan\beta d\right]}{\left[1 - j\frac{\tan\beta d}{\text{VSWR}}\right]} =$$

$$Z_r = Z_o\left[\frac{1 - j(\text{VSWR})\tan\beta d}{(\text{VSWR}) - j\tan\beta d}\right]. \tag{13-16}$$

But

$$d = \frac{\lambda}{2} + \Delta d \text{ so } \beta d = \beta\frac{\lambda}{2} + \beta\Delta d.$$

Since $\beta = \frac{2\pi}{\lambda}$,

$$\beta d = \frac{2\pi}{\lambda}\frac{\lambda}{2} + \beta\Delta d = \pi + \beta\Delta d.$$

While it is not mathematically correct let us write this as: $180° + \beta\Delta d$.

$$\tan\beta d = \tan(180° + \beta\Delta d) = \frac{\sin(180° + \beta\Delta d)}{\cos(180° + \beta\Delta d)} =$$

$$\frac{\sin 180° \cos\beta\Delta d + \cos 180° \sin\beta\Delta d}{\cos 180° \cos\beta\Delta d - \sin 180° \sin\beta\Delta d} = \frac{-\sin\beta\Delta d}{-\cos\beta\Delta d} = \tan\beta\Delta d.$$

So,

$$Z_r = Z_o\left[\frac{1 - j(\text{VSWR})\tan\beta\Delta d}{(\text{VSWR}) - j\tan\beta\Delta d}\right]. \tag{13-17}$$

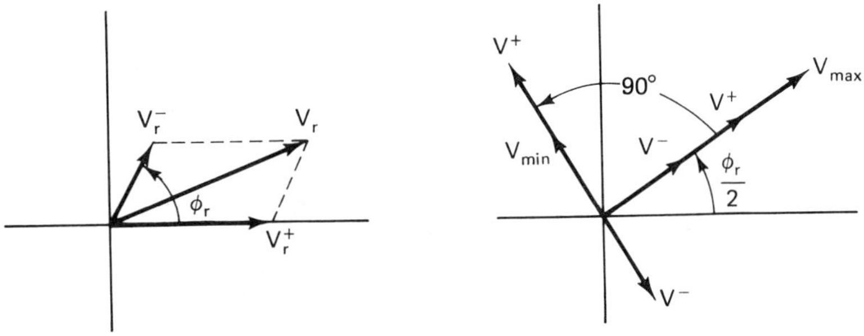

Figure 13-5. Phasor diagrams for inductive load ϕ_r is positive.

If we now go back and assume that the load is inductive (resistance in series with an inductance) we know that the angle of the voltage reflection coefficient is positive and that we have a voltage maximum first before we obtain a voltage minimum as we move down the line from the load. We know from the phasor diagram that we get a voltage maximum when $\theta = \dfrac{\phi_r}{2}$.

In Figure 13-5 we will obtain a voltage minimum when we proceed another 90° down the line from the voltage maximum. If we sketch the voltage pattern on the line we have Figure 13-6.

Now, as before, we will replace the load by a short circuit and combine the two patterns as shown in Figure 13-7.

When we replace the load by a short circuit we see that the voltage minimum now shifts to the left (towards the generator) by a

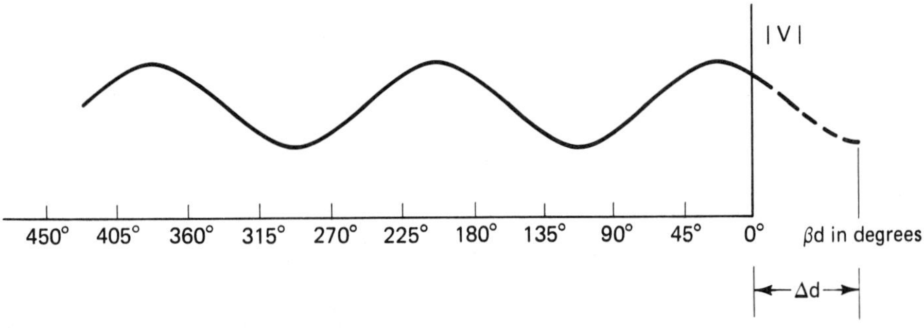

Figure 13-6. Standing wave pattern for voltage.

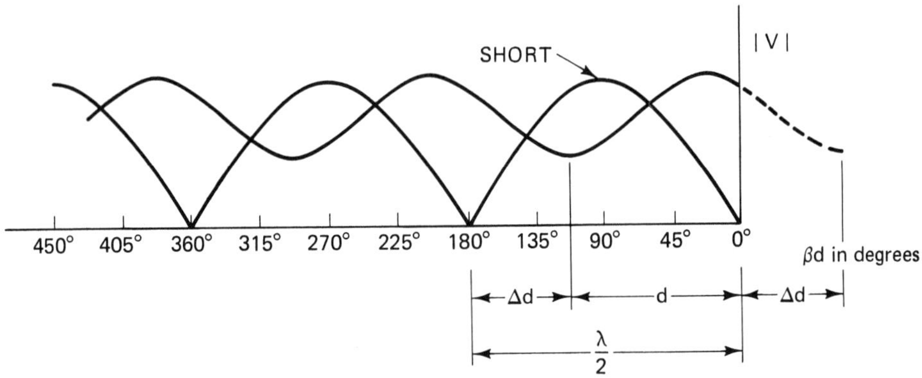

Figure 13-7. Combined pattern.

distance, Δd. Once again we will let the impedance be Z at the voltage minimum, and go back to equation (13-16).

$$Z_r = Z_o \left[\frac{1 - j(\text{VSWR})\tan \beta d}{(\text{VSWR}) - j \tan \beta d} \right].$$

In the present case we have:

$$d = \frac{\lambda}{2} - \Delta d, \text{ or } \beta d = \beta\frac{\lambda}{2} - \beta\Delta d,$$

but since $\beta = \dfrac{2\pi}{\lambda}$ we can write:

$$\beta d = \frac{2\pi}{\lambda}\frac{\lambda}{2} - \beta\Delta d = \pi - \beta\Delta d.$$

Again we will cheat a little and write this as $180° - \beta\Delta d$.

$$\tan \beta d = \tan(180° - \beta\Delta d) = \frac{\sin(180° - \beta\Delta d)}{\cos(180° - \beta\Delta d)} = \frac{\sin \beta\Delta d}{-\cos \beta\Delta d} = -\tan \beta\Delta d.$$

So

$$Z_r = Z_o \left[\frac{1 + j(\text{VSWR})\tan \beta\Delta d}{(\text{VSWR}) + j \tan \beta\Delta d} \right], \tag{13-18}$$

for an inductive load. Our results so far are:

$$Z_r = Z_o \left[\frac{1 + j(\text{VSWR})\tan \beta\Delta d}{(\text{VSWR}) + j \tan \beta\Delta d} \right],$$

for an inductive load and

$$Z_r = Z_o \left[\frac{1 - j(\text{VSWR})\tan \beta\Delta d}{(\text{VSWR}) - j \tan \beta\Delta d} \right]$$

for a capacitive load. Since

$$\frac{\beta\Delta d}{\pi} = \frac{X^\circ}{180^\circ}, \qquad X^\circ = \frac{180^\circ\ \beta\Delta d}{\pi},$$

but $\beta = \dfrac{2\pi}{\lambda}$, so

$$X^\circ = \frac{180^\circ}{\pi}\frac{2\pi}{\lambda}\Delta d = \frac{(180^\circ)(2)}{\lambda}\Delta d = (180^\circ)\frac{(\Delta d)}{\frac{\lambda}{2}}. \tag{13-19}$$

Δd is positive if V_{min} shifts towards the load (capacitive load). Δd is negative if V_{min} shifts towards the generator (inductive load). And we can now write an equation which is good for both cases:

$$Z_r = Z_o \left[\frac{1 - j(\text{VSWR})\tan X^\circ}{(\text{VSWR}) - j \tan X^\circ} \right], \tag{13-20}$$

$$\text{where } X^\circ = \frac{180^\circ\ (\pm\Delta d)}{\frac{\lambda}{2}}. \tag{13-21}$$

We use the negative sign if the voltage minimum shifts towards the generator and the positive sign if it shifts towards the load when we replace the load with a short circuit.

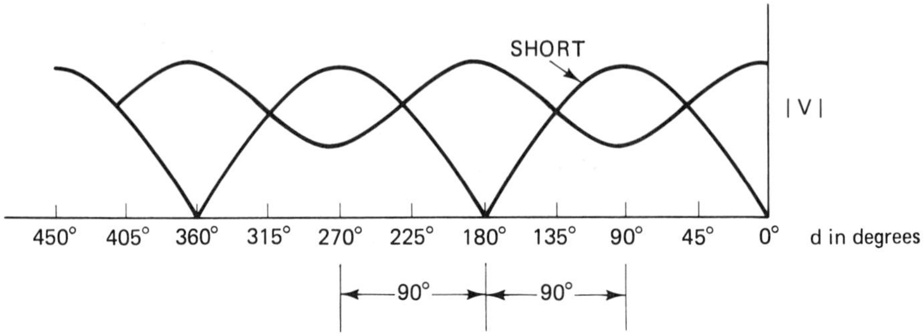

Figure 13-8. Standing wave pattern for voltage with the load resistive and greater than Z_o.

We still have two special cases to consider. Suppose the load is purely resistive and that $Z_r = R_r > Z_o = R_o$. We know that under these conditions that the angle of the voltage reflection coefficient is zero degrees and that we have a voltage maximum at the load and a voltage minimum 90 degrees from the load. Our combined pattern is given in Figure 13-8.

Now we see that when we replace the load with a short circuit the voltage minimum shifts exactly 90 degrees or $\lambda/4$ wavelengths either way. Going back to equation (13-16) we have:

$$Z_r = Z_o\left[\frac{1 - j(\text{VSWR})\tan 90^\circ}{(\text{VSWR}) - j\tan 90^\circ}\right] = Z_o\left[\frac{\dfrac{1}{\tan 90^\circ} - j(\text{VSWR})}{\dfrac{(\text{VSWR})}{\tan 90^\circ} - j}\right] =$$

$$Z_o\left[\frac{-j(\text{VSWR})}{-j}\right] = (\text{VSWR})Z_o = Z_r. \tag{13-22}$$

Our last case involves a purely resistive load that is now less than the characteristic impedance. Now the angle of the voltage reflection coefficient is 180 degrees and we have a voltage minimum at the load and every half wavelength down the line. Our combined pattern is given in Figure 13-9.

Now we see that the voltage minimum doesn't shift when we replace the load by a short circuit. When we go back to equation (13-16) with $d = 0$, we have:

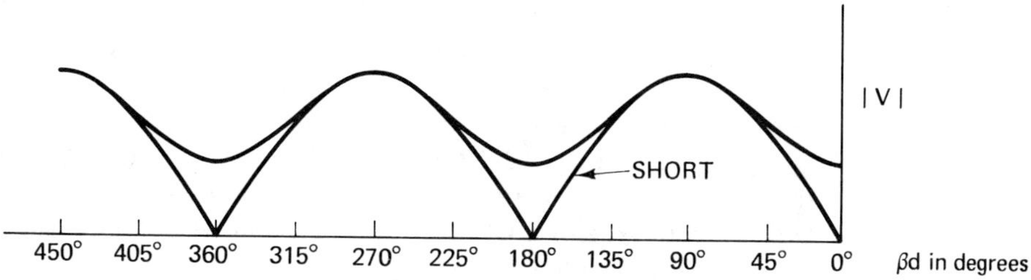

Figure 13-9. Combined pattern for voltage with the load resistive and less than Z_o.

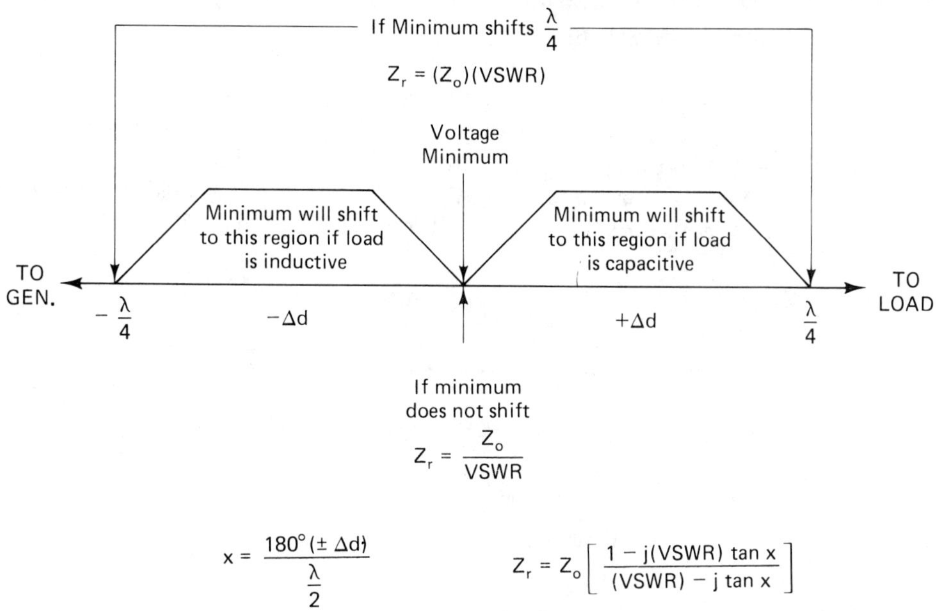

Figure 13-10. Shift of voltage minimum when load is replaced with a short circuit.

$$Z_r = Z_o \left[\frac{1 - j(\text{VSWR}) \tan 0^\circ}{(\text{VSWR}) - j \tan 0^\circ} \right] = \frac{Z_o}{(\text{VSWR})}. \qquad (13\text{-}23)$$

We have now taken care of all the possible cases and can indicate the results in Figure 13-10, which shows that we have the four possibilities discussed above. If we replace the load with a short circuit and the voltage minimum does not shift, but stays in the same place, we know that the load is purely resistive with a magnitude equal to the characteristic impedance divided by the VSWR. If the minimum shifts exactly one-quarter wavelength, we know the load is purely resistive and has a magnitude which is equal to the characteristic impedance times the VSWR. If the voltage minimum shifts to the left we use equation (13-21) to find $X°$ with Δd negative and then we use equation (13-20) to find the load impedance. If the minimum shifts to the right we use equation (13-21) to find $X°$ with Δd positive and then use equation (13-20).

13.5 Example of Finding an Unknown Load Impedance by Using the Shift of the Voltage Minimum Method

Suppose we go into the laboratory and make some measurements with the following results: $\text{Voltage}_{\text{maximum}}$ = 0.74 cm from the load. Second $\text{Voltage}_{\text{maximum}}$ = 10.74 cm from the load. $\text{Voltage}_{\text{minimum}}$ = 5.74 cm from the load. VSWR = 6.85. When the load is replaced by a short circuit, V_{minimum} = 10 cm from the load. Determine: Z_r.

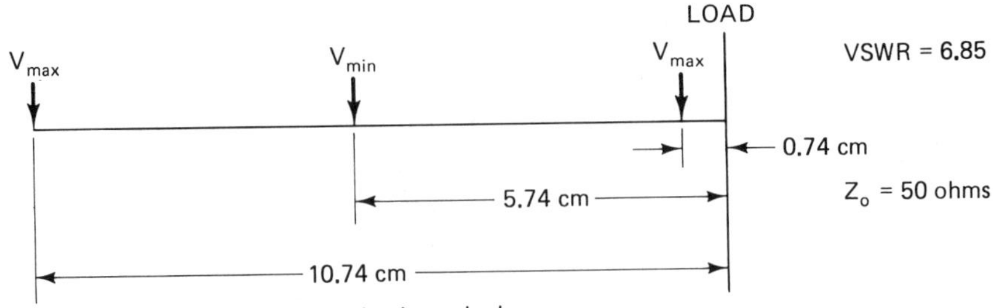

With unknown load attached

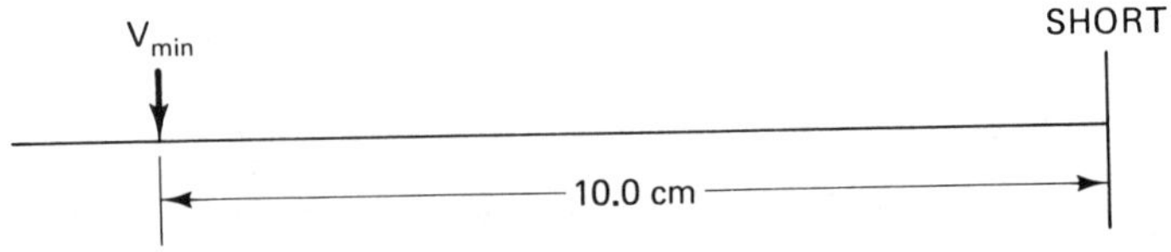

When short replaces load

Solution

After sketching out the results we see that: V_{min} shifts 10 − 5.74 = 4.26 cm to the left. Δd is − (shift to left).

$$\frac{\lambda}{2} = 10.74 - 0.74 = 10 \text{ cm.}$$

$$X^\circ = \frac{180^\circ(-\Delta d)}{\frac{\lambda}{2}} = 180^\circ \frac{(-4.26)}{10} = -76.72^\circ.$$

$$Z_r = Z_o \left[\frac{1 - j(\text{VSWR}) \tan X^\circ}{(\text{VSWR}) - j \tan X^\circ}\right] = 50 \left[\frac{1 - j6.85 \tan -76.72^\circ}{6.85 - j \tan -76.72^\circ}\right] = 180.28 \angle 56.31^\circ = 100 + j\,150 \text{ ohms.}$$

Notice that we can also use equation (13-16) directly.

$$Z_r = Z_o \left[\frac{1 - j(\text{VSWR})\tan \beta d}{(\text{VSWR}) - j \tan \beta d}\right],$$

where d is the distance to a voltage minimum. Let $\beta d = \theta$, the degree equivalent of βd.

$$Z_r = Z_o \left[\frac{1 - j(\text{VSWR}) \tan \theta}{(\text{VSWR}) - j \tan \theta}\right]. \qquad (13\text{-}24)$$

In our example we have:

$$\frac{\lambda}{2} = 10. \qquad \lambda = 20 \text{ cm, so } \theta = \frac{5.74(360^\circ)}{20} = 103.28^\circ.$$

$$Z_r = 50 \left[\frac{1 - j(6.85)\tan 103.28^\circ}{(6.85) - j \tan 103.28^\circ}\right] = 108.28 \angle 56.31^\circ = 100 + j\,150 \text{ ohms.}$$

A third solution is possible by using some equations developed earlier.

$$(3\text{-}24) \qquad |\Gamma_r| = \frac{\text{VSWR} - 1}{\text{VSWR} + 1} = \frac{6.85 - 1}{6.85 + 1} = 0.745.$$

(3-29) $$\phi = 4\pi\frac{(d_{V_{min}})}{\lambda} - \pi \pm 2n\pi, \qquad \text{Let } n = 0.$$

$$\phi = 4\pi\frac{(5.74)}{20} - \pi = 0.463 \text{ radian, or } 26.56°.$$

(10-35) $$\frac{R_r}{Z_o} = \frac{1 - |\Gamma_r|^2}{1 + |\Gamma_r|^2 - 2|\Gamma_r|\cos\phi} = \frac{1 - (0.745)^2}{1 + (0.745)^2 - 2(0.745)\cos 26.56°} =$$
$$2.04. \ R_r = 50(2.04) = 102.0 \text{ ohms.}$$

(10-36) $$\frac{X_r}{Z_o} = \frac{2|\Gamma_r|\sin\phi}{1 + |\Gamma_r|^2 - 2|\Gamma_r|\cos\phi} =$$
$$\frac{2(0.745)\sin 26.56°}{1 + (0.745)^2 - 2(0.745)\cos 26.56°} = 3.02. \qquad X_r =$$
$$50(3.02) = 151.0 \text{ ohms.}$$

So by this technique we obtain the result that $Z_r = 102 + j\,151$ ohms.

13.6 Measurements When the VSWR Is Greater than 10

Let us go back to equation (13-16).

$$Z_r = Z_o\left[\frac{1 - j(\text{VSWR})\tan\beta d}{(\text{VSWR}) - j\tan\beta d}\right] = Z_o\left[\frac{1 - j(\text{VSWR})\dfrac{\sin\beta d}{\cos\beta d}}{(\text{VSWR}) - j\dfrac{\sin\beta d}{\cos\beta d}}\right] =$$
$$Z_o\left[\frac{\cos\beta d - j(\text{VSWR})\sin\beta d}{(\text{VSWR})\cos\beta d - j\sin\beta d}\right]\left[\frac{(\text{VSWR})\cos\beta d + j\sin\beta d}{(\text{VSWR})\cos\beta d + j\sin\beta d}\right] =$$
$$Z_o\left[\frac{(\text{VSWR})\cos^2\beta d + (\text{VSWR})\sin^2\beta d}{(\text{VSWR})^2\cos^2\beta d + \sin^2\beta d}\right]$$
$$- jZ_o\left[\frac{(\text{VSWR})^2\sin\beta d\cos\beta d - \sin\beta d\cos\beta d}{(\text{VSWR})^2\cos^2\beta d + \sin^2\beta d}\right].$$

Therefore:

$$R_r = Z_o\left[\frac{(\text{VSWR})(\cos^2\beta d + \sin^2\beta d)}{(\text{VSWR})^2\cos^2\beta d + \sin^2\beta d}\right] =$$
$$\frac{Z_o(\text{VSWR})}{(\text{VSWR})^2\cos^2\beta d + \sin^2\beta d} = \frac{Z_o}{(\text{VSWR})\cos^2\beta d + \dfrac{\sin^2\beta d}{(\text{VSWR})}}.$$

If VSWR ≫ 10 we can approximate with the following equation:

$$R_r \simeq \frac{Z_o}{(\text{VSWR})\cos^2 \beta d}. \tag{13-25}$$

Also

$$X_r = -Z_o \left[\frac{(\text{VSWR})^2 \sin \beta d \cos \beta d - \sin \beta d \cos \beta d}{(\text{VSWR})^2 \cos^2 \beta d + \sin^2 \beta d}\right] = -Z_o \left[\frac{\sin \beta d \cos \beta d - \dfrac{\sin \beta d \cos \beta d}{(\text{VSWR})^2}}{\cos^2 \beta d + \dfrac{\sin^2 \beta d}{(\text{VSWR})^2}}\right].$$

If VSWR ≫ 10, we have:

$$X_r \simeq -Z_o \left[\frac{\sin \beta d \cos \beta d}{\cos^2 \beta d}\right] \simeq -Z_o \frac{\sin \beta d}{\cos \beta d} \simeq -Z_o \tan \beta d. \tag{13-26}$$

Let us represent a voltage minimum condition with the phasor diagram shown in Figure 13-11.

If we let V^+ and V^- each rotate θ degrees in their proper directions until their sum is $\sqrt{2}\, V_{min}$ at an angle of α degrees, we will have the phasor diagram shown in Figure 13-12.

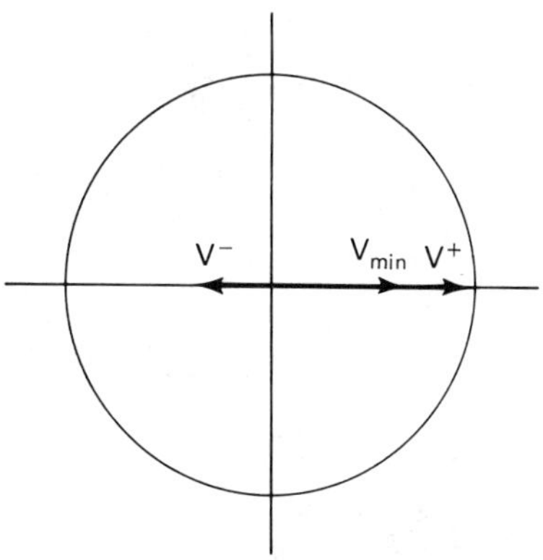

Figure 13-11. Voltage minimum.

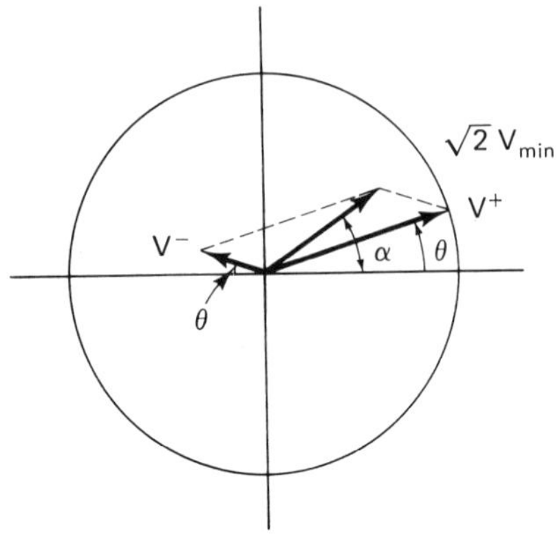

Figure 13-12. $\sqrt{2}$ voltage minimum.

$$\sqrt{2}\, V_{min} \underline{/\alpha} = |V^+| (\cos\theta + j\sin\theta) + |V^-| (-\cos\theta + j \sin\theta).$$

$$\sqrt{2}\, V_{min} \underline{/\alpha} = \cos\theta\, [|V^+| - |V^-|] + j \sin\theta\, [|V^+| + |V^+|].$$

$$\text{VSWR} = \frac{V_{max}}{V_{min}} = \frac{|V^+| + |V^-|}{|V^+| - |V^-|},$$

and $V_{min} = |V^+| - |V^-|$.

$$\frac{\sqrt{2}\, V_{min} \underline{/\alpha}}{V_{min}} = \frac{\cos\theta\, [|V^+| - |V^-|] + j \sin\theta\, [|V^+| + |V^-|]}{|V^+| - |V^-|} = \sqrt{2} \underline{/\alpha} =$$

$$\cos\theta + j \sin\theta \frac{[|V^+| + |V^-|]}{|V^+| - |V^-|}.$$

$$\sqrt{2} \underline{/\alpha} = \cos\theta + j(\text{VSWR})\sin\theta, \text{ or } |\sqrt{2} \underline{/\alpha}| = |\cos\theta + j(\text{VSWR}) \sin\theta|,$$

or

$$\sqrt{2} = \sqrt{\cos^2\theta + (\text{VSWR})^2\sin^2\theta}, \text{ or } 2 = \cos^2\theta + (\text{VSWR})^2\sin^2\theta.$$

Since $\sin^2\theta + \cos^2\theta = 1$ we can write: $\cos^2\theta = 1 - \sin^2\theta$ and our equation becomes:

$$2 = (1 - \sin^2\theta) + (\text{VSWR})^2\sin^2\theta = 1 + \sin^2\theta\, [(\text{VSWR})^2 - 1].$$

If VSWR $\gg$ 10 we have:

$$2 \simeq 1 + \sin^2\theta(\text{VSWR})^2. \qquad 1 \simeq \sin^2\theta(\text{VSWR})^2.$$

$$\sin^2\theta \simeq \frac{1}{(\text{VSWR})^2}. \qquad \sin\theta \simeq \frac{1}{\text{VSWR}} \simeq \sin \beta d'. \qquad (13\text{-}27)$$

d' is the distance between V_{min} and $\sqrt{2}\, V_{min}$.

If VSWR $\gg$ 10, $\sin \beta d' \simeq \beta d'$, so we have:

$$\beta d' \simeq \frac{1}{\text{VSWR}}, \text{ and } d' \simeq \frac{1}{\beta(\text{VSWR})}. \qquad (13\text{-}28)$$

We could have gone in the other direction as indicated in Figure 13-13.

Now the distance between the two $\sqrt{2}\, V_{min}$ points is 2 d'. Let

$$2d' = \Delta \simeq \frac{2}{\beta(\text{VSWR})}, \qquad \text{or VSWR} \simeq \frac{2}{\beta\Delta}.$$

But

$$\beta = \frac{2\pi}{\lambda}, \text{ so VSWR} \simeq \frac{2}{\frac{2\pi\Delta}{\lambda}} = \frac{\lambda}{\pi\Delta} \simeq \text{VSWR}. \qquad (13\text{-}29)$$

λ and Δ are both in meters.

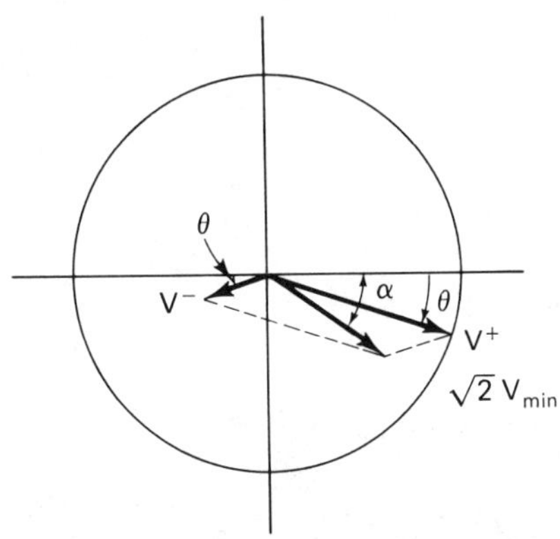

Figure 13-13. $\sqrt{2}$ voltage minimum.

The procedure in the laboratory would be to find the minimum value of the voltage and mark it on a distance scale. Then we would proceed on both sides of the minimum to values of $\sqrt{2}$ times the minimum and note the distances involved. The distance from one $\sqrt{2}\ V_{min}$ point to the other $\sqrt{2}\ V_{min}$ point would be our Δ.

13.7 Example of Finding the VSWR When the VSWR Is Greater than 10

Given:

In the laboratory we know the VSWR is greater than 10 and we make the following measurements.

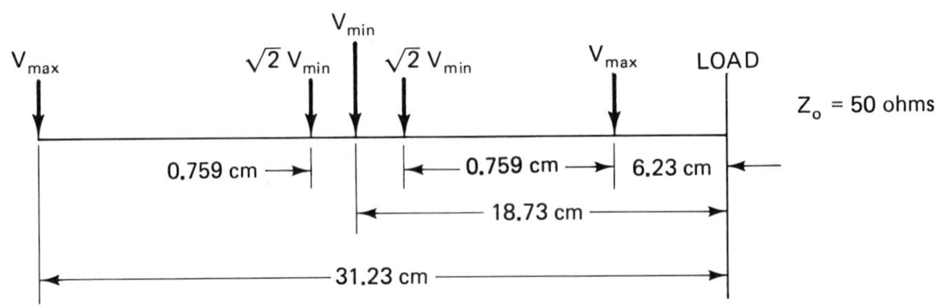

Find

VSWR and Z_r.

Solution

$$\frac{\lambda}{2} = 31.23 - 6.23 = 25 \text{ cm.} \qquad \lambda = 2(25) = 50 \text{ cm.}$$

$$\text{VSWR} \simeq \frac{\lambda}{\pi\,\Delta} \simeq \frac{50}{\pi(0.795)} \simeq 20.01.$$

$$\theta = \frac{18.73(360°)}{50} = 134.85°,$$

which is the equivalent of βd in degrees.

$$R_r \simeq \frac{Z_o}{(\text{VSWR})\cos^2\theta} \simeq \frac{50}{(20.01)(\cos 134.85°)^2} \simeq 5.02 \text{ ohms.}$$

$$X_r \simeq -jZ_o\tan\theta \simeq -j50 \tan 134.85° \simeq j50.25 \text{ ohms.}$$

$$Z_r \simeq 5.02 + j\,50.25 \text{ ohms.}$$

13.8 Chapter Summary

1. The measurement of line parameters such as R, L, C, and G along with α, β, and Z_o requires different approaches at different frequencies. At power frequencies and the middle band of frequencies the student will be using instrumentation that he has already used in previous courses. However, when very high frequencies are approached, the instrumentation and measurements change a great deal.
2. The important types of measurements involve finding R, L, C, and G. For simple configurations L and C may be calculated and R measured or calculated. By using the short-circuit and open-circuit technique Z_o can be found from:

$$Z_o = \sqrt{Z_{sc}Z_{oc}}.$$

Knowing Z_o, R, L, and C would make it possible to calculate G. The values of α and β can also be determined by the short- and open-circuit tests.

$$\alpha = \frac{1}{2\ell} \ln\left[\frac{1 + \sqrt{\frac{Z_{sc}}{Z_{oc}}}}{1 - \sqrt{\frac{Z_{sc}}{Z_{oc}}}}\right].$$

$\beta = \frac{1}{2\ell}(\phi + 2n\pi)$, where n is an integer and ϕ is the angle of

$$\ln\left[\frac{1+\sqrt{\dfrac{Z_{sc}}{Z_{oc}}}}{1-\sqrt{\dfrac{Z_{sc}}{Z_{oc}}}}\right].$$

3. One technique for determining an unknown load impedance is to measure the distance from the load to a voltage minimum and then replace the load by a short circuit and determine the shift of the voltage minimum after the short is placed on the line. Then after determining the VSWR the unknown load impedance may be found from:

$$X^\circ = 180^\circ \frac{(\pm\Delta d)}{\dfrac{\lambda}{2}}.$$

$$Z_r = Z_o\left[\frac{1 - j(\text{VSWR})\tan X^\circ}{(\text{VSWR}) - j\tan X^\circ}\right].$$

Δd is the shift of V_{min}. It is negative if the shift is toward the generator and positive if the shift is toward the load.

4. A special technique can be used when the VSWR is greater than 10 by measuring the distance between the two points on the line where the voltage is $\sqrt{2}\,V_{min}$.

$\text{VSWR} \simeq \dfrac{\lambda}{\pi\,\Delta}$, where Δ is the distance between the two points.

$$R_r \simeq \frac{Z_o}{(\text{VSWR})\cos^2\theta}.$$

$X_r \simeq -Z_o\tan\theta$, where θ is the angle equivalent of βd (the distance from the load to the voltage minimum).

5. When making any measurement at very high frequencies, the student should be very careful to read the manufacturers' instructions since some of the instrumentation at these frequencies is easily damaged and improper techniques may also cause great inaccuracies.

13.9 Review and Study Questions

1. Go back through the text and pick out those sections that deal with measurement. What topics have been covered?
2. Would it be practical to measure the standing wave on a line operating at 30 kilohertz? at 300 megahertz? at 3 gigahertz? Explain.
3. Sketch the connections for measuring the power on a three-phase, three-wire transmission line using potential and current transformers.
4. Visit a laboratory and determine the output power in watts from a number of signal generators operating at various frequencies. What is the output of signal generators operating up in the gigahertz region? What does the instruction manual say about the output impedance of the generator?
5. Give several ways of determining the unknown impedance of a load connected to a transmission line operating in the gigahertz range of frequencies.
6. Describe a technique of measuring the VSWR if the VSWR is large, say, greater than 10.
7. Sketch the standing wave pattern on a line terminated with a load impedance that is equivalent to an *R* and *L* in series. On this curve, superimpose the curve obtained when the load is replaced by a short circuit. How is the position of the voltage minimum affected by the change in loads?
8. If the load is inductive is there a voltage maximum or minimum closest to the load? Explain.
9. If the load is purely resistive is there a voltage minimum or maximum closest to the load? Explain.
10. What is a slotted line used for? How can a slotted line be used for measuring impedances?

Problems

13-1. A lossless coaxial cable is 0.953 wavelength long. It was tested by measuring the open- and short-circuit impedances. They are: $Z_{oc} = j163.542$ ohms; and $Z_{sc} = -j15.2865$ ohms. Determine Z_o for this line.

13-2. An 80-mile line was tested at 60 hertz by performing the open- and short-circuit tests. The results were as follows: Z_{oc} =

$= 1{,}335.25 \angle{-89.42^\circ}$ ohms; $Z_{sc} = 171.503 \angle{75.382^\circ}$ ohms; It is known that the wavelength is approximately 1500 miles. Determine:

(a) Z_o.
(b) α.
(c) β.
(d) R, L, G, and C.

13-3. The equivalent pye circuit of a transmission line is as follows:

$$Z_\pi = 145.595 \angle{64.785^\circ} \text{ ohms}; \quad \frac{1}{\frac{Y_\pi}{2}} = 5.61 \times 10^3 \angle{-89.787^\circ}$$

ohms.

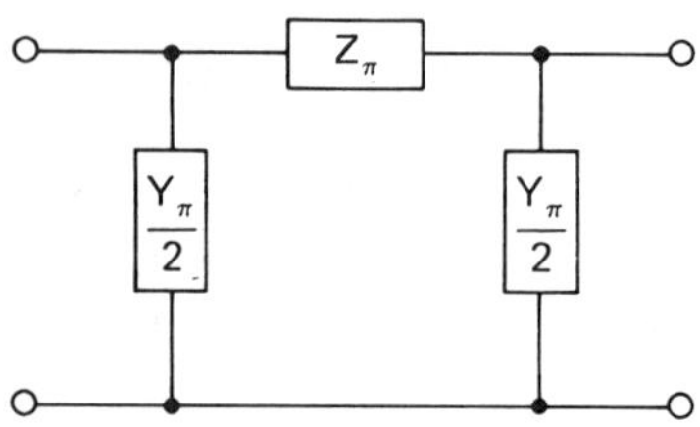

(a) Find the open circuit impedance.
(b) Find the short circuit impedance.
(c) From the results of (a) and (b) determine the following:
 (1) Z_o.
 (2) α.
 (3) β.

13-4. A test was made on a lossless line in the laboratory. The following data were obtained: $Z_o = 50$ ohms. When an unknown load impedance was replaced with a short circuit the voltage minimum shifted 0.15 wavelength to the right. If the VSWR was 3.2 when the unknown load was connected what is the value of the unknown load impedance?

13-5. An unknown load impedance is connected to a 50-ohm coaxial slotted line and the following measurements made: VSWR = 5.24. The distance between voltage minimums is 12.0 cm. When the load is replaced with a short circuit the voltage minimum shifts 2.0 cm to the left. Determine the value of the load impedance.

13-6. An unknown load impedance is connected to a 50-ohm coaxial slotted line and the following measurements made: The distance between voltage minimums = 22 cm. Total travel between two adjacent $\sqrt{2}\ V_{min}$ points is 0.6 cm. Distance to the first voltage minimum is 8 cm. Determine:

(a) The VSWR.

(b) The unknown load impedance.

14

Antennas

14.1 Introduction

One has to stretch the imagination somewhat to think of antenna systems when studying the fundamentals of transmission lines. If we think of a transmission line, however, as simply a guide for the electromagnetic fields which are transmitting energy from one point to another point, then it makes some sense to look at antennas since they too direct electromagnetic waves and the associated energy from one point to another point but without the aid of conductor-type wave guides. The study of antennas in detail is a complete course in itself; while it is fascinating our approach will be in the nature of an introduction to some of the basic concepts. This will be useful to the students who may wish to go into greater detail in the study of antennas and to the students who may find that their study of transmission lines is often associated with a laboratory where antennas serve as loads for the transmission lines being investigated. We will look at the basic concepts used when studying antennas and spend some time looking at definitions and specialized antennas. With these fundamentals, persons who end up in engineering positions that require a more complete knowledge of antenna operation may continue their studies with a minimum of difficulty when perusing the extensive literature in this area.

Since antennas are devices that convert electrical energy to electromagnetic energy we might think of them as transducers, much as a d. c. motor which is converting electrical energy to mechanical energy is called a transducer. But the process for the antenna is much more involved and requires a return to Maxwell's equations and the use of the mathematical approach along with some simplifying assumptions from time to time. Theoretically, one could start with a knowledge of how the current is varying in an antenna and develop the expression for the vector magnetic potential, $\vec{A}$, and then use the relationship that $\vec{B} = \nabla \times \vec{A}$ to determine the $\vec{B}$ field which will then allow us to find the $\vec{H}$ and $\vec{E}$ fields using our field equations. From a practical viewpoint this is not always so easy since the actual current action in an antenna is not always known and often must be approximated. However, engineers in the past have made some good guesses about the current distributions in antennas and the predicted mathematical results have checked out very closely with the measured results in the laboratory. This gave them some confidence that their educated guesses were somewhere in the ballpark and were justified on the basis of practical results.

The study of antennas is further complicated by the concept that the waves take time to move from one point to another point. We have the traveling wave concept much as we had on a conductor transmission line. So we have to take into account that the action of the wave in space is dictated by what happened in the antenna at an earlier time. This action is often referred to in terms of the *retarded potential*. It also becomes very important when one attempts to consider the case where two or more antennas are used in an array and are fed signals that may or may not be out of phase with each other. The raditaion pattern in space then becomes a function of the physical locations of the antennas and how their signals are related to each other on a time basis. The engineer is able to make the antenna system directional by properly spacing the antennas and adjusting the phase relationships between the currents in the various antennas.

The radiation patterns of antennas are also influenced by the effect of the earth; the ultimate energy that arrives at some receiving antenna may be affected, not only by the earth, but by the atmosphere, weather conditions, absorption and reflection of the waves by natural and man-made objects, the frequency of the signal and by the type of signal that was originally sent out in terms of direction, polarization and other factors. In this chapter we will look at some of the different types of radiation patterns that are usually discussed in terms of not only the $\vec{E}$ and $\vec{H}$ fields but with respect to Poynting's vector since it gives us a feel for the power being radiated by the sending antenna. Since, in engineering applications, it is nice to have some reference point to begin a discussion, we will look at some hypothetical antennas which allow us to look at real antennas on a comparison basis.

Since we are discussing the propagation of electromagnetic fields through space the student should take some time to quickly review previous work on traveling waves that is normally done when studying electromagnetic field theory. The plane wave studied in a fields course is often the practical result of a radiated wave from a real antenna because after the wave has traveled a great distance it often looks like a plane wave to the receiving antenna. If one thinks of a spherical wave traveling through space one can visualize the plane-wave effect as the spherical wave reaches its destination.

Antennas must be fed with an electrical signal that is generated and transmitted by some type of transmission line which may involve the types of transmission lines already discussed in the text, such as coaxial cables, wave guides, etc. As a circuit element, the antenna has impedance; the concept of the impedance of the antenna becomes important when the engineer must match his antenna to the connecting circuit and provide for maximum power transfer from his signal source to space through the action of the antenna. Obviously, every antenna must have an input impedance with an effective, real component that represents a resistance. The power absorbed by this equivalent resistance must represent the real power that would be transformed into electromagnetic power and sent into space if the antenna had an efficiency of 100 percent. The receiving antenna can also be thought of as having an effective impedance and an equivalent voltage source that is feeding the signal into some sort of useful device at that end. Determining the electrical characteristics of antennas is just another useful function of the antenna engineer since equivalent circuits are often used and most engineers are familiar with them from other applications.

14.2 Differential Length Antennas—Basic Ideas and Definitions

Let us start out with a very theoretical antenna which will illustrate the general procedure for dealing with the problem of determining the electromagnetic fields of any antenna. Our antenna will have a differential length and be situated at the center of the coordinate system and parallel to the z axis. We will use the spherical coordinate system with the variables, r, θ, and ϕ. The basic configuration is shown in Figure 14-1.

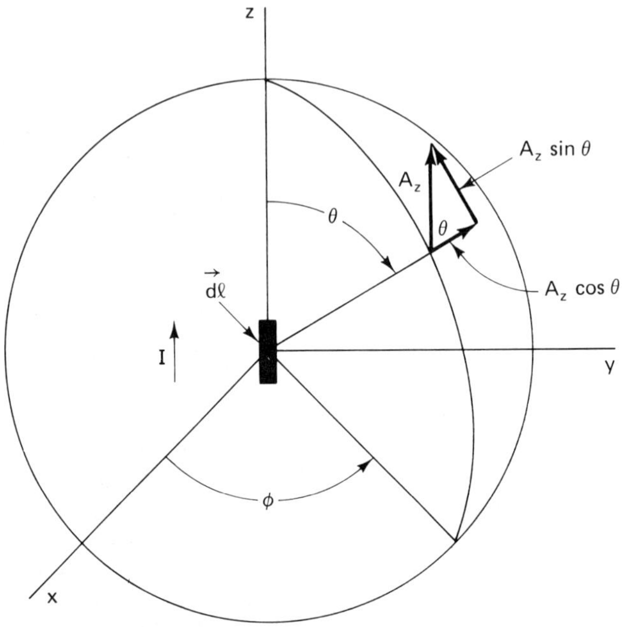

Figure 14-1. Differential antenna.

In field theory we had equation (14-1).

$$\vec{A} = \frac{\mu}{4\pi} \iiint \frac{\vec{J}\left(t - \frac{r}{c}\right)}{r} \, dvol. \tag{14-1}$$

In this equation, $\vec{A}$ is the vector magnetic potential in webers per meter, $\vec{J}$ is the current density in amperes per meter squared, and c is the velocity of light. The term $\vec{J}\left(t - \frac{r}{c}\right)$ arises because the value of $\vec{A}$ at any point in space is determined by what happened to $\vec{J}$ at an earlier time. $\vec{A}$ is usually called the retarded potential because of the time factor involved with the effect of a wave traveling through space. In our case we can also write equation (14-1) as equation (14-2).

$$\vec{A} = \frac{\mu}{4\pi} \int \frac{I\left(t - \frac{r}{c}\right) d\vec{\ell}}{r}. \tag{14-2}$$

Since we have a differential current element we can rewrite (14-2) as:

$$\vec{A} = \frac{\mu}{4\pi r} I\left(t - \frac{r}{c}\right) d\vec{\ell}. \tag{14-3}$$

As always, $\vec{A}$ takes its direction from the direction of the current, but since the current is a scalar we give the direction to $d\vec{\ell}$. In our case we are assuming that the current is in the positive z direction so we have only an A_z component for the vector magnetic potential.

Now let us assume that the current in the differential length is cosinusoidal in form.

$$\vec{A} = \frac{\mu}{4\pi r} I \cos \omega\left(t - \frac{r}{c}\right) d\vec{\ell}. \tag{14-4}$$

$$\vec{A} = \frac{\mu}{4\pi r} I \cos \left(\omega t - \frac{\omega r}{c}\right) d\vec{\ell}. \tag{14-5}$$

This looks like the form $(\omega t - \beta z)$ that we had for the transmission line where we now have:

$$\beta = \frac{\omega}{c}. \tag{14-6}$$

This, we recognize, is the form of a traveling wave and we have a delay of $\frac{\omega r}{c}$ radians, which we can think of as a phase angle delay corresponding to a time delay of $\frac{r}{c}$ seconds. To save some ink and your instructor's chalk let us define:

$$t' = t - \frac{r}{c}. \tag{14-7}$$

So now we can write:

$$A_z = \frac{\mu}{4\pi r} I \cos \omega t' \, d\ell. \tag{14-8}$$

From Figure 14-1 and using the spherical coordinate angle, θ, we can also write:

$$A_r = A_z \cos \theta. \tag{14-9}$$
$$A_\theta = -A_z \sin \theta. \tag{14-10}$$
$$A_\phi = 0. \tag{14-11}$$
$$\text{So } \vec{A} = A_z \cos \theta \, \vec{a}_r - A_z \sin \theta \, \vec{a}_\theta. \tag{14-12}$$

Now let us use a vector identity.

$$\vec{B} = \nabla \times \vec{A} = \frac{1}{r \sin \theta}\left[\frac{\partial}{\partial \theta}(A_\phi \sin \theta) - \frac{\partial A_\theta}{\partial \phi}\right]\vec{a}_r + \frac{1}{r}\left[\frac{1}{\sin \theta}\frac{\partial A_r}{\partial \phi} - \frac{\partial}{\partial r}(rA_\phi)\right]\vec{a}_\theta + \frac{1}{r}\left[\frac{\partial}{\partial r}(rA_\theta) - \frac{\partial A_r}{\partial \theta}\right]\vec{a}_\phi. \tag{14-13}$$

In our special case we have:

$$\vec{B} = \frac{1}{r}\left[\frac{\partial}{\partial r}(rA_\theta) - \frac{\partial A_r}{\partial \theta}\right]\vec{a}_\phi. \tag{14-14}$$

So

$$\vec{B} = \frac{1}{r}\left[\frac{\partial}{\partial r}(-rA_z \sin \theta) - \frac{\partial}{\partial \theta}(A_z \cos \theta)\right]\vec{a}_\phi =$$

$$\frac{1}{r}\left\{\frac{\partial}{\partial r}\left[-r \sin \theta \frac{\mu I}{4\pi r} \cos \omega\left(t - \frac{r}{c}\right) d\ell\right] - \frac{\partial}{\partial \theta}\left[\cos \theta \frac{\mu I}{4\pi r} \cos\left(t - \frac{r}{c}\right) d\ell\right]\right\}\vec{a}_\phi =$$

$$\frac{1}{r}\left[\frac{\mu I}{4\pi} d\ell\right]\left\{\frac{\partial}{\partial r}\left[-\sin \theta \cos \omega\left(t - \frac{r}{c}\right)\right] - \frac{\partial}{\partial \theta}\left[\frac{1}{r}\cos \theta \cos \omega\left(t - \frac{r}{c}\right)\right]\right\}\vec{a}_\phi =$$

$$\frac{\mu I\,d\ell}{4\pi r}\left[-\sin\theta\left(-\sin\omega\left(t-\frac{r}{c}\right)\frac{(-\omega)}{c}\right)-\frac{-\sin\theta}{r}\cos\omega\left(t-\frac{r}{c}\right)\right]\vec{a}_\phi =$$

$$\frac{\mu I\,d\ell}{4\pi}\sin\theta\left[-\frac{\omega}{rc}\sin\omega\left(t-\frac{r}{c}\right)+\frac{\cos\omega}{r^2}\left(t-\frac{r}{c}\right)\right]\vec{a}_\phi. \tag{14-15}$$

Since $\vec{B} = \mu\,\vec{H}$, $\vec{H} = \vec{B}/\mu$.

$$\vec{H} = \frac{I\,d\ell\sin\theta}{4\pi}\left[-\frac{\omega}{rc}\sin\omega t' + \frac{\cos\omega t'}{r^2}\right]\vec{a}_\phi. \tag{14-16}$$

We can now use the form of equation (14-13) to find the curl of $\vec{H}$. After some work we have:

$$\nabla\times\vec{H} = \frac{2I\,d\ell\cos\theta}{4\pi}\left[-\frac{\omega}{r^2c}\sin\omega\left(t-\frac{r}{c}\right)+\frac{\cos\omega}{r^2}\left(t-\frac{r}{c}\right)\right]\vec{a}_r$$
$$+\left[\frac{-\omega^2}{rc^2}I\,d\ell\frac{\sin\theta}{4\pi}\cos\omega\left(t-\frac{r}{c}\right)-\frac{I\,d\ell}{4\pi}\sin\theta\frac{\omega}{r^2c}\sin\omega\left(t-\frac{r}{c}\right)\right.$$
$$\left.+\frac{I\,d\ell\sin\theta}{4\pi r^3}\cos\omega\left(t-\frac{r}{c}\right)\right]\vec{a}_\theta. \tag{14-17}$$

Now using the expression from Maxwell's equations:

$$\vec{E} = \frac{1}{\varepsilon}\int(\nabla\times\vec{H})dt. \tag{14-18}$$

We obtain after some more mathematical manipulation:

$$\vec{E} = \frac{2I\,d\ell\cos\theta}{4\pi}\left[\frac{\cos\left(\omega t-\frac{\omega r}{c}\right)}{r^2c}+\frac{1}{\omega r^3}\sin\left(\omega t-\frac{\omega r}{c}\right)\right]\vec{a}_r$$
$$+\frac{I\,d\ell\sin\theta}{4\pi\varepsilon}\left[-\frac{\omega}{rc^2}\sin\omega\left(t-\frac{r}{c}\right)+\frac{1}{r^2c}\cos\omega\left(t-\frac{r}{c}\right)+\frac{1}{\omega r^3}\sin\omega\left(t-\frac{r}{c}\right)\right]\vec{a}_\theta. \tag{14-19}$$

In final condensed form we can write:

$$H_\phi = \frac{I\,d\ell\sin\theta}{4\pi}\left[-\frac{\omega\sin\omega t'}{rc}+\frac{\cos\omega t'}{r^2}\right]. \tag{14-20}$$

$$E_r = \frac{2I\,d\ell\cos\theta}{4\pi\varepsilon}\left[\frac{\cos\omega t'}{r^2c} + \frac{\sin\omega t'}{\omega r^3}\right]. \tag{14-21}$$

$$E_\theta = \frac{I\,d\ell\sin\theta}{4\pi\varepsilon}\left[-\frac{\omega\sin\omega t'}{rc^2} + \frac{\cos\omega t'}{cr^2} + \frac{\sin\omega t'}{\omega r^3}\right]. \tag{14-22}$$

$$t' = \left(t - \frac{r}{c}\right). \tag{14-23}$$

As one might suspect from looking at these equations, the total field is extremely complicated even though we have used a very simple and theoretical antenna. However, we may simplify our results by defining the "near" and "far" fields for our antenna. Equations (14-20), (14-21), and (14-22) are exact everywhere in space but we will say they describe the field components in the "near" field. Looking at the equations shows that we have some terms that decrease in magnitude very rapidly as r increases. Certainly the r^2 and r^3 terms diminish much faster than the r term since they are in the denominator. Then we can approximate for the "far" field with excellent results by neglecting the higher order terms. For the "far" field we have:

$$E_\theta \simeq \frac{I\,d\ell\sin\theta}{4\pi\varepsilon}\left[-\frac{\omega\sin\omega t'}{rc^2}\right]. \tag{14-24}$$

$$E_r \simeq 0. \tag{14-25}$$

$$H_\phi \simeq \frac{I\,d\ell\sin\theta}{4\pi}\left[-\frac{\omega\sin\omega t'}{rc^2}\right]. \tag{14-26}$$

We have arrived at the result that we can neglect E_r when we look at the "far" field by neglecting the terms containing r^2 and r^3, but, as it turns out, we can obtain the same result by considering the flow of energy in the electromagnetic field generated by our tiny antenna. To look at this aspect we must go back to Poynting's vector which is an expression for the power density in the electromagnetic field.

$$\vec{\mathcal{P}} = \vec{E} \times \vec{H}. \quad \text{(watts/meter}^2\text{)} \tag{14-27}$$

The power radiated becomes:

$$\vec{\mathcal{P}} = \vec{E} \times \vec{H} = \begin{vmatrix} \vec{a}_r & \vec{a}_\theta & \vec{a}_\phi \\ E_r & E_\theta & E_\phi \\ H_r & H_\theta & H_\phi \end{vmatrix}.$$

In our case we have:

$$\vec{\mathcal{P}} = \begin{vmatrix} \vec{a}_r & \vec{a}_\theta & \vec{a}_\phi \\ E_r & E_\theta & 0 \\ 0 & 0 & H_\phi \end{vmatrix} = E_\theta H_\phi \vec{a}_r - E_r H_\phi \vec{a}_\theta. \tag{14-28}$$

So,

$$\vec{\mathcal{P}} = \mathcal{P}_r \vec{a}_r + \mathcal{P}_\theta \vec{a}_\theta. \tag{14-29}$$

$$\mathcal{P}_r = E_\theta H_\phi. \tag{14-30}$$

$$\mathcal{P}_\theta = -E_r H_\phi. \tag{14-31}$$

However, if we look at Figure 14-2 we see that it is only the P_r component that contributes to the net outward flow of power (or energy) since it is the component that is perpendicular to the imaginary spherical surface surrounding our antenna. The P_θ component is tangent to the spherical surface and no power is moving through the surface. So we have to concentrate only on the P_r component to see what value of power moves into the space surrounding our antenna through the action of the "far" field.

$$\begin{aligned} \mathcal{P}_r &= E_\theta H_\phi \\ &= \left[\frac{I\,d\ell \sin\theta}{4\pi\varepsilon}\left\{-\frac{\omega \sin \omega t'}{rc^2} + \frac{\cos \omega t'}{cr^2} + \frac{\sin\omega t'}{\omega r^3}\right\}\right] \times \left[\frac{I\,d\ell \sin\theta}{4\pi}\left\{-\frac{\omega \sin \omega t'}{rc}\right.\right. \\ &\quad \left.\left. + \frac{\cos \omega t'}{r^2}\right\}\right] \\ &= \frac{I^2 d\ell^2 \sin^2\theta}{16\pi^2\varepsilon}\left[\frac{\omega^2 \sin^2 \omega t'}{r^2c^3} - \frac{\omega \sin \omega t' \cos \omega t'}{r^3c^2} - \frac{\omega \sin \omega t' \cos \omega t'}{r^3c^2} + \frac{\cos^2 \omega t'}{cr^4}\right. \\ &\quad \left. - \frac{\sin^2 \omega t'}{r^4c} + \frac{\sin \omega t' \cos \omega t'}{\omega r^5}\right]. \end{aligned} \tag{14-32}$$

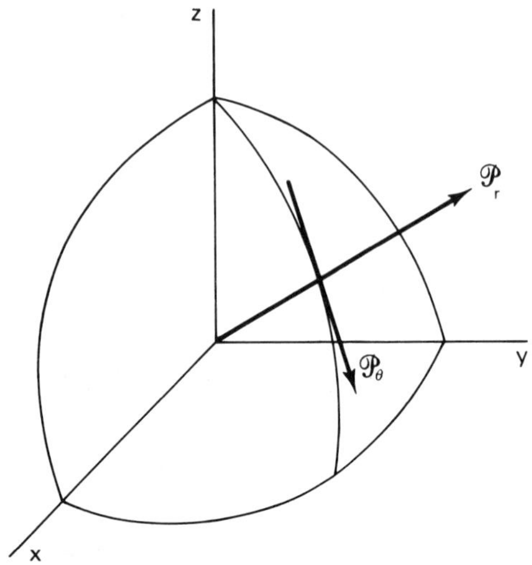

Figure 14-2. Flow of power (or energy).

Since $\sin A \cos B = \frac{1}{2}[\sin(A+B) + \sin(A-B)]$, $\sin^2 x = \frac{1}{2} - \frac{1}{2}\cos 2x$, and $\cos^2 x = \frac{1}{2} + \frac{1}{2}\cos 2x$, we can write:

$$\sin \omega t' \cos \omega t' = \frac{1}{2}[\sin(\omega t' + \omega t') + \sin(\omega t' - \omega t')] = \frac{1}{2}\sin 2\omega t'.$$

$$\mathcal{P}_r = \frac{I^2 d\ell^2 \sin^2\theta}{16\,\pi^2\varepsilon}\left[\frac{\omega^2}{r^2c^3}\left(\frac{1}{2} - \frac{1}{2}\cos 2\omega t'\right) - \frac{\omega}{r^3c^2}\left(\frac{1}{2}\sin 2\omega t'\right) - \frac{\omega}{r^3c^2}\left(\frac{1}{2}\sin 2\omega t'\right) + \frac{1}{cr^4}\left(\frac{1}{2} + \frac{1}{2}\cos 2\omega t'\right) - \frac{1}{r^4c}\left(\frac{1}{2} - \frac{1}{2}\cos 2\omega t'\right) + \frac{1}{\omega r^5}\left(\frac{1}{2}\sin 2\omega t'\right)\right].$$

$$\mathcal{P}_r = \left\{\frac{I^2 d\ell^2 \sin^2\theta}{16\pi^2\varepsilon}\right\}\left\{\left[\frac{1}{2}\frac{\omega^2}{r^2c^3} + \frac{1}{2cr^4} - \frac{1}{2cr^4}\right] + \cos 2\omega t'\left[-\frac{1}{2}\frac{\omega^2}{r^2c^3} + \frac{1}{2cr^4} + \frac{1}{2cr^4}\right] + \sin 2\omega t'\left[-\frac{1}{2}\frac{\omega}{r^3c^2}\,\frac{1}{2}\frac{\omega}{r^3c^2} + \frac{1}{2\omega r^5}\right]\right\}. \tag{14-33}$$

Equation (14-33) gives Poynting's vector as a function of time but we are usually more interested in the average power crossing our imaginary spherical surface. The average power in watts per square meter may be found, as we have done in circuits courses, from equation (14-34).

$$\mathcal{P}_{r(\text{average})} = \frac{1}{T}\int_0^T \mathcal{P}_r \, dt. \qquad (14\text{-}34)$$

However, from our previous work in circuits, we know that the time average value of the double frequency cosine and sine terms, cos $2\omega t'$ and sin $2\omega t'$ always gives a value of zero so we are able to write the answer right down.

$$\mathcal{P}_{r(\text{average})} = \frac{I^2 d\ell^2 \sin^2\theta}{16\pi^2\varepsilon}\,\frac{\omega^2}{2r^2c^3}. \qquad (\text{watts/meter}^2) \qquad (14\text{-}35)$$

By investigating equation (14-35) we see that it comes from the $\frac{1}{r}$ term in E_θ and the $\frac{1}{r}$ term in H_ϕ, so all we ever really needed was:

$$\vec{E} = \frac{-Id\ell \sin\theta \, \omega \sin\omega t'}{4\pi\varepsilon}\,\vec{a}_\theta. \qquad (14\text{-}36)$$

$$\vec{H} = -\frac{Id\ell \sin\theta \, \omega \sin\omega t'}{4\pi r c}\,\vec{a}_\phi. \qquad (14\text{-}37)$$

$$\vec{\mathcal{P}}_r = \vec{E} \times \vec{H} = \frac{I^2 d\ell^2 \sin^2\theta \, \omega^2}{16\pi\varepsilon r^2 c^3}\sin^2\omega t' \qquad (14\text{-}38)$$

Remembering that $\sin^2 x = \frac{1}{2} - \frac{1}{2}\cos 2x$, we see that equation (14-38) tells us immediately that the averge power turns out to be simply $\frac{1}{2}$ the product of the $\frac{1}{r}$ terms in the "far" field. It can be shown that this is true for the field of any antenna.

$$\mathcal{P}_{(\text{average})} = \frac{1}{2}\left[\frac{I^2 d\ell^2 \sin^2\theta \, \omega^2}{16\pi\varepsilon r^2 c^3}\right]. \qquad (\text{watts/meter}^2) \qquad (14\text{-}39)$$

Note:

$$\vec{E} = \left[- \frac{\omega I d\ell \sin\theta \sin\omega t'}{4\pi r c} \right] \frac{1}{\varepsilon_o c} \vec{a}_\theta,$$

Since $\varepsilon = \varepsilon_o$ in free space, and

$$\vec{H} = - \frac{\omega I d\ell \sin\theta \sin\omega t'}{4\pi r c} \vec{a}_\phi.$$

Therefore we can write:

$$|\vec{E}| = \frac{1}{\varepsilon_o c} |\vec{H}|, \tag{14-40}$$

and since

$$c = \frac{1}{\sqrt{\mu_o \varepsilon_o}}, \tag{14-41}$$

$$|\vec{E}| = \frac{1}{\dfrac{\varepsilon_o 1}{\sqrt{\mu_o \varepsilon_o}}} |\vec{H}| = \sqrt{\frac{\mu_o \varepsilon_o}{\varepsilon_o{}^2}} |\vec{H}| = \sqrt{\frac{\mu_o}{\varepsilon_o}} |\vec{H}|. \tag{14-42}$$

Let us define, as we did in fields, the intrinsic impedance.

$$\eta = \sqrt{\frac{\mu}{\varepsilon}}. \qquad \text{(ohms)} \tag{14-43}$$

So,

$$|\vec{E}| = \eta |\vec{H}|. \tag{14-44}$$

For free space we have:

$$\eta = \sqrt{\frac{\mu_o}{\varepsilon_o}}.$$

When we substitute the values for μ_o and ε_o we obtain $\eta = 377$ ohms, and so in free space we obtain:

$$|\vec{E}| = 377\ |\vec{H}|. \tag{14-45}$$

We observe that $\vec{E}$ and $\vec{H}$ are in time phase with each other and that $\vec{E}$ and $\vec{H}$ are at right angles to each other so that the field has the same type of properties that the plane wave studied in fields had with $\vec{E}$ and $\vec{H}$ at right angles to each other. If r is large, so that our spherical surface is large enough, the wave appears as a plane wave over a small area on the surface of the sphere.

Another important concept for any antenna is the total power radiated from the antenna in watts. Since we have the expression for the average power in watts per unit area we can find the total power radiated by integrating across the total surface area surrounding the antenna, as indicated by equation (14-46).

$$P_{(\text{power radiated})} = \oiint \mathcal{P}_{r(\text{average})} dS. \qquad (\text{watts}) \tag{14-46}$$

This equation can also be written in slightly different form as:

$$P_{(\text{power radiated})} = \oiint \frac{1}{2} (\vec{E} \times \vec{H})\ dS = \frac{1}{2} \oiint |\vec{E}|\ |\vec{H}|\ dS =$$

$$\frac{1}{2} \oiint \eta\ |\vec{H}|\ |\vec{H}|\ dS = \frac{1}{2} \oiint \eta\ |\vec{H}|^2\ dS. \tag{14-47}$$

Using a sphere, we have for our special case:

$$P_{(\text{power radiated})} = \frac{1}{2} \eta \int_{\phi=0}^{2\pi} \int_{\theta=0}^{\pi} \left[\frac{\omega I\ d\ell \sin\theta}{4\pi r c}\right]^2 (r\sin\theta d\phi)(r d\theta). \tag{14-48}$$

After going through the mathematics we obtain:

$$P_{(\text{power radiated})} = \frac{\eta \omega^2 I^2\ d\ell^2}{12\pi c^2}\ (\text{watts}). \tag{14-49}$$

Since $\beta = \frac{2\pi}{\lambda} = \frac{\omega}{c}$, we can write:

$$P_{(\text{power radiated})} = \frac{\eta}{12\pi}\beta^2 I^2 d\ell^2 \text{ (watts)}. \qquad (14\text{-}50)$$

But $\frac{I^2}{2} = \frac{I}{\sqrt{2}}\frac{I}{\sqrt{2}} = I_{rms}{}^2$. Therefore, $I^2 = 2\, I_{rms}{}^2$, so we have:

$$P_{(\text{power radiated})} = \frac{\eta}{12\pi}\left[\frac{2\pi}{\lambda}\right]^2 2I_{rms}{}^2\, d\ell^2 = \frac{2}{3}\frac{\eta}{\lambda^2} I_{rms}{}^2 d\ell^2. \qquad (14\text{-}51)$$

Since $\eta = 377 = 120\pi$ we can write:

$$P_{(\text{power radiated})} = 80\pi^2 I_{rms}{}^2 \left[\frac{d\ell}{\lambda}\right]^2. \quad \text{(watts)} \qquad (14\text{-}52)$$

Let us equate this power to the power supplied to an equivalent circuit that has the same current and a resistance which we will call the equivalent radiation resistance, $R_{(rad)}$.

$$P_{(\text{power radiated})} = 80\pi^2 I_{rms}{}^2 \left[\frac{d\ell}{\lambda}\right]^2 = I_{rms}{}^2 R_{(rad)}. \qquad (14\text{-}53)$$

So the radiation resistance of our tiny antenna becomes:

$$R_{(rad)} = 80\pi^2 \left[\frac{d\ell}{\lambda}\right]^2 \text{ ohms}. \qquad (14\text{-}54)$$

Let us assume that our tiny antenna has a length that is one hundredth of a wavelength: $d\ell = 0.01\lambda$.

$$R_{(rad)} = 80\pi^2 (0.01)^2 = 0.08 \text{ ohms}.$$

Another approach to the differential antenna would be to use phasors. Let us stop for a moment and see how this might be done since it, too, is a technique that is often helpful when dealing with antennas. We will start with the same type of figure that we had before.

We will assume as before that the current in the differential element, $d\ell$, is cosinusoidal in form.

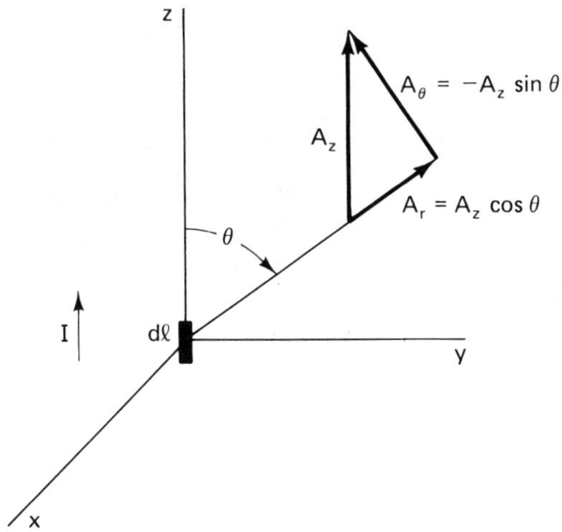

Figure 14-3. Phasors for our field.

$$I = I_o \cos \omega t. \tag{14-55}$$

Using equation (14-2) we have:

$$\vec{A} = \frac{\mu_o}{4\pi} \int \frac{I\left(t - \frac{r}{c}\right)}{r} \vec{d\ell} \text{ (at point } P\text{)}$$

or

$$\vec{A} = \frac{\mu_o}{4\pi r} I\left(t - \frac{r}{c}\right) d\ell\, \vec{a}_z. \tag{14-56}$$

To take care of the retarded potential concept let us write:

$$I' = I_o \cos\left(\omega t - \frac{r}{c}\right) = I_o \cos \omega\left(t - \frac{\omega r}{c}\right). \tag{14-57}$$

$$I' = \mathcal{R}\left[I_o e^{j(\omega t - \omega r/c)}\right] = \mathcal{R}\left[I_o e^{j\omega t} e^{-j\omega r/c}\right].$$

$$I'_{\text{phasor}} = I_o e^{-j\omega r/c}. \tag{14-58}$$

From Figure 14-3 we have for the phasor values: $A_r = A_z \cos\theta$, and $A_\theta = -A_z \sin\theta$. The phasor form of equation (14-56) is:

$$\vec{A} = \frac{\mu_o I'_{\text{phasor}} \vec{d\ell}\, \vec{a}_z}{4\pi r}. \tag{14-59}$$

To save some effort let us let $I'_{\text{phasor}} = I'_p = I_o\, e^{-j\omega r/c}$. So we can write:

$$A_r = \frac{\mu_o I_o d\ell\, e^{-j\omega r/c}}{4\pi r} \cos\theta. \tag{14-60}$$

$$A_\theta = \frac{-\mu_o I_o d\ell\, e^{-j\omega r/c}}{4\pi r} \sin\theta. \tag{14-61}$$

We now go through the same procedure that we did when we were dealing with time functions.

$$\vec{H} = \frac{\vec{B}}{\mu_o} = \frac{\nabla \times \vec{A}}{\mu_o}.$$

Skipping the details gives:

$$\vec{H} = \frac{I_o\, d\ell \sin\theta}{4\pi} e^{-j\omega r/c} \left[\frac{j\omega}{rc} + \frac{1}{r^2}\right] \vec{a}_\phi. \tag{14-62}$$

To find the E field we go back to one of Maxwell's equations and convert it to the phasor form.

$$\nabla \times \vec{H} = \vec{J} + \frac{\partial \vec{D}}{\partial t}. \qquad \nabla \times \vec{H}_p = 0 + j\omega\varepsilon_o \vec{E}_p. \tag{14-63}$$

Remember that $\vec{J}$ is assumed to be zero in free space (no current flowing).

$$\vec{E}_p = \frac{1}{j\omega\varepsilon_o}[\nabla \times \vec{H}_p]. \tag{14-64}$$

After some algebra, along with some differentiation we obtain:

$$\begin{aligned}\vec{E}_p = {} & \frac{\cos\theta I_o d\ell}{2\pi} e^{-j\omega r/c}\left[\frac{1}{r^2\varepsilon_o c} + \frac{1}{j\omega\varepsilon_o r^3}\right]\vec{a}_r + \\ & \frac{\sin\theta\, I_o d\ell}{4\pi} e^{-j\omega r/c}\left[\frac{1}{j\omega\varepsilon r^3} - \frac{\omega^2}{j\omega\varepsilon_o rc^2} + \frac{j\omega}{rcj\omega\varepsilon_o r}\right]\vec{a}_\theta.\end{aligned} \tag{14-65}$$

Summarizing our results we have:

$$H_\phi = \frac{I_o\, d\ell \sin\theta}{4\pi} e^{-j\omega r/c}\left[\frac{j\omega}{rc} + \frac{1}{r^2}\right]. \tag{14-66}$$

$$E_r = \frac{I_o\, d\ell \cos\theta}{2\pi} e^{-j\omega r/c}\left[\frac{1}{r^2\varepsilon_o c} + \frac{1}{j\omega\varepsilon_o r^3}\right]. \tag{14-67}$$

$$E_\theta = \frac{I_o\, d\ell \sin\theta}{4\pi} e^{-j\omega r/c}\left[\frac{1}{j\omega\varepsilon_o r^3} - \frac{\omega}{j\varepsilon_o rc^2} + \frac{1}{r^2 c\varepsilon_o}\right]. \tag{14-68}$$

Using the expressions:

$$\omega = 2\pi f,\; c = f\lambda,$$

and

$$e^{-j\omega r/c} = e^{-j2\pi fr/f\lambda} = e^{-j2\pi r/\lambda},$$

we obtain these equations in a slightly different form.

$$H_\phi = \frac{I_o\, d\ell \sin\theta}{4\pi} e^{-j2\pi r/\lambda}\left[\frac{j2\pi}{\lambda r} + \frac{1}{r^2}\right]. \tag{14-69}$$

$$E_r = \frac{I_o\, d\ell \cos\theta}{2\pi} e^{-j2\pi r/\lambda}\sqrt{\frac{\mu_o}{\varepsilon_o}}\left[\frac{1}{r^2} + \frac{\lambda}{j2\pi r^3}\right]. \tag{14-70}$$

$$E_\theta = \frac{I_o\, d\ell \sin\theta}{4\pi} e^{-j2\pi r/\lambda}\sqrt{\frac{\mu_o}{\varepsilon_o}}\left[\frac{j2\pi}{r\lambda} + \frac{1}{r^2} + \frac{\lambda}{j2\pi r^3}\right]. \tag{14-71}$$

The student should prove to himself that, if he converts these equations to functions of time, he will end up with the same equations that we developed when using time functions rather than phasor functions. If r is very small, say less than 1, then the $\frac{1}{r^3}$ term is very large compared to the $\frac{1}{r}$ term and so in the "near" field we can neglect the $\frac{1}{r}$ term in comparison to the $\frac{1}{r^3}$ term. However, the "far" field is the one we are really interested in so let us assume that r is very large, say 10 wavelengths or more. We would then obtain:

$$H_\phi = \frac{jI_o\, d\ell \sin\theta}{2\lambda r}\, e^{-j2\pi r/\lambda}. \tag{14-72}$$

$$E_r = 0. \tag{14-73}$$

$$E_\theta = j\sqrt{\frac{\mu_o}{\varepsilon_o}}\, \frac{I_o\, d\ell \sin\theta}{2r\lambda}\, e^{-j2\pi r/\lambda}. \tag{14-74}$$

As before, when we take $\vec{E} \times \vec{H}$ we obtain Poynting's vector in the $\vec{a}_r$ direction.

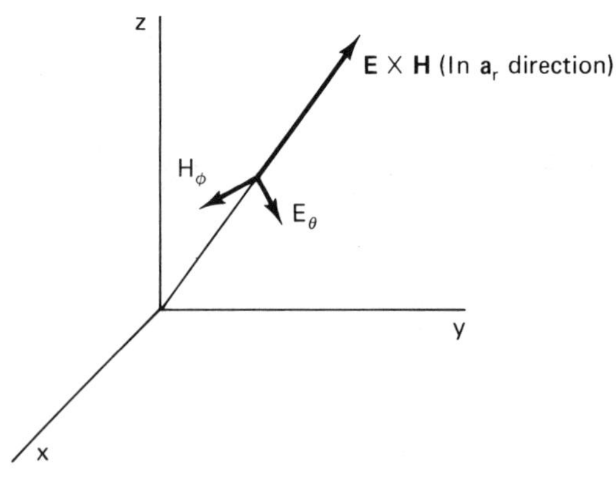

Figure 14-4. Poynting's vector.

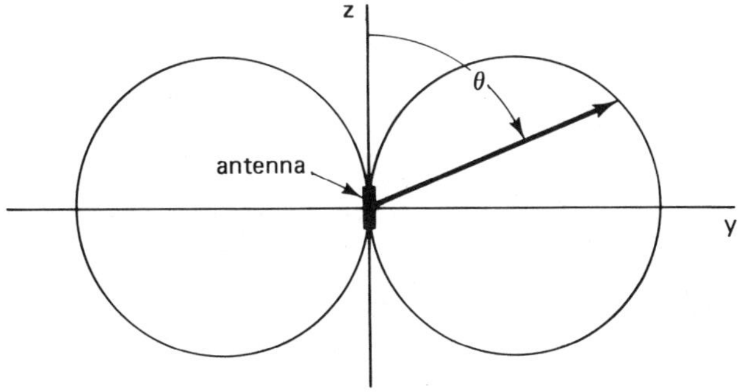

Figure 14-5. Pattern for E_θ and H_θ for differential antenna.

Both $\vec{E}$ and $\vec{H}$ vary with the sine of θ, so if we look at the pattern for either one of them we have one that is illustrated in Figure 14-5.

The radiation pattern looks like a donut without the hole which is symmetrical with the z axis. If we cut through the donut and see the pattern in the x-y plane we have a circle as indicated in Figure 14-6.

This gives a radiation pattern that is perpendicular to the miniature antenna for both the $\vec{E}$ and $\vec{H}$ components.

In the use of antennas the radiation pattern becomes important to the engineer since it gives him an understanding of how the antenna should be oriented in space for maximum energy flow from the sending antenna to the receiving antenna. For instance, in Figure 14-5 if the receiving antenna were on the z axis it would receive no signal but it would receive a maximum signal if it was situated in the x-y plane. Furthermore, the orientation of the $\vec{E}$ and $\vec{H}$ vectors in space tell the engineer how the receiving antenna should be oriented for maximum signal reception. For instance, if the $\vec{E}$ vector is horizontal to the ground (which is called horizontal polarization), the receiving antenna should also be horizontal to the ground to obtain the maximum signal. For instance, television signals are horizontally polarized.

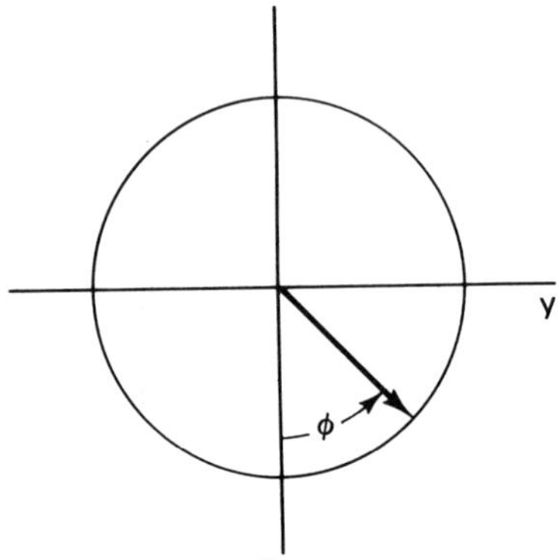

Figure 14-6. Pattern for E_θ and H_θ for differential antenna.

Let us go back to equation (14-72) and convert it to the time function.

$$H_\phi = j\frac{I_o\, d\ell \sin\theta}{2\, r\lambda}\, e^{-j2\pi r/\lambda}.$$

$$H_\phi(t) = \Re\left[j\frac{I_o\, d\ell \sin\theta\, e^{-j2\pi r/\lambda}\, e^{j\omega t}}{2r\lambda}\right] =$$

$$\Re\left[\frac{\left\{\cos\left(\omega t - \frac{j2\pi r}{\lambda}\right) + j\sin\left(\omega t - j\frac{2\pi r}{\lambda}\right)\right\} j\, I_o\, d\ell \sin\theta}{2r\lambda}\right] =$$

$$-\frac{I_o\, d\ell \sin\theta}{2r\lambda}\sin\left(\omega t - \frac{2\pi r}{\lambda}\right). \qquad (14\text{-}75)$$

Since $\lambda = \frac{c}{f}$, and $2\pi = \frac{\omega}{f}$, we can substitute and we obtain:

$$H_\phi(t) = -\frac{I_o\, d\ell \sin\theta}{2\dfrac{c}{f}r}\sin\left(\omega t - \frac{\omega}{f}r\frac{f}{c}\right) = -\frac{I_o\, d\ell \sin\theta}{2\,c\,r\dfrac{2\pi}{\omega}}\sin\left(\omega t - \frac{\omega r}{c}\right) = -\frac{I_o\, d\ell \sin\theta\ \omega}{4\pi rc}\sin\left(\omega t - \frac{\omega r}{c}\right) = -\frac{I_o\, d\ell \sin\theta\ \omega}{4\pi rc}\sin \omega t',$$

which is the same as equation (14-26).

14.3 Dipole Antennas

The dipole antenna is important from both a theoretical and practical viewpoint. They come in various lengths which are usually expressed in terms of a wavelength so that we talk about a half-wave dipole or a full-wave dipole; the folded dipole is often a basic element for a television antenna. They are also referred to in terms of the location of the input to the antenna so we talk about a center-fed dipole antenna if that is how it is connected to the line feeding it but it may be fed at different locations. A vertical antenna above the ground plane may be thought of as a dipole by using the method of images from field theory. We, however, will look briefly at a rather short dipole which is center fed and has a length which is considerably less than one wavelength; we will assume that $L \ll \lambda$, and that the current in the antenna has a linear distribution along the antenna. Obviously the current at the ends of the antenna would have to be zero. The case we are considering is illustrated in Figure 14-7(a).

In the preceding section we took the basic approach to solving the antenna problem; namely if we know something about the current we can solve for the vector magnetic potential, $\vec{A}$, and then find $\vec{B}$ using the expression that $\vec{B} = \nabla \times \vec{A}$, and from $\vec{B}$ we can proceed to find the $\vec{H}$ and $\vec{E}$ fields, Poynting's vector, radiation patterns and other information that has a practical value for the engineer. In the case of the short dipole we will approach the problem in a different manner by using the results found in Section 14.2 instead of starting from scratch. The justification is that if the antenna is short, we can assume that the contributions to the "far" field can be found by breaking the antenna length into an infinite number of infinitesimal lengths such as we used in the last section; since these tiny antennas will produce a field that is in time phase with each of the other little

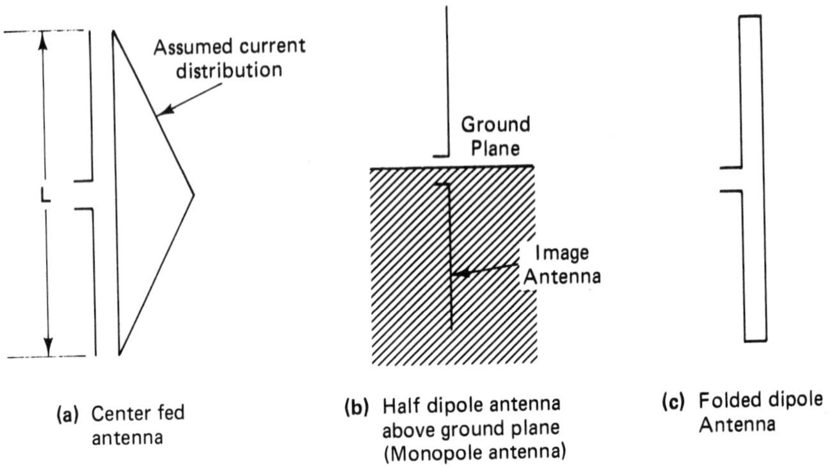

Figure 14-7. Examples of dipole antennas.

antennas we can simply evaluate the field by simple scalar addition. From equation (14-24) we know that the "far" field for a short element of antenna was:

$$E_\theta(t) = -\frac{I\,d\ell\,\sin\theta}{4\pi\varepsilon_0}\,\frac{\omega \sin \omega t'}{rc^2} = -\frac{\omega I\,d\ell\,\sin\theta}{4\pi\varepsilon_0 rc^2}\sin\omega\left(t - \frac{r}{c}\right).$$

And

$$\sin\omega\left(t - \frac{r}{c}\right) = \sin\left(\omega t - \frac{\omega r}{c}\right) = \sin(\omega t - \beta r) = I\left[\left(e^{j\omega t - \beta r}\right)\right]. \quad (14\text{-}76)$$

In equation (14-76) I stands for the imaginary part of the term in the brackets and, of course, $\beta = \frac{\omega}{c}$. So we can write our expression for $E_\theta(t)$ as:

$$E_\theta(t) = \mathcal{I}\left[-\frac{\omega I\, d\ell \sin\theta}{4\pi\varepsilon_0 r c^2}\left(e^{j\omega t - \beta r}\right)\right]. \qquad (14\text{-}77)$$

The phasor value can then be written as:

$$\vec{E} = -\frac{\omega I\, d\ell \sin\theta}{4\pi\varepsilon_0 r c^2}\, e^{-j\beta r}\, \vec{a}_\theta. \qquad (14\text{-}78)$$

Now we will find the contributions for all the little infinitesimal antennas.

$$E_\theta = \int_{-L/2}^{+L/2} -\frac{\omega I \sin\theta}{4\pi\varepsilon_0 r c^2}\, e^{-j\beta r}\, d\ell = -\frac{\omega \sin\theta\, e^{-j\beta r}}{4\pi\varepsilon_0 r c^2}\int_{-L/2}^{+L/2} I\, d\ell. \qquad (14\text{-}79)$$

We will assume a linear current distribution of the r.m.s. value of the current as shown in Figure 14-8. Let

$$\int_{-L/2}^{+L/2} I\, d\ell = I_{average}\,(L).$$

Now

$$I_{average} = \frac{1}{L}\int_{-L/2}^{+L/2} I\, dz$$

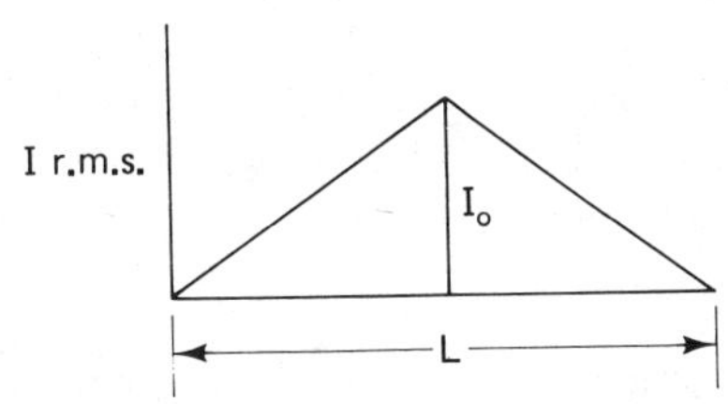

Figure 14-8. Assumed linear current distribution.

For a short dipole, it turns out that the current varies nearly linearly so we can find the average by finding the area under the curve. Area under the curve $= \frac{1}{2}$ (base) (height) for each triangle so we have:

$$\text{Area} = 2\left[\frac{1}{2}\frac{(L)}{2} I_0\right] = \frac{L\, I_0}{2}.$$

So

$$I_{\text{average}} = \frac{1}{L}\left[\frac{L\, I_0}{2}\right] = \frac{I_0}{2}. \qquad (14\text{-}80)$$

$$E_\theta = \frac{-\omega \sin\theta}{4\pi\varepsilon_0 r c^2} e^{-j\beta r} I_{\text{average}}\,(L) = -\frac{\omega \sin\theta\, e^{-j\beta r}}{4\pi\varepsilon_0 r c^2}\frac{I_0 L}{2}. \qquad (14\text{-}81)$$

And

$$H_\phi = \frac{E_\theta}{\eta} = -\frac{\omega \sin\theta\, e^{-j\beta r}}{\eta 4\pi\varepsilon_0 r c^2}\frac{I_0 L}{2}. \qquad (14\text{-}82)$$

We can find the power radiated by using equation (14-52).

$$P_{(\text{power radiated})} = 80\,\pi^2\, I_{\text{rms}}{}^2 \left[\frac{d\ell}{\lambda}\right]^2,$$

since we are assuming that $d\ell/\lambda$ is still small for our short dipole. So for our dipole antenna we have:

$$P_{(\text{power radiated})} = 80\,\pi^2 \left[\frac{L}{\lambda}\right]^2 \left[\frac{I_0}{2}\right]^2 = 20\,\pi^2 \left[\frac{L}{\lambda}\right]^2 I_{\text{rms}}{}^2. \qquad (14\text{-}83)$$

Our intuition should tell us that the half dipole (monopole) shown in Figure 14-7(b) should radiate half of this power since it is radiating only in the upper half. So for this antenna we have:

$$P_{(\text{power radiated})} = 10\,\pi^2 \left[\frac{L}{\lambda}\right]^2 I_{\text{rms}}{}^2. \text{ (watts for monopole)} \qquad (14\text{-}84)$$

Now we can find the radiation resistance for the two antennas.

$$\text{(Short dipole) } R_{(rad)} = 20\ \pi^2 \left[\frac{L}{\lambda}\right]^2 . \text{(ohms)} \qquad (14\text{-}85)$$

$$\text{(Monopole) } R_{(rad)} = 10\ \pi^2 \left[\frac{L}{\lambda}\right]^2 . \text{(ohms)} \qquad (14\text{-}86)$$

Now if we assume that the length of the antenna is one tenth of the wavelength or $L = 0.1\ \lambda$, we obtain values of 1.97 ohms and 0.987 ohms respectively. These would be very small values of impedance to match to any practical transmission line so if we used matching networks we would end up with large losses and poor efficiency.

14.4 Half-Wave Dipole

We have looked at the short dipole but let us now look at a dipole which has some real practical use in engineering work. We have seen that if we know about the current in the antenna, we can determine the vector magnetic potential, $\vec{A}$, and proceed from it to other items of use to us. In order to look at the half-wave dipole let us go back to our transmission line results to try to find out about the current distribution in our antenna. For an open-circuit line we had a current distribution as shown in Figure 14-9.

Now, since our center-fed, dipole antenna looks something like an open circuit, let us bend over the curves in Figure 14-9, giving us the picture shown in Figure 14-10.

This makes some sense since the dipole looks like an open-circuited transmission line. A better justification for the assumption

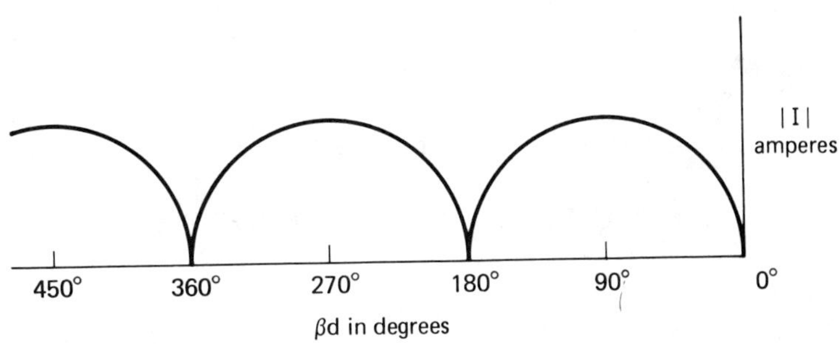

Figure 14-9. Current distribution for open-circuit transmission line.

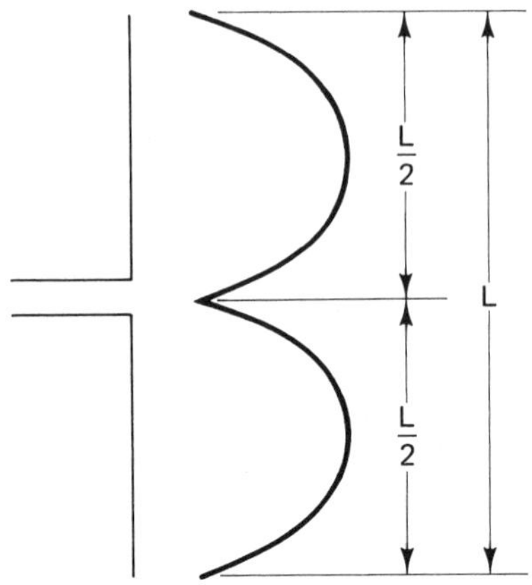

Figure 14-10. Sinusoidal current distribution for dipole antenna.

of the current distribution is that the end results we will obtain give results that are quite close to what we can measure in the laboratory; therefore, the actual current distribution must be relatively close to that indicated in Figure 14-10. Now let us express the current distribution mathematically as:

$$I = I_{max} \sin\beta\left(\frac{L}{2} - z\right), \ z > 0. \tag{14-87}$$

$$I = I_{max} \sin\beta\left(\frac{L}{2} + z\right), \ z < 0. \tag{14-88}$$

The assumed current distribution is indicated in Figure 14-11.

Going back to equation (14-2) we have:

$$\vec{A} = \frac{\mu_0}{4\pi} \frac{I\left(t - \frac{r}{c}\right)}{r} \vec{d\ell}.$$

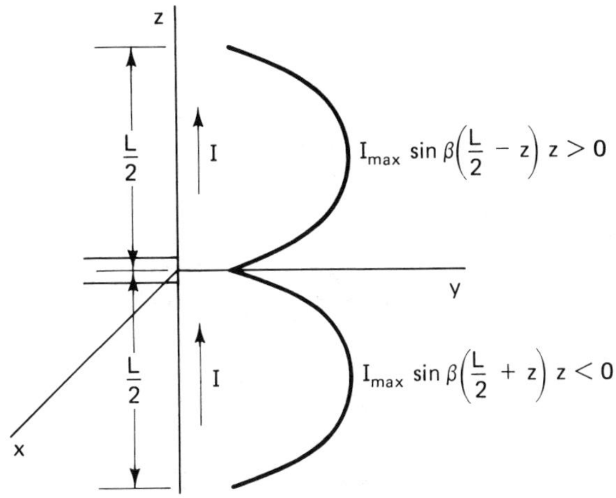

Figure 14-11. Current distribution for dipole antenna.

Now, since I is in the z direction we can write the solution for the phasor value of vector magnetic potential:

$$A_z = \frac{\mu_0}{4\pi} \int_{-L/2}^{0} \frac{I_{max} \sin \beta\left(\frac{L}{2} + z\right)}{r} e^{-j\beta r}\, dz + \frac{\mu_0}{4\pi} \int_{0}^{L/2} \frac{I_{max} \sin\beta\left(\frac{L}{2} - z\right)}{r} e^{-j\beta r}\, dz. \quad (14\text{-}89)$$

In Figure 14-12 we will assume that we are dealing with the "far" field so that the two lines are approximately parallel.

In terms of r (the distance from dz) we have:

$$r = r_0 - z \cos \theta. \quad (14\text{-}90)$$

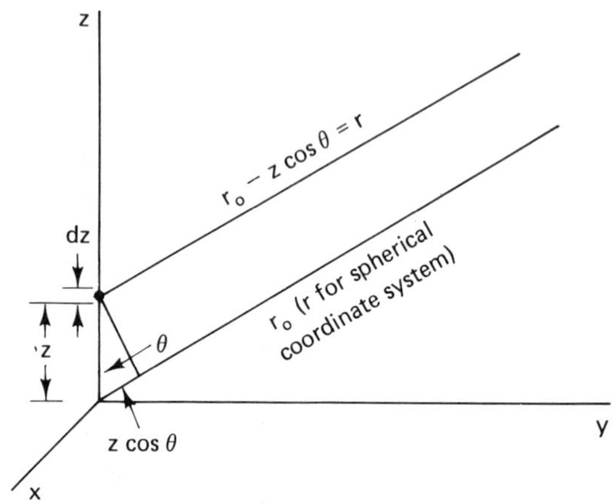

Figure 14-12. Distances used for dipole antenna.

$$A_z = \frac{\mu_0 I_{max}}{4\pi} \int_{-L/2}^{0} \frac{\sin\beta\left(\frac{L}{2} + z\right) e^{-j\beta(r_0 - z\cos\theta)}}{r_0 - z\cos\theta}\, dz + \frac{\mu_0 I_{max}}{4\pi} \int_{0}^{L/2} \frac{\sin\beta\left(\frac{L}{2} - z\right) e^{-j\beta(r_0 - z\cos\theta)}}{r_0 - z\cos\theta}\, dz. \tag{14-91}$$

$$A_z = \frac{\mu_0 I_{max}}{4\pi} e^{-j\beta r_0} \left[\int_{-L/2}^{0} \frac{\sin\beta\left(\frac{L}{2} + z\right) e^{j\beta z\cos\theta}}{r_0 - z\cos\theta}\, dz + \int_{0}^{L/2} \frac{\sin\beta\left(\frac{L}{2} - z\right) e^{j\beta z\cos\theta}}{r_0 - z\cos\theta}\, dz \right]. \tag{14-92}$$

When we get far enough away from the dipole so that the two lines are essentially parallel and $r \gg z$ we have:

$$r_0 - z\cos\theta \simeq r_0 \simeq r. \tag{14-93}$$

So we can approximate the value of A_z as equation (14-94).

$$A_z \simeq \frac{\mu_0 I_{max}}{4\pi r} e^{-j\beta r}\left[\int_{-L/2}^{0} \sin\beta\left(\frac{L}{2} + z\right) e^{j\beta z\cos} dz + \int_{0}^{L/2} \sin\beta\left(\frac{L}{2} - z\right) e^{j\beta z\cos} dz\right]. \quad (14\text{-}94)$$

In order to integrate, let $z = -z_1$, $dz = -dz_1$. When $z = -L/2$, $z_1 = +L/2$. Looking at just the first integral we can change it as follows:

$$\int_{L/2}^{0} \sin\beta\left(\frac{L}{2} - z_1\right) e^{-j\beta z_1\cos\theta} (-dz_1)$$

$$= -\int_{L/2}^{0} \sin\beta\left(\frac{L}{2} - z_1\right) e^{-j\beta z_1\cos\theta} (dz_1).$$

Now if we go back and let $z_1 = z$, $dz_1 = dz$, and when we change the limits we have:

$$\int_{0}^{L/2} \sin\beta\left(\frac{L}{2} - z\right) e^{-j\beta z\cos} dz. \quad (14\text{-}95)$$

So we can rewrite equation (14-94) as:

$$A_z \simeq \frac{\mu_0 I_{max} e^{-j\beta r}}{4\pi r}\left[\int_{0}^{L/2} \sin\beta\left(\frac{L}{2} - z\right) e^{-j\beta z\cos\theta} dz + \int_{0}^{L/2} \sin\beta\left(\frac{L}{2} - z\right) e^{j\beta z\cos\theta} dz\right]. \quad (14\text{-}96)$$

$$A_z \simeq \frac{\mu_0 I_{max} e^{-j\beta r}}{4\pi r}\left[\int_{0}^{L/2} \sin\beta\left(\frac{L}{2} - z\right) (2) \left\{\frac{e^{j\beta z\cos\theta} + e^{-j\beta z\cos\theta}}{2}\right\} dz\right] =$$

$$\frac{\mu_0 I_{max} e^{-j\beta r}}{2\pi r}\int_{0}^{L/2} \sin\beta\left(\frac{L}{2} - z\right)\cos(\beta z\cos\theta) dz. \quad (14\text{-}97)$$

The integration is fun but let us leave it so your instructor can impress you with his mathematical ability. We let:

$$M = \frac{L}{2} \text{ (half the length of the dipole antenna).} \qquad (14\text{-}98)$$

So we have:

$$A_z \simeq \frac{\mu_0 I_{max} e^{-j\beta r}}{2\pi r\beta}\left[\frac{\cos(M\beta\cos\theta) - \cos\beta M}{\sin^2\theta}\right]. \qquad (14\text{-}99)$$

Going back to equation (14-12):

$\vec{A} = A_z \cos\theta\, \vec{a}_r - A_z \sin\theta\, \vec{a}_\theta$, and the expressions that $\vec{B} = \nabla \times \vec{A}$, and $\vec{H} = \frac{\vec{B}}{\mu_0}$, we obtain the equations:

$$H_\phi \simeq \frac{jI_{max} e^{-j\beta r}}{2\pi r}\left[\frac{\cos(\beta M\cos\theta) - \cos\beta M}{\sin\theta}\right]. \qquad (14\text{-}100)$$

$$E_\theta \simeq \eta\, H_\phi = \frac{j\eta I_{max} e^{-j\beta r}}{2\pi r}\left[\frac{\cos(\beta M\cos\theta) - \cos\beta M}{\sin\theta}\right]. \qquad (14\text{-}101)$$

Since $\beta = \frac{2\pi}{\lambda}$, and for our half-wave dipole $\frac{L}{2} = \frac{\lambda}{4} = M$, we can write:

(Half-wave dipole)

$$E_\theta \simeq \frac{j\eta I_{max} e^{-j\beta r}}{2\pi r}\,\frac{\cos\left(\frac{\pi}{2}\cos\theta\right)}{\sin\theta}. \qquad (14\text{-}102)$$

$$H_\phi \simeq \frac{jI_{max} e^{-j\beta r}}{2\pi r}\,\frac{\cos\left(\frac{\pi}{2}\cos\theta\right)}{\sin\theta}. \qquad (14\text{-}103)$$

The pattern for both E_θ and H_ϕ varies with the factor given in equation (14-104)

$$\frac{\cos\left(\frac{\pi}{2}\cos\theta\right)}{\sin\theta}. \qquad (14\text{-}104)$$

Let us let the first part of equation (14-102) be equal to unity and look at the pattern we obtain.

$$\frac{j\eta I_{max} e^{-j\beta r}}{2\pi r} = 1. \qquad (14\text{-}105)$$

When we do this, our pattern for $\vec{E}$ is given by equation (14-104). It is shown in Figure 14-13.

We see the pattern resembles the one we obtained for the short dipole because we still have the donut shape but it becomes a flattened donut. When we evaluate Poynting's vector around the surface of a sphere surrounding the antenna we obtain:

$$P_{(rad)} = 36.558\, I_{max}^2 = 73.116\, I_{rms}^2. \text{ (watts)} \qquad (14\text{-}106)$$

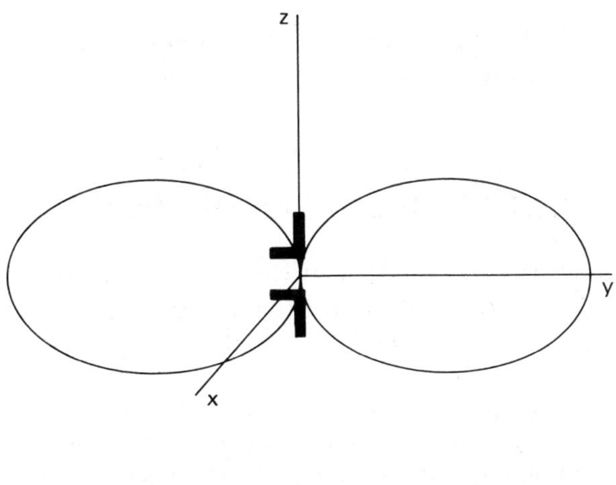

Figure 14-13. Pattern for E_θ and H_ϕ for half-wave dipole.

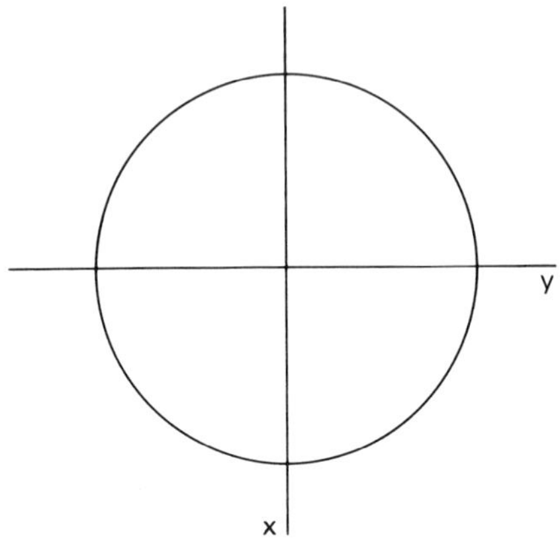

Figure 14-14. Pattern for E_θ and H_ϕ in x-y plane for half-wave dipole.

Equating this to our equivalent resistance we find that the radiation resistance becomes:

$$R_{\text{radiation}} = 73.116 \text{ ohms.} \tag{14-107}$$

If we have half of the half-wave dipole mounted above the ground plane (monopole) we would obtain:

$$R_{\text{radiation}} = 36.5 \text{ ohms.} \tag{14-108}$$

Again, if we look at our pattern in the x-y plane we have a circular pattern as shown in Figure 14-14.

14.5 Antenna Patterns and Definitions

We have indicated the patterns of several antennas in terms of E_θ (or H_ϕ) and plotted them several different ways. Since our antennas were orientated so that they had their major axis parallel to and on the z axis and were located at the origin, the patterns did not vary with ϕ but they did vary with θ. The equations indicate that the pattern does not vary with r, only the magnitude. Since we are talking about a three-dimensional pattern we are able to show only a part of the pattern in two dimensions. For the dipole we just looked at we can choose some plane such as the y-z plane to see what happens as a function of θ and the x-y plane to see what the pattern is as a function of ϕ. For our dipole the pattern was not a function of ϕ and we ended up with a circle of constant radius.

The three patterns commonly used by the antenna engineer are:

1. Field intensity.
2. Power density.
3. Radiation intensity.

Before looking at these patterns let us look at the concept of a solid angle. We will take a sphere as shown in Figure 14-15, with a radius of r meters.

If we take some surface area on the sphere we can define a solid angle subtended by the volume starting at the origin as:

$$\Omega(\text{angle in steradians}) = \frac{S(\text{area})}{r^2}. \tag{14-109}$$

Notice that a steradian is dimensionless:

$$\left[\frac{\text{meter}^2}{\text{meter}^2}\right]$$

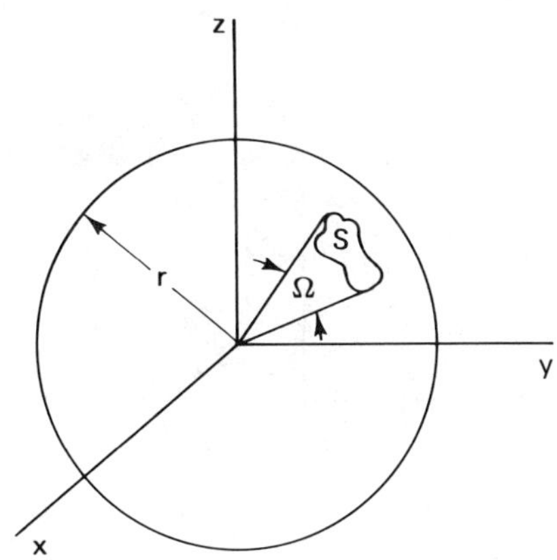

Figure 14-15. Solid angle.

For a whole sphere we would have:

$$\Omega = \frac{4\pi r^2}{r^2} = 4\pi \text{ steradians.}$$

So there are 4π steradians in the whole solid angle subtended by the total surface of a sphere. It is convenient when talking about steradians to consider a unit sphere so the angle is simply the area on the surface of the unit sphere. Notice also that the units of r^2 in equation (14-109) turn out to be:

$$r^2 = \frac{S}{\Omega} = \frac{\text{meter}^2}{\text{steradian}}. \quad (14\text{-}110)$$

We have already looked at some field intensity patterns. Normally they are plotted in terms of the E field. For instance, if we have a form for E such as $E_\theta = A \sin\theta$, where A is some constant, we obtain the donut-without-the-hole type of pattern.

If we take Poynting's vector and take the time average to find the average real power in watts per meter squared and plot the results we have a plot of the power density pattern. It is really a plot of:

$$P_{\text{density}} = \frac{1}{2}\mathcal{R}\,(\vec{E} \times \vec{H}^*) \quad (14\text{-}111)$$

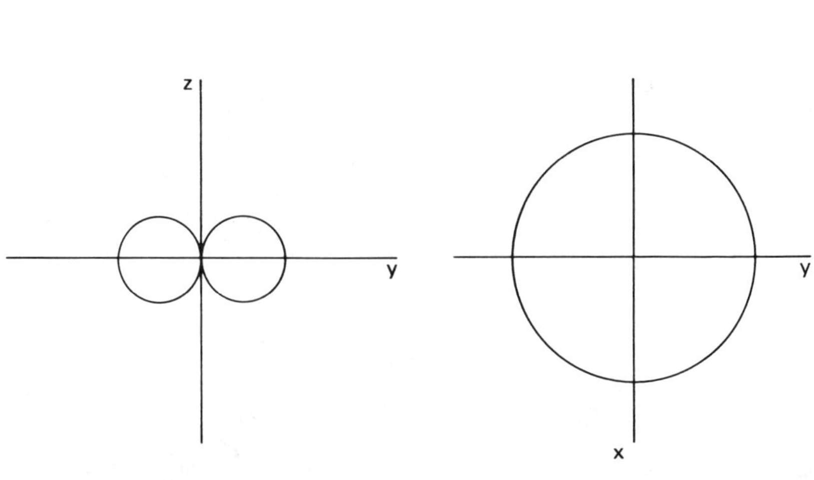

Figure 14-16. Pattern for E_θ and H_ϕ field when $E_\theta = A \sin\theta$.

In equation (14-111) we are dealing with the phasor values. Since we need to consider only the variation with the variables and not the actual magnitudes to obtain the shape of the pattern, let us assume that we have the following:

$$E_\theta = A \sin \theta \text{ and } H_\phi = \frac{E_\theta}{\eta} = \frac{A}{\eta} \sin \theta.$$

When we go through with the steps indicated in (14-111) we have a term for the power density such as:

$$P_{\text{density}} = B \sin^2 \theta.$$

A plot of the power density is given in Figure 14-17.

It is customary when plotting the power density patterns to indicate the half-power points. In our example the half-power points would occur when $\theta = 45°$. The angle between the half-power points is called the "beam width." In our example the beam width is 90°.

The third pattern is the radiation pattern and is a plot of the radiation intensity. By definition:

$$\mathcal{R}_{\text{intensity}} = r^2 P_{\text{density}}. \qquad (14\text{-}112)$$

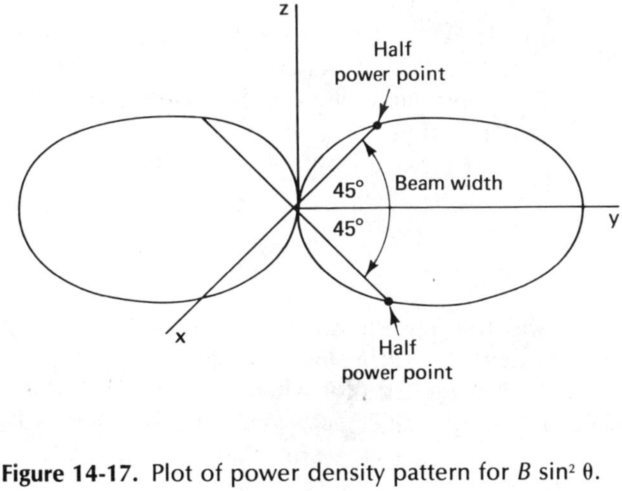

Figure 14-17. Plot of power density pattern for $B \sin^2 \theta$.

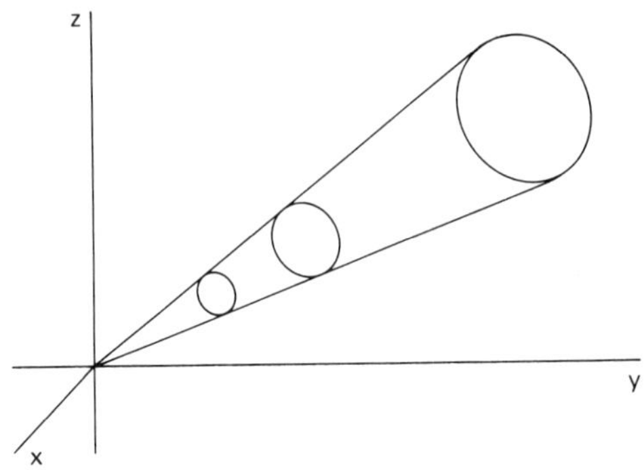

Figure 14-18. Solid angle from source at origin.

It has units of:

$$\frac{\text{meters}^2}{\text{steradian}}\,\frac{\text{watts}}{\text{meter}^2} = \frac{\text{watts}}{\text{steradian}}.$$

It thus becomes a measure of the power in watts that is going through space in a solid angle. If we think of a conical shape, as indicated in Figure 14-18, we see that as we move farther from the source (antenna), the watts contained in the solid angle are spread over a larger and larger area so the power density in watts per square meter is decreasing.

If we take a unit sphere with $r^2 = 1$ we have:

$$\mathcal{R}_{\text{intensity}} = (1)^2\, P_{\text{density}}$$

and the pattern would be the same as the power density pattern. Since we are normally interested only in the shapes of the three curves we can plot what is commonly called the "normalized" patterns. In our case we would have: $E_{\theta(\text{normalized})} = \sin\theta$; $P_{\text{density(normalized)}} = \sin^2\theta$; $\mathcal{R}_{\text{intensity(normalized)}} = \sin^2\theta$.

14.6 Special Antenna Types

In order to make our introduction to antennas as fundamental as possible we should briefly look at some useful concepts that will allow us to better understand the literature if we need additional information.

An *isotropic* antenna is a theoretical antenna that does not exist in the real world (only approached in practice) but possesses some characteristics that are often useful when dealing with antennas. It is defined as an antenna where the radiation intensity is the same in all directions. If we take our expression for finding the power radiated from an antenna we have:

(14-46) $$P_{(\text{power radiated})} = \oiint \mathcal{P}_{r(\text{average})} dS. \text{ (watts)}$$

$$P_{(\text{power radiated})} = \oiint \mathcal{P}_r \, r^2 \sin\theta \, d\theta \, d\phi. \quad (14\text{-}113)$$

For the isotropic antenna, P_r is independent of θ and ϕ, so we have:

$$P_{(\text{power radiated})} = \mathcal{P}_{r(\text{average})} \oiint dS = P_r(4\pi r^2). \quad (14\text{-}114)$$

So

$$\mathcal{P}_r = \frac{P}{4\pi r^2} = \frac{1}{2}\mathcal{R}(\vec{E} \times \vec{H}^*). \quad (14\text{-}115)$$

$$\mathcal{R}_{\text{intensity}} = r^2\mathcal{P}_r = r^2\frac{P}{4\pi r^2} = \frac{P}{4\pi}. \text{ (watts/steradian)} \quad (14\text{-}116)$$

For the isotropic antenna the radiation intensity is constant and given by equation (14-116).

Let us go back to equation (14-113) and substitute for P_r.

$$P_{(\text{power radiated})} = \oiint \frac{\mathcal{R}_{\text{intensity}}}{r^2} r^2 \sin\theta \, d\theta \, d\phi = \int_{\theta=0}^{\pi}\int_{\phi=0}^{2\pi} \mathcal{R}_{\text{intensity}} \sin\theta \, d\theta \, d\phi =$$

$$\oiint \mathcal{R}_{\text{intensity}} \, d\Omega = 4\pi\mathcal{R}_{\text{intensity}}. \quad (14\text{-}117)$$

In equation (14-117), $d\Omega = \sin\theta \, d\theta \, d\phi =$ the solid angle in steradians which has the units of square radians.

So we find that the total power radiated is given by the integral of the radiation intensity, $\mathcal{R}$, over a solid angle of 4π steradians. For our isotropic antenna we have:

$$P_{\text{(power radiated)}} = 4\pi\, \mathcal{R}_{\text{intensity}}. \tag{14-118}$$

There are two more important concepts that are used when discussing antennas. They are directivity and gain.

Directivity is defined as the ratio of the maximum radiation intensity to the average radiation intensity.

$$D = \frac{\mathcal{R}_{\text{max}}}{\mathcal{R}_{\text{ave}}}. \tag{14-119}$$

We can find the average value of the radiation intensity as:

$$\mathcal{R}_{\text{ave}} = \frac{1}{4\pi} \oiint \mathcal{R}\, d\Omega. \tag{14-120}$$

$$4\pi\, \mathcal{R}_{\text{ave}} = \oiint \mathcal{R}\, d\Omega = P_{\text{(power radiated)}},$$

from equation (14-117), so

$$\mathcal{R}_{\text{ave}} = \frac{P_{\text{(power radiated)}}}{4\pi}. \tag{14-121}$$

So we have:

$$D = \frac{\mathcal{R}_{\text{max}}\,(4\pi)}{\mathcal{R}_{\text{ave}}\,(4\pi)} = \frac{4\pi \mathcal{R}_{\text{max}}}{P_{\text{power radiated)}}}. \tag{14-122}$$

The directivity can thus be written as the ratio of 4π times the maximum radiation intensity to the total power radiated.

The gain of an antenna is the ratio of the maximum radiation intensity for the given antenna to the maximum radiation intensity of some reference antenna with the same power input to both antennas. It is sometimes called the power gain.

$$G = \frac{\mathcal{R}_{\text{max(for antenna under study)}}}{\mathcal{R}_{\text{max (for some reference antenna)}}}. \tag{14-123}$$

The gain is a relative quantity which involves a reference antenna and takes the efficiencies of both antennas into account.

$$\text{Efficiency} = \frac{\text{power radiated}}{\text{power input into antenna}}. \tag{14-124}$$

It has become common practice to use either the isotropic or the dipole antenna as the reference when discussing the gain of an antenna. The gain, using an isotropic antenna as the reference, would be:

$$G_{\text{isotropic}} = \frac{\mathcal{R}_{\text{max}}}{\mathcal{R}_{\text{max (isotropic antenna)}}}. \qquad (14\text{-}125)$$

We will assume the isotropic antenna is lossless, and, of course, the radiation intensity is the same in all directions by definition.

$$(\text{Isotropic antenna})\ \mathcal{R}_{\text{max}} = \text{constant} = \mathcal{R}_{\text{average}}. \qquad (14\text{-}126)$$

When lossless,

$$P_{\text{input}} = P_{\text{radiated}} = 4\pi\ R_{\text{average}}. \qquad (14\text{-}127)$$

For the antenna we are looking at, let us define R_{max}, as the radiation intensity if the antenna is 100 percent efficient. Then:

$$\mathcal{R}_{\text{max}} = \eta \mathcal{R}_{\text{max}}',$$

where η is the efficiency. So,

$$G_{\text{isotropic}} = \frac{\eta\ \mathcal{R}_{\text{max}}'}{R_{\text{max (isotropic antenna)}}} = \eta\ D. \qquad (14\text{-}128)$$

If the antenna under consideration is lossless, then equation (14-128) gives:

$$G_{\text{isotropic}} = \frac{\mathcal{R}_{\text{max}}'}{\mathcal{R}_{\text{max}}} = \frac{\mathcal{R}_{\text{max}}'}{\mathcal{R}_{\text{ave}}} = D. \qquad (14\text{-}129)$$

So if the given antenna can be considered lossless the gain over the lossless isotropic antenna is the same as the directivity, D. Since many antennas have low losses there is often very little difference between the gain over the lossless isotropic antenna and the directivity.

A second antenna used for discussing the gain of an antenna is a lossless dipole.

$$G_{\text{dipole}} = \frac{\mathcal{R}_{\max}}{\mathcal{R}_{\max(\text{lossless dipole})}}. \tag{14-130}$$

Often the gain and directivity are specified in decibels.

$$G(db) = 10 \log G. \tag{14-131}$$

$$D(db) = 10 \log D. \tag{14-132}$$

Let us look at an example of these definitions. Suppose we have the following information about our antenna: Input power = 100 watts; $\eta = 95\%$; $\mathcal{R}_{\max} = 220$ watts/steradian.

$$\mathcal{R}_{\text{ave}} = \frac{P_{\text{radiated}}}{4\pi} = \frac{(0.95)\,(100)}{4\pi} = 7.56 \text{ watts/steradian}$$

$$D = 4\pi \frac{\mathcal{R}_{\max}}{P_{\text{radiated}}} = \frac{4\pi(220)}{(0.95)(100)} = 37.05.$$

Or

$$D = 10 \log 37.05 = 15.69\ db.$$

$$\mathcal{R}_{\text{isotropic}} = \frac{P_{\text{in}}}{4\pi} = \frac{100}{4\pi} = 7.96 \text{ watts/steradian}.$$

$$G_{\text{isotropic}} = \frac{(0.95)\,(220)}{7.96} = 26.256.$$

Or

$$G_{\text{isotropic}} = 10 \log 26.256 = 14.19\ db.$$

Polarization is another term used in describing the action of an antenna field. It basically tells about the orientation of the electric field vector, $\vec{E}$. For example, in fields courses the student usually studies a plane wave with an x component of $\vec{E}$ and a y component of $\vec{H}$ moving in the z direction or with a similar orientation in space. The $\vec{E}$ vector could be written as:

$$\vec{E} = E_{\max} \cos(\omega t - \beta z)\, \vec{a}_x.$$

We would describe the wave as being polarized in the x direction. Similarly if we have a wave traveling through space with the $\vec{E}$ vector parallel to the ground we would say the wave is horizontally polarized. When $\vec{E}$ is vertical to the ground we say it is vertically polarized.

It is possible to have an infinite variety of polarizations. For instance, if:

$$\vec{E} = E_x\cos(\omega t - \beta z)\,\vec{a}_x + E_y\cos(\omega t - \beta z + \phi)\,\vec{a}_y. \quad (14\text{-}133)$$

Suppose we look at the case where $\phi = 0$ and take the point in space where $z = 0$.

$$\vec{E} = E_x \cos \omega t\,\vec{a}_x + E_y \cos \omega t\,\vec{a}_y.$$

Furthermore, let us assume that $|E_x| = |E_y|$. We can set up a table for the x and y components as a function of time.

ωt	*x component*	*y component*
0	1.0	0
$\frac{\pi}{2}$	0.0	−1.0
π	−1,0	0
$\frac{3\pi}{2}$	0	1.0

Figure 14-19 shows what happens to the vector addition of the two components as time goes on. We obtain a rotating vector value for $\vec{E}$ and the tip of $\vec{E}$ traces a circle. We would say that this wave is circularly polarized. If E_x and E_y do not have the same magnitudes we would end up with an elliptically polarized wave. The tip of the $\vec{E}$ vector would trace out an ellipse.

14.7 Chapter Summary

1. Theoretically, the fields propagated by an antenna may be found if one knows how the current is related to position and time in the antenna. Knowing about the current allows one to find the vector potential, $\vec{A}$, and from it one can find $\vec{B}$, $\vec{H}$, and $\vec{E}$.

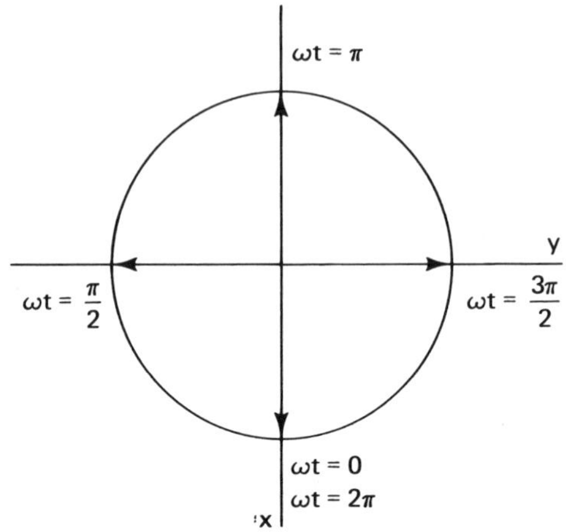

Figure 14-19. Circularly polarized wave.

2. $\vec{A}$, known as the retarded vector magnetic potential, is the result of the finite time it takes for the fields to move through space. The potential at a point in space, at time t_2, results because of what happened at the antenna at an earlier time, t_1, and is related to the distance traveled and the velocity of the wave.
3. The differential antenna is a theoretical antenna which not only illustrates the principles of handling antenna theory but can be used to find the fields of finite sized antennas. Its basic components are:

$$H_\phi = \frac{I\,d\ell \sin\theta}{4\pi}\left[\frac{-\omega \sin \omega t'}{rc} + \frac{\cos \omega t'}{r^2}\right].$$

$$E_r = \frac{2\,I\,d\ell \cos\theta}{4\pi\varepsilon_0}\left[\frac{\cos \omega t'}{r^2 c} + \frac{\sin \omega t'}{\omega r^3}\right].$$

$$E_\theta = \frac{I\,d\ell \sin\theta}{4\pi\varepsilon_0}\left[\frac{-\omega \sin \omega t'}{rc^2} + \frac{\cos \omega t'}{cr^2} + \frac{\sin \omega t'}{\omega r^3}\right].$$

$$t' = \left(t - \frac{r}{c}\right).$$

However, in the "far" field the important terms are:

$$E_\theta \simeq \frac{I\,d\ell \sin\theta}{4\pi\varepsilon_0}\left[\frac{-\omega \sin \omega t'}{rc^2}\right].$$

$$H_\phi \simeq \frac{I\,d\ell \sin\theta}{4\pi}\left[\frac{-\omega \sin \omega t'}{rc}\right].$$

$$E_r \simeq 0.$$

4. Poynting's vector, $\vec{E} \times \vec{H}$, gives the power flow in watts per square meter for the fields generated by the antenna. If it is in terms of time it gives the instantaneous power. By integrating over a full period and dividing by the time period one may find the time average power.

$$\mathcal{P}_{r(\text{average})} = \frac{1}{T}\int_0^T \mathcal{P}_r(t)\,dt.$$

5. For the differential antenna the time average value of Poynting's vector is:

$$\mathcal{P}_{r(\text{average})} = \frac{I^2\,d\ell^2 \sin^2\theta}{16\pi^2\varepsilon_0}\,\frac{\omega^2}{2r^2c^3}.$$

This equation comes from the $\frac{1}{r}$ terms in both E_θ and H_ϕ. The terms in $\frac{1}{r^2}$ and $\frac{1}{r^3}$ do not contribute to the average power moving away from the antenna in the "far" field.

6. In free space the propagated waves travel at the speed of light.

7. The radiation pattern for $\vec{E}$ for the differential antenna is a function of the angle θ only.

$$E_\theta = (\text{constant}) \sin\theta.$$

8. The power radiated may be equated to a radiation resistance by the equation:

$$P_{(\text{power radiated})} = I_{\text{rms}}^{\,2}\,R_{(\text{rad})}.$$

$R_{(rad)}$ is the equivalent resistance that consumes the same real power from the source as the power radiated by the antenna.

9. The radiation resistance for the differential antenna is very, very small and from a practical viewpoint it would be difficult to match a differential antenna to a real transmission line.
10. In this chapter we discussed a short dipole and a regular dipole antenna. The short dipole has a radiation pattern that is similar to the differential antenna.

$$E_\theta = (\text{constant}) \sin \theta.$$

This gives a pattern that looks like a donut without the hole.

11. The regular dipole with a length, L, and with $M = \frac{L}{2}$ has a field component given in the form:

$$E_\theta = (\text{constant}) \left[\frac{\cos(\beta M \cos\theta) - \cos \beta M}{\sin \theta} \right].$$

12. If the regular dipole is one-half wavelength long the pattern becomes:

$$E_\theta = (\text{constant}) \left[\frac{\cos \frac{\pi}{2} \cos \theta}{\sin \theta} \right].$$

The radiation resistance for the regular half-wave dipole antenna is 73.116 ohms.

13. A steradian is a unit of solid angle which is the solid angle subtended by the volume starting at the center of a sphere and ending at the surface of the sphere. It is defined in terms of the area subtended on the surface.

$$\Omega \text{ (steradian)} = \frac{S(\text{area})}{r^2}.$$

The r is the radius of the sphere. For a unit sphere the solid angle is simply the area subtended on the surface. The total (and maximum) solid angle is 4π steradians.

14. There are three types of radiation patterns commonly used: (a) Field intensity pattern; (b) Power density pattern; (c) Radiation intensity pattern. The field intensity pattern is normally a plot of E_θ. The power density pattern is a plot of the power radiated in watts per steradian. They can be defined as:

$$P_{\text{density}} = \frac{1}{2}\,\mathcal{R}(\vec{E} \times \vec{H}) \text{ watts per square meter,}$$

where $\mathcal{R}$ stands for the real part of Poynting's vector, $\vec{E} \times \vec{H}$.

$$\mathcal{R}_{\text{intensity}} = r^2\, P_{\text{density}} \text{ watts per steradian.}$$

15. An isotropic antenna is one where the radiation intensity is the same in all directions. The maximum and the average value are the same. For this antenna we can write:

$$\mathcal{R}_{\text{intensity}} = \frac{P}{4\pi}. \qquad \mathcal{P}_r = \frac{P}{4\pi r^2}.$$

16. Directivity is defined as:

$$D = \frac{\mathcal{R}_{\text{max}}}{\mathcal{R}_{\text{ave}}}.$$

It is sometimes defined in terms of decibels as:

$$D(db) = 10 \log D.$$

17. Gain is defined as:

$$G = \frac{\mathcal{R}_{\text{max (for antenna under study)}}}{\mathcal{R}_{\text{max (for reference antenna)}}}.$$

The reference antenna is usually either a lossless isotropic or a lossless half-wave dipole. Sometimes it also is defined in terms of decibels as:

$$G(db) = 10 \log G.$$

18. Polarization of radiated waves refers to the orientation of the $\vec{E}$ vector.

14.8 Review and Study Questions

1. What is the retarded potential? Why does it exist?
2. What are the units of the following?
   (a) Vector magnetic potential, $\vec{A}$?
   (b) Magnetic flux density vector, $\vec{B}$?
   (c) Current density vector, $\vec{J}$?
   (d) Electric field intensity vector, $\vec{E}$?
   (e) Magnetic field intensity vector, $\vec{H}$?
3. What components of $\vec{E}$ and $\vec{H}$ are necessary in order to have energy flow in the $\vec{a}_r$ direction? What components would be necessary in order to have the energy flow in the $\vec{a}_z$ direction?
4. How is the direction of $\vec{A}$ related to the current flow in an antenna? How is the magnitude of $\vec{A}$ related to the magnitude of the current when we are concerned about the magnitude of $\vec{A}$ at a distance of one-half wavelength from the current element?
5. What is a differential antenna? Why is it important to the engineer?
6. What are the units of $\frac{\omega}{c}$?
7. What is Poynting's vector? What does it tell us?
8. How do we find the time average value of a time function?
9. What is the time average power of a differential antenna?
10. Define the "near" field. What are the most important mathematical terms for the "near" field of a differential antenna?
11. What is the "far" field? What are the most important mathematical terms for the "far" field of a differential antenna?
12. How does the field of a differential antenna vary as a function of ϕ? of θ? of r?
13. How is the total power radiated from an antenna found?
14. Define the radiation resistance.
15. Why can we neglect the E_r term for our differential antenna?
16. How does the radiation resistance of a half-wave dipole compare with the radiation resistance of a short dipole and a differential antenna?
17. What justification do we have for assuming a sinusoidal type current distribution in a half-wave dipole antenna?

18. Give the differences between the following three patterns: (a) Field intensity pattern; (b) Power density pattern; (c) Radiation intensity pattern. Give the units of each. Why is polar graph paper of use to us for these plots?
19. How many steradians are contained in the upper hemisphere of the earth?
20. Define directivity.
21. Define gain.
22. What is an isotropic antenna? Why is it important to the antenna engineer?
23. What is polarization? How can a wave be elliptically polarized?

Problems

14-1. A differential antenna is connected to a 50 ohm coaxial cable. Determine the VSWR on the cable if the differential antenna has the following lengths:
(a) 0.01 wavelength.
(b) 0.005 wavelength.

14-2. The field intensity 6 miles from a differential antenna is 1,300 microvolts per meter. Determine:
(a) The field strength 32 miles from the antenna.
(b) The field strength 64 miles from the antenna.

14-3. An antenna is located in Chicago, Illinois. Determine the time it takes for a propagated wave to travel to:
(a) Minneapolis, Minnesota (405 miles).
(b) A rocket ship piloted by Buck Rogers and located 6.5 million miles away.

14-4. Sketch the radiation pattern for E_θ in the y-z plane and the x-y plane if the antenna is located at the origin and has the following values for E_θ:
(a) $E_\theta = 10 \cos \theta$, volts per meter.
(b) $E_\theta = 10 \sin \theta$, volts per meter.
(c) $E_\theta = 10 \sin \theta \cos \theta$, volts per meter.
(d) $E_\theta = 10 \sin^2 \theta$, volts per meter.

14-5. For a certain antenna it is found that the phasor value of the components of $\vec{A}$ are as follows:

$$A_r = \frac{\mu_0 I d\ell \cos \theta \, e_{-j\omega r/c}}{4\pi r} \text{ and } A_\theta = -\frac{\mu_0 I \, d\ell \sin \theta \, e_{-j\omega r/c}}{4\pi r}.$$

(a) Determine the phasor value of $\vec{H}$.
(b) Determine the phasor value of E_r.

14-6. A dipole is one wavelength long.
(a) Determine the expression for E_θ.
(b) If the antenna is located at the origin sketch the pattern for E_θ in the y-z plane.
(c) Sketch the radiation pattern for E_θ in the x-y plane.

14-7. A dipole is one-quarter wavelength long.
(a) Determine the expression for E_θ.
(b) If the antenna is located at the origin, sketch the pattern for E_θ in the y-z plane.
(c) Sketch the radiation pattern for E_θ in the x-y plane.

14-8. The normalized value of $\vec{E}$ for a certain dipole is given as:

$$E_{\theta(\text{normalized})} = \frac{\cos\left(\frac{\pi}{2}\cos\theta\right).}{\sin\theta}.$$

(a) Draw the field pattern for this antenna in polar coordinates for the y-z plane.

14-9. For an isotropic antenna situated in free space, Poynting's vector is given as $6/r^2\,\vec{a}_r$ average watts per meter squared.
(a) Determine the total power radiated in all directions from the antenna.
(b) If the antenna current is 0.9 amperes, r.m.s., determine the radiation resistance of the antenna.

14-10. For a certain antenna the far-field components are given as follows:

$$\vec{H} = \frac{I\ell\sin\theta}{2\lambda r}\sin\omega\left(t - \frac{v}{c}\right)\vec{a}_\phi.$$

$$\vec{E} = -\sqrt{\frac{\mu_0}{\varepsilon_0}}\frac{I\ell\sin\theta}{2\lambda r}\sin\omega\left(t - \frac{r}{c}\right)\vec{a}_\theta.$$

(a) Find the time-average power density radiated into free space.
(b) Find the time-average power radiated into free space.

14-11. A certain antenna has the following radiation intensity pattern:

$$\mathcal{R}_{\text{intensity}} = 2\sin\theta \text{ watts per steradian.}$$

(a) Sketch the radiation pattern in the y-z plane.
(b) Find the total power radiated from the antenna.
(c) If an isotropic source radiated the same total power what would be the maximum (and average) radiation, $R_{\text{intensity}}$ for the isotropic antenna?
(d) What is the directivity of this antenna?
(e) What is the gain of this antenna over a lossless isotropic antenna if the antenna is 98 percent efficient?

14-12. A certain antenna has a maximum radiation intensity of 6 watts per steradian in the + x direction. The average radiation intensity is 2.4 watts per steradian.
(a) What is the directivity of the antenna?
(b) If the antenna has a radiation efficiency of 90 percent what is the gain of the antenna?

14-13. Repeat problem 14-12 if the maximum radiation intensity is 10 watts per steradian in the same direction.

14-14. Calculate the directivity for a source that has the following radiation intensity pattern: $\mathcal{R}_{\text{intensity}} = \mathcal{R}_{\text{max}} \sin\theta \sin^3\phi$, when $0 \leq \theta \leq \pi$, and $0 \leq \phi \leq \pi$, and $\mathcal{R}_{\text{intensity}} = 0$ everywhere else in space.

14-15. Repeat problem 14-14 if: $\mathcal{R}_{\text{intensity}} = \mathcal{R}_{\text{max}} \sin^2\theta \sin^3\phi$, when $0 \leq \theta \leq \pi$, and $0 \leq \phi \leq \pi$, and $\mathcal{R}_{\text{intensity}} = 0$ everywhere else in space.

References

Aluminum Association: *Aluminum Electrical Conductor Handbook* (2nd ed). The Aluminum Association, New York, 1982.

Angelakos, D. J., & Everhart, T. E. *Microwave Communications*. McGraw-Hill, New York, 1968.

Chipman, R. A. *Theory and Problems of Transmission Lines*. Schaum's Outline Series, McGraw-Hill, New York, 1968.

Dwight, H. G. *Electrical Elements of Power Transmission Lines*. Macmillan, New York, 1954.

Dwight, H. B. *Tables of Integrals and Other Mathematical Data*. Macmillan, New York, 1934.

Everitt, W. L. *Communication Engineering* (2nd ed). McGraw-Hill, New York, 1937.

Hayt, W. H., Jr. *Engineering Electromagnetics* (4th ed). McGraw-Hill, New York, 1981.

Johnson, W. C. *Transmission Lines and Networks*. McGraw-Hill, New York, 1950.

Kimbark, E. W. *Electrical Transmission of Power and Signals*. Wiley, New York, 1950.

King, D. D. *Measurements at Centimeter Wavelength*. Boston Technical Publishers, Inc., Boston, 1965.

King, R. W., Mimno, H. R., & Wing, A. H. *Transmission Lines, Antennas and Wave Guides*, McGraw-Hill, New York, 1945.

Knowlton, A. E. (Ed.). *Standard Handbook for Electrical Engineers* (9th ed). McGraw-Hill, New York, 1957.

Kraus, J. D. *Antennas*. McGraw-Hill, New York, 1950.

Lance, A. L. *Introduction to Microwave Theory and Measurements.* McGraw-Hill, New York, 1964.

Lathi, B. P. *Signals, Systems and Communication.* Wiley, New York, 1965.

Magnusson, P. C. *Transmission Lines and Wave Propagation.* Allyn & Bacon, Rockleigh, N. J., 1965.

McLachlan, N. W. *Bessel Functions for Engineers.* Oxford University Press, Fair Lawn, N. J., 1961.

Moore, R. K. *Traveling Wave Engineering,* McGraw-Hill, New York, 1960.

Okress, E. C. (Ed.). *Microwave Power Engineering, Vol. 1: Generation, Transmission, Rectification.* Academic Press, New York, 1968.

Paris, D. T. & Hurd, F. K. *Basic Electromagnetic Theory.* McGraw-Hill, New York, 1969.

Popovic, B. D. *Introductory Engineering Electromagnetics.* Addison Wesley, Reading, Mass., 1971.

Sarbacher, R. I., & Edson, W. A. *Hyper and Ultrahigh Frequency Engineering,* John Wiley, New York, 1947.

Schelkunoff, S. A., & Friis, H. T. *Antenna Theory and Practice,* Wiley, New York, 1952.

Seshadri, S. R. *Fundamentals of Transmission Lines and Electromagnetic Fields* Addison Wesley, Reading, Mass., 1971.

Skilling, H. H. *Electric Transmission Lines.* McGraw-Hill, New York, 1951.

Skilling, H. H. *Fundamentals of Electric Waves* (2nd ed.). Wiley, New York, 1948.

Stewart, J. L. *Circuit Analysis of Transmission Lines.* Wiley, New York, 1958.

Westinghouse Electric Corporation. *Electrical Transmission and Distribution Reference Book.* Westinghouse, 777 Penn Center Blvd., Pittsburgh, Pennsylvania, 1964.

Wolff, E. A. *Antenna Analysis.* Wiley, New York, 1967.

Woodruff, L. F. *Principles of Electric Power Transmission* (2nd ed.). Wiley, New York, 1948.

Zahn, M. *Electromagnetic Field Theory: A Problem Solving Approach.* Wiley, 1979.

Appendix

I. Symbols Used Throughout the Text

Note: One symbol may stand for several different physical entities. The reader should determine the meaning of a symbol within the material being studied.

$\| \ \|$	Magnitude of quantity in the brackets.
a	Radius of a round conductor or the largest inside dimension of a rectangular wave guide in units of length.
$\vec{a}_x, \vec{a}_y, \vec{a}_z$	Unit vectors in the Cartesian coordinate system.
$\vec{a}_\rho, \vec{a}_\phi, \vec{a}_z$	Unit vectors in the cylindrical coordinate system.
$\vec{a}_r, \vec{a}_\theta, \vec{a}_\phi$	Unit vectors in the spherical coordinate system.
$\vec{A}$	Vector magnetic potential in webers per meter.
α(alpha)	Attenuation constant in nepers per unit length or angle of Z_r/Z_o in radians or degrees.
β(beta)	Phase constant in radians per unit length.
$\vec{B}$	Magnetic flux density in webers per meter squared.
b	Inner radius of the outer conductor of a coaxial cable or the smallest inside dimension of a rectangular wave guide in units of length.

γ(gamma)	Propagation constant equal to the (attenuation constant) + j (phase constant).
Γ_r(Gamma)	Voltage reflection coefficient at the receiving end.
Γ_s(Gamma)	Voltage reflection coefficient at the sending end.
Γ_r(current)	Current reflection coefficient at the receiving end.
Γ_s(current)	Current reflection coefficient at the sending end.
$\Gamma_d(\Gamma_x)$	Voltage reflection coefficient at the point d (or x) on the transmission line.
c	Outer radius of the outer conductor of a coaxial cable in units of length or the velocity of light in meters per second.
CP	Complex power. $VI^* = P + jQ$.
d	Distance from the receiving end of the transmission line to a point on the line in units of length.
$\vec{D}$	Electric flux density in coulombs per meter squared.
D	The geometric distance in units of length or the directivity of an antenna or the distance between the centers of two conductors in units of length.
Δ(Delta)	Incremental quantity such as Δx.
∇	Vector operator delta.
δ(Delta)	Skin depth in units of distance.
ε(Epsilon)	Permittivity in farads per meter.
$\varepsilon_{o'}$	Permittivity of free space.
ε_r	Relative permittivity (dielectric constant).
e	Base of natural logarithms.
$\vec{E}$	Electric field intensity in volts per meter.
$\vec{E}_p$	Phasor value of $\vec{E}$.
f	Frequency in hertz.

f_c	Cutoff frequency of wave guide in hertz.
G	Conductivity in mhos or mhos per unit length or the gain or the power gain of an antenna.
$G_{isotropic}$	Gain of an antenna using a lossless isotropic antenna as the reference antenna.
G_{dipole}	Gain of an antenna using a dipole antenna as the reference antenna.
η(Eta)	Intrinsic impedance in ohms or the efficiency defined as the output over the input.
θ(Theta)	Angle of voltage reflection coefficient or angle of impedance or admittance in radians or degrees.
θ_s	Angle between sending end voltage and sending end current in radians or degrees.
θ_r	Angle between the receiving end voltage and the receiving end current in degrees or radians.
Θ(Theta)	$ZY = \gamma\ell$.
$\vec{H}$	Magnetic field intensity in amperes per meter.
$\vec{H}_p$	Phasor value of $\vec{H}$.
$\mathcal{I}$	Imaginary part of a complex number.
$\mathcal{I}_{\mathcal{M}}$	Imaginary part of some quantity.
$\vec{J}$	Current density in amperes per meter squared.
J_o	Bessel function of the first kind, order zero.
K_o	Bessel function of the second kind, order zero.
ℓ	Length of a transmission line in units of length.
L	Inductance in henrys or in henrys per unit length.
λ(Lambda)	Wavelength in units of length.
λ_c	Cutoff wavelength of wave guide.
λ_g	Wavelength in the waveguide.
μ(Mu)	Permeability in henrys per meter.
μ_o	Permeability of free space.

π(Pi)	3.14159265.
ρ(Rho)	Volume charge density in coulombs per meter squared or the resistivity in ohm–meters or one of the coordinates in the cylindrical coordinate system.
P	Real power in watts.
P_{density}	The time average value of Poynting's vector in watts per meter squared.
P_r	Real power on a transmission line in watts.
$P_{\text{power radiated}}$	Power radiated from an antenna (crossing a sphere surrounding the antenna) in watts.
$\mathcal{P}$	Poynting's vector in watts per meter squared.
$\mathcal{P}_{\text{(average)}}$	The time average value of $\mathcal{P}$ in watts per meter squared $= P_{\text{density}}$.
$\mathcal{P}_r$	Component of Poynting's vector in the $\vec{a}_r$ direction in spherical coordinates.
$\mathcal{P}_{r\text{(average)}}$	Time average value of $\mathcal{P}_r = \mathcal{P}_{\text{ave}}$.
Q	Imaginary power (volt–amperes reactive) in vars.
r	Subscript to indicate the receiving end of a transmission line such as v_r, V_r, i_r, I_r.
R	Resistance in ohms or in ohms per unit length.
$\mathcal{R}$	Real part of a complex number.
$R_{\text{(rad)}}$	Radiation resistance of an antenna in ohms.
$\mathcal{R}_{\text{intensity}}$	Radiation intensity (power in a solid angle) in watts per steradian.
s	Subscript to indicate the sending end of a transmission line such as v_s, V_s, i_s, I_s.
t	Time in units of time.
Σ(Sigma)	Summation of items.
σ(Sigma)	Conductivity in mhos per meter.
Φ(Phi)	Magnetic flux in webers.

ϕ(Phi)	Angle of the voltage reflection coefficient or the angle of the propagation constant in radians or degrees.
v_p	Phase velocity in length units per time units.
v_g	Group velocity in length units per time units.
ψ(Psi)	Electric flux in coulombs.
Ω(Omega)	Angle in steradians or impedance in ohms.
$\mho$ (Inverted Omega)	Admittance in mhos.
VSWR	Voltage Standing wave ratio.
ω(Omega)	Radian frequency or angular velocity in radians per second.
x	Distance from the sending end of a transmission line to a point on the line.
y	Admittance in mhos or mhos per unit length.
Y	Admittance per unit length times length.
Y_o	Characteristic admittance of a transmission line.
z	Impedance in ohms or ohms per unit length.
z	Impedance per unit length times length.
Z_o	Characteristic impedance of a transmission line.

II. Conversion Factors

1 centimeter = 0.393701 inches = 32.808×10^{-3} feet.

1 meter = 39.3701 inches = 100 centimeters = 3.2808 feet.

1 kilometer = 0.62137 miles.

1 inch = 1,000 mils = 2.54 centimeters = 25.4×10^{-3} meters.

1 foot = 30.48 centimeters = 304.8×10^{-3} meters.

1 mile = 5,280 feet = 1.60934 kilometers = 1.60934×10^{3} meters.

1 mil = 1×10^{-3} inches.

1 radian = 57.29578 degrees.

1 degree = 17.453×10^{-3} radians.

1 neper = 8.6858890 decibels.

1 decibel = 0.1151292 nepers.

1 steradian = 0.07957747 solid angle.

1 solid angle = 12.5663706 steradians.

III. Physical Constants

(a) $\varepsilon_o = (8.854187818 \pm 0.000000071) \times 10^{-12} \simeq \dfrac{10^{-9}}{36\pi} = 8.8419 \times 10^{-12}$ farads per meter.

(b) $\mu_o = 4\pi \times 10^{-7}$ henrys per meter.

(c) $c = (2.997925 \pm 0.000002) \times 10^8 \simeq 3 \times 10^8$ meters per second.

(d) $\pi = 3.141592653589793$.

IV. Prefixes for SI Units

p(pico)	10^{-12}
n(nano)	10^{-9}
μ(micro)	10^{-6}
m(milli)	10^{-3}
K(kilo)	10^{3}
M(mega)	10^{6}
G(giga)	10^{9}
T(tera)	10^{12}

V. Field Equations

(a) $\vec{\nabla} \times \vec{H} = \vec{J} + \dfrac{\partial \vec{D}}{\partial t}$. (d) $\vec{\nabla} \cdot \vec{B} = 0$.

(b) $\nabla \times \vec{E} = -\dfrac{\partial \vec{B}}{\partial t}$. (e) $\vec{B} = \mu\vec{H}$.

(c) $\nabla \cdot \vec{D} = \rho$. (f) $\vec{D} = \varepsilon\vec{E}$.

(g) $\varepsilon = \varepsilon_r \varepsilon_o$. (i) $\vec{\mathcal{P}} = \vec{E} \times \vec{H}$.

(h) $\mu = \mu_r \mu_o$.

VI. Reference Equations

(a) $\sin x = \dfrac{e^{jx} - e^{-jx}}{2j}$

(b) $\cos x = \dfrac{e^{jx} + e^{-jx}}{2}$.

(c) $e^{jx} = \cos x + j \sin x$.

(d) $e^{-jx} = \cos x - j \sin x$.

(e) $\cosh x = \dfrac{e^{x} + e^{-x}}{2}$.

(f) $\sinh x = \dfrac{e^{x} - e^{-x}}{2}$.

(g) $e^{x} = \cosh x + \sinh x$.

(h) $e^{-x} = \cosh x - \sinh x$.

(i) $\cosh^2 x - \sinh^2 x = 1$.

(j) $\sinh(x \pm jy) = \sinh x \cos y \pm j \cosh x \sin y$.

(k) $\cosh(x \pm jy) = \cosh x \cos y \pm j \sinh x \sin y$.

(l) $\tanh(x + jy) = \dfrac{\sinh 2x + j \sin 2y}{\cosh 2x + \cos 2y}$.

(m) $\tanh(x - jy) = \dfrac{\sinh 2x - j \sin 2y}{\cosh 2x + \cos 2y}$.

(n) $\sinh x = x + \dfrac{x^3}{3!} + \dfrac{x^5}{5!} + \dfrac{x^7}{7!} + \ldots$

(o) $\cosh x = 1 + \frac{x^2}{2!} + \frac{x^4}{4!} + \frac{x^6}{6!} + \ldots$

(p) $\tanh x = x - \frac{x^3}{3} + \frac{2x^5}{15} - \frac{17x^7}{315} + \frac{62x^9}{2835} - \ldots \quad x^2 < \frac{\pi^2}{4}.$

(q) $\frac{d \sinh x}{dx} = \cosh x.$

(r) $\frac{d \cosh x}{dx} = \sinh x.$

(s) $\frac{d \tanh x}{dx} = \operatorname{sech}^2 x.$

(t) $\sinh(x \pm y) = \sinh x \cosh y \pm \cosh x \sinh y.$

(u) $\cosh(x \pm y) = \cosh x \cosh y \pm \sinh x \sinh y.$

(v) $\tanh \frac{x}{2} = \frac{\cosh x - 1}{\sinh x}.$

(w) $\tanh x = \frac{\sinh x}{\cosh x} = \frac{e^x - e^{-x}}{e^x + e^{-x}}.$

(x) $\sinh(-x) = -\sinh x.$

(y) $\cosh(-x) = \cosh x.$

(z) $\tanh(-x) = -\tanh x.$

VII. Answers to Selected Problems.

Chapter 1

1-2 (a) Yes, if $\alpha = 0$ and $\beta = \omega\sqrt{LC}$; (b) Yes, if $\beta = \omega\sqrt{LC}$; (c) Yes, if $v^p = \frac{1}{\sqrt{LC}}$.

1-4 (a) 237.6 ohms/mile; (b) 3.22 henrys/mile; (c) 609.57×10^{-12} farads/foot; (d) 2.796×10^{-9} mhos/meter.

1-6 (b) $(4 + j2) \times 10^{-3}$ mhos; (c) $R = 250$ ohms, $C = 21.22 \times 10^{-9}$ farads.

1-8 (a) 0.00237 nepers/mile; (b) 0.198 radians/mile; (c) $632.54 \angle -0.68°$ ohms; (d) 31.64 miles; (e) 315,394 miles/second.

1-10 (a) 0.23 inches.

1-12 (a) $200.2 \angle 1.114°$ ohms; (b) $4{,}000 \angle 90°$ ohms; (c) $4{,}000{,}100 \angle 90°$ ohms.

1-14 (a) $1.34 \times 10^3 \angle -38.09°$ ohms.

1-16 $\dfrac{\partial i}{\partial x} = 2.371 \times 10^{-6} \cos(10^3 t - 46.378°)$ amperes/meter.

Chapter 2

2-2 (a) $j15.46$ ohms; (b) $j46.57$ ohms; (c) $j705.1$ ohms; (d) $-j45.8$ ohms.

2-4 (a) $-j161.6$ ohms; (b) $-j53.67$ ohms; (c) $-j3.55$ ohms; (d) $j54.57$ ohms.

2-6 (a) $0.333 \angle -90°$ amperes; (b) $141.42 \angle -90°$ volts; (c) 0 amperes; (d) $j300$ ohms.

2-8 (a) $0.333 \angle 90°$ amperes; (b) 0 volts; (c) $0.47 \angle 180°$ amperes; (d) $-j300$ ohms.

2-10 infinite ohms.

2-12 (a) $1.0 \angle -90°$ amperes; (b) 0 volts; (c) $1.414 \angle -90°$ amperes; (d) $100 \angle 90°$ ohms.

2-14 $50 \angle -90°$ ohms.

2-16 (a) 14.32×10^9 hertz; (b) 1.5×10^8 meters/second; (c) 1.047×10^{-2} meters; (d) $150 \angle -600x$ volts.

Chapter 3

3-2 $11.41 \angle 0°$ volts; $18.3 \times 10^{-3} \angle 90°$ amperes.

3-4 $0.0692 \angle 0°$ amperes; -0.0908 amperes.

3-6 1.796; 2.142 gigahertz.

3-8 1/8 wavelength, 300 ohms; 3/8 wavelength, 33.333 ohms.

3-10 (b) $3.418 \angle 19.44^\circ$ volts, $9.48 \times 10^{-3} \angle -36.869^\circ$; (c) 4.

3-12 (b) $6.94 \angle 26.565^\circ$ volts, $0.098 \angle -18.435^\circ$ amperes; (c) 2.62.

3-14 (a) $45.307 \angle -17.68^\circ$ ohms; (b) 17.5 centimeters.

3-16 (a) 3 gigahertz; (b) infinity; (c) $106.25 \angle 90^\circ$ ohms.

3-18 (a) $1 \angle -90^\circ$ amperes; (b) 0 volts; (c) 1.414 amperes; (d) $100 \angle 90^\circ$ ohms; (e) 0 watts.

3-20 (a) 4; (b) 333.3×10^6 hertz; (c) $12.5 \angle 0^\circ$ ohms; (d) $0.6 \angle 180^\circ$.

3-22 (a) $0.333 \angle 90^\circ$ amperes; (b) 141.42 volts; (c) 0 amperes; (d) $300 \angle 90^\circ$ ohms.

3-24 (a) $99.93 \angle -33.7^\circ$ volts; (b) $44.66 \angle -63.425^\circ$ ohms.

Chapter 4

4-2 (a) 0.5 + j1; (b) 0.48 − j0.59; (c) 0 + j0.25; (d) 0.4 −j0.2.

4-4 $438 \angle 90^\circ$ ohms, $0.228 \angle -90^\circ$ amperes.

4-6 (a) $26.58 \angle -48.8^\circ$ ohms; (b) 3.35; (c) $0.541 \angle 12.5^\circ$.

4-8 (a) $87.24 \angle 34.96^\circ$ ohms; (b) 0.419; (c) 44°; (d) 2.39; (e) 15.55 cm.; (f) 20.75 ohms.

4-10 (a) 0 + j106.5 ohms; (b) 3×10^9 hertz; (c) infinity.

4-12 (a) $0.445 \angle -117.5^\circ$; (b) $0.56 \angle 69^\circ$; (c) $0.37 \angle -162^\circ$.

4-14 $33.0 \angle 90^\circ$.

4-16 $196.97 \angle -23.96$

4-18 $41.0 \angle 0^\circ$ ohms.

4-20 (a) infinite ohms; (b) 0 ohms; (c) $50 \angle 90^\circ$ ohms.

4-22 (a) 87.5 + j0 ohms; (b) 28.5 + j0 ohms.

4-24 $297.06 \angle -6.76^\circ$ ohms.

4-26 (a) $100 \angle 0^\circ$ ohms; (b) 0.124 wavelengths; (c) $25 \angle 0^\circ$, 0.374 wavelengths.

4-28 (a) 10; (b) 333.33×10^6 hertz; (c) 5 + j0; (d) $0.82 \angle 180^\circ$; (e) $27.28 \angle 76.22^\circ$ ohms.

4-30 (a) $100 \angle 0^\circ$; (b) 2, 1.

Chapter 5

5-2 (a) $1 \angle -53.13^\circ$ amperes; (b) $121.25 \angle 28.07^\circ$ volts; (c) $0.728 \angle 75.25^\circ$ amperes; (d) 60 watts.

5-4 (a) $101.07 \angle -35.23^\circ$ volts; (b) $1.86 \angle 49.02^\circ$ amperes; (c) $0.639 \angle -53.67^\circ$ amperes; (d) 61.29 watts.

5-6 (a) $86.85 \angle 34.65^\circ$ ohms; (b) 2.39; (c) $107.95 \angle -24.80^\circ$ volts.

5-8 (a) $1.61 \angle 46.73^\circ$ amperes; (b) $160.96 \angle 46.73^\circ$ volts; (c) $0.0299 \angle -30.02^\circ$ mhos; (d) $2.99 \angle -30.02^\circ$ amperes; (e) 259.0 watts.

5-10 (a) $100 \angle -36.869^\circ$ ohms; (b) $0.5 \angle -126.869^\circ$ amperes.

5-12 (a) $-j0.229$ amperes; (b) $170.13 \angle 180^\circ$ volts; (c) $j435.9$ ohms.

5-14 (a) $0.932 \angle -52.32^\circ$ amperes; (b) $147.45 \angle -33.88^\circ$ volts; (c) 130.44 watts; (d) $2.70 \angle 49.97^\circ$ amperes.

5-16 (a) $86.83 \angle 34.62^\circ$ ohms; (b) $41.66 \angle 24.79^\circ$ volts; (c) $0.62 \angle -66.78^\circ$ amperes; (d) $0.518 \angle -34.62^\circ$ amperes .

5-18 First stub at 0.1657 wavelengths from load with a stub length of 0.342 wavelengths, second stub at 0.465 wavelengths from the load with a stub length of 0.157 wavelengths.

5-20 $9.62 \angle -24.56^\circ$ ohms, provide a stub with a length of 0.074 wavelengths placed 0.25 wavelengths from load, alternate stub would be 0.426 wavelengths long and placed 0.375 wavelengths from the load.

5-22 stub is 0.0914 wavelengths long and placed 0.1775 wavelengths from the load.

Chapter 6

6-2 (a) 3.76, 3.626; (b) $1.098 \angle -8.39^\circ$, $0.618 \angle -32.93^\circ$; (c) $1.29 \angle 49.86^\circ$, $2.89 \angle 63.94^\circ$; (d) -0.416, $j0.909$.

6-4 (a) $0.703 \angle 99.546^\circ$; (b) $158.52 \angle -20.82^\circ$ ohms; (c) $0.0315 \angle 20.82^\circ$ amperes; (d) 0.147 watts; (e) $1{,}407 \angle -133.99^\circ$ volts; (f) $9.995 \times 10^{-3} \angle -178.99^\circ$ amperes; (g) 9.9×10^{-3} watts, 6.72%.

6-6 (a) $1{,}131.0 \angle -4.28^\circ$ volts; (b) $1.738 \angle 87.05^\circ$ amperes; (c) 0.051; (d) 89.42 watts.

6-8 (a) 5.0316×10^{-3} nepers/kilometer, 22.1168×10^{-3} radians/kilometer, $644.63 \angle -12.706^\circ$ ohms; (b) $24.93 \angle -41.58^\circ$ volts, $0.124 \angle -41.58^\circ$ amperes, $0.1152 \angle -30.38^\circ$ amperes, 3.108 watts, 4.969 watts.

6-10 (a) $1.55 \angle 12.706^\circ$ amperes; (b) $904.26 \angle -25.34^\circ$ volts; (c) $1.40 \angle -12.64^\circ$ amperes; (d) 1,513.28 watts; (e) 1,237.4 watts, 81.69%.

6-12 (a) $39{,}587.9 \angle -4.28^\circ$ volts; (b) 0 amperes; (c) $60.84 \angle 87.05^\circ$ amperes; (d) $82.86 \angle 75.38^\circ$ amperes, (e) 87.86%.

6-14 (a) 0 volts; (b) $210.75 \angle -61.0^\circ$ amperes; (c) $186.33 \angle -57.2^\circ$ amperes.

Chapter 7

7-2 (a) $155{,}158.59 \angle 10.637^\circ$ volts (268,742.5 volts, line-to-line); (b) $191.78 \angle -6.20^\circ$ amperes; (c) 0.957; (d) 5,439.9 kilowatts.

7-4 $0.144 + j0.7065$ ohms per mile per phase, $0 + j6.004 \times 10^{-6}$ mhos per mile per phase.

7-6 (a) $39{,}592.7 \angle -4.3^\circ$ volts; (b) $60.84 \angle 87.05^\circ$ amperes; (c) 109,585.8 watts; (d) 0.0514 leading.

7-8 $40 + j159.99$ ohms, $-j2{,}777.77$ ohms, $34.48 + j157.17$ ohms, $-j2{,}751.08$ ohms, $155{,}158.4 \angle 10.634^\circ$ volts, $191.56 \angle -6.21^\circ$ amperes, 0.957, 5,336.9 kilowatts.

7-10 $20 + j79.99$ ohms, $-j1{,}388.8$ ohms, $20.39 + j80.72$ ohms, $-6.9 - j1{,}415.48$ ohms.

7-12 (a) $0.89 \angle 3.3^\circ$; (b) $0.9629 \angle 1.02^\circ$.

7-14 (a) $108.65 \angle 66.07^\circ$ amperes; (b) $131{,}383.66 \angle 6.29^\circ$ volts; (c) 92.77% compared to 40.88% with lower voltage.

Chapter 8

8-2 (a) 300 volts in series with 200 and 100 ohms, 300 volts in series with 200 and 50 ohms; (b) 100, 55.55, 60.493, 59.945, 60.006 volts; (c) 0, 1.33, 1.185, 1.2016, 1.1998 amperes; (d) 1.2 amperes, 60 volts.

8-4 (a) 0, 66.66, 74.07, 74.89, 74.98 volts; (b) 0, 1.333, 1.48, 1.497, 1.499 amperes; (c) 1, 1.444, 1.493, 1.4993, 1.4999 amperes; (d) 100, 77.77, 75.308, 75.034, 75.003 volts; (e) 0, 100, 66.96, 77.77, 74.67, 75.308, 74.89, 75.034, 74.986, 75.003 volts; (f) 75 volts, 1.5 amperes, 75 volts, 1.5 amperes.

8-6 (a) 1.333, 1, 1, 1, 1, 1, 1 amperes; (b) 0, 1, 1, 1, 1, 1 amperes; (c) 0, 125, 125, 125, 125, 125 volts; (d) 0, 1.333, 1, 1, 1, 1 amperes; (e) 125 volts, 125 volts, 1 ampere, 1 ampere.

8-8 (a) 0, 150, 150, 150, 150 volts, 112.5, 150, 150, 150, 150 volts; (b) 0, 0.75, 0.75, 0.75, 0.75 amperes, 1.125, 0.75, 0.75, 0.75, 0.75 amperes; (c) 0, 112.5, 150, 150, 150, 150, 150, 150 volts; (d) 150 volts, 0.75 amperes, 150 volts, 0.75 amperes.

Chapter 9

9-2 (a) $213.97 \angle -14.55°$ ohms, 16.9×10^{-3} nepers/kilometer, 65.06×10^{-3} radians/kilometer; (b) 96.57×10^{6} meters/second; (c) 96.57×10^{3} meters.

9-4 (a) 1.913×10^{-6} henrys/meter; (b) 6.128×10^{-12} farads/meter; (c) 889.48×10^{-9} nepers/meter, 1.567×10^{-6} radians/meter, $780.25 \angle -29.57°$ ohms, 2.409×10^{8} meters/second.

9-6 (a) 2.74×10^{-6} henrys/meter; (b) $4.19\ 5 \times 10^{-12}$ farads/meter; (c) 7.806×10^{-6} nepers/meter, 7.910×10^{-6} radians/meter, $7.026 \times 10^{3} \angle -44.619°$ ohms, 47.65×10^{6} meters/second.

9-8 25.32 ohms.

9-10 72.76°C.

9-12 (a) 0.159886230, same; (b) −2.788980132, book −2.788980155; (c) 3 terms, 6 terms.

9-14 114,768.49.

9-16 (a) 0.0952; (b) 132.0; (c) 388.96.

9-18 (a) $0.912 \angle 65.43°$ ohms/mile; (b) $5.08 \times 10^{-6} \angle 90°$ mhos/mile; (c) $106.70 \angle 65.43°$ ohms, $595.35 \times 10^{-6} \angle 90°$ mhos; (d) 458.34×10^{-6} nepers/mile, 2.104×10^{-3} radians/mile, $423.35 \angle -12.284°$ ohms.

9-20 (a) $0.8402 \angle 78.53°$ ohms/mile; (b) $5.123 \times 10^{-6} \angle 90°$ mhos/mile; (c) $98.31 \angle 78.53°$ ohms, $599.46 \times 10^{-6} \angle 90°$ mhos;

(d) 207.218×10^{-6} nepers/mile, 2.064×10^{-3} radians/mile, $404.95 \angle -5.731°$ ohms.

Chapter 10

10-2 (a) 2.55 henrys/mile; (b) 16.66×10^3 miles/second, 19.93×10^3 miles/second, 20.0×10^3 miles/second; (c) 20×10^3 miles/second.

10-4 (a) $0.7567 \angle 163.56°$; (b) $0.754 \angle -16.43°$; (c) $0.746 \angle 163.56°$.

10-6 (a) $300 \angle 0°$ ohms; (b) $0.5 \angle 0°$.

10-8 $45.307 \angle -17.68°$ ohms.

10-10 (a) 4; (b) 333.33×10^6 hertz; (c) $200 \angle 0°$ ohms; (d) $0.6 \angle 0°$.

10-12 (a) $297.68 \angle -6.36°$ ohms; (b) $29.97 \angle -1.59°$ volts.

Chapter 11

11-2 (a) 1.2139 decibels; (b) 139.76×10^{-6} nepers/meter.

11-4 (a) 34.65 miles; (b) 3.01 decibels.

11-6 (a) 5.42×10^{-3} nepers/kilometer, 22.35×10^{-3} radians/kilometer, $647.85 \angle -13.52°$ ohms; (b) $0.4425 \angle 0°$ volts, $683.0 \times 10^{-6} \angle 13.527°$ amperes, $0.154 \angle -145.869°$ amperes; (c) 15.00 watts; (d) 293.88×10^{-6} watts; (e) 66.413×10^{-3} watts; (f) 66.413×10^{-3} watts, 0 watts, 0 watts.

11-8 (a) 2.697 watts; (b) 2.697 watts; (c) $0.5927 \angle -46.846°$; (d) 0.3513; (e) 1.46 watts.

Chapter 12

12-2 (a) 16.156 gigahertz; (b) 0.01857 meters, 1.857 centimeters; (c) 0.05199 meters; (d) 0.01749 meters; (e) 8.917×10^8 meters/second; (f) 1.009×10^8 meters/second.

12-4 (a) 47.57 radians/meter; (b) 0.1442 meters; (c) 2.0794×10^9 hertz; (d) 0.13208 meters; (e) 4.067×10^8 meters/second; (f) 2.2127×10^8 meters/second; (g) 0.097 meters.

12-6 (a) 6,561.6 megahertz, 13,123 megahertz.

12-8 0.2567×10^9 hertz.

12-10 (a) 5.65 meters; (b) 0.02133 meters; (c) 79.544×10^9 meters/second; (d) 1.1314×10^6 meters/second.

Chapter 13

13-2 (a) $478.53 \angle -7.018°$ ohms; (b) 326.95×10^{-9} nepers/meter; (c) 2.656×10^{-6} radians/meter; (d) 499.8×10^{-6} ohms/mile, 5.303×10^{-3} henrys/mile, 0 mhos/mile, 23.872×10^{-9} farads/mile.

13-4 $64.828 \angle -53.93°$ ohms.

13-6 (a) 23.342; (b) $110.186 \angle -83.53°$ ohms.

Chapter 14

14-2 (a) 260 microvolts per meter; (b) 121.875 microvolts per meter.

14-4 Sketches.

14-6 (a) $\dfrac{j\eta I_{max} e^{-j\beta r}}{2\pi r} \dfrac{\cos(\pi \cos \theta) + 1}{\sin \theta}$; (b) and (c) Sketches.

14-8 Sketch.

14-10 $\dfrac{1}{2} \sqrt{\dfrac{\mu_o}{\varepsilon_o}} \dfrac{I^2 \ell^2 \sin^2 \theta}{4 \lambda^2 r^2}$ watts per meter squared; (b) $\sqrt{\dfrac{\mu_o}{\varepsilon_o}} \dfrac{(\pi)}{3} \dfrac{I^2 \ell^2}{\lambda^2}$ watts.

14-12 (a) 2.5; (b) 2.25.

14-14 6.

INDEX

C

D

E

F

G

H

I

K

L

M

N

T

U

V

W